RELATIVITY MADE RELATIVELY EASY

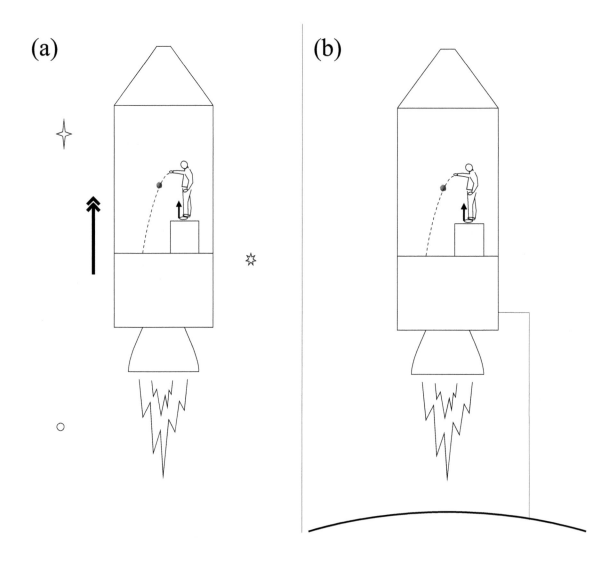

(a)

(b)

Relativity Made Relatively Easy

ANDREW M. STEANE

OXFORD
UNIVERSITY PRESS

OXFORD
UNIVERSITY PRESS

Great Clarendon Street, Oxford, OX2 6DP,
United Kingdom

Oxford University Press is a department of the University of Oxford.
It furthers the University's objective of excellence in research, scholarship,
and education by publishing worldwide. Oxford is a registered trade mark of
Oxford University Press in the UK and in certain other countries

First Edition published in 2012

Impression: 1

British Library Cataloguing in Publication Data

Data available

Library of Congress Cataloging in Publication Data

Library of Congress Control Number: 2012941966

ISBN 978–0–19–966285–2 (hbk)
      978–0–19–966286–9 (pbk)

Printed and bound in Great Britain by
CPI Group (UK) Ltd, Croydon, CR0 4YY

Dedication

This book is dedicated to Derek Stacey and David Lucas: two physicists from whose generosity, encouragement and example I have greatly benefited.

Preface

The aim of this book is to help people understand and enjoy the world in which we all live. It is written for the undergraduate student of physics, and is intended to teach. The text presents an extensive study of Special Relativity, and a (gentle, but exact) introduction to General Relativity. It is not intended to be the first introduction to Special Relativity for most students, although for a bright student it could function as that. Therefore basic ideas such as time dilation and space contraction are recalled but not discussed at length. However, I think it is also beneficial to have a thorough discussion of those concepts at as simple a level as possible, so I have provided one in another book called *The Wonderful World of Relativity*. The present book is self-contained and does not require knowledge of the first one, but a more basic text such as *The Wonderful World* or something similar is recommended as a preparation for this book.

The book has two more specific aims. The first is to allow an undergraduate physics course to extend somewhat further and wider in this area than has traditionally been the case, while ensuring that the mainstream of students can still handle the material; the second is to show how physics 'works' more generally and to act as a prelude to advanced topics such as classical and quantum field theory. The title *Relativity Made Relatively Easy* is therefore playful, yet serious. The text aims to make manageable what would otherwise be regarded as hard; to make derivations as simple as possible and physical ideas as transparent as possible. It is intended to teach, and therefore little prior mathematical knowledge is assumed. Although spacetime and relativity are the main themes, physical ideas such as fields and flow, symmetry and stress are expounded along the way. These ideas connect to other areas such as hydrodynamics, electromagnetism, and particle physics. The present volume covers Special Relativity thoroughly except for spinors and Lagrangian density methods for fields, and it introduces General Relativity with the minimum of mathematical apparatus required to acquire correct ideas and quantitative results for static metrics. The affine connection (Christoffel symbol), for example, is not needed in order to achieve this. A second volume will extend the treatment of General Relativity somewhat more thoroughly, and will also introduce cosmology, spinors and some field theory (which explains the occasional mention of 'volume 2' in the present text).

Although many universities are now extending their core coverage of Relativity, and the present volume is intended to meet that need, a few sections go further than most undergraduate courses will want to go. These are intended to fill the gap between undergraduate and graduate study, and to offer general reading for the professional physicist.

The exercises are an integral part of the text, and are of three types. Some are examples to build familiarity, some introduce formulae or results that fill out or complete the main text, and some build physical intuition and a sound grasp of the big ideas.

Sections or chapters with a * in the heading can be omitted or skimmed at first reading.

Acknowledgements

I have, of course, learned Relativity mostly from other people. All writers in this area have learned from the pioneers of the subject, especially Einstein, Lorentz, Maxwell, Minkowski, and Poincaré. I am also indebted to tutors such as N. Stone and W. S. C. Williams at Oxford University, and to authors who have preceded me—especially texts by (in alphabetical order) J. Binney, T.-P. Cheng, A. Einstein, R. Feynman, A. P. French, R. d'Inverno, J. D. Jackson, H. Muirhead, W. Rindler, F. Rohrlich, R. Shankar, E. F. Taylor, J. A. Wheeler, and W. S. C. Williams. I thank A. Barr for some comments and suggestions.

Several sections of the book compress, extend, clarify or combine treatments from other authors. I have not traced the ancestry of all the material, but have indicated cases where I quote an approach or argument that I think might not be regarded as common currency. I have also developed many ideas independently, but naturally after a century of discussion, a significant proportion of such ideas must be rediscoveries.

Einstein emphasized the need to think of a reference frame in physical and not abstract terms, as a physical entity made of rods and clocks. As a student I resisted this idea, feeling that a more abstract idea, liberated from mere matter, must be superior. I was wrong. The whole point of Relativity is to see that abstract notions of space and time are superfluous and misleading.

Feynman offered very useful guidance on how to approach things simply while retaining rigour. Readers familiar with his, Leighton, and Sands' 'Lectures on Physics' will recognize its influence on Chapter 10. I have filled in mathematical methods in order to allow the student to tackle example calculations, and to make Chapter 11 possible.

I learned a significant number of detailed points from Rindler's work, and my contribution has been to clarify where possible. Appendix D and section 16.2.1 re-present arguments I found in his book, with more comfort and explanation for the reader, and section 13.1 follows his line of argument, which I could not improve.

Most of the exercises are either original or significantly reconfigured, or are standard short problems. However, some have been largely copied from elsewhere—a § indicates those from W. Rindler, and §§ those from W. S. C. Williams. I thank these authors for their permission. Wolfgang Rindler has produced many excellent exercises in several books; I thank him for allowing a selection of them to adorn this book.

I thank my family, who accepted the compromises on home life which I made in order to work on the book.

Finally, my special thanks go to the Physics Department of Oxford University, its 'Ion Traps and Quantum Computing' research group, and David Lucas and Charles de Bourcy. The University allowed an atmosphere of academic freedom and trust that made it possible to devote time to this book, the research group, under David's guidance, accepted my long absences in a generous spirit, and Charles kindly read and checked part I of the manuscript with great care and insight. Any remaining defects are my own responsibility.

Contents

Part I

The relativistic world

Basic ideas

1

The primary purpose of this chapter is to offer a way in for readers completely unfamiliar with Special Relativity, and to recall the main ideas for readers who have some preliminary knowledge of the subject. For the former category, appendix A contains some of the basic arguments that will not be repeated in the main text (and that can be found in introductory texts such as *The Wonderful World of Relativity*). The right moment to turn to that appendix, if you need to, is after completing section 1.2 of this chapter.

In order to discuss space and time without being vague, it is extremely helpful to introduce the notion of a *reference body*. This is usually called an '*inertial frame of reference*', but this phrase is in some respects unfortunate. The phrase 'frame of reference' is used in an abstract way in everyday language, but in physics we mean something more concrete: a large rigid physical object which could, in principle, exist in the vicinity of any system whose evolution we wish to discuss. Such a 'reference body' clarifies what we mean when we talk of distance and time. By 'distance' we mean the number of particles or rods of the reference body between two places. The reference body keeps track of time as well, since the particles making it can be imagined to be tiny regular clocks (think of an atom with an internal vibration, for example). By 'time' at any given place we mean the number of repetitions of some such regularly repeating process ('clock') at that place.

'Frame of reference' and 'reference body' are synonyms in physics. Most people like to think of a frame of reference as having the form of a scaffolding of ideally thin and rigid rods, with clocks attached. One might also think of it as a large brick (but one with the unusual property that others things can move through it unimpeded). It is a mistake to try to be too abstract here. Although the scaffolding or rigid body is not necessarily present, our reasoning about distance and time must be consistent with the fact that such a body might in principle be present in any region of spacetime, and used to define those concepts.

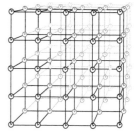

Fig. 1.1

An **inertial** frame of reference is one in which Newton's First Law of Motion holds. If a particle not subject to any forces always has a constant velocity relative to some frame of reference (as determined by distance and time measurements furnished by nearby parts of the frame), no matter when or where the particle sets out, then that frame is an inertial frame of reference. It follows immediately that all the parts of an inertial frame of reference have the same velocity relative to one such particle as it moves past, and by considering many freely moving

Fig. 1.2

particles at different places and times one may infer that all the parts of an inertial frame move along together, having the same velocity and zero acceleration relative to any other such frame, assuming that distance relationships obey Euclidean geometry.

An *observer* is a reasoning being who could in principle be situated at rest in some given frame of reference. We use the word 'observe' to mean *not* what the observer directly sees, but what he or she can deduce to be the case at each time and place in his/her reference frame. For example, suppose two explosions occur, and an observer is located closer to one than to the other in his own reference frame. If such an observer receives light-flashes from the two explosions simultaneously, then he 'observes' (i.e., deduces) that the explosions were not simultaneous in his reference frame.

1.1 Newtonian physics

Let us briefly survey the connection between inertial reference frames according to classical physics, as developed by Galileo, Newton, and others.

A crucial idea, first presented at length by Galileo, is the idea that the behaviour of physical systems is the same in any given inertial reference frame, irrespective of whether that frame may be in uniform motion with respect to others. For example, it is possible to play table tennis in a carriage of a moving railway train without noticing the motion of the train (as long as the rails are smooth and the train has constant velocity). There is no need to adjust one's calculations of the trajectory of the ball or the choice of force to apply using the bat: all the behaviour is the same as it would be in a motionless train. This idea, which we shall state more carefully in a moment, is called the Principle of Relativity; it is obeyed by both classical and relativistic physics.

When we analyze the motions of bodies it is useful to introduce a coordinate system (in both space and time), which is simply a way of noting positions and times relative to a reference body (= inertial frame of reference). An *event* is a point in space and time. It is useful to know, for any given event, how the coordinates of the event relative to one reference body relate to the coordinates of the same event relative to another reference body. If reference frames S and S' have all their axes aligned, but frame S' moves along the positive x direction relative to S at speed v, then we say the reference frames are in *standard configuration* (Fig. 1.3). The coordinates of any given event, as determined in two reference frames in standard configuration, are related, *according to Newtonian physics*, by

$$t' = t,$$
$$x' = x - vt,$$
$$y' = y,$$
$$z' = z. \tag{1.1}$$

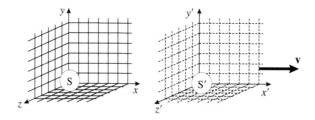

Fig. 1.3 Two reference frames (=reference bodies) in standard configuration. S′ moves in the x direction relative to S, with its axes aligned with those of S. The picture shows the situation at the moment (defined in S) when the axes of S′ have just swept past those of S. The whole reference frame of S′ is in motion together at the same velocity $\mathbf{v}$ relative to S. Equally, the frame of S is in motion at velocity $-\mathbf{v}$ relative to S′.

This set of equations is called the *Galilean transformation*. It can also be written in matrix notation as

$$
\begin{pmatrix} t' \\ x' \\ y' \\ z' \end{pmatrix} = \begin{pmatrix} 1 & 0 & 0 & 0 \\ -v & 1 & 0 & 0 \\ 0 & 0 & 1 & 0 \\ 0 & 0 & 0 & 1 \end{pmatrix} \begin{pmatrix} t \\ x \\ y \\ z \end{pmatrix},
\tag{1.2}
$$

or

$$
\begin{pmatrix} t' \\ \mathbf{r}' \end{pmatrix} = \mathcal{G} \begin{pmatrix} t \\ \mathbf{r} \end{pmatrix},
\tag{1.3}
$$

where

$$
\mathcal{G} \equiv \begin{pmatrix} 1 & 0 & 0 & 0 \\ -v & 1 & 0 & 0 \\ 0 & 0 & 1 & 0 \\ 0 & 0 & 0 & 1 \end{pmatrix}.
\tag{1.4}
$$

The inverse Galilean transformation is

$$
\begin{pmatrix} t \\ x \\ y \\ z \end{pmatrix} = \begin{pmatrix} 1 & 0 & 0 & 0 \\ v & 1 & 0 & 0 \\ 0 & 0 & 1 & 0 \\ 0 & 0 & 0 & 1 \end{pmatrix} \begin{pmatrix} t' \\ x' \\ y' \\ z' \end{pmatrix}.
\tag{1.5}
$$

which can also be written

$$
\begin{pmatrix} t \\ \mathbf{r} \end{pmatrix} = \mathcal{G}^{-1} \begin{pmatrix} t' \\ \mathbf{r}' \end{pmatrix}.
\tag{1.6}
$$

The reader is invited to verify this—that is, check that the matrix given in eqn (1.5) is indeed the inverse of $\mathcal{G}$.

Matrix notation makes it easy to check things such as the effect of transforming from one reference frame to another and then to a third. For example, the net effect of transforming to another frame and then back to the first is given by $\mathcal{G}^{-1}\mathcal{G}$, which is, of course, the identity matrix.

1.2 Special Relativity

1.2.1 The Postulates of Special Relativity

Turning now to Special Relativity, we shall find that the Principle of Relativity is still obeyed, but the Galilean transformation fails.

The Main Postulates of Special Relativity are

Postulate 1, 'Principle of Relativity': *The motions of bodies included in a given space are the same among themselves, whether that space is at rest or moves uniformly forward in a straight line.*

Postulate 2, 'Light Speed Postulate':
Version A: *There is a finite maximum speed for signals.*
Version B: *There is an inertial reference frame in which the speed of light in vacuum is independent of the motion of the source.*

The Principle of Relativity (Postulate 1) is obeyed by classical physics; the Light Speed Postulate is not. The Principle of Relativity can also be stated as:

The laws of physics take the same mathematical form in all inertial frames of reference.

In Postulate 2, either version A or version B is sufficient on its own to allow Special Relativity to be developed. Version A does not mention light; which makes it clear that Special Relativity underlies all theories in physics, not just electromagnetism. For this reason version A is preferred. However, we will preserve the practice of calling this postulate the 'Light Speed Postulate' because in vacuum, far from material objects, light-waves move at the maximum speed for signals. With this piece of information about light, one can use either version to derive the other.

Einstein used version B of the Light Speed Postulate. It is often stated as 'the speed of light is independent of the motion of the source.' In this statement the fact that motion can only ever be relative motion is taken for granted, and it is a statement about what is observed in any reference frame. In our version B we chose to make a slightly more restricted statement (choosing just one reference frame), merely because it is interesting to hone ones assumptions down to the smallest possible set. By combining this with Postulate 1 it immediately follows that all reference frames will have this property.

The Light Speed Postulate can be tested very accurately by astronomical observations. For instance, in a binary star system each star orbits the centre of mass. If the emitted light had a speed depending on the motion of the source, then it would propagate to Earth at a speed which varies with the time of emission. For example, light emitted at one point on the orbit would catch up and possibly overtake light emitted at another point. This would be observable as multiple images appearing in detectors on Earth, or as some more modest change in the detected pattern of fluctuation (e.g. of Doppler effect or intensity). No such effects have ever been detected. A typical approach would be to test the claim that the speed of light emitted by a source of velocity v is given by $c + kv$, where k is a constant to be determined. By using X-ray sources one can avoid complicating issues arising from scattering of visible light by the interstellar medium, and by a clever combination of position and Doppler measurements (see exercise 1.5) one can determine the upper bound $k < 10^{-9}$ using data from binary systems situated at a distance of order 10^4 light-years from Earth and having an orbital period of a few days.

In order to make clear what is assumed and what is derived, it is useful to add two further postulates to the list:

Postulate 0, 'Euclidean geometry': *The rules of Euclidean geometry apply to all spatial measurements within any given inertial reference frame.*

Postulate 3, 'Conservation of momentum': *Internal interactions among the parts of an isolated system cannot change the system's total momentum, where momentum is a vector function of rest mass and velocity.*

Postulate 0 (Euclidean geometry) is obeyed by Special Relativity but not by General Relativity. Postulate 3 (conservation of momentum) allows the central elements of dynamics to be deduced, including the famous formula $E = mc^2$ (which cannot be derived from the Main Postulates alone)[1].

[1] One can replace the statement of Postulate 3 by a statement about translational symmetry; see chapter 14.

1.2.2   Central ideas about spacetime

Recall that a 'point in spacetime' is called an *event*. This is something happening at an instant of time at a point in space, with infinitesimal time duration and spatial extension. For an example, tap the tip of a pencil once on a table top, or click your fingers.

A *particle* is a physical object of infinitesimal spatial extent, which can exist for some extended period of time. The line of events which gives the location of the particle as a function of time is called its *worldline*; see Fig. 1.4.

If two events have coordinates (t_1, x_1, y_1, z_1) and (t_2, x_2, y_2, z_2) in some reference frame, then the quantity

$$s^2 = -c^2(t_2 - t_1)^2 + (x_2 - x_1)^2 + (y_2 - y_1)^2 + (z_2 - z_1)^2 \qquad (1.7)$$

is called the *squared spacetime interval* between them. Note the crucial minus sign in front of the first term. We emphasize it by writing eqn (1.7) as

$$s^2 = -c^2 \Delta t^2 + \Delta x^2 + \Delta y^2 + \Delta z^2. \qquad (1.8)$$

If $s^2 < 0$ then the time between the events is sufficiently long that a particle or other signal (moving at speeds less than c) could move from one event to the other. Such a pair of events is said to be separated by a *time-like interval*. If $s^2 > 0$ then the time between the events is too short for any physical influence to move between them. This is called a *spacelike interval*. If $s^2 = 0$ then we have a *null interval*, which means that a light pulse or other light-speed signal could move directly from one event to the other.

Although the parts t_i, x_i, y_i, z_i needed to calculate an interval will vary from one reference frame to another, we will find in chapter 2 that the net result, s^2, is **independent of reference frame**: all reference

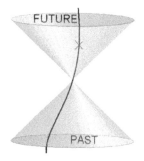

Fig. 1.4 A spacetime diagram showing a worldline and a light-cone (past and future branches). Time runs in the vertical direction on the diagram, and one spatial dimension has been suppressed. The cross ($\times$) marks an example event. The apex of the cone is another event.

Fig. 1.5 The set of events at fixed interval from the origin forms a hyperboloid of revolution. The figure shows such a surface for the case where the interval from the origin is spacelike. (a) shows the surface as traced out by a set of hyperbolae, and (b) shows the same surface and illustrates the interesting fact that it can also be constructed from a set of exactly straight lines (cf. section 3.2).

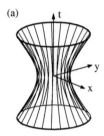

(a)

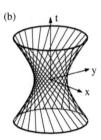

(b)

frames agree on the value of this quantity. This is similar to the fact that the length of a vector is unchanged by rotations of the vector. A quantity whose value is the same in all reference frames is called a *Lorentz invariant* (or Lorentz scalar). **Lorentz invariants play a central role in Special Relativity.**

The set of events with a null spacetime interval from any given event lie on a cone called the *light-cone* of that event; see Fig. 1.4. The part (or 'branch') of this cone extending into the past is made of the worldlines of photons that form a spherical pulse of light collapsing onto the event, and the part extending into the future is made of the worldlines of photons that form a spherical pulse of light emitted by the event. The cone is an abstraction: the incoming and outgoing light-pulses do not have to be there. The past part of the light-cone surface of any event A is called the *past light-cone* of A, and the future part of the surface is called the *future light-cone* of A. The whole of the future cone (i.e., the body of the cone as well as the surface) is called the *absolute future* of A, and consists of all events which could possibly be influenced by A (in view of the Light Speed Postulate). The whole of the past cone is called the *absolute past* of A, and consists of all events which could possibly influence A. The rest of spacetime, outside either branch of the light-cone, can neither influence nor be influenced by A. It consists of all events with a spacelike separation from A.

The set of events at fixed interval from the origin satisfies the equation

$$-c^2t^2 + x^2 + y^2 + z^2 = L \qquad (1.9)$$

where L is a constant. If $L > 0$ the interval is spacelike; if $L < 0$ the interval is time-like. This is the equation of a hyperboloid of revolution; see Figs 1.5 and 1.6.

The single most basic insight into spacetime that Einstein's theory introduces is the *relativity of simultaneity*: two events that are simultaneous in one reference frame are not necessarily simultaneous in another. In particular, if two events happen simultaneously at different spatial locations in reference frame F, then they will not be simultaneous in any reference frame moving relative to F with a non-zero component of velocity along the line between the events. An example is furnished by 'Einstein's train'; see Fig. 1.7.

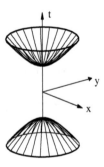

Fig. 1.6 The set of events at a fixed time-like interval from the origin.

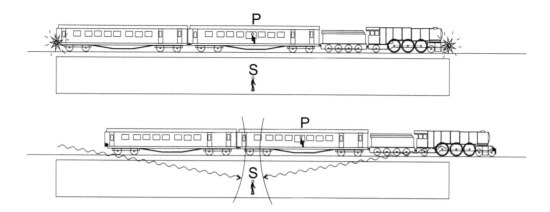

Fig. 1.7 Einstein's train. A fast-moving train is moving past a platform. Firecrackers are placed at the two ends of the train, each triggered to explode when it reaches the corresponding end of the platform. Suppose an observer S standing still in the middle of the platform receives flashes of light from the two explosions simultaneously. He infers that the explosions were simultaneous (they both occured at a time $\frac{1}{2}L_0/c$ before he received the signals, where L_0 is the length of the platform according to his measurements). From this he can also infer that the train has the same length as the platform. S also finds that the front flash reaches a passenger P seated in the middle of the train before the back flash does. For example, when the flashes arrive at S one has already passed P, while the other has not. It follows that the passenger himself experiences the flashes separated by a finite interval of time. P considers that the flashes travelled *equal* distances (relative to him) to reach him, since he is seated in the middle of the train, and one flash occurred at the front, one at the back: there are scorch marks on the train to prove it. Since the speed of light is a universal constant, P must infer that the explosions were *not* simultaneous in his reference frame: the front one occurred before the back one. Hence simultaneity is not absolute: it depends on a reference frame. (P may also infer that the train is longer than the platform, according to his measurements.)

By careful argument from the postulates one can connect timing and spatial measurements in one inertial reference frame to those in any other inertial reference frame in a precise, quantitative way. In the next chapter we will introduce the *Lorentz transformation* to do this in general. Arguments for some simple cases are a useful way into the subject, and are summarized in appendix A.

1.3   Matrix methods

By writing down the Galilean transformation using a matrix, we have already assumed that the reader has some idea what a matrix is and how it is used. However, in case matrices are unfamiliar we will here summarize the matrix mathematics we shall need. This will not substitute for a more lengthy course of mathematical training, but it may be a useful reminder.

A matrix is a table of numbers. We will only need to deal with real matrices (with rare exceptions), so the numbers are real numbers. In an '$n \times m$' matrix the table has n rows and m columns. Here is a 2×3 matrix, for example:

$$\begin{pmatrix} 1.2 & -3.6 & 8 \\ 2 & 4.5 & 2 \end{pmatrix}. \tag{1.10}$$

If either n or m is 1 then we have a vector; if both are 1 than we have a scalar.

A vector of 1 row is called a row vector; a vector of 1 column is called a column vector:

$$\text{row vector: } (1, -3, 2), \qquad \text{column vector: } \begin{pmatrix} 1 \\ -3 \\ 2 \end{pmatrix}.$$

The sum of two matrices, written as $A + B$, is only defined (so it is only a legal operation) when A and B have the same shape: that is, the two matrices have the same number of rows n, and they also have the same number of columns m (but n does not have to equal m). $A + B$ is then defined to mean the matrix formed from the sums of the corresponding components of A and B. To be precise, if M_{ij} refers to the element of matrix M in the ith row and jth column, then the matrix sum is defined by

$$M = A + B \qquad \Leftrightarrow \qquad M_{ij} = A_{ij} + B_{ij}.$$

This rule applies to vectors and scalars too, since they are special cases of matrices, and it agrees with the familiar rule for summing vectors: add the components.

The product of two matrices, written as AB, is only defined (so it is only a legal operation) when A and B have appropriate shapes: the number of *columns* in the first matrix has to equal the number of *rows* in the second matrix. For example, a 2×3 matrix can multiply a 3×5 matrix, but it cannot multiply a 2×3 matrix. The product is defined by the mathematical rule

$$M = AB \qquad \Leftrightarrow \qquad M_{ij} = \sum_k A_{ik} B_{kj}. \qquad (1.11)$$

It is important to note that this rule is **not commutative**: AB is not necessarily the same as BA. The rule is important in order to have a precise definition, but the use of subscripts and the sum can leave the operation obscure until one tries a few examples. It amounts to the following. You have to work your way through the elements of M one by one. To obtain the element of M on the ith row and jth column, take the ith row of A and the jth column of B. Regard these as two vectors and evaluate their scalar product: that is, 'dive' the row of A onto the column of B, multiply corresponding elements, and then sum. The result is the value of M_{ij}.

The only way to become familiar with matrix multiplication is by practice. By applying the rule, you will find that if a $k \times n$ matrix multiplies a $n \times m$ matrix then the result is a $k \times m$ matrix. This is a very useful check to keep track of what you are doing.

The whole point of matrix notation is that much of the time we can avoid actually carrying out the element-by-element multiplications and additions. Instead we manipulate the matrix symbols. For example, if $A + B = C$ and $A - B = D$ then we can deduce that $C + D = 2A$

without needing to carry out any element-by-element analysis. The following mathematical results apply to matrices (as the reader can show by applying the rules developed above):

$$A + B = B + A$$
$$A + (B + C) = (A + B) + C$$
$$(AB)C = A(BC)$$
$$A(B + C) = AB + AC.$$

We shall mostly be concerned with square matrices and with vectors. The square matrices will be mostly 4×4, so they can be added and multiplied to give other 4×4 matrices. A square matrix can multiply a *column* vector, giving a result that is a column vector (since a 4×4 matrix multiplying a 4×1 matrix gives a 4×1 matrix). For example:

$$\begin{pmatrix} 1.2 & -3.6 & 8 & 2 \\ 2 & 4.5 & 2 & 0.5 \\ -1 & 5 & 1 & -0.5 \\ -2 & 3.2 & 3 & -5 \end{pmatrix} \begin{pmatrix} 1 \\ -3 \\ 2 \\ -4 \end{pmatrix} = \begin{pmatrix} 20 \\ -9.5 \\ -12 \\ 14.4 \end{pmatrix}.$$

A square matrix can be multiplied from the left by a *row* vector, giving a result that is a row vector (since a 1×4 matrix multiplying a 4×4 matrix gives a 1×4 matrix).

Matrix inverse

Many, but not all, square matrices have an *inverse*. This is written M^{-1} and is defined by

$$MM^{-1} = M^{-1}M = I \qquad (1.12)$$

where I is the *identity matrix*, consisting of ones down the diagonal and zeros everywhere else. For example, in the 4×4 case it is

$$I = \begin{pmatrix} 1 & 0 & 0 & 0 \\ 0 & 1 & 0 & 0 \\ 0 & 0 & 1 & 0 \\ 0 & 0 & 0 & 1 \end{pmatrix}.$$

The identity matrix has no effect when it multiplies another matrix: $IM = MI = M$ for all M. Inverses of non-square matrices can also be defined, but we shall not need them.

There is no definition of a 'division' operation for matrices (in the sense of one matrix 'divided by' another), but often multiplication by the inverse achieves what might be regarded as a form of division. For example, if $AB = C$ and A has an inverse, then by premultiplying both sides by A^{-1} we obtain $A^{-1}AB = A^{-1}C$, and therefore $B = A^{-1}C$ (by using the fact that $A^{-1}A = I$ and $IB = B$).

The inverse of a 2×2 matrix is easy to find:

$$M = \begin{pmatrix} a & b \\ c & d \end{pmatrix} \Leftrightarrow M^{-1} = \frac{1}{ad - bc} \begin{pmatrix} d & -b \\ -c & a \end{pmatrix}.$$

Here the inverse exists when $ad - bc \neq 0$, and you can check that it satisfies eqn (1.12).

There is also a general rule on how to find the inverse of a matrix of any size; you should consult a mathematics textbook when you need it.

The inverse of a product is the product of the inverses, but you have to reverse the order:

$$(AB)^{-1} = B^{-1}A^{-1}. \tag{1.13}$$

Proof: $(AB)(B^{-1}A^{-1}) = A(B^{-1}B)A^{-1} = AA^{-1} = I$, and you are invited to show by a similar method that $(B^{-1}A^{-1})(AB) = I$.

Transpose and scalar product

The *transpose* of a matrix, written M^T, is the matrix obtained by swapping the rows and columns. To be precise:

$$A = M^T \qquad \text{means} \qquad A_{ij} = M_{ji}.$$

For example, the transpose of the matrix displayed in eqn (1.10) is

$$\begin{pmatrix} 1.2 & 2 \\ -3.6 & 4.5 \\ 8 & 2 \end{pmatrix}.$$

The transpose of a row vector is a column vector, and the transpose of a column vector is a row vector.

The following results are useful:

$$(A + B)^T = A^T + B^T \tag{1.14}$$

$$(AB)^T = B^T A^T \tag{1.15}$$

$$(A^T)^{-1} = (A^{-1})^T \tag{1.16}$$

Note the order reversal in eqn (1.15). You can easily prove this result using eqn (1.11). Then eqn (1.16) follows, since if M is the inverse of A^T then we must have $A^T M = I$, taking the transpose of both sides gives $M^T A = I^T = I$ and hence $M^T = A^{-1}$ and the result follows.

The product of a row vector and a column vector of the same length is often useful because it is simple: it is a 1×1 matrix—in other words, a scalar. If we start with a pair of column vectors $\mathbf{u}$ and $\mathbf{v}$ of the same size, then we can obtain such a scalar by

$$\mathbf{u}^T \mathbf{v}. \tag{1.17}$$

This comes up often, so it is given a name: it is called the *scalar product* or *inner product* of the vectors. (The inner product of a pair of row vectors would be $\mathbf{u}\mathbf{v}^T$.) You can calculate it by multiplying corresponding components and summing. For example, if $\mathbf{u}$ has components u_1, u_2, u_3 and $\mathbf{v}$ has components v_1, v_2, v_3 then

$$\mathbf{u}^T \mathbf{v} = u_1 v_1 + u_2 v_2 + u_3 v_3. \tag{1.18}$$

Most science or mathematics students will meet the scalar product first in the context of vector analysis in space, where one is typically dealing with three-component vectors representing things such as displacement, velocity, and force. In this context it can be convenient not to be too concerned whether the vectors are row or column vectors, and so the dot notation is introduced: the scalar product is written $\mathbf{u} \cdot \mathbf{v}$. In Relativity we will be dealing with 4-component vectors in time and space, and for them we will introduce a special meaning for the dot notation and for the phrase 'scalar product'.

1.4 Spacetime diagrams

Figs 1.4, 1.5, and 1.6 are all examples of *spacetime diagrams*. A spacetime diagram shows events and worldlines and other information in a natural way, and when used correctly is a great help to understanding Relativity. By convention, time is usually shown vertically on the diagram. It is then most easy to show either two spatial dimensions (in a perspective or projected view) or a single spatial dimension. The coordinate axes in Figs 1.5 and 1.6 show the coordinates for one inertial reference frame which has been given a privileged status on the diagram. It is the reference frame S associated with a body whose worldline is vertical on the diagram. The *same* region of spacetime can be discussed equally well in terms of the coordinate system of any other reference frame S′, moving with respect to the first. Such a coordinate system can be shown by adding to the diagram a worldline of a particle fixed in S′, and a line of simultaneity for S′; see Fig. 1.8. (For the line of simultaneity, consult Fig. A.2 in the appendix.) These lines are the time axis t' and position axis x' for the second frame. Events separated by an interval parallel to t' on the diagram are at the same position in S′ (they are separated only by time in S′). Events separated by an interval parallel to x' on the diagram are simultaneous in S′ (they are separated only by space in S′).

To obtain the time and distance between events, in either frame, from such a diagram, the axes must be calibrated. We can always choose the origin of frame S′ so that the event $\{t = 0,\, x = 0\}$ occurs at $\{t' = 0,\, x' = 0\}$. Consider the event E on the time axis of S′, one unit of time in frame S′ after $t' = 0$. This event is at $\{t' = 1,\, x' = 0\}$, and at some $\{t, x\}$ related by $x = vt$ where v is the relative velocity of the frames. The spacetime interval between the origin and E is invariant, i.e., the same in both frames, so

$$-c^2 t'^2 = -c^2 t^2 + x^2 = -t^2 (c^2 - v^2)$$

Fig. 1.8 Spacetime diagram showing coordinate axes for two observers in relative motion in one spatial dimension. The light-cone symbol in the corner indicates that this is a spacetime diagram (if this were in any doubt), and also indicates the orientation of the diagram and the directions of light-like worldlines. Note that this diagram is completely flat: it shows motion in just one spatial dimension (e.g. beads sliding along a wire, trains running along a track), and is not a perspective view of a three-dimensional diagram (which could be used to show motion in two spatial dimensions).

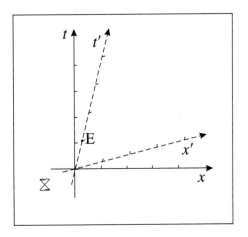

Fig. 1.9 Spacetime diagram showing coordinate axes for two observers in relative motion in one spatial dimension, with an example event A and a further worldline. The distance between A and the worldline, in any given frame, is indicated by the length of a line extending from A to the worldline, oriented parallel to a line of simultaneity for that frame: i.e., the x-axis and x'-axis respectively.

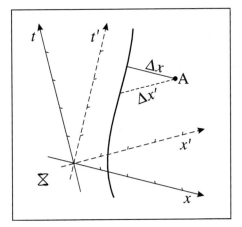

therefore $t = \gamma t'$ where $\gamma = (1 - v^2/c^2)^{-1/2}$ is the Lorentz factor. It follows that one unit of 'primed time' along the t' takes up more than one unit of 'unprimed time' along the t axis (since $\gamma > 1$).

When treating a single pair of reference frames it can be convenient to take care of this calibration issue by choosing a more symmetric configuration of the diagram. Suppose the frames are S and S'. There always exists an intermediate frame F such that S and S' move with equal and opposite velocities relative to F. If we take the point of view of F for the purpose of constructing the diagram, then the coordinate axes of S and S' will be placed as shown in Fig. 1.9, and this is convenient because now both sets of axes get the same calibration, because both suffer the same time dilation and Lorentz contraction relative to F. This makes it easy to compare distance or time measurements in one frame with corresponding measurements in the other. For example, the distance between the curved worldline and the event marked A on Fig. 1.9, as observed in the two frames, is indicated by the two straight line segments marked Δx and $\Delta x'$. One can see at a glance that $\Delta x < \Delta x'$ in this example.

Exercises

(1.1)

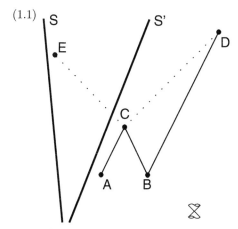

The spacetime diagram in the figure shows the worldlines of two inertial observers S and S′, and various other events. Find the sequence of the labeled events, first in the rest frame of S, and then in the rest frame of S′. Is causality always respected?

(1.2) Using a spacetime diagram, or otherwise, prove that

  (i) the temporal order of two events is the same in all reference frames if and only if they are separated by a time-like interval,

  (ii) there exists a reference frame in which two events are simultaneous if and only if they are separated by a space-like interval.

(1.3) **Bondi's k factor**. Hermann Bondi proposed the following neat argument. Suppose two observers A and B move uniformly along a line at relative speed v. Clocks held by the observers are set to zero when they meet. At event A the first observer sends a light-pulse to the second, who receives it at event B and immediately returns it, the return pulse arriving at event C.

  (i) Show that

$$t_B = \frac{t_A}{1 - v/c} \quad \text{and} \quad t_C = t_B(1 + v/c)$$

    where t_A, t_B, t_C are the times of the three events as observed in frame A.

  (ii) Let t'_B be the time of the reflection event, as registered on a clock held by B, and define a

factor $k \equiv t'_B/t_A$. This is the ratio of reception time observed by the receiver to emission time observed by the emitter, when a light-signal is sent between the parties. Since the whole situation is linear, k is independent of time. Since the situation is symmetric, we must find the same factor k for the return signal: i.e., $k = t_C/t'_B$. Use this information to find k in terms of v.

  (iii) Hence find t_B/t'_B in terms of v (time dilation), and also explain how k is related to the Doppler effect.

(1.4) A rod of rest length L_0 lies on the x axis of some reference frame S, and moves at speed v along that axis. At some moment, the motion is reversed, such that all particles in the rod change their velocity from $+v$ to $-v$ simultaneously in S. Find the initial and final lengths of the rod, in frame S and in a frame S′ in standard configuration with S. (Note that both frames are inertial; their motion does not change!) Using a spacetime diagram, find the sequence of events in S′: which end of the rod starts to move first, relative to S′?

(1.5) Consider using observations of a binary star system to test the Light Speed Postulate, as follows. At distance D from Earth a small X-ray-emitting star follows a circular orbit of radius r around a large companion, at a speed $v_0 \ll c$. The distance of the small star from Earth is then $D + r \sin \omega t$, and its velocity component towards Earth is $v = v_0 \cos \omega t$ where $v_0 = r\omega$. In a model where the speed of light is given by $c + kv$, where v is the velocity of the source and $k \ll 1$ is a constant, show that X-rays emitted at time t are received on Earth at time

$$t_E = t + T + \frac{r}{c} \sin \omega t - \frac{T v_0}{c} k \cos \omega t$$

to lowest order in k, where $T = D/c$. Hence show that

$$\frac{dt_E}{dt} = 1 + (V/c) \cos(\omega t + \phi)$$

where $V = v_0(1 + (kT\omega)^2)^{1/2}$ and $\tan \phi = kT\omega$. Explain why the observed Doppler effect λ_E/λ

is given by dt_E/dt (e.g. consider two successive emitted wavefronts, with $v \ll c$ so that time dilation is negligible). The X-ray intensity varies as a function of time owing to eclipsing by the large companion. By comparing the detected intensity curve with the detected Doppler shift curve, one can measure the offset phase ϕ. Find an upper bound on k if this phase is determined to be $\phi < 0.06$ rad for the source Her-X1, which lies at a distance 2×10^4 light-years and has an orbital period of 1.7 days[2].

(1.6) Find the velocity at which you should move relative to Holbein's picture *The Ambassadors* in order that the image of the skull should not be stretched in your rest frame.

---

[2] K. Breeher, Phys. Rev. Lett. **39**, 1051 (1877).

The Lorentz transformation

2

In a first introduction to Special Relativity (such as appendix A), the reasoning is kept as direct as possible. Simple physical scenarios are used to deduce basic mathematical results. Now we will introduce a more algebraic approach. This is needed in order to generalize and to proceed. In particular, it will save a lot of trouble in calculations involving a change of reference frame, and we will learn how to formulate laws of physics so that they obey the Main Postulates of the theory.

2.1 Introducing the Lorentz transformation

The Lorentz transformation, for which this chapter is named, is the coordinate transformation which replaces the Galilean transformation presented in eqn (1.1).

Let S and S′ be reference frames allowing coordinate systems (t, x, y, z) and (t', x', y', z') to be defined. Let their corresponding axes be aligned, with the x and x' axes along the line of relative motion, so that S′ has velocity v in the x direction in reference frame S. Also, let the origins of coordinates and time be chosen so that the origins of the two reference frames coincide at $t = t' = 0$. Hereafter we refer to this arrangement as the 'standard configuration' of a pair of reference frames. In such a standard configuration, if an event has coordinates (t, x, y, z) in S, then its coordinates in S′ are given by

$$t' = \gamma(t - vx/c^2) \tag{2.1}$$

$$x' = \gamma(-vt + x) \tag{2.2}$$

$$y' = y \tag{2.3}$$

$$z' = z \tag{2.4}$$

where the **Lorentz factor** $\gamma = \gamma(v) = 1/(1 - v^2/c^2)^{1/2}$. This set of simultaneous equations is called the Lorentz transformation; we will derive it from the Main Postulates of Special Relativity in section 2.1.1.

By solving for (t, x, y, z) in terms of (t', x', y', z') you can easily derive the inverse Lorentz transformation:

$$t = \gamma(t' + vx'/c^2) \tag{2.5}$$

$$x = \gamma(vt' + x') \tag{2.6}$$

$$y = y' \tag{2.7}$$

$$z = z' \tag{2.8}$$

This can also be obtained by replacing v by $-v$ and swapping primed and unprimed symbols in the first set of equations. This is how it must turn out, since if S$'$ has velocity $\mathbf{v}$ in S, then S has velocity $-\mathbf{v}$ in S$'$, and both are equally valid inertial frames.

Let us immediately extract from the Lorentz transformation the phenomena of time dilation and Lorentz contraction. For the former, simply choose two events at the same spatial location in S, separated by time τ. We may as well choose the origin, $x = y = z = 0$, and times $t = 0$ and $t = \tau$ in frame S. Now apply eqn (2.1) to the two events. We find that the first event occurs at time $t' = 0$, and the second at time $t' = \gamma\tau$, so the time interval between them in frame S$'$ is $\gamma\tau$: i.e., longer than in the first frame by the factor γ. This is time dilation.

For Lorentz contraction one must consider not two events but two worldlines. These are the worldlines of the two ends, along the x direction, of some object fixed in S. Place the origin on one of these worldlines, and then the other end lies at $x = L_0$ for all t, where L_0 is the rest length. Now consider these worldlines in the frame S$'$ and choose the time $t' = 0$. At this moment the worldline passing through the origin of S is also at the origin of S$'$: i.e., at $x' = 0$. Using the Lorentz transformation, the other worldline is found at

$$t' = \gamma(t - vL_0/c^2), \qquad x' = \gamma(-vt + L_0). \tag{2.9}$$

Since we are considering the situation at $t' = 0$ we deduce from the first equation that $t = vL_0/c^2$. Substituting this into the second equation we obtain $x' = \gamma L_0(1 - v^2/c^2) = L_0/\gamma$. Thus in the primed frame at a given instant the two ends of the object are at $x' = 0$ and $x' = L_0/\gamma$. Therefore the length of the object is reduced from L_0 by a factor γ. This is Lorentz contraction.

For relativistic addition of velocities, eqn (A.11), consider a particle moving along the x' axis with speed u in frame S$'$. Its worldline is given by $x' = ut'$. Substituting in eqn (2.6) we obtain $x = \gamma(vt' + ut') = \gamma^2(v + u)(t - vx/c^2)$. Solve for x as a function of t, and one obtains $x = wt$ with w as given by eqn (A.11).

For the Doppler effect, consider a sequence of wavefronts emitted from the origin of S at times $0, t_0, 2t_0, \ldots$. The first wavefront is detected in S$'$ at $t' = 0$. The next wavefront has the worldline $x = c(t - t_0)$, and the worldline of the origin of S$'$ is $x = vt$. These two lines intersect at $x = vt = c(t - t_0)$, hence $t = t_0/(1 - v/c)$. Now use the Lorentz transformation (2.5) with $x' = 0$ to find the time of the reception event in

Table 2.1 Useful relations involving γ. $\beta = v/c$ is the speed in units of the speed of light. $dt/d\tau$ relates the time between events on a worldline to the proper time, for a particle of speed v. dt'/dt relates the time between events on a worldline for two reference frames of relative velocity $\mathbf{v}$, with $\mathbf{u}$ the particle velocity in the unprimed frame. If two particles have velocities $\mathbf{u}, \mathbf{v}$ in some reference frame, then $\gamma(w)$ is the Lorentz factor for their relative velocity.

$$\gamma$$

$$\beta = \sqrt{1 - 1/\gamma^2}, \qquad \gamma^2 v^2 = (\gamma^2 - 1)c^2 \tag{2.10}$$

$$\frac{d\gamma}{dv} = \gamma^3 v/c^2, \qquad \frac{d}{dv}(\gamma v) = \gamma^3 \tag{2.11}$$

$$\frac{dt}{d\tau} = \gamma, \qquad \frac{dt'}{dt} = \gamma_v(1 - \mathbf{u} \cdot \mathbf{v}/c^2) \tag{2.12}$$

$$\gamma(w) = \gamma(u)\gamma(v)(1 - \mathbf{u} \cdot \mathbf{v}/c^2) \tag{2.13}$$

S′: $t' = t/\gamma = t_0/\gamma(1 - v/c)$. This gives the period of the wave observed in S′. After inverting to convert times into frequencies, we obtain eqn (A.10), for S′ moving away from the source.

To summarize:

The Postulates of Relativity, taken together, lead to a description of spacetime in which the notions of simultaneity, time duration, and spatial distance are well-defined in each inertial reference frame, but their values, for a given pair of events, can vary from one reference frame to another. In particular, objects evolve more slowly and are contracted along their direction of motion when observed in a reference frame relative to which they are in motion.

A good way to think of the Lorentz transformation is to regard it as a kind of 'translation' from the t, x, y, z 'language' to the t', x', y', z' 'language'. The basic results given above serve as an introduction, to increase our confidence with the transformation and its use. In this and the next chapter we will use it to treat more general situations, such as addition of non-parallel velocities, the Doppler effect for light emitted at a general angle to the direction of motion, and other phenomena.

Table 2.1 summarizes some useful formulae related to the Lorentz factor $\gamma(v)$. Derivations of eqns (2.12) and (2.13) will be presented in section 2.5, while the derivation of the others is left as an exercise for the reader.

Why not start with the Lorentz transformation?

Question: 'The Lorentz transformation allows all the basic results of time dilation, Lorentz contraction, Doppler effect, and addition of velocities to be derived quite readily. Why not start with it, and avoid all the trouble of the slow step-by-step arguments presented in introductory treatments of Relativity?'

Answer: The cautious step-by-step arguments are needed in order to understand the results, and the character of spacetime. Only then is the physical meaning of the Lorentz transformation clear. We can present things quickly now, because spacetime, time dilation, and space contraction are discussed in the previous chapter and in appendix A (and at greater length in *The Wonderful World*). Such a discussion has to take place somewhere. The derivation of the Lorentz transformation given in section 2.1.1 can seem like mere mathematical trickery unless we maintain a firm grasp on what it all means.

2.1.1   Derivation of Lorentz transformation

To derive the Lorentz transformation from first principles, one may reason as follows.

We seek a transformation of coordinates such that the coordinate systems in two inertial reference frames S and S′ will lead to results consistent with the Principle of Relativity and the Speed of Light Postulate. We shall assume that the transformation is *linear*, (i.e., the equations for t', x', y', z' only contain terms linear in t, x, y, z, not higher powers or products of them); if we find a linear transformation then the assumption will be proved to have been justified.

Fig. 2.1

After adopting the standard configuration of the axes of the inertial frames, we can derive by symmetry arguments the equations $y' = y$ and $z' = z$, (i.e., the absence of any transverse effect). Let us assume there is an effect on transverse distance, in order to prove a contradiction. Take a cylindrical pipe of diameter D_0 in its rest frame, cut it in two, and let the two segments approach one another along their common axis (Fig. 2.1). Let S be the rest frame of the second pipe segment. In this frame the other segment is moving. Let D_1, D_2 be the diameters of the two pipe segments as observed in S. Clearly, $D_2 = D_0$. If $D_1 < D_0$, then pipe 1 will pass inside pipe 2. However, by the Principle of Relativity we must then find that in the rest frame S′ of pipe 1, $D_2' < D_0 = D_1'$, so pipe 2 passes inside pipe 1. This is a contradiction, because which pipe is inside is an absolute property.[1] Therefore $D_1 < D_0$ cannot be true. Hence we must have $D_1 \geq D_0$. However, the case $D_1 > D_0$ again leads to a contradiction. It follows that $D_1 = D_0$. One can extend this argument to the whole of the rectangular meshes in the yz and $y'z'$ planes in Fig. 1.3.

[1] If you doubt this, replace one pipe with a wide metal sheet having a hole of proper diameter D_0. The remaining pipe cannot both pass through and not pass through this hole. A change of reference frame *has no influence whatsoever on events*; it influences merely how distances and times between events are measured.

It remains to find the equations relating t' and x' to (t, x, y, z). Using methods such as radar signalling to establish simultaneity (appendix A), it is not hard to prove that events in a plane at any given t, x are also simultaneous in S′, so t' depends only on t and x. Also, we seek a solution where the axes of S and S′ remain aligned as the reference frames move, so x' depends only on t and x. Therefore we can restrict the last part of the derivation to one spatial dimension, and we seek a pair of equations having the general form

$$\left.\begin{aligned} t' &= at + bx \\ x' &= dt + ex \end{aligned}\right\} \qquad (2.14)$$

where a, b, d, e are constant coefficients to be discovered.

We have four unknowns, and we can find four equations for them as follows:

(1) Since the reference frame S$'$ moves with speed v along the x direction in S, the point $x' = 0$ must move as $x = vt$:

$$0 = dt + evt$$

$$\Rightarrow d = -ev. \tag{2.15}$$

(2) Similarly, the point $x = 0$ in S moves as $x' = -vt'$ in S$'$; therefore $-vt' = dt$ when $t' = at$, hence

$$d = -av. \tag{2.16}$$

Combining eqns (2.15) and (2.16) we have $a = e$.

(3) A light-speed signal in S must also have the speed of light in S$'$, so $x = ct$ must give $x' = ct'$:

$$\left.\begin{array}{l} t' = at + bct, \\ ct' = dt + ect \end{array}\right\} \Rightarrow ac + bc^2 = d + ec$$

$$\Rightarrow d = bc^2. \tag{2.17}$$

(4) So far we have established that the general form is

$$t' = a(t - vx/c^2),$$
$$x' = a(-vt + x). \tag{2.18}$$

It remains to find an expression for a. This can be done by applying the Principle of Relativity. First manipulate eqns (2.18) so as to obtain t and x in terms of t' and x':

$$t = \frac{1}{a(1 - v^2/c^2)}(t' + vx'/c^2),$$

$$x = \frac{1}{a(1 - v^2/c^2)}(vt' + x'). \tag{2.19}$$

(An easy way to obtain this is to express (2.18) using a 2×2 matrix and pre-multiply by the inverse of the matrix.) Now argue (by the Principle of Relativity) that the second set of equations must be the same as the first set, except for a change in the sign of v and swapping primed and unprimed symbols. This implies that

$$a = \frac{1}{a(1 - v^2/c^2)} \tag{2.20}$$

Hence $a = \gamma$, and we have derived the Lorentz transformation.

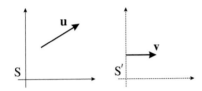

Fig. 2.2 A particle has velocity **u** in frame S. Frame S′ moves at velocity **v** relative to S, with its spatial axes aligned with those of S.

2.2   Velocities

Let reference frames S, S′ be in standard configuration with relative velocity v, and suppose a particle moves with velocity **u** in S (see Fig. 2.2). What is the velocity **u**′ of this particle in S′?

For the purpose of the calculation we can, without loss of generality, put the origin of coordinates on the worldline of the particle. Then the trajectory of the particle is $x = u_x t, y = u_y t, z = u_z t$. Applying the Lorentz transformation, we have

$$x' = \gamma(-vt + u_x t)$$

$$y' = u_y t$$

$$z' = u_z t \tag{2.21}$$

for points on the trajectory, with

$$t' \;=\; \gamma(t - v u_x t/c^2). \tag{2.22}$$

This gives $t = t'/\gamma(1 - u_x v/c^2)$, which, when substituted in the equations for x', y', z', implies

$$u'_x = \frac{u_x - v}{1 - u_x v/c^2}, \tag{2.23}$$

$$u'_y = \frac{u_y}{\gamma(1 - u_x v/c^2)}, \tag{2.24}$$

$$u'_z = \frac{u_z}{\gamma(1 - u_x v/c^2)}. \tag{2.25}$$

Writing

$$\mathbf{u} = \mathbf{u}_{\parallel} + \mathbf{u}_{\perp} \tag{2.26}$$

where $\mathbf{u}_{\parallel}$ is the component of **u** in the direction of the relative motion of the reference frames, and $\mathbf{u}_{\perp}$ is the component perpendicular to it, the result is conveniently written in vector notation:

$$\mathbf{u}'_{\parallel} = \frac{\mathbf{u}_{\parallel} - \mathbf{v}}{1 - \mathbf{u} \cdot \mathbf{v}/c^2}, \qquad \mathbf{u}'_{\perp} = \frac{\mathbf{u}_{\perp}}{\gamma_v (1 - \mathbf{u} \cdot \mathbf{v}/c^2)}. \tag{2.27}$$

These equations are called the equations for the 'relativistic transformation of velocities' or 'relativistic addition of velocities'. The subscript on the γ symbol acts as a reminder that it refers to $\gamma(v)$ not $\gamma(u)$. If **u** and **v** are the velocities of two particles in any given reference frame, then **u**′ is their relative velocity (think about it!).

When **u** is parallel to **v** we regain eqn (A.11).

When **u** is perpendicular to **v** we have $\mathbf{u}'_{\parallel} = -\mathbf{v}$ and $\mathbf{u}'_{\perp} = \mathbf{u}/\gamma_v$. The latter can be interpreted as an example of time dilation (in S′ the particle takes a longer time to cover a given distance). For this case, $u'^2 = u^2 + v^2 - u^2 v^2/c^2$.

Sometimes it is useful to express the results as a single vector equation. This is easily done using $\mathbf{u}_{\parallel} = (\mathbf{u} \cdot \mathbf{v})\mathbf{v}/v^2$ and $\mathbf{u}_{\perp} = \mathbf{u} - \mathbf{u}_{\parallel}$, giving:

$$\mathbf{u}' = \frac{1}{1 - \mathbf{u}\cdot\mathbf{v}/c^2}\left[\frac{1}{\gamma_v}\mathbf{u} - \left(1 - \frac{\mathbf{u}\cdot\mathbf{v}}{c^2}\frac{\gamma_v}{1+\gamma_v}\right)\mathbf{v}\right]. \qquad (2.28)$$

It will be useful to have the relationship between the gamma factors for $\mathbf{u}', \mathbf{u}$ and $\mathbf{v}$. One can obtain this by squaring eqn (2.28) and simplifying, but the algebra is laborious. A much better way is to use an argument via invariant quantities. This will be presented in section 2.6; the result is given in eqn (2.13). That equation also serves as a general proof that the velocity addition formulae never result in a speed $w > c$ when $u, v \le c$. For, if $u \le c$ and $v \le c$ then the right-hand side of (2.13) is real and non-negative, and therefore $\gamma(w)$ is real, hence $w \le c$.

Let θ be the angle between $\mathbf{u}$ and $\mathbf{v}$, then $u_\parallel = u\cos\theta$, $u_\perp = u\sin\theta$, and from eqn (2.27) we obtain

$$\tan\theta' = \frac{u'_\perp}{u'_\parallel} = \frac{u\sin\theta}{\gamma_v(u\cos\theta - v)}. \qquad (2.29)$$

This is the way a direction of motion transforms between reference frames. In the formula, $\mathbf{v}$ is the velocity of frame S′ relative to frame S. (We shall present a quicker derivation of this formula in section 2.5.3 by using a 4-vector.) The classical (Galilean) result would give the same formula, but with $\gamma = 1$. Therefore, the distinctive effect of the Lorentz transformation is to 'throw' the velocity forward more than one might expect (as well as to prevent the speed exceeding c).

Fig. 2.3 presents some examples of eqn (2.29). If in an explosion in reference frame S′, particles are emitted in all directions with the same speed u', then in frame S the particle velocities are directed in a cone

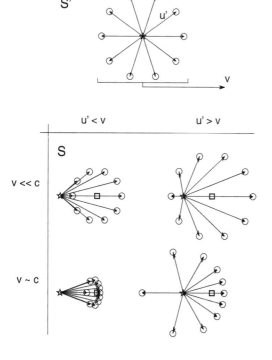

Fig. 2.3 Transformation of velocities. An isotropic explosion in frame S′ produces particles all moving at speed u' in S′, and a fragment is left at the centre of the explosion (top diagram). The fragment and frame S′ move to the right at speed v relative to frame S. The lower four diagrams show the situation in frame S. The ∗ shows the location of the explosion event. The square shows the present position of the central fragment, the circles show positions of the particles, and the arrows show the velocities of the particles. The left diagrams show examples with $u' < v$, the right with $u' > v$. The top two diagrams show the case $u', v \ll c$. Here the particles lie on a circle centred at the fragment, as in classical physics. The bottom diagrams show examples with $v \sim c$, thus bringing out the difference between the relativistic and the classical predictions. The lower right shows $u' = c$: 'headlight effect' for photons. The photons lie on a circle centred at the position of the explosion (not the fragment), but more of them move forward than backward.

Fig. 2.4 Two illustrative collision experiments. The lines show particle tracks, and the dashed ellipses indicate part of the cylindrical detector. The central circle indicates where a collision has taken place. (a) shows a case where the collisions products emerge roughly isotropically, and (b) shows a case where most particles emerge in three jets. It is highly likely that the explanation for (b) is the emission of a small number of short-lived fast-moving particles, each of which gave rise to one jet.

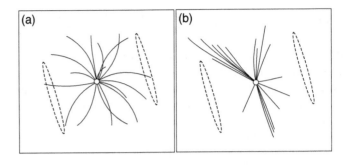

angled forwards along the direction of propagation of S′ in S, for $v > u'$, and mostly in such a cone for $v < u'$. This is not completely unlike the classical prediction (shown in the top two diagrams of Fig. 2.3), but the 'collimation' into a narrow beam is more pronounced in the relativistic case. A practical result of this is illustrated in Fig. 2.4. When a fast-moving particle decays in flight, the products are emitted roughly isotropically in the rest frame of the decaying particle, and therefore in any other frame they move in a directed 'jet' along the line of motion of the original particle. Such jets are commonly observed in particle accelerator experiments (Fig. 2.4b). They are a signature of the presence of a short-lived fast-moving particle that gave rise to the jet.

Is it acceptable to set $c = 1$?

It is a common practice to set $c = 1$ for convenience when doing mathematical manipulations in Special Relativity. Then one can leave c out of the equations, which reduces clutter and can make things easier. When you need to calculate a specific number for comparison with experiment, you must either put back all the cs into your final equations, or remember that the choice $c = 1$ is consistent only when the units of distance and time (and all other units that depend on them) are chosen appropriately. For example, one could work with seconds for time, and light-seconds for distance. (One light-second is equal to 299792458 metres.) The only problem with this approach is that you must apply it consistently throughout. To identify the positions where c or a power of c appears in an equation, one can use dimensional analysis, but when one has further quantities also set equal to 1, this can require some careful thought. Alternatively you can make sure that all the units you use (including mass, energy, etc.) are consistent with $c = 1$.

Some authors like to take this further, and argue that Relativity teaches us that there is something basically wrong about assigning different units to time and distance. We recognise that the height and width of any physical object are just different uses of essentially the same type of physical quantity—namely, spatial distance—so the ratio of height to width is a dimensionless number. One might want to argue that similarly, temporal and spatial separation are just different uses of essentially the same quantity—namely, separation in spacetime—so the ratio of distance to time (what we call speed) should be regarded as dimensionless.

Ultimately this is a matter of taste. Clearly time and space are intimately related, but they are not quite the same: there is no way that a proper time could be mistaken for, or regarded as, a rest length, for example. My preference is to regard the statement 'set $c = 1$' as shorthand for 'set $c = 1$ distance-unit per time-unit'. In other words, I do not regard speed as dimensionless, but I recognise that to choose 'natural units' can be convenient. 'Natural units' are units where c has the value '1 speed-unit'.

2.3   Lorentz invariance and 4-vectors

It is possible to continue by finding equations describing the transformation of acceleration, and then introducing force and its transformation. However, a much better insight into the whole subject is gained if we learn a new type of approach in which time and space are handled together.

Question: Can we derive Special Relativity directly from the invariance of the interval? Do we have to prove that the interval is Lorentz-invariant first?

Answer: This question addresses an important technical point. It is good practice in physics to look at things in more than one way. A good way to learn Special Relativity is to take the Postulates as the starting point, and derive everything from there. This is approach adopted in *The Wonderful World of Relativity* and also in this book. Therefore, you can regard the logical sequence as 'postulates $\Rightarrow$ Lorentz transformation $\Rightarrow$ invariance of interval and other results.' However, it turns out that the spacetime interval alone, if we *assume* its frame-independence, is sufficient to derive everything else! This more technical and mathematical argument is best assimilated after one is already familiar with Relativity. Therefore we are not adopting it at this stage, but some of the examples in this chapter serve to illustrate it. In order to proceed to General Relativity it turns out that the clearest line of attack is to assume by postulate that an invariant interval can be defined by combining the squares of coordinate separations, and then derive the nature of spacetime from that and some further assumptions about the impact of mass–energy on the interval. This leads to 'warping of spacetime', which we observe as a gravitational field.

First, let us arrange the coordinates t, x, y, z into a vector of four components. It is good practice to make all the elements of such a '4-vector' have the same physical dimensions, so we let the first component be ct, and define

$$\mathsf{X} \equiv \begin{pmatrix} ct \\ x \\ y \\ z \end{pmatrix}. \tag{2.30}$$

We will always use a capital letter and the plain font as in 'X' for 4-vector quantities. For the familiar '3-vectors' we use a bold Roman font as in '**x**', and mostly but not always a small letter. You should think of 4-vectors as column vectors not row vectors, so that the Lorentz transformation equations can be written[2]

$$\mathsf{X}' = \Lambda \mathsf{X} \tag{2.31}$$

with

$$\Lambda \equiv \begin{pmatrix} \gamma & -\gamma\beta & 0 & 0 \\ -\gamma\beta & \gamma & 0 & 0 \\ 0 & 0 & 1 & 0 \\ 0 & 0 & 0 & 1 \end{pmatrix} \tag{2.32}$$

where

$$\beta \equiv \frac{v}{c}. \tag{2.33}$$

The right-hand side of eqn (2.31) represents the product of a 4×4 matrix Λ with a 4×1 vector X, using the standard rules of matrix multiplication. You should check that eqn (2.31) correctly reproduces eqns (2.1) to (2.4).

The inverse Lorentz transformation is obviously

$$\mathsf{X} = \Lambda^{-1}\mathsf{X}' \tag{2.34}$$

(just multiply both sides of eqn (2.31) by Λ^{-1}), and one finds

$$\Lambda^{-1} = \begin{pmatrix} \gamma & \gamma\beta & 0 & 0 \\ \gamma\beta & \gamma & 0 & 0 \\ 0 & 0 & 1 & 0 \\ 0 & 0 & 0 & 1 \end{pmatrix}. \tag{2.35}$$

It should not surprise us that this is simply Λ with a change of sign of β. You can confirm that $\Lambda^{-1}\Lambda = I$ where I is the identity matrix.

When we want to refer to the components of a 4-vector, we use the notation

$$\mathsf{X}^\mu = \mathsf{X}^0, \mathsf{X}^1, \mathsf{X}^2, \mathsf{X}^3, \qquad \text{or} \qquad \mathsf{X}^t, \mathsf{X}^x, \mathsf{X}^y, \mathsf{X}^z, \tag{2.36}$$

where the zeroth component is the 'time' component, ct for the case of X as defined by eqn (2.30), and the other three components are the 'spatial' components, x, y, z for the case of (2.30). The reason for placing the indices as superscripts rather than subscripts will emerge later.

[2] The Lorentz transformation is commonly indicated by either a calligraphic Roman letter $\mathcal{L}$, or by the Greek capital Lambda, Λ. We adopt the latter in order that $\mathcal{L}$ can be used for other things such as a Lagrangian. This is also helpful to connect to research literature where the use of Λ is almost universal.

2.3.1   Rapidity

Define a parameter ρ by

$$\tanh(\rho) = \frac{v}{c} = \beta, \qquad (2.37)$$

then

$$\cosh(\rho) = \gamma, \quad \sinh(\rho) = \beta\gamma, \quad \exp(\rho) = \left(\frac{1+\beta}{1-\beta}\right)^{1/2}, \qquad (2.38)$$

so the Lorentz transformation is

$$\Lambda = \begin{pmatrix} \cosh\rho & -\sinh\rho & 0 & 0 \\ -\sinh\rho & \cosh\rho & 0 & 0 \\ 0 & 0 & 1 & 0 \\ 0 & 0 & 0 & 1 \end{pmatrix}. \qquad (2.39)$$

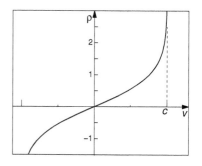

Fig. 2.5 Rapidity versus v.

The quantity ρ is called the *hyperbolic parameter* or the *rapidity*. This form makes some types of calculation easy. For example, consider a group of three references frames S, S′, S″ all moving colinearly, with ρ_u the rapidity with which frame S′ moves relative to S, and ρ_v the rapidity with which frame S″ moves relative to S′. Then for any 4-vector V,

$$\mathsf{V}'' = \Lambda_v \mathsf{V}' = \Lambda_v \Lambda_u \mathsf{V}.$$

It follows that the transformation from frame S to S″ is $\Lambda = \Lambda_v \Lambda_u$. In matrix form this is

$$\Lambda = \begin{pmatrix} \cosh\rho_v & -\sinh\rho_v & 0 & 0 \\ -\sinh\rho_v & \cosh\rho_v & 0 & 0 \\ 0 & 0 & 1 & 0 \\ 0 & 0 & 0 & 1 \end{pmatrix} \begin{pmatrix} \cosh\rho_u & -\sinh\rho_u & 0 & 0 \\ -\sinh\rho_u & \cosh\rho_u & 0 & 0 \\ 0 & 0 & 1 & 0 \\ 0 & 0 & 0 & 1 \end{pmatrix}$$

$$= \begin{pmatrix} (\cosh\rho_v \cosh\rho_u + \sinh\rho_v \sinh\rho_u) & (-\cosh\rho_v \sinh\rho_u - \sinh\rho_v \cosh\rho_u) & 0 & 0 \\ \multicolumn{4}{c}{\dots \text{ etc. } \dots} \end{pmatrix}$$

$$= \begin{pmatrix} \cosh(\rho_v + \rho_u) & -\sinh(\rho_v + \rho_u) & 0 & 0 \\ -\sinh(\rho_v + \rho_u) & \cosh(\rho_v + \rho_u) & 0 & 0 \\ 0 & 0 & 1 & 0 \\ 0 & 0 & 0 & 1 \end{pmatrix}$$

where in the intermediate step some of the working is omitted, since it is more useful to see the pattern than to write out all the terms. From the form of the result it is clear that for relative motion *all in the same direction*, the rapidities simply add

$$\rho_w = \rho_v + \rho_u \qquad (2.40)$$

where ρ_w is the rapidity with which S″ moves relative to S. This makes rapidity a useful tool for studying straight-line motion; cf. section 4.2.4. If the relative motion of one pair of frames is in a direction different from that of the other pair, then the calculation is a lot more complicated, and no such simple result emerges. For motion all in a single direction, however, one can use eqn (2.40) as an alternative way to deduce the addition of velocities formula $w = (u + v)/(1 + uv/c^2)$ by using the standard result for the hyperbolic tangent of a sum:

$$\tanh(\rho_u + \rho_v) = \frac{\tanh \rho_u + \tanh \rho_v}{1 + \tanh \rho_u \tanh \rho_v}.$$

Example   A rocket engine is programmed to fire in bursts such that each time it fires, the rocket achieves a velocity increment of u—meaning that in the inertial frame where the rocket is at rest before the engine fires, its speed is u after the engine stops. Calculate the speed w of the rocket relative to its starting rest frame after n such bursts, all colinear.

Solution
Define the rapidities ρ_u and ρ_w by $\tanh \rho_u = u/c$ and $\tanh \rho_w = w/c$, then by eqn (2.40) we have that ρ_w is given by the sum of n increments of ρ_u: i.e., $\rho_w = n\rho_u$. Therefore, $w = c \tanh(n\rho_u)$. (This can also be written $w = c(z^n - 1)/(z^n + 1)$ where $z = \exp(2\rho_u)$.)

You can readily show that the Lorentz transformation can also be written in the form

$$\begin{pmatrix} ct' + x' \\ ct' - x' \\ y' \\ z' \end{pmatrix} = \begin{pmatrix} e^{-\rho} & & & \\ & e^{\rho} & & \\ & & 1 & \\ & & & 1 \end{pmatrix} \begin{pmatrix} ct + x \\ ct - x \\ y \\ z \end{pmatrix}. \tag{2.41}$$

The form (2.39) can be regarded as a 'rotation' through an imaginary angle $i\rho$ if we also multiply the zeroth component of 4-vectors by i.

2.4   Lorentz-invariant quantities

Under a Lorentz transformation a 4-vector changes, but not out of all recognition. In particular, a 4-vector has a size or 'length' that is not affected by Lorentz transformations. This is like 3-vectors, which preserve their length under rotations, but the 'length' has to be calculated in a specific way.

To find our way to the result we need, first recall how the length of a 3-vector is calculated. For $\mathbf{r} = (x, y, z)$ we would have $r \equiv |\mathbf{r}| \equiv \sqrt{x^2 + y^2 + z^2}$. In vector notation this is

$$|\mathbf{r}|^2 = \mathbf{r} \cdot \mathbf{r} = \mathbf{r}^T \mathbf{r} \tag{2.42}$$

where the dot represents the scalar product, and in the last form we assumed $\mathbf{r}$ is a column vector, and $\mathbf{r}^T$ denotes its transpose: i.e. a row vector. Multiplying that 1×3 row vector onto the 3×1 column vector in the standard way results in a 1×1 'matrix'—in other words a scalar, equal to $x^2 + y^2 + z^2$.

The 'length' of a 4-vector is calculated similarly, but with a crucial sign that enters in because time and space are not exactly the same as each other. For the 4-vector X given in eqn (2.30) you are invited to check that the combination

$$-(\mathsf{X}^0)^2 + (\mathsf{X}^1)^2 + (\mathsf{X}^2)^2 + (\mathsf{X}^3)^2 \qquad (2.43)$$

is 'Lorentz-invariant'. That is,

$$-c^2 t'^2 + x'^2 + y'^2 + z'^2 = -c^2 t^2 + x^2 + y^2 + z^2, \qquad (2.44)$$

cf. eqn (1.7). In matrix notation, this quantity can be written

$$-c^2 t^2 + x^2 + y^2 + z^2 = \mathsf{X}^T g \mathsf{X} \qquad (2.45)$$

where

$$g = \begin{pmatrix} -1 & 0 & 0 & 0 \\ 0 & 1 & 0 & 0 \\ 0 & 0 & 1 & 0 \\ 0 & 0 & 0 & 1 \end{pmatrix}. \qquad (2.46)$$

More generally, if A is a 4-vector, and $\mathsf{A}' = \Lambda \mathsf{A}$, then we have

$$\mathsf{A}'^T g \mathsf{A}' = (\Lambda \mathsf{A})^T g (\Lambda \mathsf{A})$$
$$= A^T (\Lambda^T g \Lambda) A, \qquad (2.47)$$

(where we used eqn (1.15)). Therefore $\mathsf{A}'^T g \mathsf{A}' = \mathsf{A}^T g \mathsf{A}$ as long as

$$\Lambda^T g \Lambda = g. \qquad (2.48)$$

You should now check that g as given in eqn (2.46) indeed satisfies this matrix equation. This proves that for *any* quantity A that transforms in the same way as X, the scalar quantity $\mathsf{A}^T g \mathsf{A}$ is 'Lorentz-invariant', meaning that it does not matter which reference frame is chosen for the purpose of calculating it, as the answer will always come out the same.

g is called 'the metric' or 'the metric tensor'. A generalized form of it plays a central role in General Relativity. In order to consider Lorentz transformations of all kinds (for relative motion in any direction, with or without rotation and inversion) one may regard g as the prior, given quantity, and then the important eqn (2.48) is the defining property of Lorentz transformations Λ in general. This will be explored further in chapter 6.

In the case of the spacetime displacement (or 'interval') 4-vector X, the invariant 'length' we are discussing is the spacetime interval s previewed in eqn (1.7), taken between the origin and the event at X. As we mentioned in eqn (A.4), in the case of time-like intervals the invariant interval length is c times the proper time. To see this, calculate the interval in the reference frame where the X has no spatial part: i.e. $x = y = z = 0$. Then it is obvious that $\mathsf{X}^T g \mathsf{X} = -c^2 t^2$ and the time t is the *proper* time between the origin event 0 and the event at X, because it is the time in the frame where O and X occur at the same position.

Time-like intervals have a negative value for $s^2 \equiv -c^2 \Delta t^2 + (\Delta x^2 + \Delta y^2 + \Delta z^2)$, so taking the square root would produce an imaginary number. However, the significant quantity is the proper time given by $\tau = (-s^2)^{1/2}/c$, which is real, not imaginary. In algebraic manipulations, mostly it is not necessary to take the square root in any case. For

Table 2.2 A selection of useful 4-vectors. Some have more than one name. Their definition and use is developed in the text. The Lorentz factor γ is γ_u: i.e. it refers to the speed u of the particle in question in the given reference frame. $\dot{\gamma}$ is used for $d\gamma/dt$ and $W = dE/dt$. The last column gives the invariant squared 'length' of the 4-vector, but is omitted in those cases where it is less useful in analysis. Above the line are time-like 4-vectors; below the line the acceleration is space-like, and the wave vector may be space-like or time-like.

symbol	definition	components	name(s)	invariant
X	X	$(ct, \mathbf{r})$	4-displacement, interval	$-c^2\tau^2$
U	$dX/d\tau$	$(\gamma c, \gamma\mathbf{u})$	4-velocity	$-c^2$
P	$m_0 U$	$(E/c, \mathbf{p})$	energy-momentum, 4-momentum	$-m_0^2 c^2$
F	$dP/d\tau$	$(\gamma W/c, \gamma\mathbf{f})$	4-force, work-force	
J	$\rho_0 U$	$(c\rho, \mathbf{j})$	4-current density	$-c^2\rho_0^2$
A	A	$(\varphi/c, \mathbf{A})$	4-vector potential	
A	$dU/d\tau$	$\gamma(\dot{\gamma}c, \dot{\gamma}\mathbf{u} + \gamma\mathbf{a})$	4-acceleration	a_0^2
K	$\square\phi$	$(\omega/c, \mathbf{k})$	wave vector	

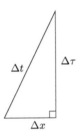

Fig. 2.6 The relationship between $\Delta\tau$, Δt, and Δx can be seen this way (with $c = 1$ and $\Delta y = \Delta z = 0$). The diagram may be thought of as half a 'light-pulse clock'; cf. Fig. A.4.

intervals lying on the surface of a light-cone the 'length' is zero, and these are called **null** intervals.

Table 2.2 gives a selection of 4-vectors and their associated Lorentz-invariant 'length-squared'. These 4-vectors and the use of invariants in calculations will be developed as we proceed. The terminology 'time-like', 'null', and 'space-like' is extended to all 4-vectors in an obvious way, according as $(A^0)^2$ is greater than, equal to, or less than $(A^1)^2 + (A^2)^2 + (A^3)^2$. Note that a 'null' 4-vector is not necessarily zero; rather, it is a 'balanced' 4-vector, poised on the edge between time-like and space-like.

It is helpful to have a mathematical definition of what we mean in general by a 4-vector. The definition is: *a 4-vector is any set of four scalar quantities that transform in the same way as (ct, x, y, z) under a change of reference frame.* Such a definition is useful because it means that we can infer that the basic rules of vector algebra apply to 4-vectors. For example, the sum of two 4-vectors A and B, written A + B, is evaluated by summing the corresponding components, just as is done for 3-vectors. Standard rules of matrix multiplication apply, such as $\Lambda(A + B) = \Lambda A + \Lambda B$. A small change in a 4-vector, written for example dA, is itself a 4-vector.

You can easily show that eqn (2.48) implies that $A^T g B$ is Lorentz-invariant for any pair of 4-vectors A, B. This combination is essentially a form of scalar product, so for 4-vectors we define

$$A \cdot B \equiv A^T g B. \tag{2.49}$$

That is, a central dot operator appearing between two 4-vector symbols is defined to be shorthand notation for the combination $A^T g B$. The result is a scalar, and it is referred to as the 'scalar product' of the 4-vectors. In terms of the components it is

$$-A^0B^0 + (A^1B^1 + A^2B^2 + A^3B^3).$$

A 'vector product' or 'cross product' can also be defined for 4-vectors, but it requires a 4×4 matrix to be introduced. This will be deferred until chapter 12.

4-vector notation; metric signature

Unfortunately there is more than one convention concerning notation for 4-vectors. There are two issues: the order of components, and the sign of the metric. For the former, the notation adopted in this book is the one that is most widely used now, but in the past authors have sometimes preferred to put the time component last instead of first, and then number the components 1 to 4 instead of 0 to 3. Also, sometimes you find $i = \sqrt{-1}$ attached to the time component. This is done merely to allow the invariant length-squared to be written $\sum_\mu (A^\mu)^2$, and the i^2 factor then takes care of the sign. One reason to prefer the introduction of the g matrix (eqn (2.46)) to the use of i is that it allows the transition to General Relativity to proceed more smoothly.

The second issue is the sign of g. When discussing General Relativity, the most common practice in writing the Minkowski metric g is the one adopted in this book. However, within purely Special Relativistic treatments another convention is common, and is widely adopted in the particle physics community. This is to define g with the signs $1, -1, -1, -1$ down the diagonal: i.e. the negative of the version we adopt here. As long as one is consistent, either convention is valid—but beware: changing convention will result in a change of sign of all scalar products. For example, we have $P \cdot P = -m^2c^2$ for the energy-momentum 4-vector, but the other choice of metric would give $P \cdot P = m^2c^2$. The number of positive and negative signs in the metric is called the *signature*. This can also be deduced from the trace of the metric (the sum of the diagonal elements) if the number of dimensions is given. Our metric has signature $+2$, and the other choice has signature -2. The reason that $1, -1, -1, -1$ is preferred by many authors is that it makes time-like vectors have positive 'size', and most of the important basic vectors are time-like (see table 2.2). However, the reasons to prefer $-1, 1, 1, 1$ outweigh this, in my opinion. They are

(1) It can be confusing to use $(+1, -1, -1, -1)$ in General Relativity.

(2) Expressions like $U \cdot P$ ought to remind us of $\mathbf{u} \cdot \mathbf{p}$.

(3) It is more natural to take the 4-gradient as $(-\partial/\partial ct, \partial/\partial x, \partial/\partial y, \partial/\partial z)$, since then it more closely resembles the familiar 3-gradient.

The 4-gradient (item 3) will be introduced in chapter 6, and its relation to the metric will be explained in chapter 12.

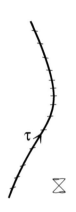

Fig. 2.7

2.5   Basic 4-vectors

2.5.1   Proper time

Consider a worldline, such as the one shown in Fig. 2.7. We would like to describe events along this line, and if possible we would like a description that does not depend on a choice of frame of reference. This is just like the desire to do classical (Newtonian) mechanics without picking any particular coordinate system: in Newtonian mechanics it is achieved by using 3-vectors. In Special Relativity, we use 4-vectors. We also need a parameter to indicate which event we are talking about, i.e. 'how far' along the worldline it is. In Newtonian mechanics this job was done by the time, because that was a universal among reference frames connected by a Galilean transformation. In Special Relativity we use the *proper time* τ. By this we mean the integral of all the little infinitesimal bits of proper time 'experienced' by the particle along its history. This is a suitable choice because this proper time is Lorentz-invariant, agreed among all reference frames.

This basic role of proper time is a central idea of the subject.

In Newtonian mechanics a particle's motion is described by using a position 3-vector **r** that is a function of time, so **r**(t). This is shorthand notation for three functions of t; the time t serves as a parameter. In relativity when we use a 4-vector to describe the worldline of some object, you should think of it as a function of the proper time along the worldline, so $\mathsf{X}(\tau)$. This is a shorthand notation for four functions of τ; the proper time τ serves as a parameter.

Let X be the displacement 4-vector describing a given worldline. This means that its components in any reference frame S give $ct, x(t), y(t), z(t)$ for the trajectory relative to that frame. Two close-together events on the worldline are (ct, x, y, z) and $(c(t + dt), x + dx, y + dy, z + dz)$. The proper time between these events is

$$d\tau = \frac{1}{c} \left(c^2 dt^2 - dx^2 - dy^2 - dz^2 \right)^{1/2} \tag{2.50}$$

$$= dt \left(1 - u^2/c^2 \right)^{1/2} \tag{2.51}$$

where $\mathbf{u} = (dx/dt, dy/dt, dz/dt)$ is the velocity of the particle in S. We thus obtain the important relation

$$\frac{dt}{d\tau} = \gamma \tag{2.52}$$

for neighbouring events on a worldline, where the γ factor is the one associated with the velocity of the particle in the reference frame in which t is calculated.

Eqn (2.12ii) concerns the time between events on a worldline as observed in two frames, neither of which is the rest frame. The worldline is that of a particle having velocity $\mathbf{u}$ in the frame S, with $\mathbf{v}$ the velocity of S′ relative to S. To derive the result, let $(t, \mathbf{r}) = (t, \mathbf{u}t)$ be the coordinates in S of an event on the worldline of the first particle; then the Lorentz transformation gives

$$t' = \gamma_v(t - vx/c^2) = \gamma_v(t - \mathbf{u} \cdot \mathbf{v}t/c^2).$$

Differentiating with respect to t, with all the velocities held constant, gives eqn (2.12ii).

2.5.2   Velocity, acceleration

We have a 4-vector for spacetime displacement, so it is natural to ask whether there is a 4-vector for velocity, defined as a rate of change of the 4-displacement of a particle. To construct such a quantity, we note first of all that for 4-vector X, a small change $d\mathsf{X}$ is itself a 4-vector. To obtain a 'rate of change of X' we should take the ratio of $d\mathsf{X}$ to a small time interval. But take care: if we want the result to be a 4-vector then the small time interval had better be Lorentz-invariant. Fortunately there is a Lorentz-invariant time interval that naturally presents itself: the proper time along the worldline. We thus arrive at the definition

$$\text{4-velocity} \quad \mathsf{U} \equiv \frac{d\mathsf{X}}{d\tau}. \tag{2.53}$$

The 4-velocity 4-vector has a direction in spacetime pointing along the worldline.

If we want to know the components of the 4-velocity in any particular frame, we use eqn (2.52):

$$\mathsf{U} \equiv \frac{d\mathsf{X}}{d\tau} = \frac{d\mathsf{X}}{dt}\frac{dt}{d\tau} = (\gamma_u c,\ \gamma_u \mathbf{u}). \tag{2.54}$$

The notation $(\ldots, \ldots)$ is a list of elements; it is shown horizontal on the page in order to save space, but U should be understood to be a column vector when it is used in matrix expressions such as $\mathsf{U}' = \Lambda \mathsf{U}$. The invariant length or size of the 4-velocity is just c (this is obvious if you calculate it in the rest frame, but for practice you should do the calculation in a general reference frame too). This size is not only Lorentz-invariant (that is, the same in all reference frames) but also constant (that is, not changing with time), even though U can change with time (it is the 4-velocity of a general particle undergoing any form of motion, not just inertial motion). In units where $c = 1$, a 4-velocity is a unit vector.

4-acceleration is defined as one would expect by $\mathsf{A} = d\mathsf{U}/d\tau = d^2\mathsf{X}/d\tau^2$, but now the relationship to a 3-vector is more complicated:

$$\mathsf{A} \equiv \frac{d\mathsf{U}}{d\tau} = \gamma\frac{d\mathsf{U}}{dt} = \gamma\left(\frac{d\gamma}{dt}c,\ \frac{d\gamma}{dt}\mathbf{u} + \gamma\mathbf{a}\right) \tag{2.55}$$

[3] $u = (\mathbf{u} \cdot \mathbf{u})^{1/2} \Rightarrow du/dt = (1/2)$
$(u^2)^{-1/2}(\mathbf{u} \cdot \mathbf{a} + \mathbf{a} \cdot \mathbf{u}) = \mathbf{u} \cdot \mathbf{a}/u,$
or use $(d/dt)(u_x^2 + u_y^2 + u_z^2)^{1/2}$.

where $\gamma = \gamma(u)$ and $\mathbf{a}$ is the 3-acceleration. Using $d\gamma/dt = (d\gamma/du)$ (du/dt) with the γ relation (2.11) and[3] $du/dt = (\mathbf{u} \cdot \mathbf{a})/u$, we find

$$\frac{d\gamma}{dt} = \gamma^3 \frac{\mathbf{u} \cdot \mathbf{a}}{c^2}. \tag{2.56}$$

Therefore

$$\mathsf{A} = \gamma^2 \left(\frac{\mathbf{u} \cdot \mathbf{a}}{c} \gamma^2, \frac{\mathbf{u} \cdot \mathbf{a}}{c^2} \gamma^2 \mathbf{u} + \mathbf{a} \right). \tag{2.57}$$

In the rest frame of the particle this expression simplifies to

$$\mathsf{A} = (0, \mathbf{a}_0) \tag{2.58}$$

where we write $\mathbf{a}_0$ for the acceleration observed in the rest frame. If one takes an interest in the scalar product $\mathsf{U} \cdot \mathsf{A}$, one may as well evaluate it in the rest frame, and thus one finds that

$$\mathsf{U} \cdot \mathsf{A} = 0. \tag{2.59}$$

That is, *the 4-acceleration is always orthogonal to the 4-velocity.* This makes sense, because the magnitude of the 4-velocity should not change: it remains a unit vector. 4-velocity is time-like, and 4-acceleration is space-like and orthogonal to it. This does not, of course, imply that 3-acceleration is orthogonal to 3-velocity (though it can be, but usually is not)[4].

[4] The transformation of 3-acceleration is best obtained by differentiating eqn (2.27) and using eqn (2.12), yielding

$\mathbf{a}'_\parallel = \mathbf{a}_\parallel (\gamma\alpha)^{-3},$

$\mathbf{a}'_\perp = (\mathbf{a}_\perp + (\mathbf{a} \cdot \mathbf{v})\mathbf{u}_\perp/\alpha c^2)(\gamma\alpha)^{-2},$

where $\alpha = 1 - \mathbf{u} \cdot \mathbf{v}/c^2$.

Using the Lorentz-invariant length-squared of A one can relate the acceleration in any given reference frame to the acceleration in the rest frame $\mathbf{a}_0$:

$$\gamma^4 \left(-\left(\frac{\mathbf{u} \cdot \mathbf{a}}{c}\right)^2 \gamma^4 + \left(\frac{\mathbf{u} \cdot \mathbf{a}}{c^2} \gamma^2 \mathbf{u} + \mathbf{a}\right)^2 \right) = a_0^2. \tag{2.60}$$

This simplifies to

$$a_0^2 = \gamma^4 a^2 + \gamma^6 (\mathbf{u} \cdot \mathbf{a})^2/c^2 \quad = \gamma^6 (a^2 - (\mathbf{u} \wedge \mathbf{a})^2/c^2) \tag{2.61}$$

where we give two versions for the sake of convenience in later discussions. As a check, you can obtain the first version from the second by using the triple product rule.

$\mathbf{a}_0$ is the *proper acceleration*, which is the acceleration relative to a frame in which the particle is momentarily at rest. $\mathbf{a}$ is the acceleration relative to a frame of our choosing—for example, the laboratory frame—in which the particle has velocity $\mathbf{u}$.

Eqn (2.61) gives $a_0 = \gamma^2 a$ and $a_0 = \gamma^3 a$ for the cases of $\mathbf{a}$ orthogonal and parallel to $\mathbf{u}$, respectively. In the first case the factor γ^2 comes from two factors of time dilation, and in the second case there is a further factor of γ because of Lorentz contraction. To see this, consider a pair of neighbouring events A,B on the worldline, separated by proper time $d\tau$. At event A the particle is at rest in some frame; at B the velocity of the particle in that frame is $\mathbf{a}_0 d\tau$ (to first order in $d\tau$), and the distance travelled from A is $dl_0 = \frac{1}{2}a_0 d\tau^2$. In another frame moving perpendicular to $\mathbf{a}_0$, the time interval between A and B is dilated but the distance is the same, leading to

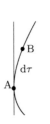

Fig. 2.8

$$a = \frac{2dl_0}{(\gamma d\tau)^2} = \frac{a_0}{\gamma^2}, \tag{2.62}$$

whereas in a frame moving parallel to $\mathbf{a}_0$ one has also contraction of the distance, leading to

$$a = \frac{2dl_0/\gamma}{(\gamma d\tau)^2} = \frac{a_0}{\gamma^3}. \tag{2.63}$$

For example, for circular motion the proper acceleration a_0 is γ^2 times larger than the acceleration $(a = u^2/r)$ in the laboratory, and for straight-line motion it is γ^3 times larger.

Straight-line motion at constant a_0 is motion at constant $\gamma^3 a$. Using the gamma relation (2.11ii), this is motion at constant $(d/dt)(\gamma v)$: in other words, constant rate of change of momentum—i.e. constant force. This will be discussed in detail in section 4.2.4. For such motion the acceleration in the original rest frame falls in proportion to $1/\gamma^3$ as γ increases, which is just enough to maintain a_0 at a constant value.

Addition of velocities: a comment

In section 2.5.2 we showed that the velocity 4-vector describing the motion of a particle has a constant magnitude or 'length', equal to c. It is a unit vector when $c = 1$ unit. This means that one should treat with caution the sum of two velocity 4-vectors:

$$\mathsf{U}_1 + \mathsf{U}_2 =? \tag{2.64}$$

Although the sum on the left hand side is mathematically well-defined, the sum of two 4-velocities does *not* make another 4-velocity, because the sum of two time-like unit vectors is not a unit vector.

The idea of adding velocity vectors comes from classical physics, but if one pauses to reflect one soon realizes that it is not the same sort of operation as, for example, adding two displacements. A displacement in spacetime added to another displacement in spacetime corresponds directly to another displacement. For the case of time-like displacements, for example, it could represent a journey from event A to event B, followed by a journey from event B to event C (where each journey has a definite start and finish time as well as position). Hence it makes sense to write

$$\mathsf{X}_1 + \mathsf{X}_2 = \mathsf{X}_3. \tag{2.65}$$

Adding velocity 4-vectors, however, gives a quantity with no ready physical interpretation. It is a bit like forming a sum of temperatures: one can add them up, but what does it mean? In the classical case the sum of 3-vector velocities makes sense because the velocity of an object C relative to another object A is given by the vector sum of the velocity of C relative to B and the velocity of B relative to A. In Special Relativity velocities do not sum like this, and one must instead use the velocity transformation equations (2.27).

2.5.3 Momentum, energy

Supposing that we would like to develop a 4-vector quantity that behaves like momentum, the natural thing to do is to try multiplying a 4-velocity by a mass. We must ensure that the mass we choose is Lorentz-invariant—which is easy: just use the rest mass. Thus we arrive at the definition

$$\text{4-momentum } \mathsf{P} \equiv m_0 \mathsf{U} = m_0 \frac{d\mathsf{X}}{d\tau}. \qquad (2.66)$$

P, like U, points along the worldline. Using eqn (2.12) we can write the components of P in any given reference frame as

$$\mathsf{P} = \gamma m_0 \frac{d\mathsf{X}}{dt} = (\gamma_u m_0 c, \gamma_u m_0 \mathbf{u}) \qquad (2.67)$$

for a particle of velocity $\mathbf{u}$ in the reference frame.

In chapter 5, relativistic expressions for 3-momentum and energy will be developed. The argument can also be found in many introductory texts. One obtains the important expressions

$$E = \gamma m_0 c^2, \qquad \mathbf{p} = \gamma m_0 \mathbf{u} \qquad (2.68)$$

for the energy and 3-momentum of a particle of rest mass m_0 and velocity $\mathbf{u}$. It follows that the 4-momentum can also be written as

$$\mathsf{P} = (E/c, \mathbf{p})$$

and for this reason P is also called the energy-momentum 4-vector.

In the present chapter we have obtained this 4-vector quantity purely by mathematical argument, and we can call it 'momentum' if we chose. The step of claiming that this quantity has a conservation law associated with it is a further step; it is a statement of physical law. This will be presented in chapter 5.

The relationship

$$\frac{\mathbf{p}}{E} = \frac{\mathbf{u}}{c^2} \qquad (2.69)$$

(which follows from eqn (2.68)) can be useful for obtaining the velocity if the momentum and energy are known.

We used the symbol m_0 for rest mass in the formulae above. This was for the avoidance of all doubt, so that it is clear that this is a rest mass and not some other quantity such as γm_0. Since rest mass is Lorentz-invariant, however, it is by far the most important mass-related concept, and for this reason the practice of referring to γm_0 as 'relativistic mass' is mostly unhelpful, and is best avoided. Therefore, we shall never use the symbol m to refer to γm_0. This frees us from the need to attach a subscript zero, and throughout this book the symbol m will only ever refer to rest mass.

Invariant, covariant, conserved

Invariant or 'Lorentz-invariant' means *the same in all reference frames*

Covariant is a technical term applied to some 4-vector quantities, and is used to mean 'invariant' when it is the mathematical form of an equation (such as $\mathsf{F} = d\mathsf{P}/d\tau$) that is invariant.

Conserved means 'not changing with time' or 'the same before and after'.

Rest mass is Lorentz-invariant but not conserved. Energy is conserved but not Lorentz-invariant.

2.5.4   The direction change of a 4-vector under a boost

The simplicity of the components in $\mathsf{P} = (E/c, \mathbf{p})$ makes P a convenient 4-vector to work with in many situations. For example, to obtain the formula (2.29) for the transformation of a direction of travel, we can use the fact that P is a 4-vector. Suppose a particle has 4-momentum P in frame S. The 4-vector nature of P means that it transforms as $\mathsf{P}' = \Lambda \mathsf{P}$ so,

$$E'/c = \gamma(E/c - \beta p_x),$$

$$p'_x = \gamma(-\beta E/c + p_x),$$

$$p'_y = p_y,$$

and since the velocity is parallel to the momentum we can find the direction of travel in frame S$'$ by $\tan \theta' = p'_y/p'_x$:

$$\tan \theta' = \frac{p_y}{\gamma(-vE/c^2 + p_x)} = \frac{u_y}{\gamma_v(-v + u_x)} = \frac{u \sin \theta}{\gamma_v(u \cos \theta - v)}, \quad (2.70)$$

where we used eqn (2.69). This is valid for any 4-vector, if we take it that u refers to the ratio of the spatial to the temporal part of the 4-vector, multiplied by the speed of light.

Fig. 2.9 gives a graphical insight into this result (see the caption for the argument). The diagram can be applied to any 4-vector, but since it can be useful when considering collision processes, an energy-momentum 4-vector is shown for illustrative purposes.

In the case of a null 4-vector (e.g., P for a zero-rest-mass particle) another form is often useful:

$$\cos \theta' = \frac{cp'_x}{E'} = \frac{\gamma(-\beta E/c + p \cos \theta)}{\gamma(E/c - \beta p \cos \theta)} = \frac{\cos \theta - \beta}{1 - \beta \cos \theta} \quad (2.71)$$

where we used $E = pc$.

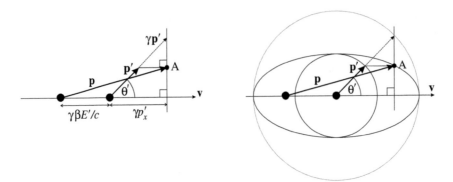

Fig. 2.9 A graphical method for obtaining the direction in space of a 4-vector after a Lorentz 'boost': i.e.. a change to another reference frame whose axes are aligned with the first. (Note that this is neither a spacetime diagram nor a picture in space; it is purely a mathematical construction). Let frame S′ be in standard configuration with S. $\mathbf{p}'$ is a momentum vector in S′. The point A on the diagram is located such that its y position agrees with p'_y, and its x position is $\gamma p'_x$ from the foot of $\mathbf{p}'$. $\mathbf{p}$ is the momentum vector observed in frame S. It is placed so that its foot is at a distance $\gamma \beta E'/c$ to the left of the foot of $\mathbf{p}'$, and it extends from there to A. It is easy to check that it thus has the correct x and y components as given by Lorentz transformation of $\mathbf{p}'$. The interest is that one can show that when θ' varies while maintaining p' fixed, the point A moves around an ellipse. Therefore, the right-hand diagram shows the general pattern of the relationship between $\mathbf{p}$ and $\mathbf{p}'$.

2.5.5 Force

We now have at least two ways in which force could be introduced:

$$\mathsf{F} \overset{?}{=} m_0 \mathsf{A} \quad \text{or} \quad \mathsf{F} \overset{?}{=} \frac{\mathrm{d}\mathsf{P}}{\mathrm{d}\tau}. \tag{2.72}$$

Both of these are perfectly well-defined 4-vector equations; but they are not the same, because the rest mass is not always constant. We are free to choose either, because the relation is a *definition* of 4-force, and we can define things how we wish. However, some definitions are more useful than others, and there is no doubt about which one permits the most elegant theoretical description of the large quantity of available experimental data. It is the second:

$$\mathsf{F} \equiv \frac{\mathrm{d}\mathsf{P}}{\mathrm{d}\tau}. \tag{2.73}$$

The reason why this is the most useful way to define 4-force is related to the fact that P is conserved.

We have

$$\mathsf{F} = \frac{\mathrm{d}\mathsf{P}}{\mathrm{d}\tau} = \left(\frac{1}{c}\frac{\mathrm{d}E}{\mathrm{d}\tau}, \frac{\mathrm{d}\mathbf{p}}{\mathrm{d}\tau} \right),$$

where $\mathbf{p}$ is the relativistic 3-momentum $\gamma m_0 \mathbf{u}$. To work with F in practice it will often prove helpful to adopt a particular reference frame and study its spatial and temporal components separately. To this end we *define* a vector $\mathbf{f}$ by

$$\mathbf{f} \equiv \frac{\mathrm{d}\mathbf{p}}{\mathrm{d}t} \tag{2.74}$$

which is called the force or 3-force. Then we have

$$\mathsf{F} = \frac{d\mathsf{P}}{d\tau} = \gamma \frac{d\mathsf{P}}{dt} = \gamma \frac{d}{dt}(E/c, \mathbf{p}) = (\gamma W/c,\, \gamma \mathbf{f}). \qquad (2.75)$$

where $W = dE/dt$ can be recognized as the rate of doing work by the force.

2.5.6   Wave vector

Another 4-vector appears in the analysis of wave motion. It is the wave-4-vector (or '4-wave-vector')

$$\mathsf{K} = (\omega/c,\ \mathbf{k}) \qquad (2.76)$$

where ω is the angular frequency of the wave, and $\mathbf{k}$ is the spatial wave-vector, which points in the direction of propagation and has size $k = 2\pi/\lambda$ for wavelength λ. We shall postpone the proof that K is a 4-vector until chapter 6. We introduce it here because it offers the most natural way to discuss the general form of the Doppler effect, for a source moving in an arbitrary direction. Note that the waves described by $(\omega/c,\ \mathbf{k})$ could be any sort of wave motion, not just light-waves. They could be waves on water, or pressure waves, etc. The 4-wave-vector can refer to any quantity a whose behaviour in space and time takes the form

$$a = a_0 \cos(\mathbf{k} \cdot \mathbf{r} - \omega t)$$

where the wave amplitude a_0 is a constant. The phase of the wave is

$$\phi = \mathbf{k} \cdot \mathbf{r} - \omega t = \mathsf{K} \cdot \mathsf{X}.$$

Since ϕ can be expressed as a dot product of 4-vectors, it is a Lorentz-invariant quantity.[5]

2.6   The joy of invariants

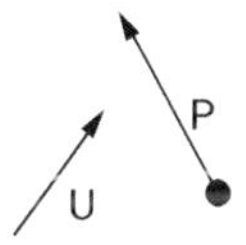

Suppose an observer whose 4-velocity is U observes a particle having 4-momentum P. What is the energy E_O of the particle relative to the observer?

   This is an eminently practical question, and we should like to answer it. One way would be to express P in component form in some arbitrary frame, and Lorentz-transform to the rest frame of the observer. However, do not try it! You should learn to think in terms of 4-vectors, and not go to components if you do not need to.

   We know that the quantity we are looking for must depend on both U and P, and it is a scalar. Therefore, let us consider $\mathsf{U} \cdot \mathsf{P}$. This is such a scalar, and has physical dimensions of energy. Evaluate it in the rest frame of the observer: there $\mathsf{U} = (c, 0, 0, 0)$, so we obtain minus c times the zeroth component of P in that frame: i.e. the particle's energy

[5] In chapter 6 we start by showing that ϕ is invariant without mentioning K, and then define K as its 4-gradient.

Fig. 2.10

in that frame, which is the very thing we wanted. In symbols, this is $U \cdot P = -E_O$. Now bring in the fact that $U \cdot P$ is Lorentz-invariant. This means that nothing was overlooked by evaluating it in one particular reference frame, and it will always give E_O. We are done: *the energy of the particle relative to the observer is* $-U \cdot P$.

This calculation illustrates a very important technique called **the method of invariants**. The idea has been stated beautifully by Hagedorn:

'If a question is of such a nature that its answer will always be the same, no matter in which inertial frame one starts, it must be possible to formulate the answer entirely with the help of those invariants which one can build with the available 4-vectors. One then finds the answer in a particular inertial frame which one can choose freely and in such a way that the answer is there obvious or most easy. One looks then how the invariants appear in this particular system, expresses the answer to the problem by these same invariants, and one has found at the same time the general answer.'

He continues to add that it is worthwhile to devote some time to thinking this through until one has understood that there is no hocus-pocus or guesswork and that the method is completely safe. I agree!

Example   For any isolated system of particles there exists a reference frame in which the total 3-momentum is zero. Such a frame is called the CM (centre-of-momentum) frame. For a system of two particles of 4-momenta P_1, P_2, what is the total energy in the CM frame?

Solution
We have three invariants to hand: $P_1 \cdot P_1 = -m_1^2 c^2$, $P_2 \cdot P_2 = -m_2^2 c^2$, and $P_1 \cdot P_2$. Other invariants, such as $(P_1 + P_2) \cdot (P_1 + P_2)$, can be expressed in terms of these three. Let S' be the CM frame. In the CM frame the total energy is obviously $E_1' + E_2'$. We want to write this in terms of invariants. In the CM frame we have, by definition, $\mathbf{p}_1' + \mathbf{p}_2' = 0$. This means that $(P_1' + P_2')$ has zero momentum part, and its energy part is the very thing we have been asked for. Therefore, the answer can be written as

$$E_{\text{tot}}^{\text{CM}} = E_{\text{tot}}' = c\sqrt{-(P_1' + P_2') \cdot (P_1' + P_2')}$$

$$= c\sqrt{-(P_1 + P_2) \cdot (P_1 + P_2)}, \tag{2.77}$$

where the last step uses the invariant nature of the scalar product. We now have the answer we want in terms of the given 4-momenta, and it does not matter in what frame ('laboratory frame') they may have been specified.

The method of invariants provides a very convenient way to derive the equation (2.13) relating the Lorentz factors for different 3-velocities. We consider the quantity $U \cdot V$ where U and V are the 4-velocities of particles moving with velocities $\mathbf{u}$, $\mathbf{v}$ in some frame. Then, using eqn (2.54) twice,

$$\mathsf{U} \cdot \mathsf{V} = \gamma_u \gamma_v (-c^2 + \mathbf{u} \cdot \mathbf{v}).$$

Let $\mathbf{w}$ be the relative 3-velocity of the particles, which is equal to the velocity of one particle in the rest frame of the other. In the rest frame of the first particle its velocity would be zero, and that of the other particle would be $\mathbf{w}$. Evaluating $\mathsf{U} \cdot \mathsf{V}$ in that frame gives

$$\mathsf{U}' \cdot \mathsf{V}' = -\gamma_w c^2.$$

Now use the fact that $\mathsf{U} \cdot \mathsf{V}$ is Lorentz-invariant. This means that the above two expressions are equal:

$$\gamma_w c^2 = \gamma_u \gamma_v (c^2 - \mathbf{u} \cdot \mathbf{v}).$$

This is eqn (2.13). (See exercise 2.6 for another method.)

2.7 Summary

The main ideas of this chapter have been the Lorentz transformation, 4-vectors, and Lorentz-invariant quantities, especially proper time. To help keep your thoughts on track you should consider the space-time displacement X and the energy-momentum P to be the 'primary' 4-vectors—those most important to remember. They have the simplest expression in terms of components (see table 2.2): their expressions do not involve γ. For wave motion, the 4-wave-vector is the primary quantity.

The next most simple 4-vectors are 4-velocity U and 4-force F.

Exercises

(2.1) Show that

    (i) for any time-like vector Y there exists a frame in which its spatial part is zero,

    (ii) any vector orthogonal to a time-like vector must be space-like,

    (iii) with one exception, any vector orthogonal to a null vector is space-like, and describe the exception.

(2.2) Show that

    (i) the instantaneous 4-velocity of a particle is parallel to the worldline,

    (ii) if a pair of events is simultaneous in the rest frame of some observer, then the 4-displacement between them is orthogonal to that observer's worldline.

    (iii) if the 4-displacement between any two events is orthogonal to an observer's worldline, then the events are simultaneous in the rest frame of that observer.

(2.3) A stack of synchronized clocks is formed in frame S', such that all their hour and minute hands turn at the same rate about a common axis, aligned with the x' axis. Describe the stack of clocks at some instant in frame S.

(2.4) In a given inertial frame S, two particles are shot out from a point in orthogonal spatial directions with equal speeds v. At what rate does the distance between the particles increase in S? What is the speed of each particle relative to the other? Which of these quantities can exceed c?

(2.5) Two particles move along the x axis at speeds $0.8c$ and $0.5c$, the faster setting out two metres behind the slower. After how many seconds do they collide?

(2.6) Two particles have velocities $\mathbf{u}$, $\mathbf{v}$ in some reference frame. The Lorentz factor for their relative velocity $\mathbf{w}$ is given by

$$\gamma(w) = \gamma(u)\gamma(v)(1 - \mathbf{u} \cdot \mathbf{v}/c^2).$$

Prove this twice, by using each of the following two methods.

(i) In the given frame, the worldline of the first particle is $\mathsf{X} = (ct, \mathbf{u}t)$. Transform to the rest frame of the other particle to obtain $t' = \gamma_v t (1 - \mathbf{u} \cdot \mathbf{v}/c^2)$. Obtain dt'/dt and hence the required result.

(ii) Use the invariant $\mathsf{U} \cdot \mathsf{V}$ (as in the text).

(2.7) Show that if two particles have velocities $\mathbf{u}$, $\mathbf{v}$ relative to a given frame, then the speed of one particle relative to the other is

$$w = \frac{c\sqrt{c^2(\mathbf{u}-\mathbf{v})^2 - u^2v^2 + (\mathbf{u} \cdot \mathbf{v})^2}}{c^2 - \mathbf{u} \cdot \mathbf{v}}$$

(2.8) Show from eqns (2.27) that when $\mathbf{u}$ is perpendicular to $\mathbf{v}$, $\gamma(u') = \gamma(u)\gamma(v)$.

(2.9) §In frame S a guillotine blade in the (x, y) plane falls in the negative y direction towards a block level with the x axis and centred at the origin. The angle of the edge of the blade is such that the point of intersection of blade and block moves at a speed greater than c in the positive x direction. In some frame S' in standard configuration with S, this point moves in the *opposite* direction along the block. Now suppose that when the centre of the blade arrives at the block, the whole blade instantaneously evaporates in frame S (for example, it could be vapourized by a very powerful laser beam incident from the z direction). A piece of paper placed on the block is therefore cut on the

negative x-axis only. Explain this in S'. (Hint: the spacetime diagram.)

(2.10) How many boosts by $0.5c$ are required to reach the speed $0.99c$? (where each boost results in speed $0.5c$ in whatever is the rest frame before the boost, and $0.99c$ is the final speed relative to the initial rest frame). (Hint: rapidity.)

(2.11) Evaluate the result of two Lorentz transformations in succession, for relative motion all along the same direction, without using rapidity, and confirm that the result is consistent with eqn (A.11). (This exercise merely shows how to do something 'the hard way'.)

(2.12) Use the Lorentz transformation of the energy-momentum 4-vector to re-derive the velocity transformation equations (2.27), as follows. First obtain $p'_x = \gamma_v(-vE/c^2 + p_x)$, with $E = \gamma_u m_0 c^2$ and $p_x = \gamma_u m_0 u_x$. Therefore

$$m_0 \frac{dx'}{d\tau} = \gamma_v \gamma_u m_0 (u_x - v).$$

Next, make use of

$$\frac{dx'}{dt'} = \frac{dx'}{d\tau} \frac{d\tau}{dt} \frac{dt}{dt'}$$

(cf. exercise 2.6).

(2.13) A particle has energy 1 joule and momentum 10^{-9} kg m/s. Find its speed.

(2.14) Let P_1 and P_2 be the 4-momenta of two particles. Show that

$$\mathsf{P}_1 \cdot \mathsf{P}_2 = -m_1 m_2 c^2 \gamma_u \qquad (2.78)$$

where γ_u is the Lorentz factor of the relative speed of the particles.

(2.15) A pair of pions are emitted from a point with equal speeds v in opposite directions in frame S, and subsequently decay after proper lifetimes τ_1, τ_2, respectively. Find the distance between the decay events (i) in frame S and (ii) in the frame in which the decay events are simultaneous.

Moving light sources

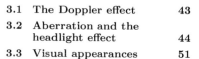

3.1 The Doppler effect

Suppose a wave source in frame S' emits a plane wave of angular frequency ω_0 in a direction making angle θ_0 with the x' axis and lying in the $x'y'$ plane. Then the wave 4-vector in S' is $\mathsf{K}' = (\omega_0/c, k_0 \cos\theta_0, k_0 \sin\theta_0, 0)$. (Here we adopt the subscript zero to indicate values in the frame where the source is at rest).

Applying the inverse Lorentz transformation, the wave 4-vector in S is

$$
\begin{pmatrix} \omega/c \\ k\cos\theta \\ k\sin\theta \\ 0 \end{pmatrix} = \begin{pmatrix} \gamma & \gamma\beta & 0 & 0 \\ \gamma\beta & \gamma & 0 & 0 \\ 0 & 0 & 1 & 0 \\ 0 & 0 & 0 & 1 \end{pmatrix} \begin{pmatrix} \omega_0/c \\ k_0\cos\theta_0 \\ k_0\sin\theta_0 \\ 0 \end{pmatrix} = \begin{pmatrix} \gamma(\omega_0/c + \beta k_0 \cos\theta_0) \\ \gamma(\beta\omega_0/c + k_0 \cos\theta_0) \\ k_0\sin\theta_0 \\ 0 \end{pmatrix} \quad (3.1)
$$

Therefore (extracting the first line, and the ratio of the next two):

$$
\omega = \gamma\omega_0\left(1 + \frac{k_0}{\omega_0}v\cos\theta_0\right), \quad (3.2)
$$

$$
\tan\theta = \frac{\sin\theta_0}{\gamma(\cos\theta_0 + v(\omega_0/k_0)/c^2)}. \quad (3.3)
$$

Eqn (3.2) is the Doppler effect. We did not make any assumption about the source, so this result describes waves of all kinds, not just light.

For light-waves one has $\omega_0/k_0 = c$, so

$$
\omega = \gamma(1 + \beta\cos\theta_0)\omega_0. \quad (3.4)
$$

For $\theta_0 = 0$ we have the 'longitudinal Doppler effect' for light:

$$
\frac{\omega}{\omega_0} = \gamma(1 + v/c) = \left(\frac{1 + v/c}{1 - v/c}\right)^{1/2}.
$$

Another standard case is the 'transverse Doppler effect', observed when $\theta = \pi/2$: i.e., when the received light travels perpendicularly to the velocity of the source *in the reference frame of the receiver* (Note that this is not the same as $\theta_0 = \pi/2$). From eqn (3.3) this occurs when $\cos\theta_0 = -v/c$, so

$$
\frac{\omega}{\omega_0} = \gamma(1 - v^2/c^2) = \frac{1}{\gamma}.
$$

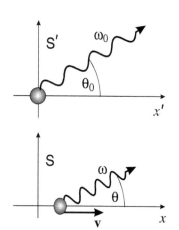

Fig. 3.1

This can be interpreted as an example of time dilation: the process of oscillation in the source is slowed down by a factor γ. It is a qualitatively different prediction from the classical case (where there is no transverse effect) and so represents a direct test of Special Relativity. In practice the most accurate tests combine data from a variety of angles, and a comparison of the frequencies observed in the forward and back longitudinal directions allows the classical prediction to be ruled out, even when the source velocity is unknown.

It can be useful to have the complete Doppler effect formula in terms of the angle θ in the laboratory frame. This is most easily done by considering the invariant $\mathsf{K} \cdot \mathsf{U}$, where U is the 4-velocity of the source. In the source rest frame this evaluates to $-(\omega_0/c)c = -\omega_0$. In the 'laboratory' frame S it evaluates to

$$(\omega/c, \, \mathbf{k}) \cdot (\gamma c, \gamma \mathbf{v}) = \gamma(-\omega + \mathbf{k} \cdot \mathbf{v}) = -\gamma \omega \left(1 - \frac{kv}{\omega} \cos \theta \right).$$

Equating the two expressions, we have

Doppler effect

$$\frac{\omega}{\omega_0} = \frac{1}{\gamma(1 - (v/v_p) \cos \theta)}. \tag{3.5}$$

where $v_p = \omega/k$ is the phase velocity in the laboratory frame. The transverse effect is easy to 'read off' from this formula (as is the effect at any θ). This version, and its straightforward derivation from $\mathsf{K} \cdot \mathsf{U}$, is the most useful form of the Doppler effect formula.

The transverse Doppler effect has to be taken into account in high-precision atomic spectroscopy experiments. In an atomic vapour the thermal motion of the atoms results in 'Doppler broadening'—a spread of observed frequencies, limiting the attainable precision. For atoms at room temperature, the speeds are of the order of a few hundred metres per second, giving rise to longitudinal Doppler shifts of the order of hundreds of MHz for visible light. To avoid this, a collimated atomic beam is used, and the transversely emitted light is detected. For a sufficiently well-collimated beam, the remaining contribution to the Doppler broadening is primarily from the transverse effect. In this way the experimental observation of time dilation has become commonplace in atomic spectroscopy laboratories, as well as in particle accelerators.

3.2   Aberration and the headlight effect

The change in direction of travel of waves (especially light-waves) when the same wave is observed in one of two different inertial frames is called *aberration*. The new name should not be taken to imply that there is anything new here, beyond what we have already discussed. It is just an example of the change in direction of a 4-vector. The name arose historically because changes in the direction of rays in optics were referred to as 'aberration'.

The third line of eqn (3.1) reads $k \sin \theta = k_0 \sin \theta_0$. For light-waves the phase velocity is an invariant, so this can be converted into

$$\omega \sin \theta = \omega_0 \sin \theta_0. \tag{3.6}$$

This expresses the relation between Doppler shift and aberration.

Returning to eqn (3.1) and taking the ratio of the first two lines, one has, for the case $\omega_0/c = k_0$ (e.g., light-waves):

$$\cos \theta = \frac{\cos \theta_0 + v/c}{1 + (v/c) \cos \theta_0}. \tag{3.7}$$

By solving this for $\cos \theta_0$ you can confirm that the formula for $\cos \theta_0$ in terms of $\cos \theta$ can be obtained as usual by swapping 'primed' for unprimed symbols and changing the sign of v (where here the 'primed' symbols are indicated by a subscript zero).

Consider light emitted by a point source fixed in S'. In any given time interval t in S, an emitted photon[1] moves through ct in the direction θ, while the light-source moves through vt in the x-direction; see Fig. 3.2. Consider the case $\theta_0 = \pi/2$; for example, a photon emitted down the y' axis. For example, there might be a pipe laid along the y' axis and the photon travels down it (Fig. 3.3). Observed in the other frame, such a pipe will be parallel to the y axis, and the photon will still travel down it. In time t the photon travels through distance ct in a direction to be discovered, while the pipe travels through a distance vt in the x direction. Therefore, for this case, $c \cos \theta = v$, in agreement with eqn (3.7). A source that emitted isotropically in its rest frame would emit half the light into the directions $\theta_0 \le \pi/2$. The receiver would then observe this light to be directed into a cone with half-angle $\cos^{-1} v/c$—i.e., *less than $\pi/2$*; see Fig. 3.4. This 'forward beaming' is called the **headlight effect** or searchlight effect.

The lower right part of Fig. 2.3 gives an example of the headlight effect. If in an explosion in reference frame S', photons or light-pulses are emitted in all directions, then in frame S the velocities are directed mostly in a cone angled forwards along the direction of propagation of S' in S.

The full headlight effect involves both the direction and the intensity of the light. To understand the intensity (i.e., energy crossing unit area in unit time) consider Fig. 3.5, which shows a plane pulse of light propagating between two mirrors (such as in a laser cavity, for example). We consider a pulse which is rectangular in frame S', and long enough so that it is monochromatic to good approximation, and wide enough so that diffraction can be neglected. Let the pulse length be n wavelengths: i.e., $n\lambda_0$ in frame S'. Imagine a small antenna which detects the pulse as it passes by. Such an antenna will register n oscillations. This number n must be frame-independent. It follows that the length of the pulse in frame S is $n\lambda$.

We shall now prove an interesting property of the propagation of such a rectangular light-pulse: namely, *the area of the wavefronts is Lorentz-invariant*. This follows from the fact that *null worldlines are lines of*

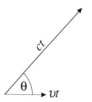

Fig. 3.2

[1] We use the word 'photon' for convenience here. It does not mean the results depend on a particle theory for light. It suffices that the waves travel in straight lines: i.e., along the direction of the wave vector. The 'photon' here serves as a convenient way to keep track of the motion of a given wavefront in vacuum.

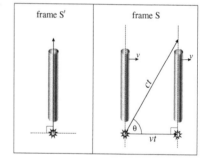

Fig. 3.3 If a photon travels down a given pipe in one reference frame then it will do so in all reference frames. In particular, if the pipe is at rest relative to the source and oriented at right angles to the relative motion of source and observer, then we can use this to deduce quickly the direction of the photon in the observer's frame S, since such a pipe is merely contracted not rotated, as shown. Hence one finds $\cos \theta = v/c$.

Fig. 3.4 The headlight effect for photons. An ordinary incandescent light-bulb is a good approximation to an isotropic emitter in its rest frame: half the power is emitted into each hemisphere. In any frame relative to which the light-bulb moves at velocity **v**, the emission is not isotropic but preferentially in the forward direction. The light appearing in the forward hemisphere of the rest frame is emitted in the general frame into a cone in the forward direction of half-angle $\cos^{-1} v/c$ (so $\sin\theta = 1/\gamma$). Its energy is also boosted. The remainder of the emitted light fills the rest of the full solid angle (the complete distribution is given in eqns (3.13) and (3.14)).

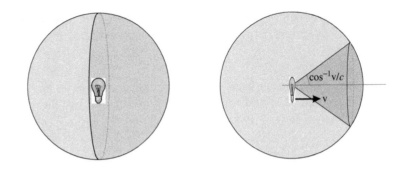

constant interval. To be precise, all events on a null worldline are at the same spacetime interval from some given event B, where B can be positioned anywhere (in some arbitrarily chosen reference frame) but must occur at the right time. This seems surprising, but the proof is simple: see Fig. 3.6.

The straight lines shown in Fig. 1.5b are examples of this fact, and we now see that they are all null lines. With this in mind, the proof can be furnished another way by the use of 4-vectors (exercise 3.8).

Now consider the wavefronts shown in Fig. 3.5. Each end of a wavefront follows a trajectory like the one shown in Fig. 3.6, and each wavefront is an example of the line AB. Focus attention on the $\phi = 0$ wavefront and ignore all the others (if you prefer, suppose we are considering a short pulse of light, or two photons travelling abreast). For convenience, place the origin (of time as well as space coordinates) at B and suppose the phase is zero there. In some other frame (S′), the trajectory followed by the other end of the wavefront (the end passing through A) is given by linear functions $x'(t')$, $y'(t')$, $z'(t')$ which satisfy

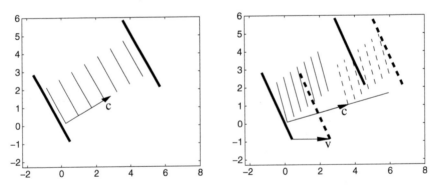

Fig. 3.5 The effect of a change of reference frame on a plane wave. The diagrams show a pulse of light propagating between a pair of mirrors: for example, the mirrors of a laser cavity. The left diagram shows the situation in S′, the rest frame of the mirrors. The right diagram shows the mirrors and wavefronts at two instants of time in frame S (full lines show the situation at $t = 0$, dashed lines show the situation at a later time t). In this frame the laser cavity suffers a Lorentz contraction and the pulse length is reduced by a larger factor. The wavefronts are no longer perpendicular to the mirror surfaces. The angles are such that the centre of each wavefront still arrives at the centre of the right mirror, and after reflection will meet the oncoming left mirror at its centre also. The width of the wavefronts is the same in the two frames (see text).

$-c^2t'^2 + r'^2 = L^2$, where $r' = (x'^2 + y'^2 + z'^2)^{1/2}$, by using the fact that all events along the worldline are at the same interval from the origin B, and this interval is Lorentz-invariant. It follows that at $t' = 0$: i.e., when the wavefront in question arrives at B, the other end of the wavefront is at $r' = L$. *Therefore the width of the wavefront is Lorentz-invariant.* The height of the wavefronts (that is, their length in the direction perpendicular to the relative motion of S and S') is obviously invariant, so we deduce that the area (and shape) of the wavefronts is invariant. QED.

This fact has some further nice implications: it means that a circular disk-shaped pulse of light has a circular (not elliptical) shape in all frames (and with the same radius), and a pattern of photons all propagating abreast will 'look the same' (i.e., form the same image on a flat piece of card placed perpendicular to the beam) no matter what the motion of the card relative to other things (this configuration is called a *super-snapshot*).

Since the area is invariant, and the length transforms as the wavelength, it follows that the volume of the pulse transforms in the same way as its wavelength. Now, the intensity I of a plane wave is proportional to the energy per unit volume u. We have, therefore:

$$\frac{I}{I_0} = \frac{u}{u_0} = \frac{E/\lambda}{E_0/\lambda_0} \tag{3.8}$$

where E is the energy of the pulse. Such a pulse of light can be regarded as an isolated system having zero rest mass and a well-defined energy-momentum 4-vector P describing its total energy and momentum. This statement is non-trivial, and will be re-examined in chapters 8 and 16. Since $P \cdot P = 0 = K \cdot K$ and $\mathbf{p}$ is in the same direction as $\mathbf{k}$, we find that the 4-vectors P and K are in the same spacetime direction, so their components transform similarly. To be precise, $E/E_0 = \omega/\omega_0$. It follows that, for a plane wave, the intensity transforms as the square of the frequency:

$$\frac{I}{I_0} = \frac{u}{u_0} = \frac{\omega^2}{\omega_0^2}. \tag{3.9}$$

(This result can be obtained more directly by tensor methods; see eqn (16.51)). For the forward direction the Doppler effect gives $\omega > \omega_0$; then eqn (3.9) predicts an intensity increase even for a plane wave. This forms the second part of the 'headlight effect'. It means that not only is there a steer towards forward directions, but also an increase in intensity of the plane wave components that are emitted in a forward direction.

The headlight effect can be observed in high-energy collision experiments, and also in astrophysics. Some astrophysical sources emit fast-moving jets of material, which in turn glow. Owing to the headlight effect the observed emission from such jets is mostly along the line of the jet. Owing to the expansion of the universe, distant galaxies are moving away from us. The light-emission from each galaxy is roughly isotropic in its rest frame, so owing to the headlight effect the light is mostly 'beamed'

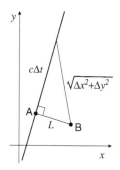

Fig. 3.6 The thick line shows the trajectory of a photon relative to some reference frame S. Event A is some arbitrary event on the worldline. Let B be an event that is simultaneous with A and on a perpendicular from A, in the frame S. Then all events on the worldline have the same interval from B. For, by Pythagoras' theorem, $(\Delta x^2 + \Delta y^2) = L^2 + c^2 \Delta t^2$. Hence the squared interval from B to any event on the worldline is L^2.

one frame

another frame

Fig. 3.7

away from us, making the galaxies appear dimmer. This helps to resolve *Olber's paradox*, concerning why the sky is dark at night.

The headlight effect is put to good use in X-ray sources based on 'synchrotron radiation'. When a charged particle accelerates, its electric field must distort, with the result that it emits electromagnetic waves (see chapter 8). In the case of electrons moving in fast circular orbits, the centripetal acceleration results in radiation called synchrotron radiation. In the rest frame of the electron at any instant, the radiation is emitted symmetrically about an axis along the acceleration vector (i.e., about an axis along the radius vector from the centre of the orbit), and has maximum intensity in the plane perpendicular to this axis. In the laboratory frame two effects come into play: the Doppler effect results in frequency shifts up to high frequency for light emitted in the forward direction, and the headlight effect ensures that most of the light appears in this direction. The result is a narrow beam, almost like a laser beam, of hard X-rays or gamma-rays. This beam is continually swept around a circle, so a stationary detector will receive pulses of X-rays or gamma-rays. (See section 8.3 for more information).

So far we have examined the headlight effect by finding the direction and energy of any given particle or ray. Another important quantity is a measure of how much light is emitted into any given small range of directions. This is done by imagining a sphere around the light-source, and asking how much light falls onto a given region of the sphere.

In spherical polar coordinates, the element of solid angle is (Fig. 3.8)

$$d\Omega = \sin\theta d\theta d\phi. \tag{3.10}$$

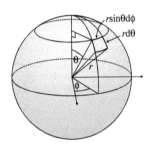

Fig. 3.8 A small change in θ and ϕ gives the arc lengths shown; these arcs meet at right angles, so the area is $r^2 \sin\theta d\theta d\phi$; the solid angle is defined as the ratio of this area to r^2 (so the full solid angle of a sphere is 4π).

When we transform between frames, taking the axis of polar coordinates along the relative motion of the frames, we shall find that $\theta' \neq \theta$ but $\phi' = \phi$ (the latter follows from cylindrical and mirror symmetry, or, equally, from the simple form of the transverse part of the Lorentz transformation). It follows that $d\phi' = d\phi$. Hence the solid angle transforms as

$$\frac{d\Omega'}{d\Omega} = \frac{\sin\theta' d\theta' d\phi'}{\sin\theta d\theta d\phi} = \frac{-d(\cos\theta')d\phi'}{-d(\cos\theta)d\phi} = \frac{d\cos\theta'}{d\cos\theta}. \tag{3.11}$$

(where $d(\cos\theta)$ is simply a way of writing df if $f(\theta) \equiv \cos\theta$). Reverting now to the subscript zero notation, we have $\theta' \equiv \theta_0$ and $\Omega' \equiv \Omega_0$. The relationship between $\cos\theta$ and $\cos\theta_0$ is given by eqn (3.7). For simplicity of calculation, write this in the form $f = (f_0 + \beta)/(1 + \beta f_0)$, where $f = \cos\theta$ and $f_0 = \cos\theta_0$, and then it is easy to evaluate df/df_0. Hence we find that *for a group of light-rays propagating outwards from a point*, the solid angle taken up by the rays transforms as

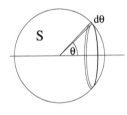

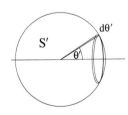

Fig. 3.9 Photons emitted into $d\theta$ in frame S are found propagating in the range of angles $d\theta'$ in frame S'.

$$\frac{d\Omega}{d\Omega_0} = \frac{1}{\gamma^2(1 + \beta\cos\theta_0)^2} = \left(\frac{\omega_0}{\omega}\right)^2 \tag{3.12}$$

where the final step used eqn (3.4).

If the source emits isotropically in its rest frame, then the number of photons going into any given solid angle in any given time is proportional to that solid angle. The significance of eqn (3.12) is that those same photons fill a different solid angle in the other frame—a smaller one for directions ahead of the source, a larger one for directions behind. In other words, there is a brightening in the forward direction: more photons per unit solid angle (or per unit area at any given distance). More generally, the number of photons per unit solid angle is written $dN/d\Omega$, which can be a function of θ and ϕ, and we have that the enhancement factor when one examines the same source from the perspective of another frame is given by

$$\frac{dN}{d\Omega} = \frac{dN}{d\Omega_0}\frac{d\Omega_0}{d\Omega} = \frac{dN}{d\Omega_0}\left(\frac{\omega}{\omega_0}\right)^2 \tag{3.13}$$

since the number N of photons emitted is invariant. For example, the enhancement factor for emission into a small solid angle in the directly forward direction (at $\theta = \theta_0 = 0$) is $\gamma^2(1+\beta)^2 = (1+\beta)/(1-\beta)$.

For isotropic emission in the rest frame one has $dN/d\Omega_0 = N/(4\pi)$ if N is the total number of photons emitted into all directions, but eqn (3.13) does not need to assume this.

The simplicity of the final result (3.13) is owing to eqn (3.12). It is remarkable: the angles are so arranged that the solid angle transforms in the same way as the square of the frequency. There is no very simple reason for this, but a moderately intuitive argument runs as follows. Consider a given emission event E at the origin and a detection event D at $(t_0, \mathbf{r}_0)$ in the rest frame of the source. Since light travels at speed c we must have $r_0 = ct$, so the 4-vector $\mathsf{X} = (ct_0, \mathbf{r}_0)$ is null. Applying a Lorentz transformation, X transforms just the same way as the 4-wave-vector K, which is also null. Therefore the distance r between E and D, evaluated in other frames, varies from one frame to another in the same way as the wave vector k, and hence (using the universal phase velocity c for light-waves) in the same way as the frequency ω. Now consider two detectors, both present at event D, and presenting the same cross-sectional area, but moving with different velocities. We will prove shortly that they will intercept the same group of rays from E. The one observing the higher frequency finds that the emission event was further away, and therefore that detector presents a smaller solid angle at the emission point, in proportion to $1/r^2$. Therefore the solid angle filled by any given group of propagating photons varies as $d\Omega/d\Omega_0 = (r_0/r)^2 = (\omega_0/\omega)^2$, which is eqn (3.12).

The argument assumed that the detectors at D intercept the same bundle of ray directions from E, independent of the state of motion of the detector. To be precise, if we suppose that the source emitted N photons in all directions in a short burst, then all identical detectors at D receive the same fraction of those N, assuming that each detector has the same proper size and is oriented so as to receive the light at normal incidence in its rest frame. This is obvious for detectors moving directly towards or away from the source (no transverse contraction), but more generally it is a non-trivial result of the behaviour indicated in Figs 3.7 and 3.5. Each detector 'chops out' a certain segment of a flat light-pulse

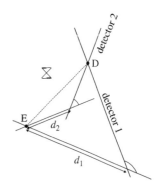

Fig. 3.10 The detector moving towards the source considers that the emission event E was further away.

Fig. 3.11 The Doppler effect and the headlight effect combine in this image of waves emitted by a moving oscillating source. The image shows an example where the emission is isotropic in the rest frame of the source, and the phase velocity is c. Each wavefront is circular, but more bunched up and brighter in the forward direction.

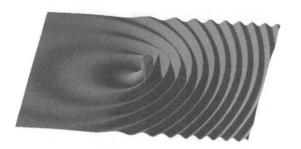

washing over it. It chops out that same segment in the rest frame of the source.

Eqn (3.13) concerns the number of particle velocities or ray directions per unit solid angle, not the flux of energy per unit solid angle. To obtain the latter we need to combine eqns (3.13) and (3.9). The emission into all direction can always be expressed as a set of plane waves; eqn (3.13) shows that for a point source the density (per unit solid angle) of plane wave components transforms as ω^2, and eqn (3.9) states that the intensity of each plane wave transforms as ω^2. It follows that, for a monochromatic source that emits isotropically in its rest frame, the flux of energy per unit solid angle transforms as

$$\frac{\mathrm{d}\mathcal{P}}{\mathrm{d}\Omega} = \left(\frac{\omega^4}{\omega_0^4}\right) \frac{\mathrm{d}\mathcal{P}_0}{\mathrm{d}\Omega_0}. \tag{3.14}$$

This fourth power relationship is a strong dependence. For v close to c, eqn (3.2) gives $\omega \simeq 2\gamma\omega_0$ for emission in the forward direction. At $\gamma \simeq 100$, for example, the brightness in the forward direction is enhanced approximately a billion-fold.

3.2.1   Stellar aberration

'Stellar aberration' is the name for the change in direction of light arriving at Earth from a star, owing to the relative motion of the Earth and the star. Part of this relative motion is constant (over large time-scales) so gives a fixed angle change: we cannot tell it is there unless we have further information about the position or motion of the star. However, part of the angle change varies, owing to the changing direction of motion of the Earth in the course of a year, and this small part can be detected by sufficiently careful observations. Before carrying out a detailed calculation, let us note the expected order of magnitude of the effect. At $\theta_0 = \pi/2$ we have $\cos\theta = v/c$, therefore $\sin(\pi/2 - \theta) = v/c$. For $v \ll c$ this shows that the angle $\pi/2 - \theta$ is small, so we can use the small angle approximation for the sin function, giving $\theta \simeq \pi/2 - v/c$. Indeed, since the velocities are small, one does not need Relativity to calculate the effect. Over the course of six months the angle observed in the rest frame of the Earth is expected to change by about $2v/c \simeq 0.0002$ radians, which is $0.01°$ or about 40 seconds of arc.

It is to his credit that in 1727 James Bradley achieved the required stability and precision in observations of the star γ Draconis. In the course of a year he recorded angle changes in the light arriving down a telescope fixed with an accuracy of a few seconds of arc, and thus he clearly observed the aberration effect. In fact his original intention was to carry out triangulation using the Earth's orbit as a baseline, and thus deduce the distance to the star. The triangulation or 'parallax' effect is also present, but it is much smaller than aberration for stars sufficiently far away. Bradley's observed angle changes were not consistent with parallax (the maxima and minima occured at the wrong points in the Earth's orbit), and he correctly inferred that they were related to the velocity not the position of the Earth.

In the rest frame of the star, it is easy to picture the aberration effect: as the light 'rains down' on the Earth, the Earth with the telescope on it moves across; see Fig. 3.12. Clearly, if a ray of light entering the top of the telescope is to reach the bottom of the telescope without hitting the sides, the telescope must not point straight at the star; it must be angled forward slightly into the 'shower' of light.

In the rest frame of the Earth, we apply eqn (3.7) supposing S′ to be the rest frame of the star. θ is the angle between the received ray and the velocity vector of the star in the rest frame of the Earth. First consider the case where the star does not move relative to the Sun, then v in the formula is the speed of the orbital motion of the Earth. Since this is small compared to c, one may use the binomial expansion $(1 - (v/c)\cos\theta)^{-1} \simeq 1 + (v/c)\cos\theta$ and then multiply out, retaining only terms linear in v/c, to obtain

$$\cos\theta' \simeq \cos\theta - \frac{v}{c}\sin^2\theta. \qquad (3.15)$$

This shows that the largest difference between θ' and θ occurs when $\sin\theta = \pm 1$. This happens when Earth's velocity is at right angles to a line from the Earth to the star. For a star directly above the plane of Earth's orbit, the size of the aberration angle is constant and the star appears to move around a circle of angular diameter $2v/c$; for a star at some other inclination the star appears to move around an ellipse of (angular) major axis $2v/c$.

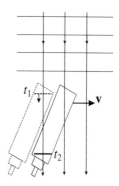

Fig. 3.12 Stellar aberration pictured in the rest frame of the star. The light 'rains down' in the vertical direction, while the telescope fixed to planet Earth moves across. The horizontal lines show wavefronts. The thicker dashed wavefront shows the position at time t_1 of a portion of light that entered the telescope (dashed) a short time ago. In order that it can arrive at the bottom of the telescope, where the *same* bit of light is shown by a bold full line, it is clear that the telescope must be angled into the 'shower' of light. (To be precise, the bold line shows where the light would go if it were not focused by the objective lens of the telescope. The ray passing through the centre of a thin lens is undeviated, so the focused image appears centred on that ray.) This diagram suffices to show that a tilt of the telescope is needed, and in particular, if the telescope later moves in the opposite direction then its orientation must be changed if it is to be used to observe the same star.

3.3   Visual appearances*

This section can be omitted at first reading—not because it is difficult (it is not), but because it is of relatively small importance. No further results in forthcoming chapters will require it.

When we discuss the situation in a given reference frame, we usually mean the dispositions, velocities, accelerations, etc. at some instant of time throughout that frame. However, another question that can naturally arise is 'what do things look like?'—where we mean 'look like' quite literally: where does the light received during some small time interval by an observer located at some particular place appear to have come from? This is the same question as one asks when considering the concept of a virtual image in optics. When high-speed motion is involved, to answer the question it is necessary to take into account the travel time of the light from source to observer. For example, if a train is approaching while I stand at the platform, the light I see now from the back of the train must have set off before the light I see now from the front of the train. Therefore the visual appearance of the train is 'stretched' (a purely classical phenomenon). The size of this visual stretching is similar to the Lorentz contraction, and in some arrangements exactly the same!

Example   Consider a cube of proper side L_0 oriented horizontally and passing horizontally from left to right at height h above a strip of photographic film. Light from the Sun falls perpendicularly onto the film. The film is exposed briefly at some moment in its own rest frame, so as to register the shadow of the cube. Find the length of the shadow.

Solution

Let us calculate in the rest frame of the film throughout. The cube is Lorentz-contracted by a factor γ along its direction of motion. The shadow of the lower face clearly fills a length L_0/γ of the film. The travel times of light from the lower and upper faces of the cube to the film are h/c and $(h + L_0)/c$ respectively. Therefore the shadow of the upper face trails behind that of the lower face by a distance $L_0 v/c$. The total length of the observed shadow is therefore $L_0(1/\gamma + v/c)$.

In this example the cube's shadow is longer, not shorter, than L_0. If one considers an image formed by light-rays emitted by the cube and arriving perpendicularly onto the film (a so-called 'super-snapshot') then the registered image is the same as would be obtained from a non-moving cube of the *same* size L_0 but rotated through $\sin^{-1} v/c$ (exercise for the reader). In this sense one may say that the cube 'appears' to be rotated and not contracted (and therefore an intelligent observer would deduce that the cube was, at any instant in his frame, contracted and not rotated!).

For the case of observation of a sphere, the following neat demonstration is due to Penrose.

Write the angular transformation eqn (3.7) in the form[2]

$$\tan\tfrac{1}{2}\theta = \sqrt{\frac{1 - v/c}{1 + v/c}}\, \tan\tfrac{1}{2}\theta_0. \tag{3.16}$$

Now consider an observer at a particular place O. All that the observer sees can be mapped onto a sphere around him (the very 'sphere of the heavens' we sometimes find ourselves imagining when we look at the stars); see Fig. 3.14. Let him arrange for it to be further mapped onto the tangent plane shown in Fig. 3.14, by stereographic projection from A. We are doing this merely as a mathematical device to help understand what different observers will see. The projected image has a vertical dimension proportional to $\tan\tfrac{1}{2}\theta$.

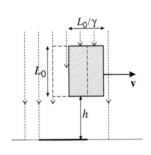

Fig. 3.13

[2] This can be obtained by using either the trigonometric identity $\cos\theta = (1 - t^2)/(1 + t^2)$ where $t = \tan(\theta/2)$, or the identity $t = \sin\theta/(1 + \cos\theta)$.

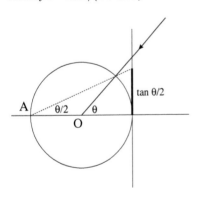

Fig. 3.14

Now suppose there is a spherical object moving in the sky, emitting light, and consider two observers who momentarily meet at O. Depending on their motion, both observers receive light propagating at angles transformed from θ_0, but according to eqn (3.16) the stereographic projection images they each find will only differ by a scale factor. Let one observer be at rest in the rest frame of the spherical object. It is obvious that he will see a circular shape on his sky. Now, *stereographic projection maps circles to circles*, so he will also see a circular image on his vertical 'cinema screen'. It follows that the other observer, no matter what his motion, will also see a circular stereographic projection (since it merely differs by a scale factor), and therefore *he sees a circular shape in his sky*. (Any pictures painted on the surface of the sphere will nonetheless appear distorted.) We conclude that, despite Lorentz contraction (or rather, *because of length contraction*, since without it the image would be stretched), the boundary of a moving spherical object will always present a circular visual appearance, to all observers. It does not 'look' contracted—which implies that it must actually be contracted.

To undertake a more general consideration of visual appearances, consider a point source moving along the line $y = y_0$ at constant velocity v. Its x-position is given by $x(t) = x_0 + vt$, where x_0 is a constant. A photon emitted by the source at time $t = t_s$ and propagating to the origin arrives there at time

$$t = t_s + \frac{1}{c}\left(y_0^2 + (x_0 + vt_s)^2\right)^{1/2}, \qquad (3.17)$$

and propagates along a line making an angle θ to the y-axis given by

$$\tan\theta = (x_0 + vt_s)/y_0. \qquad (3.18)$$

Eqns (3.17) and (3.18) allow one to reconstruct the appearance of a general moving object, since such an object can be considered to form a set of point sources with various values of (x_0, y_0). The image formed at any given time by a small imaging apparatus located at the origin is constructed from rays arriving there simultaneously.

An appearance of superluminal motion

Suppose that a quasar 1 billion light-years from Earth explodes, emitting material which moves outwards in two lobes. The explosion is recorded by observers on Earth, and one year later an image is taken of the (brightly glowing) lobes, using a powerful telescope; Fig. 3.15. Suppose the angular separation of one lobe from the location of the explosion, as recorded on the image, is then found to be 1 milli-arcsecond (approximately 5×10^{-9} radians). It would appear, then, that the lobe has moved a distance $5 \times 10^{-9} \times 10^9 = 5$ light-years in one year. This naive calculation is said to imply a 'visual appearance of faster-than-light motion'. Of course, to an intelligent observer there is no such appearance, because he would not accept the calculation: the travel time of the light

view at some date

$\Delta x = 5$ ly

viewed a year later

Fig. 3.15 Astronomical observations. Surprising or not?

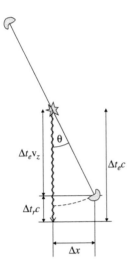

Fig. 3.16 The situation near the star when the lobe arrived at Δx. The bright light-pulse from the explosion is on its way. The lobe is about to emit the light which will contribute to the second image above. The transverse velocity of the lobe is $\Delta x / \Delta t_e$, not $\Delta x / \Delta t_r$.

has not been properly taken into account. What is going on is that the lobes are not moving purely in the transverse direction to the line of sight from Earth, but they have a component of velocity along the line of sight; Fig. 3.16. If this component is v_z then you can easily check that the difference in reception times Δt_r is related to the difference in emission times Δt_e by

$$\Delta t_r = \Delta t_e \left(1 - \frac{v_z}{c}\right).$$

Clearly $\Delta t_e > \Delta t_r$ if the source has a component of velocity towards us. The correct conclusion in the above example is that v_z was positive and the lobe took more than one year to travel between the observed positions.

One cannot fully deduce the direction of motion of the lobe from the given observations, but one can constrain it by imposing the condition $v < c$, where v is its speed. If the lobe velocity makes an angle θ with the line of sight, then the observed 'apparent transverse speed' is given by

$$\frac{\Delta x}{\Delta t_r} = \frac{\Delta x}{\Delta t_e} \frac{\Delta t_e}{\Delta t_r} = \frac{v \sin\theta}{1 - v \cos\theta / c}.$$

Exercises

(3.1) Two photons travel along the x-axis of S, with a constant distance L between them. Find the distance between them as observed in S'. How is this result related to the Doppler effect?

(3.2) The emission spectrum from a source in the sky is observed to have a periodic fluctuation, as shown in the data displayed in Fig. 3.17.

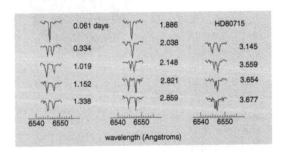

Fig. 3.17 Spectra of light received from an astronomical object at specific times during an observation period of a few days.

It is proposed that the source is a binary star system. Explain how this could give rise to the data. Extract an estimate for the component of orbital velocity in the line of sight. Assuming the stars have equal mass, estimate also the distance between them and their mass.

(3.3) §§Excited ions in a fast beam emit light on a given internal transition. The wavelength observed in the direction parallel to the beam is 359.5 nm, the wavelength observed in the direction perpendicular to the beam in the laboratory is 474.4 nm. Find the wavelength in the rest frame of the ions, and the speed of the ions in the laboratory. [*Ans.* 406.3 nm, 0.2422c]

(3.4) A light-source moves relative to an observer S at speed v. Show that for light emitted at a small angle θ_0 to the forward direction in the source frame, the angle of the ray observed by S is

$$\theta \simeq \theta_0 \sqrt{\frac{c - v}{c + v}}.$$

(3.5) A particle moves with $\gamma \gg 1$ and emits light at a small angle θ to its line of flight in the laboratory. Show that the angle of emission (not necessarily small) in the particle's rest frame is given approximately by

$$\cos^{-1}\frac{1-\gamma^2\theta^2}{1+\gamma^2\theta^2}.$$

(3.6) Given a radioactive source that emits neutrinos isotropically, how could a narrow beam of neutrinos be obtained?

(3.7) $\S\S$A galaxy with a negligible speed of recession from Earth has an active nucleus. It has emitted two jets of hot material with the same speed v in opposite directions, at an angle θ to the direction to Earth. A spectral line in singly-ionised Mg (proper wavelength $\lambda_0 = 448.1$ nm) is emitted from both jets. Show that the wavelengths $\lambda_\pm$ observed on Earth from the two jets are given by

$$\lambda_\pm = \lambda_0\gamma(1 \pm (v/c)\cos\theta)$$

(you may assume the angle subtended at Earth by the jets is negligible). If $\lambda_+ = 420.2$ nm and $\lambda_- = 700.1$ nm, find v and θ.

In some cases the receding source is difficult to observe. Suggest a reason for this. [*Ans.* 0.6c, 65.4°]

(3.8) (i) An arbitrary null worldline can be specified by $\mathsf{X} = \mathsf{X}_0 + \alpha\mathsf{K}$ where X_0 is constant, K is null and α is a parameter. Show that all events on the worldline are at the same interval from any given event X_B in the plane $\mathsf{K}\cdot\mathsf{X}_B = \mathsf{K}\cdot\mathsf{X}_0$.

(ii) Relate the quantities in part (i) to the behaviour of a wavefront of a plane light-wave. Hence show that the area of such wavefronts is Lorentz invariant. (This is arguably a neater method to the one given before eqn (3.8).)

(3.9) **Moving mirror.** A plane mirror moves uniformly with velocity $\mathbf{v}$ in the direction of its normal in a frame S. An incident light-ray has angular frequency ω_i and is reflected with angular frequency ω_r. Show that

$$\omega_i \sin\theta_i = \omega_r \sin\theta_r$$

where θ_i, θ_r are the angles of incidence and reflection, and

$$\frac{\tan(\theta_i/2)}{\tan(\theta_r/2)} = \frac{1+v/c}{1-v/c}.$$

Visual appearances

(3.10) Two observers move towards a small distant light-source. At the event when one observer is overtaken by the other, he finds that the source looks twice as wide as it does to the other. Show that the relative speed of the observers is $3c/5$. (Hint: consider a single ray emitted from the edge of the source and observed by both observers.)

(3.11) §A rocket flies through a circular hoop, along the axis, at speed v. How far past the hoop is the rocket when the hoop appears to the pilot to be exactly in the lateral direction? Is the same answer obtained in Galilean physics?

(3.12) It is sometimes claimed that according to Special Relativity it is possible to 'see round corners'. Is this true? (Hint: the answer is *no*!) The claim is related to the change of angle of propagation of light from one frame to another. For example, light emitted from the back of a fast train at a shallow angle to the surface (i.e., at a large angle to the direction of propagation of the train) will be 'thrown forward' and found to be propagating somewhat in the forward direction in another frame. A person standing on a platform and watching the train pass from his left to right will therefore first see the back face of the last carriage when he looks somewhat to his left: i.e., in a direction he might find surprising. This is the basis of the 'seeing round a corner' claim. Investigate this by calculation and diagrams. Does the light propagate in a straight line? Does the corner of the train get in the way of the light?

(3.13) A rod of proper length L_0 lies along the line $y = h$ and moves along that line at speed v. Define the 'apparent length' of the rod as the length of a stationary rod lying at $y = h$ which would have the same appearance (to a viewer at the origin) as the moving rod. Establish that the apparent length is γL_0 at $t = 0$, and tends to $L_0\alpha$ and L_0/α at $t = \pm\infty$, respectively, where $\alpha = ((1 - v/c)/(1 + v/c))^{1/2}$.

Dynamics

We are now ready to carry out the sort of calculation one often meets in mechanics problems: the motion of a particle subject to a given force, and the study of collision problems through conservation laws.

Since the concept of force is familiar in classical mechanics, we shall start with that, treating problems where the force is assumed to be known, and we wish to derive the motion. However, since we are also interested in exploring the foundations of the subject, one should note that most physicists would agree that the notion of conservation of momentum is prior to, or underlies, the notion of force. In other words, force is to be understood as a useful way to keep track of the tendency of one body to influence the momentum of another when they interact. We *define* the 3-force $\mathbf{f}$ as equal to $d\mathbf{p}/dt$, where $\mathbf{p} = \gamma_v m_0 \mathbf{v}$ is the 3-momentum of the body on which it acts. This proves to be a useful idea, because there are many circumstances where the force can also be calculated in other ways. For example, for a spring satisfying Hooke's law we would have $\mathbf{f} = -k\mathbf{x}$, where $\mathbf{x}$ is the extension, and in electromagnetic fields we would have $\mathbf{f} = q(\mathbf{E} + \mathbf{v} \wedge \mathbf{B})$, etc. Therefore it makes sense to study cases where the force is given and the motion is to be deduced. However, the whole argument relies on the definition of momentum, and the reason momentum is defined as $\gamma_v m_0 \mathbf{v}$ is that this quantity satisfies a conservation law, which we shall discuss in the next chapter.

In the first section we introduce some general properties of the 4-force. We then treat various examples using the more familiar language of 3-vectors. This consists of various applications of the relativistic 'second law of motion' $\mathbf{f} = d\mathbf{p}/dt$.

4.1 Force

Let us recall the definition of 4-force (eqn (2.73)):

$$\mathsf{F} \equiv \frac{d\mathsf{P}}{d\tau} = \left(\frac{1}{c}\frac{dE}{d\tau}, \frac{d\mathbf{p}}{d\tau} \right) = \left(\frac{\gamma}{c}\frac{dE}{dt}, \gamma\mathbf{f} \right). \tag{4.1}$$

where $\mathbf{f} \equiv d\mathbf{p}/dt$. Suppose a particle of 4-velocity U is subject to a 4-force F. Taking the scalar product, we obtain the Lorentz-invariant quantity

$$\mathsf{U} \cdot \mathsf{F} = \gamma^2 \left(-\frac{dE}{dt} + \mathbf{u}\cdot\mathbf{f} \right). \tag{4.2}$$

One expects that this should be something to do with the rate of doing work by the force. Because the scalar product of two 4-vectors is Lorentz-invariant, one can calculate it in any convenient reference frame and obtain an answer that applies in all reference frames. So let us calculate it in the rest frame of the particle ($\mathbf{u} = 0$), obtaining

$$\mathsf{U} \cdot \mathsf{F} = -c^2 \frac{\mathrm{d}m_0}{\mathrm{d}\tau}, \tag{4.3}$$

since in the rest frame $\gamma = 1, \dot{\gamma} = 0, E = m_0 c^2$ and $\mathrm{d}t = \mathrm{d}\tau$. We now have the result all in terms of Lorentz-invariant quantities, and we obtain an important basic property of 4-force:

When $\mathsf{U} \cdot \mathsf{F} = 0$, *the rest mass is constant.*

A force which does not change the rest mass of the object on which it acts is called a *pure force*. The work done by a pure force goes completely into changing the kinetic energy of the particle. In this case we can set eqn (4.2) equal to zero, thus obtaining

$$\frac{\mathrm{d}E}{\mathrm{d}t} = \mathbf{f} \cdot \mathbf{u} \qquad [\text{ for pure force, } m_0 \text{ constant} \tag{4.4}$$

This is just like the classical relation between force and rate of doing work. An important example of a pure force is the force exerted on a charged particle by electric and magnetic fields. Fundamental forces that are non-pure include the strong and weak forces of particle physics.

A 4-force which does not change a body's velocity is called *heat-like*. Such a force influences the rest mass (for example by feeding energy into the internal degrees of freedom of a composite system such as a spring or a gas).

In this chapter we will study equations of motion only for the case of a pure force. The next chapter will include general forces (not necessarily pure), studied through their effects on momenta and energies.

4.1.1   Transformation of force

We introduced the 4-force on a particle by the sensible definition $\mathsf{F} = \mathrm{d}\mathsf{P}/\mathrm{d}\tau$. Note that this statement makes Newton's second law a definition of force, rather than a statement about dynamics. Nonetheless, just as in classical physics, a physical claim is being made: we claim that there will exist cases where the size and direction of the 4-force can be established by other means, and then the equation can be used to find $\mathrm{d}\mathsf{P}/\mathrm{d}\tau$. We also made the equally natural definition $\mathbf{f} = \mathrm{d}\mathbf{p}/\mathrm{d}t$ for 3-force. However, we are then faced with the fact that a Lorentz factor γ appears in the relationship between F and $\mathbf{f}$: see eqn (4.1). This means that the transformation of 3-force, under a change of reference frame, depends not only on the 3-force $\mathbf{f}$ *but also on the velocity of the particle on which it acts*. The latter may also be called the velocity of the 'point of action of the force'.

Let $\mathbf{f}$ be a 3-force in reference frame S, and let $\mathbf{u}$ be the 3-velocity in S of the particle on which the force acts. Then, by applying the

Lorentz transformation to $\mathsf{F} = (\gamma_u W/c, \gamma_u \mathbf{f})$, where $W = dE/dt$, one obtains

$$\frac{\gamma_{u'}}{c}\frac{dE'}{dt'} = \gamma_v\gamma_u\left((dE/dt)/c - \beta f_\|\right),$$

$$\gamma_{u'}\mathbf{f}'_\| = \gamma_v\gamma_u\left(-\boldsymbol{\beta}(dE/dt)/c + \mathbf{f}_\|\right),$$

$$\gamma_{u'}\mathbf{f}'_\perp = \gamma_u\mathbf{f}_\perp, \tag{4.5}$$

where $\mathbf{u}'$ is related to $\mathbf{u}$ by the velocity transformation formulae (2.27). With the help of eqn (2.13) relating the γ factors, one obtains

$$\mathbf{f}'_\| = \frac{\mathbf{f}_\| - (\mathbf{v}/c^2)dE/dt}{1 - \mathbf{u}\cdot\mathbf{v}/c^2}, \qquad \mathbf{f}'_\perp = \frac{\mathbf{f}_\perp}{\gamma_v(1 - \mathbf{u}\cdot\mathbf{v}/c^2)}. \tag{4.6}$$

These are the transformation equations for the components of $\mathbf{f}'$ parallel and perpendicular to the relative velocity of the reference frames, when in frame S the force $\mathbf{f}$ acts on a particle moving with velocity $\mathbf{u}$. (Note the similarity with the velocity transformation equations, owing to the similar relationship with the relevant 4-vector).

For the case of a pure force, it is useful to substitute eqn (4.4) into (4.6i), giving

$$\mathbf{f}'_\| = \frac{\mathbf{f}_\| - \mathbf{v}(\mathbf{f}\cdot\mathbf{u})/c^2}{1 - \mathbf{u}\cdot\mathbf{v}/c^2} \qquad [\text{ if } m_0 = \text{const.} \tag{4.7}$$

Unlike in classical mechanics, $\mathbf{f}$ is not invariant between inertial reference frames. However, a special case arises when m_0 is constant and the force is parallel to the velocity $\mathbf{u}$. Then the force is the same in all reference frames whose motion is also parallel to $\mathbf{u}$. This is easily proved by using eqn (4.7) with $\mathbf{f}\cdot\mathbf{u} = fu$, $\mathbf{u}\cdot\mathbf{v} = uv$ and $\mathbf{f}_\perp = 0$. Alternatively, simply choose S to be the rest frame ($\mathbf{u} = 0$) so one has $dE/dt = 0$, and then transform to any frame S' with $\mathbf{v}$ parallel to $\mathbf{f}$. The result is $\mathbf{f}' = \mathbf{f}$ for all such S'.

The transformation equations also tell us some interesting things about forces in general. Consider, for example, the case $\mathbf{u} = 0$: i.e., $\mathbf{f}$ is the force in the rest frame of the object on which it acts. Then eqn (4.6) says $\mathbf{f}'_\perp = \mathbf{f}_\perp/\gamma$: i.e., the transverse force in another frame is smaller than the transverse force in the rest frame. Since transverse area contracts by this same factor γ, we see that the force per unit area is independent of reference frame.

Suppose that an object is put in tension by forces that are just sufficient to break it in the rest frame. In frames moving perpendicular to the line of action of such forces, the tension force is reduced by a factor γ, and yet the object still breaks. Therefore the breaking strength of material objects is smaller when they move! We will see how this comes about for the case of electromagnetic forces in chapter 7.

The **Trouton–Noble experiment** nicely illustrates the relativistic transformation of force; see Fig. 4.1.

Next, observe that if $\mathbf{f}$ is independent of $\mathbf{u}$, then $\mathbf{f}'$ does depend on $\mathbf{u}$. Therefore, independence of velocity is not a Lorentz-invariant property.

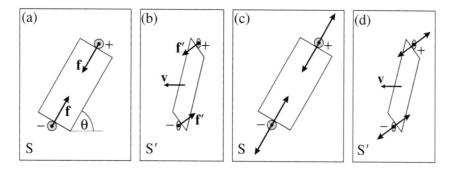

Fig. 4.1 The Trouton–Noble experiment Suppose that two opposite charges are attached to the ends of a non-conducting rod, so that they attract one another. Suppose that in frame S the rod is at rest, and oriented at angle θ to the horizontal axis. The forces exerted by each particle on the other are equal and opposite, directed along the line between them and of size f (fig. (a)). Now consider the situation in a reference frame S$'$ moving horizontally with speed v. The rod is Lorentz-contracted horizontally (the figure shows an example with $\gamma = 2.294$). The force transformation equations (4.6) state that in S$'$ the force is the same in the horizontal direction, but reduced in the vertical direction by a factor γ, as shown. Therefore the forces $\mathbf{f}'$ are not along the line between the particles in S$'$ (fig. (b)). Is there a net torque on the system? This torque, if it existed, would allow the detection of an absolute velocity, in contradiction of the Principle of Relativity. The answer (supplied by Lorentz (1904)) is given by figures (c) and (d), which indicate the complete set of forces acting on each particle, including the reaction from the surface of the rod. These are balanced, in any frame, so there is no torque. (There are also balanced stresses in the material of the rod (not shown), placing it in compression.) In 1901 (i.e., before Special Relativity was properly understood) Fitzgerald noticed that the energy of the electromagnetic field in a capacitor carrying given charge would depend on its velocity and orientation (see exercise 16.10 of chapter 7), implying that there would be a torque tending to orient the plates normal to the velocity through the 'aether'. The torque was sought experimentally by Trouton and Noble in 1903, with a null result. The underlying physics is essentially the same as for the rod with charged ends, but the argument in terms of field energy is more involved, because there is a flow of energy and momentum in the field, discussed in chapter 16.

A force which does not depend on the particle velocity in one reference frame transforms into one that does in another reference frame. This is the case, for example, for electromagnetic forces. It is a problem for Newton's law of gravitation, however, which we deduce is not correct.

To determine the velocity-dependence of $\mathbf{f}'$ in terms of the velocity in the primed frame, i.e. $\mathbf{u}'$, use the velocity transformation equation (2.27i) to write

$$\frac{1}{1 - \mathbf{u}\cdot\mathbf{v}/c^2} = \gamma_v^2(1 + \mathbf{u}' \cdot \mathbf{v}/c^2). \qquad (4.8)$$

4.2   Motion under a pure force

For a pure force we have $dm_0/dt = 0$, and so eqn (2.74) is

$$\mathbf{f} = \frac{d}{dt}(\gamma m_0 \mathbf{u}) = \gamma m_0 \mathbf{a} + m_0 \frac{d\gamma}{dt}\mathbf{u}, \qquad (4.9)$$

$$\frac{dK}{dt} = \mathbf{f}\cdot\mathbf{u}. \qquad (4.10)$$

We continue to use $\mathbf{u}$ for the velocity of the particle, so $\gamma = \gamma(u)$, and we rewrote eqn (4.4) in order to display all the main facts in one place, with $K \equiv E - m_0 c^2$ the kinetic energy. The most important thing to notice

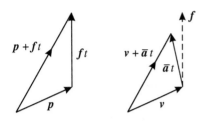

Fig. 4.2 Force and acceleration are usually not parallel. The diagram at left shows the change in momentum from $\mathbf{p}$ to $\mathbf{p_f} = \mathbf{p} + \mathbf{f}t$ when a constant force $\mathbf{f}$ acts for time t. The diagram at right shows what happens to the velocity. The initial velocity is parallel to the initial momentum $\mathbf{p}$, and the final velocity is parallel to the final momentum $\mathbf{p_f}$, but the proportionality constant γ has changed, because the size of v changed. As a result, the change in the velocity vector is not parallel to the line of action of the force. Thus the acceleration is not parallel to $\mathbf{f}$. (The figure shows $\bar{\mathbf{a}}t$ where $\bar{\mathbf{a}}$ is the mean acceleration during the time t; the acceleration is not constant in this example.)

is that the relationship between force and kinetic energy is the familiar one, but acceleration is not parallel to the force, except in special cases such as constant speed (leading to $d\gamma/dt = 0$) or $\mathbf{f}$ parallel to $\mathbf{u}$. Let us see why.

Force is defined as a quantity relating primarily to *momentum*, not *velocity*. When a force pushes on a particle moving in some general direction, the particle is 'duty-bound' to increase its momentum components, each in proportion to the relevant force component. For example, the component of momentum perpendicular to the force, $\mathbf{p_\perp}$, should *not* change. Suppose the acceleration, and hence the velocity change, were parallel with the force. This would mean that the component of velocity perpendicular to the force remains constant. However, in general the speed of the particle does change, leading to a change in γ, so this would result in a change in $\mathbf{p_\perp}$, which is not allowed. We deduce that when the particle speeds up it must redirect its velocity so as to reduce the component perpendicular to $\mathbf{f}$, and when the particle slows down it must redirect its velocity so as to increase the component perpendicular to $\mathbf{f}$. Fig. 4.2 gives an example.

There are two interesting ways to write the $d\gamma/dt$ part. First, we have $E = \gamma m_0 c^2$, so when m_0 is constant we should recognise $d\gamma/dt$ as dE/dt up to constants:

$$\frac{d\gamma}{dt} = \frac{1}{m_0 c^2}\frac{dE}{dt} = \frac{\mathbf{f}\cdot\mathbf{u}}{m_0 c^2}, \tag{4.11}$$

using eqn (4.4), so

$$\mathbf{f} = \gamma m_0 \mathbf{a} + \frac{\mathbf{f}\cdot\mathbf{u}}{c^2}\mathbf{u}. \tag{4.12}$$

This is a convenient form with which to examine the components of $\mathbf{f}$ parallel and perpendicular to the velocity $\mathbf{u}$. For the parallel component one has $\mathbf{f}\cdot\mathbf{u} = f_\| u$ and thus

$$f_\| = \gamma m_0 a_\| + f_\| u^2/c^2 \qquad \Rightarrow \qquad f_\| = \gamma^3 m_0 a_\|.$$

The perpendicular component can then be found using

$$\mathbf{f_\perp} = \mathbf{f} - f_\|\hat{\mathbf{u}} = \left(\gamma m_0 \mathbf{a} + f_\|\frac{u^2}{c^2}\hat{\mathbf{u}}\right) - f_\|\hat{\mathbf{u}}$$

$$= \gamma m_0 \mathbf{a} - f_\|\hat{\mathbf{u}}/\gamma^2 = \gamma m_0(\mathbf{a} - a_\|\hat{\mathbf{u}}) = \gamma m_0 \mathbf{a_\perp}.$$

To summarize,

$$f_\| = \gamma^3 m_0 a_\|, \qquad f_\perp = \gamma m_0 a_\perp. \tag{4.13}$$

These relationships are not special cases; they are true for any motion (unlike eqns (2.63), (2.62)). Since any force can be resolved into longitudinal and transverse components, eqn (4.13) provides one way to find the acceleration, which would be hard to do directly from eqn (4.9). Sometimes people like to use the terminology 'longitudinal mass' $\gamma^3 m_0$ and 'transverse mass' γm_0. This can be useful, but we will not adopt it. The main point is that there is a greater inertial resistance to velocity changes (whether an increase or a decrease) along the direction of motion, compared to the inertial resistance to picking up a velocity component transverse to the current motion (and both exceed the inertia of the rest mass).

One can also use eqn (2.56) in eqn (4.9), giving

$$\mathbf{f} = \gamma m_0 \left(\mathbf{a} + \gamma^2 \frac{\mathbf{u}\cdot\mathbf{a}}{c^2}\mathbf{u}\right) = \gamma^3 m_0 \left((1 - u^2/c^2)\mathbf{a} + \frac{\mathbf{u}\cdot\mathbf{a}}{c^2}\mathbf{u}\right). \quad (4.14)$$

This allows one to obtain the longitudinal and transverse acceleration without an appeal to work and energy.

The 'instantaneous rest frame'

The notion of an 'instantaneous rest frame' has to be correctly understood in the case of a particle undergoing acceleration. It does *not* refer to an accelerating reference frame, but to a sequence of inertial reference frames.

For a particle undergoing any type of motion, one can always talk about the 'rest frame' of the particle for any given event A on the particle's worldline. This is the inertial reference frame in which the velocity of the particle is zero at event A. Let us call this frame S_A. This reference frame is moving inertially—at constant velocity—whether or not the particle is. If the particle is accelerating at event A, then it is at rest in S_A only momentarily. That does not mean S_A is a non-inertial frame, it just means that immediately before and after A the frame S_A is not the particle's rest frame.

We can imagine a continuous set of inertial reference frames moving around in any region of space. Each has constant velocity. As a given particle accelerates through the space, its rest frame is now one member of the set, now another. When we speak of 'the instantaneous rest frame' it means whichever inertial frame in the set is the rest frame at the event under consideration. One may then naturally extend this idea to a sequence of events, and then the phrase 'the instantaneous rest frame' refers to a sequence of inertial frames; it is purely and simply a shorthand phrase for 'the sequence of instantaneous rest frames'.

4.2.1 Linear motion and rapidity

For a particle undergoing straight-line motion, the rapidity is often a useful quantity to consider. This is because for this case there is a simple relationship between rapidity in one frame and another.

Consider a particle accelerating along a line. That is, the velocity is aligned with the acceleration, and the acceleration is always in the same direction (but not necessarily of constant size). Let S_A be the instantaneous rest frame at some event A (see box above). At A the particle has zero velocity in the frame under consideration, and in the next small time interval $d\tau$ it acquires a velocity $dv = a_0 d\tau$, where a_0 is the proper acceleration and $d\tau$ is the proper time, since both quantities are being evaluated in the rest frame. Hence the rapidity ρ_0 increases from zero to $\tanh^{-1}(a_0 d\tau/c) \simeq a_0 d\tau/c$. Therefore

$$\frac{d\rho_0}{d\tau} = \frac{a_0}{c} \qquad (4.15)$$

where the equation applies in frame S_A for events in the vicinity of A. Now recall from the discussion in section 2.3.1 that, for velocity changes all in the same direction, rapidities add. It follows that the rapidity of the particle, as observed in any other frame S moving along the line of acceleration, is $\rho = \rho_A + \rho_0$, where ρ_A is the rapidity of frame S_A as observed in S. I insist again that S_A is an inertial frame, not an accelerating one, so ρ_A is constant. Hence

$$\frac{d\rho}{d\tau} = \frac{d\rho_A}{d\tau} + \frac{d\rho_0}{d\tau} = \frac{d\rho_0}{d\tau} = \frac{a_0}{c} \qquad (4.16)$$

where $a_0 = a_0(\tau)$ is in general a function of proper time. This equation applies for events in the vicinity of A, but now we can argue that it does not matter which event A was chosen. Therefore, *for linear motion, the rate of change of rapidity with respect to proper time is equal to the proper acceleration divided by c.*

4.2.2   Hyperbolic motion: the 'relativistic rocket'

Suppose that the engine of a rocket is programmed in such a way as to maintain a constant proper acceleration. This means that the rate of expulsion of rocket fuel should reduce in proportion to the remaining rest mass, so that the acceleration measured in the instantaneous rest frames stays constant. In an interstellar journey, a (not too large) constant proper acceleration might be desirable in order to offer the occupants of the rocket a constant 'artificial gravity'. For this reason, motion at constant proper acceleration is sometimes referred to as the case of a 'relativistic rocket'.

If the proper acceleration is constant, then eqn (4.16) can be easily solved for ρ as a function of τ:

$$\rho = \frac{a_0}{c}\tau + \text{const.} \qquad (4.17)$$

This gives the rapidity relative to a frame S of our chosing, as a function of proper time. If we choose the frame S to be that in which the rocket is at rest at proper time zero, then the constant of integration is zero and we have (using the definition of rapidity, eqn (2.37))

$$v = c \tanh \rho = c \tanh(a_0 \tau/c). \qquad (4.18)$$

(We are reverting to $\mathbf{v}$ rather than $\mathbf{u}$ for the particle velocity.) This is the speed of the rocket, relative to the initial rest frame S, expressed as a function of proper time on board the rocket.

We would also like to know the speed in frame S as a function of time t in that frame. To this end, recall $\gamma = \cosh\rho$ (eqn (2.38)) and use

$$\frac{\mathrm{d}t}{\mathrm{d}\tau} = \gamma = \cosh(a_0\tau/c). \tag{4.19}$$

Integrating this gives

$$t = (c/a_0)\sinh\rho \tag{4.20}$$

where we settled the constant of integration by choosing $t = 0$ at $\tau = 0$. This formula can be used to carry out an exact calculation of the relative aging of the twins in the twin paradox (see example below).

We can now get v as a function of t by using

$$\tanh\rho = \frac{\sinh\rho}{\cosh\rho} = \frac{\sinh\rho}{(1 + \sinh^2\rho)^{1/2}},$$

hence

$$v(t) = \frac{a_0 t}{(1 + a_0^2 t^2/c^2)^{1/2}}. \tag{4.21}$$

The motion at constant proper acceleration has the following intriguing property. According to either eqn (4.18) or (4.21), the velocity relative to frame S tends to a constant value (the speed of light) as time goes on. Therefore the acceleration relative to frame S falls to zero. The particle always finds itself to have the same constant acceleration in its own rest frame, yet its acceleration relative to any given inertial frame, such as the initial rest frame, dies away to zero as the particle speed approaches c. It is like the Alice and the Red Queen in Lewis Carroll's *Through the looking glass*, forever running to stand still. The particle accelerates and accelerates, and yet only approaches a constant velocity.

For a further comment on constant proper acceleration, see the end of section 2.8.

Next we investigate the distance travelled. We have

$$\frac{\mathrm{d}x}{\mathrm{d}\tau} = \frac{\mathrm{d}x}{\mathrm{d}t}\frac{\mathrm{d}t}{\mathrm{d}\tau} = (c\tanh\rho)\cosh\rho = c\sinh(a_0\tau/c)$$

using eqns (4.18) and (4.19), hence

$$(x - b) = (c^2/a_0)\cosh\rho \tag{4.22}$$

where b is a constant of integration. Combining eqn (4.20) with (4.22) gives

$$(x - b)^2 - c^2 t^2 = (c^2/a_0)^2. \tag{4.23}$$

This is the equation of a hyperbola—see Fig. 4.3—and for this reason motion at constant proper acceleration is called 'hyperbolic motion'. It should be contrasted with the 'parabolic motion' (in spacetime) that is obtained for classical motion at constant acceleration. It is also useful to notice that $(x - b)^2 - c^2 t^2$ is the invariant spacetime interval between

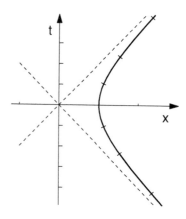

Fig. 4.3 Spacetime diagram showing the worldline of a particle undergoing constant *proper acceleration*. That is, if at any event A on the worldline one picks the inertial reference frame S_A whose velocity matches that of the particle at A, then the acceleration of the particle at A, as observed in frame S_A, has a value a_0, independent of A. The worldline is a hyperbola on the diagram; see eqn (4.23). The asymptotes are at the speed of light. The motion maintains a fixed spacetime interval from the event where the asymptotes cross (cf. Chapter 9). The tick-marks on the worldline indicate the elapse of proper time. This type of motion can be produced by a constant force acting parallel to the velocity.

Table 4.1 A summary of results for straight-line motion at constant proper acceleration a_0 ('hyperbolic motion'). The hyperbolic angle θ has been introduced for convenience. If the origin is chosen so that $b = 0$ then some further simplifications are obtained, such as $x = \gamma x_0$, $v = c^2 t/x$. Note that the Lorentz factor γ increases linearly with distance covered.

$\theta = a_0 \tau/c,$	$\rho = \theta$	(4.24)
$t = (c/a_0)\sinh\theta,$	$x = b + (c^2/a_0)\cosh\theta$	(4.25)
$v = c\tanh\theta,$	$\gamma = \cosh\theta = a_0(x-b)/c^2$	(4.26)
$\tau = c\theta/a_0,$	$\gamma^3 a = a_0$	(4.27)

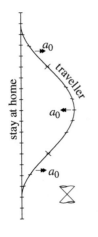

Fig. 4.4 Twin paradox. The journey of the travelling twin can be formed from three pieces of hyperbolic motion.

the event $(t = 0, x = b)$ and the location (t, x) of the particle at any instant. Thus the motion maintains a constant interval from a certain event situated off the worldline. This event is singled out by the initial conditions and the size of the acceleration.

The various results we have calculated are summarized in table 4.1.

Example   *Twin paradox.* A space rocket travels away from Earth with constant proper acceleration $a = 9.8$ ms^{-2} while 10 years pass on board the rocket. It then reverses its motor and slows for 10 proper years. When it comes to rest at a distant star, how far is it from Earth (in the rest frame of Earth)? If it then returns by the same method, how much time will have passed on Earth between initial departure and final return?

Solution

To do this calculation it is convenient to notice that in units of years and light-years, the acceleration 9.8 m/s^2 (which is the typical acceleration due to gravity on Earth) is almost 1 (in fact it is 1.0323 light-years/year^2). During the first 10 proper years, from eqn (4.22) we learn that the rocket travels $\cosh 10 \simeq 11\,000$ light-years from Earth. During the next part of its journey the rocket moves just as if at had set out from the star and accelerated towards Earth, but with time reversed. Such a time reversal does not affect the distance travelled, therefore the distance is again 11 000 light-years, making a total of 22 000 light-years. From eqn (4.20) we learn that during the initial acceleration, $\sinh 10 \simeq 11\,000$ years pass on Earth, so the total time on Earth between departure and final return is 44 000 years.

This example shows how to estimate the range of interstellar voyages at given acceleration. It also illustrates that explorers can thus 'travel into the future', but with no return ticket into the past. An important application of all the above formulae is to the design of linear particle accelerators, where a constant proper acceleration is a reasonable first approximation to what can be achieved. The situation of constant proper acceleration has many further fascinating properties and is discussed at length in chapter 9 as a prelude to General Relativity.

4.2.3   4-vector treatment of hyperbolic motion

If we make the most natural choice of origin, so that $b = 0$ in eqn (4.23), then the equations for x and t in terms of θ combine to make the 4-vector displacement

$$\mathsf{X} = (ct,\ x) = x_0(\sinh\theta,\ \cosh\theta) \tag{4.28}$$

where $x_0 = c^2/a_0$ and we suppressed the y and z components which remain zero throughout. We then obtain

$$\mathsf{U} = \frac{\mathrm{d}\mathsf{X}}{\mathrm{d}\tau} = \frac{\mathrm{d}\mathsf{X}}{\mathrm{d}\theta}\frac{\mathrm{d}\theta}{\mathrm{d}\tau} = c(\cosh\theta,\ \sinh\theta) \tag{4.29}$$

$$\text{and}\quad \dot{\mathsf{U}} = \frac{\mathrm{d}\mathsf{U}}{\mathrm{d}\tau} = a_0(\sinh\theta,\ \cosh\theta) = \frac{a_0^2}{c^2}\mathsf{X}. \tag{4.30}$$

$$\Rightarrow \qquad \ddot{\mathsf{U}} \propto \mathsf{U} \tag{4.31}$$

where the dot signifies $\mathrm{d}/\mathrm{d}\tau$. We shall now show that this relationship between 4-velocity and rate of change of 4-acceleration can be regarded as the defining characteristic of hyperbolic motion.

Suppose we have motion that satisfies (4.31): i.e.,

$$\dot{\mathsf{A}} = \alpha^2 \mathsf{U} \tag{4.32}$$

where α is a constant. Consider $\mathsf{A}\cdot\mathsf{A}$, and recall $\mathsf{A}\cdot\mathsf{A} = a_0^2$ (from eqn (2.58)). Differentiating with respect to τ gives

$$\frac{\mathrm{d}}{\mathrm{d}\tau}(a_0^2) = 2\dot{\mathsf{A}}\cdot\mathsf{A} = 2(\alpha^2\mathsf{U})\cdot\mathsf{A} = 0$$

where we used eqn (4.32) and then the general fact that 4-velocity is perpendicular to 4-acceleration (eqn (2.59)). It follows that a_0 is constant. Hence eqn (4.32) implies motion at constant proper acceleration.

The constant α is related to the proper acceleration. To find out how, consider $\mathsf{A}\cdot\mathsf{U} = 0$. Differentiating with respect to τ gives

$$\dot{\mathsf{A}}\cdot\mathsf{U} + \mathsf{A}\cdot\dot{\mathsf{U}} = 0 \qquad \Rightarrow \qquad \dot{\mathsf{A}}\cdot\mathsf{U} = -a_0^2 \tag{4.33}$$

(using eqn (2.58)). This is true for any motion, not just hyperbolic motion. Applying it to the case of hyperbolic motion, eqn (4.32), we find $-\alpha^2 c^2 = -a_0^2$ hence $\alpha = a_0/c$.

Eqn (4.32) is a second-order differential equation for U, and it can be solved straightforwardly using exponential functions. Upon substituting in the boundary conditions $\mathsf{U} = (c, \mathbf{0})$ and $\dot{\mathsf{U}} = (0, \mathbf{a}_0)$ at $\tau = 0$ one obtains a cosh function for U^0 and a sinh function for the spatial part, the same as we already found in the previous section.

To do the whole calculation starting from the 4-vector equation of motion

$$\mathsf{F} = m_0\frac{\mathrm{d}\mathsf{U}}{\mathrm{d}\tau} \tag{4.34}$$

(valid for a pure force) we need to know what F gives the motion under consideration. Clearly it must be, in component form, $(\gamma\mathbf{f}\cdot\mathbf{v}/c,\ \gamma\mathbf{f})$ in some reference frame, but we would prefer a 4-vector notation which does not rely on any particular choice of frame. The most useful way to write it turns out to be

$$F = \mathbb{F}g U/c \qquad (4.35)$$

where g is the metric and

$$\mathbb{F} = \begin{pmatrix} 0 & f_0 & 0 & 0 \\ -f_0 & 0 & 0 & 0 \\ 0 & 0 & 0 & 0 \\ 0 & 0 & 0 & 0 \end{pmatrix} \qquad (4.36)$$

(for a force along the x direction)[1], with constant f_0. Substituting this into eqn (4.34) we obtain

$$\left. \begin{array}{c} f_0 U^1/c = m_0 \dot{U}^0 \\ f_0 U^0/c = m_0 \dot{U}^1 \end{array} \right\}$$

where the superscripts label the components of U. This pair of simultaneous first-order differential equations may be solved in the usual way, by differentiating the second and substituting into the first, to find

$$\ddot{U}^1 = \left(\frac{f_0}{m_0 c} \right)^2 U^1.$$

This is one component of eqn (4.32), whose solution we discussed above.

4.2.4   Motion under a constant force

The phrase 'constant force' might have several meanings in a relativistic calculation. It could mean constant with respect to time in a given inertial frame or to proper time along a worldline, and it might refer to the 3-force or the 4-force. In this section we will study the case of motion of a particle subject to a 3-force whose size and direction is independent of time and position in a given reference frame.

The reader might wonder why we are not treating a constant 4-force. The reason is that this would be a somewhat unrealistic scenario. If the 4-force is independent of proper time then all parts of the energy-momentum 4-vector increase together, and this means the combination $E^2 - p^2 c^2$ must be changing, so we do not have a pure force. It is not impossible, but it represents a non-simple (and rather artificial) situation. If the 4-force on a particle is independent of reference frame time then its spatial part must be proportional to $1/\gamma_v$ where v is the speed of the particle in the reference frame. Again, it is not impossible but it is rather unusual or artificial.

The case of a 3-force $\mathbf{f}$ that is independent of position and time in a given reference frame, on the other hand, is quite common. It is obtained, for example, for a charged particle moving in a static uniform electric field. Its treatment is very simple for a particle starting from rest:

[1] The matrix $\mathbb{F}$ is introduced in eqn (4.36) merely to show how to write the equation of motion we need. In chapter 7 we shall learn that $\mathbb{F}$ is a contravariant second-rank tensor, but you do not to worry about that for now.

$$\frac{d\mathbf{p}}{dt} = \mathbf{f} \qquad \Rightarrow \qquad \mathbf{p} = \mathbf{p_0} + \mathbf{f}t$$

since $\mathbf{f}$ is constant. If $\mathbf{p_0} = 0$ then the motion is in a straight line with $\mathbf{p}$ always parallel to $\mathbf{f}$, and by solving the equation $p = \gamma m_0 v = ft$ for v one finds

$$v = \frac{ft}{\sqrt{m_0^2 + f^2 t^2/c^2}}. \tag{4.37}$$

This result is plotted in Fig. 4.5.

Example   An electron is accelerated from rest by a static uniform electric field of strength 1000 V/m. How long does it take (in the initial rest frame) for the electron's speed to reach 0.99c?

Solution
The equation $\mathbf{f} = q\mathbf{E}$ for the force due to an electric field is valid at all speeds. Therefore we have $f = 1.6 \times 10^{-16}$ N and the time is $t = \gamma m_0 v/f \simeq 12 \,\mu$s.

In section 4.1.1 (the transformation equations for force) we saw that in this case ($\mathbf{f}$ parallel to $\mathbf{v}$) the force is the same in all reference frames moving in the same direction as the particle. That is, if we were to evaluate the force in other reference frames moving parallel to the particle velocity, then we would find the same force. In particular, we might take an interest in the reference frame in which the particle is momentarily at rest at some given time—the 'instantaneous rest frame'. We would find that the force on the particle in this new reference frame is the same as in the first one, and therefore at the moment when the particle is at rest in the new reference frame, it has the very same acceleration that it had in the original rest frame when it started out! Such a particle always finds itself to have the same acceleration in its own rest frame, so we have an example of motion at constant proper acceleration—the 'hyperbolic motion' already treated in section 4.2.2. A particle momentarily at rest has $\gamma = 1$ so, using eqn (4.9), the relationship between the force and the proper acceleration is simply

$$f = m_0 a_0. \tag{4.38}$$

By substituting this into eqn (4.37) one finds that that equation is identical to eqn (4.21). All the properties of the motion follow as before, summarized in table 4.1.

When the initial velocity is not along the line of the constant force, the proper acceleration is not constant (exercise 4.7).

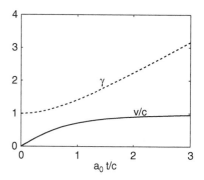

Fig. 4.5 Speed (full curve) and Lorentz factor (dashed curve) as a function of time for straight-line motion under a constant force. The product of these two curves is a straight line.

4.2.5   Circular motion

Another very simple case is obtained when $d\gamma/dt = 0$, i.e. motion at constant speed. From eqn (4.11) this happens when the force remains perpendicular to the velocity. An example is the force on a charged particle moving in a magnetic field: then

$$\mathbf{f} = q\mathbf{v} \wedge \mathbf{B} = \gamma m_0 \mathbf{a}. \tag{4.39}$$

The solution of the equation of motion proceeds exactly as in the classical (low velocity) case, except that a constant factor γ appears wherever the rest mass appears. For an initial velocity perpendicular to $\mathbf{B}$ the resulting motion is circular[2]. The particle moves at speed v around a circle of radius

$$r = \frac{\gamma m_0 v}{qB} = \frac{p}{qB}. \tag{4.40}$$

In particle physics experiments, a standard diagnostic tool is to record the track of a particle in a uniform magnetic field of known strength. This equation shows that if the charge q is also known, then the particle's momentum can be deduced directly from the radius of the track. The equation is also crucial for the design of ring-shaped particle accelerators using magnetic confinement. It shows that to maintain a given ring radius r, the strength of the magnetic field has to increase in proportion to the particle's momentum, not its speed. In modern accelerators the particles move at close to the speed of light anyway, so v is essentially fixed at $\simeq c$, but this does not free us from the need to build ever more powerful magnetic field coils if we want to confine particles of higher energy.

The period and angular frequency of the motion are

$$T = \frac{2\pi r}{v} = 2\pi \frac{\gamma m_0}{qB}, \qquad \omega = \frac{qB}{\gamma m_0}. \tag{4.41}$$

The classical result that the period is independent of the radius and speed is lost. This makes the task of synchronizing applied electric field pulses with the motion of the particle (in order to accelerate the particle) more technically demanding. It required historically the development of the 'synchrotron' from the 'cyclotron'.

Combined electric and magnetic fields will be considered in chapter 13.

4.2.6 Motion under a central force

The case of a *central* force is that in which the force experienced by a particle is always directed towards or away from one point in space (in a given inertial frame). This is an important basic case, partly because in the low-speed limit it arises in the '2-body problem', where a pair of particles interact by a force directed along the line between them. In that case the equations can be simplified by separating them into one equation for the relative motion, and another for the motion of the centre of mass of the system. This simplification is possible because one can adopt the approximation that the field transmits cause and effect instantaneously between the particles, with the result that the force on one particle is always equal and opposite to the force on the other. In the case of high speeds this cannot be assumed. If two particles interact at a distance it must be because they both interact locally with a third party—for example, the electromagnetic field—and the dynamics of the field cannot be ignored. We shall look into this more fully in chapter 13. The main conclusion for our present discussion is that the '2-body' problem is really a '2-body plus field' problem and has no simple solution.

Nevertheless, the idea of a central force remains important and can be a good model when one particle interacts with a very much heavier particle and energy loss by radiation is small—for example, a planet orbiting the Sun. Then the acceleration of the heavy particle can be neglected, and in the rest frame of the heavy particle the other particle experiences, to good approximation, a central force. This can also be used to find out approximately how an electron orbiting an atomic nucleus would move if it did not emit electromagnetic waves.

Consider, then, a particle of rest mass[3] m and position vector $\mathbf{r}$ subject to a force

$$\mathbf{f} = f(r)\hat{\mathbf{r}}. \tag{4.42}$$

Introduce the **3-angular momentum**

$$\mathbf{L} \equiv \mathbf{r} \wedge \mathbf{p}. \tag{4.43}$$

By differentiating with respect to time one finds

$$\dot{\mathbf{L}} = \dot{\mathbf{r}} \wedge \mathbf{p} + \mathbf{r} \wedge \dot{\mathbf{p}} = \mathbf{r} \wedge \mathbf{f}, \tag{4.44}$$

(since $\mathbf{p}$ is parallel to $\dot{\mathbf{r}}$ and $\dot{\mathbf{p}} = \mathbf{f}$) which is true for motion under any force (and is just like the classical result). For the case of a central force one has conservation of angular-momentum:

$$\frac{\mathrm{d}\mathbf{L}}{\mathrm{d}t} = 0 \quad \Longrightarrow \quad \mathbf{L} = \text{const.} \tag{4.45}$$

It follows from this that the motion remains in a plane (the one containing the vectors $\mathbf{r}$ and $\mathbf{p}$), since if $\dot{\mathbf{r}}$ were ever directed out of that plane then $\mathbf{L}$ would necessarily point in a new direction. Adopting plane polar coordinates $\{r, \phi\}$ in this plane we have

$$\mathbf{p} = \gamma m\mathbf{v} = \gamma m(\dot{r}, r\dot{\phi}) = (p_r, \gamma mr\dot{\phi}). \tag{4.46}$$

Therefore

$$L = \gamma mr^2\dot{\phi}. \tag{4.47}$$

(the angular momentum vector being directed normal to the plane). Using $\mathrm{d}t/\mathrm{d}\tau = \gamma$, it is useful to convert this to the form

$$\frac{\mathrm{d}\phi}{\mathrm{d}\tau} = \frac{L}{mr^2}. \tag{4.48}$$

Note also that $p^2 = p_r^2 + L^2/r^2$, which is like the classical result.

Let E be the energy of the particle, in the sense of its rest energy plus kinetic energy, then using $E^2 - p^2c^2 = m^2c^4$ we obtain

$$p_r^2 = \frac{E^2}{c^2} - \frac{L^2}{r^2} - m^2c^2. \tag{4.49}$$

[3] In this section we drop the subscript zero on m; it always means rest mass.

To make further progress it is useful to introduce the concept of *potential energy* V. This is defined by

$$V(\mathbf{r}) \equiv - \int_{\mathbf{r}_0}^{\mathbf{r}} \mathbf{f} \cdot d\mathbf{r}'. \tag{4.50}$$

Such a definition is useful when the integral around any closed path is zero so that V is single-valued. When this happens the force is said to be *conservative*. Using eqn (4.10) (valid for a pure force) we then find that during any small displacement $d\mathbf{r}$ the kinetic energy lost by the particle is equal to the change in V:

$$dK = (\mathbf{f} \cdot \mathbf{u})dt = \mathbf{f} \cdot d\mathbf{r} = -dV. \tag{4.51}$$

It follows that the quantity

$$\mathcal{E} \equiv E + V \tag{4.52}$$

is a constant of the motion. In classical mechanics V is often called 'the potential energy of the particle', and then $\mathcal{E}$ is called 'the total energy of the particle'. However, strictly speaking V is not a property of the particle: it makes no contribution whatsoever to the energy possessed by the particle, which remains $E = \gamma m c^2$. V is just a mathematical device introduced in order to identify a constant of the motion. Physically it could be regarded as the energy owned not by the particle but by the *other* system (such as an electric field) with which the particle is interacting.

We can write eqn (4.52) in two useful ways:

$$\gamma m c^2 + V = \text{const} \tag{4.53}$$

$$\text{and} \quad p_r^2 c^2 + \frac{L^2 c^2}{r^2} + m^2 c^4 = (\mathcal{E} - V)^2 \tag{4.54}$$

(using eqn (4.49)). Since for a given force, V is a known function of r, the first equation enables the Lorentz factor for the total speed to be obtained at any given r for given initial conditions. Using the angular momentum (also fixed by initial conditions) one can then also find $\dot{\phi}$ and hence $\dot{r}$.

Eqn (4.54) is a differential equation for r as a function of time (since $p_r = \gamma m \dot{r}$). It is easiest to seek a solution as a function of proper time τ, since

$$\frac{dr}{d\tau} = \frac{dr}{dt}\frac{dt}{d\tau} = \dot{r}\gamma = \frac{p_r}{m}$$

so we have

$$\frac{1}{2}m\left(\frac{dr}{d\tau}\right)^2 = \tilde{\mathcal{E}} - V_{\text{eff}}(r) \tag{4.55}$$

where

$$\tilde{\mathcal{E}} \equiv \frac{\mathcal{E}^2 - m^2 c^4}{2mc^2}, \tag{4.56}$$

$$V_{\text{eff}}(r) \equiv \frac{V(r)(2\mathcal{E} - V(r))}{2mc^2} + \frac{L^2}{2mr^2}. \tag{4.57}$$

Eqn (4.55) has precisely the same form as an equation for classical motion in one dimension in a potential $V_{\text{eff}}(r)$. Therefore we can immediately deduce the main qualitative features of the motion.

Inverse-square-law force

Consider, for example, an inverse-square-law force, such as that arising from Coulomb attraction between opposite charges. Writing $\mathbf{f} = -\alpha \hat{\mathbf{r}}/r^2$ and therefore $V = -\alpha/r$, we have

$$V_{\text{eff}} = \frac{1}{2mc^2}\left(\frac{L^2 c^2 - \alpha^2}{r^2} - \frac{2\alpha\mathcal{E}}{r}\right). \tag{4.58}$$

The second term gives an attractive $1/r$ potential well that dominates at large r. If the first term is non-zero then it dominates at small r and gives either a barrier or an attractive well, depending on the sign. Thus there are two cases to consider:

$$\text{(i) } L > L_c, \quad \text{(ii) } L \leq L_c, \quad \text{where} \quad L_c \equiv \frac{\alpha}{c}. \tag{4.59}$$

(i) For large angular momentum, the 'centrifugal barrier' is sufficient to prevent the particle approaching the origin, just as in the classical case. There are two types of motion: unbound motion (or 'scattering') when $\tilde{\mathcal{E}} > 0$, and bound motion when $\tilde{\mathcal{E}} < 0$, in which case r is constrained to stay in between turning points at $V_{\text{eff}}(r) = \tilde{\mathcal{E}}$.

(ii) For small angular momentum, something qualitatively different from the classical behaviour occurs: when $L \leq L_c$ the motion has no inner turning point and the particle is 'sucked in' to the origin. The motion conserves L and therefore is a spiral in which $\gamma \to \infty$ as $r \to 0$. In this limit the approximation that the particle or system providing the central force does not itself accelerate is liable to break down; the main point is that a Coulomb-law scattering centre can result in a close spiralling collision even when the incident particle has non-zero angular momentum. In Newtonian physics this type of behaviour would require an attractive force with a stronger dependence on distance. For a scattering process in which the incident particle has momentum p_i at infinity, and impact parameter b, the angular momentum is $L = bp_i$. All particles with impact parameter below $b_c = L_c/p_i$ will suffer a spiralling close collision. The collision cross-section for this process is

$$\pi b_c^2 = \frac{\pi\alpha^2}{c^2 p_i^2} = \frac{\pi\alpha^2}{E^2 - m^2 c^4}.$$

This is very small in practice. For example, for an electron moving in the Coulomb potential of a proton, $b_c \simeq 1.4 \times 10^{-12}$ m when the incident

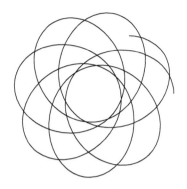

Fig. 4.6 Example orbit of a fast-moving particle in a $1/r$ potential. Only part of the orbit is shown; its continuation follows the same pattern.

kinetic energy is 1 eV. Using the Newtonian formula for gravity to approximate conditions in the solar system, one obtains $b_c \simeq GM/vc$, where $M \simeq 2 \times 10^{30}$ kg is the solar mass and $v \ll c$ is the speed of an object such as a comet. b_c exceeds the radius of the Sun when the incident velocity (far from the Sun) is below 640 m/s.

In the case $L > L_c$ and $\tilde{\mathcal{E}} < 0$, where there are stable bound orbits, a further difference from the classical motion arises. The classical $1/r$ potential leads to elliptical orbits, in which the orbit closes on itself after a single turn. This requires that the distance from the origin oscillates in step with the movement around the origin, so that after r completes one cycle between its turning points, ϕ has increased by 2π. There is no reason why this synchrony should be maintained when the equation of motion changes, and in fact it is not. The orbit has the form of a rosette; see Fig. 4.6. In order to deduce this, we can turn eqn (4.55) into an equation for the orbit, as follows. First differentiate with respect to τ, to obtain

$$m\frac{d^2 r}{d\tau^2} = -\frac{dV_{\text{eff}}}{dr} \tag{4.60}$$

where we cancelled a factor of $dr/d\tau$ which is valid except at the stationary points. Apply this to the case of an inverse-square-law force, for which the effective potential is given in eqn (4.58):

$$\frac{d^2 r}{d\tau^2} = \frac{L^2 - \alpha^2/c^2}{m^2 r^3} - \frac{\alpha\mathcal{E}}{m^2 c^2 r^2}. \tag{4.61}$$

Although this equation can be tackled by direct integration, the best way to find the orbit is to make two changes of variable. Using eqn (4.48), derivatives with respect to τ can be expressed in terms of derivatives with respect to ϕ. Then one changes variable from r to $u = 1/r$, obtaining

$$\frac{d^2 r}{d\tau^2} = -\left(\frac{L}{m}\right)^2 u^2 \frac{d^2 u}{d\phi^2} \tag{4.62}$$

and therefore eqn (4.61) becomes

$$\frac{d^2 u}{d\phi^2} = -\left(1 - \frac{\alpha^2}{L^2 c^2}\right) u + \frac{\alpha\mathcal{E}}{L^2 c^2}. \tag{4.63}$$

This is the equation for simple harmonic motion. Hence the orbit is given by

$$r(\phi) = \frac{1}{u} = \frac{1}{A\cos(\tilde{\omega}(\phi - \phi_0)) + \alpha\mathcal{E}/(L^2 c^2 - \alpha^2)} \tag{4.64}$$

where A and ϕ_0 are constants of integration, and

$$\tilde{\omega} = \sqrt{1 - \frac{\alpha^2}{L^2 c^2}}. \tag{4.65}$$

The radial motion completes one period when ϕ increases by $2\pi/\tilde{\omega}$. A Newtonian calculation gives $\tilde{\omega} = 1$, which means that the orbit closes (forming an ellipse). Note that the departure from the Newtonian prediction is largest at small L, not large L. For the relativistic case far from the critical angular momentum: i.e., $L \gg \alpha/c$, one has $\tilde{\omega} \simeq 1 - \alpha^2/2L^2c^2$. Therefore, when r returns to its minimum value (so-called *perihelion* in the case of planets orbiting the Sun) ϕ has increased by 2π plus an extra bit equal to

$$\delta\phi = \frac{\pi\alpha^2}{L^2c^2}. \tag{4.66}$$

The location of the innermost point of the orbit shifts around (or 'precesses') by this amount per orbit. For the case of an electron orbiting a proton, the combination α/Lc is equal to the fine structure constant when $L = \hbar$, and this motion was used by Sommerfeld to construct a semi-classical theory for the observed fine structure of hydrogen (subsequently replaced by the correct quantum treatment). For the case of gravitational attraction to a spherical mass, the result (4.66) is about six times smaller than the precession predicted by General Relativity.

4.2.7 (An)harmonic motion*

Another important basic problem in mechanics is motion in a quadratic potential well. In classical mechanics this gives simple harmonic motion: i.e., sinusoidal oscillation with a period independent of the amplitude.

The relativistic problem is more complicated because we have the non-linear equation of motion

$$-kx = \frac{d}{dt}(\gamma m_0 v) \quad \left[= \gamma^3 m_0 \frac{d^2x}{dt^2} \right] \tag{4.67}$$

where k is the 'spring constant'.

We can get some immediate insight by using energy methods We introduce potential energy once again, by eqn (4.50), and the motion is conservative: the 'total energy' $\mathcal{E} \equiv m_0c^2 + K + V$ is a constant of the motion, with

$$V(x) = \frac{1}{2}kx^2 = \frac{1}{2}m_0\omega_0^2 x^2, \tag{4.68}$$

where $\omega_0 = \sqrt{k/m_0}$ is the angular frequency of oscillations in the classical (low-velocity) limit. This gives

$$\mathcal{E} = \gamma m_0 c^2 + kx^2/2. \tag{4.69}$$

This immediately tells us γ as a function of x:

$$\gamma = \gamma_0 - \frac{\omega_0^2 x^2}{2c^2} \tag{4.70}$$

where $\gamma_0 = \mathcal{E}/(m_0c^2)$ is another way of expressing the conserved total energy. Thus γ as a function of x forms an inverted parabola. The maximum excursion is when the speed falls to zero: i.e., $\gamma = 1$. Putting this into eqn (4.70) gives the amplitude of the motion:

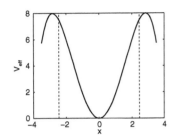

Fig. 4.7 The effective potential V_{eff} (eqn (4.74)) for the example case $\gamma_0 = 4$, in units such that $m_0\omega_0^2 = 1$ and $\omega_0/c = 1$. The dashed lines show the positions of maximum excursion of the particle, eqn (4.71). The turning points in V_{eff} are at $x = \pm(c/\omega_0)\sqrt{2\gamma_0}$, which is further from the origin than $x_{\max}$, so the motion never goes over the hump and escapes. However, when γ_0 is large (ultra-relativistic case) the motion gets close to the turning points and therefore a large proportion of the proper time is spent there. This makes sense, since this is where the particle is moving slowly.

$$x_{\max} = \frac{c}{\omega_0}\sqrt{2(\gamma_0 - 1)}. \tag{4.71}$$

In the low-velocity limit, of course, all the usual classical results are obtained. In the extreme high-velocity limit, $\gamma_0 \gg 1$, the amplitude is large and the particle spends most of the time (in the lab frame) travelling at $v \simeq c$, apart from short turn-around periods at the maximum excursion. Therefore, in this limit one quarter-period of the oscillatory motion is approximately $x_{\max}/c$, so the period is $T \simeq 4\sqrt{2\gamma_0}/\omega_0$.

To obtain more information it is more convenient to find x as a function of proper time rather than coordinate time. That is, we express eqn (4.67) as

$$-kx = \frac{dp}{d\tau}\frac{d\tau}{dt} = \frac{m_0}{\gamma}\frac{d^2x}{d\tau^2} \tag{4.72}$$

where we used $p = m_0 dx/d\tau$ and the familiar (I hope) $dt/d\tau = \gamma$. Substituting expression (4.70) for γ in terms of x, we can convert this into a differential equation for $x(\tau)$:

$$\frac{d^2x}{d\tau^2} = -\gamma_0\omega_0^2 x + \frac{\omega_0^4}{2c^2}x^3. \tag{4.73}$$

This is like an equation for classical motion in an effective potential well:

$$V_{\text{eff}} = \frac{1}{2}\gamma_0 m_0\omega_0^2 x^2 - \frac{m_0\omega_0^4}{8c^2}x^4. \tag{4.74}$$

Thus the relativistic motion as a function of proper time looks like classical motion in an anharmonic well formed by a combination of quadratic and quartic terms. Fig. 4.7 shows the shape of the effective potential well for an example case, with some general remarks in the caption.

The differential equation has an analytic solution in terms of elliptic integrals.[4] To convert from proper time back to coordinate time t one uses eqn (4.70) again, now writing the left-hand side as $dt/d\tau$ and regarding it as a differential equation for t as a function of τ (using the solution of eqn (4.73) for $x(\tau)$ on the right-hand side).

[4] W. Moreau, R. Easther, and R. Neutze, Am. J. Phys. **62**, 531 (1994).

Exercises

(4.1) Show that $\mathbf{f} = m_0\mathbf{a}$ in the instantaneous rest frame.

(4.2) Does Special Relativity place any bounds on the possible sizes of forces or accelerations?

(4.3) Obtain the transformation equations for 3-force, by starting from the Lorentz transformation of energy-momentum, and then differentiating with respect to t'. (Hint: argue that the relative velocity

$\mathbf{v}$ of the reference frames is constant, and use or derive the expression (2.12ii) for dt/dt'.)

(4.4) In the twin paradox, the travelling twin leaves Earth on board a spaceship undergoing motion at constant proper acceleration of 9.8 m/s². After 5 years of proper time for the spaceship, the direction of the rockets are reversed so that the spaceship accelerates towards Earth for 10 proper years. The rockets are then reversed again to allow the spaceship to slow and come to rest on Earth after a further 5 years of spaceship proper time. How much does the travelling twin age? How much does the stay-at-home twin age?

(4.5) A particle moves hyperbolically with proper acceleration a_0, starting from rest at $t = 0$. At $t = 0$ a photon is emitted towards the particle from a distance c^2/a_0 behind it. Prove that the photon never catches up with the particle, and furthermore, in the instantaneous rest frames of the particle, the distance to the photon is always c^2/a_0.

(4.6) Show that the motion of a particle in a uniform magnetic field is in general helical, with the period for a cycle independent of the initial direction of the velocity. (Hint: what can you learn from $\mathbf{f} \cdot \mathbf{v}$?)

(4.7) **Constant force**. Consider motion under a constant force, for a non-zero initial velocity in an arbitrary direction, as follows.

(i) Write down the solution for $\mathbf{p}$ as a function of time, taking as initial condition $\mathbf{p}(0) = \mathbf{p}_0$.

(ii) Show that the Lorentz factor as a function of time is given by $\gamma^2 = 1 + \alpha^2$, where $\alpha = (\mathbf{p}_0 + \mathbf{f}t)/mc$.

(iii) You can now write down the solution for $\mathbf{v}$ as a function of time. Do so.

(iv) Now restrict attention to the case where $\mathbf{p}_0$ is perpendicular to $\mathbf{f}$. Taking the x-direction along $\mathbf{f}$ and the y-direction along $\mathbf{p}_0$, show that the trajectory is given by

$$x = \frac{c}{f}\left(m^2c^2 + p_0^2 + f^2t^2\right)^{1/2} + \text{const} \quad (4.75)$$

$$y = \frac{cp_0}{f}\log\left(ft + \sqrt{m^2c^2 + p_0^2 + f^2t^2}\right)$$

$$+ \text{const} \quad (4.76)$$

(v) Explain (without carrying out the calculation) how the general case can then be treated by a suitable Lorentz transformation.

Note that the calculation as a function of proper time is best accomplished another way; see chapter 13.

(4.8) For motion under a pure (rest mass preserving) inverse square law force $\mathbf{f} = -\alpha \mathbf{r}/r^3$, where α is a constant, derive the energy equation $\gamma mc^2 - \alpha/r = \text{constant}$.

5 The conservation of energy-momentum

So far we have discussed energy and momentum by introducing the definitions (2.68) without explaining where they come from. Historically, in 1905 Einstein first approached the subject of force and acceleration by finding the equation of motion of a charged particle subject to electric and magnetic fields, assuming the charge remained constant and the Maxwell and Lorentz force equations were valid, and that Newton's second law applied in the particle's rest frame. He could then use the theory he himself developed to understand what must happen in other frames, and hence derive the equation of motion for a general velocity of the particle. Subsequently, Planck pointed out that the result could be made more transparent if one understood the 3-momentum to be given by $\gamma m_0 \mathbf{v}$. A significant further development took place in 1909 when Lewis and Tolman showed that this definition was consistent with momentum conservation in all reference frames. Nowadays, we can side-step these arguments by proceeding straight to the main result using 4-vector methods. However, when learning the subject the Lewis and Tolman argument remains a useful way in, so we shall present it first.

5.1 Elastic collision, following Lewis and Tolman

The concept of conservation of momentum brings a great deal of insight in Newtonian physics, so it is natural to ask what role it plays, if any, once we accept that space and time are non-Galilean. Is momentum conservation still valid in Special Relativity? The question must be asked because the quantity $m_0 \mathbf{v}$ (the product of rest mass and velocity) is *not* conserved—its sum over the members of an isolated set of interacting particles can change when the particles interact. However, we can seek an alternative conserved property as follows. We propose that there might exist a vector property involving rest mass and velocity that is conserved, and then we use the Principle of Relativity and the Light Speed Postulate to find out as much as we can about such a property (without making any other assumptions). It turns out that this is sufficient to specify the property uniquely! In other words, *if* there is a conserved quantity, then it has to be of a specific type. It will then require experiments to check whether the conservation actually holds.

This idea can be stated more precisely as follows. For a particle of rest mass m_0 and velocity $\mathbf{v}$, define the property

$$\mathbf{p} \equiv m_0 \alpha(v)\mathbf{v}, \tag{5.1}$$

where $\alpha(v)$ is some function of speed to be determined. Propose that the function $\alpha(v)$ is universal—the same for all particles and physical conditions—and propose that $\sum_i \mathbf{p}_i$ is a conserved quantity when the sum is taken over the parts i of an isolated system. We can choose any name we like for our newly invented quantity $\mathbf{p}$, and we choose to call it 'momentum'. Next, we examine a physical process of our own choosing (it will be a collision) in one frame, and use the Lorentz transformation to deduce the velocities in another reference frame. We will thus learn what momentum conservation in one frame implies about the motion in another. By requiring conservation in the second frame also, we shall have two sets of equations for a single process. These equations can be used to determine the function α (or to discover that there is no possible solution).

The type of process we shall examine is an elastic collision between two particles of the same rest mass. By 'elastic' we mean that there is no change of rest mass of either particle during the collision; we only need to argue that such collisions are possible.

There always exists a frame F relative to which the initial velocities are equal and opposite, and we can always choose axes such that the x axis of F bisects the angle between the initial and final velocities, so that the collision looks like Fig. 5.1. This choice automatically guarantees the conservation of the x component of momentum during the collision, because this is unchanged for both particles. It remains to consider the y component.

We will not need to examine the momenta in frame F any further. Instead we take an interest in two other frames: one moving to the left (along the negative x direction) and keeping pace with the first particle, the other moving to the right and keeping pace with the second particle; see Fig. 5.2. Let the relative speed of these two reference frames be v (this speed is related to the speeds of the particles in F, but we will not need to know what the relation is). In the first (left-going) reference frame the lower particle simply moves up and down at some speed u. It follows by symmetry that in the second reference frame the upper

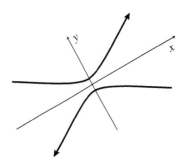

Fig. 5.1 A general collision of two identical particles initially moving towards one another with identical speeds. No matter what angles are involved, we can always choose a set of axes oriented as shown, relative to the trajectories, in order to simplify the analysis.

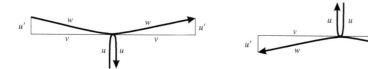

Fig. 5.2 The same collision as in Fig. 5.1, but viewed from two different reference frames—one moving left keeping pace with the lower particle, and one moving right keeping pace with the upper particle. The relative speed of these two frames is v. From the overall symmetry one can deduce that the vertical speed of the upper particle in frame 1 is the same as the vertical speed of the lower particle in frame 2, etc. w is a total speed; u' and v are its components in the horizontal and vertical directions.

particle must move down and up at that same speed u. Let the vertical component of the speed of the other particle be u' in each case (this also applies in both reference frames, by symmetry). This information has been indicated in Fig. 5.2.

Now invoke the proposed definition of momentum, eqn (5.1). In the first reference frame the momentum of the lower particle is $m_0\alpha(u)u$ vertically upwards before the collision, and $m_0\alpha(u)u$ downwards after the collision, so it undergoes a net change of $2m_0\alpha(u)u$.

The other particle has total momentum $m_0\alpha(w)\mathbf{w} = m_0\alpha(w)$ $(\mathbf{v} + \mathbf{u}')$ before the collision. Its horizontal and vertical components are $(m_0\alpha(w))v$ and $(m_0\alpha(w))u'$ respectively. Note that $\alpha(w)$ appears in both of these formulae, not $\alpha(v)$ or $\alpha(u')$. After the collision, the horizontal component remains the same but the vertical component reverses. Therefore the net change is in the vertical direction and is equal to $2m_0\alpha(w)u'$.

Now we assert conservation of momentum:

$$2m_0\alpha(u)u = 2m_0\alpha(w)u' \qquad \Rightarrow \qquad \alpha(u)u = \alpha(w)u'. \quad (5.2)$$

To find the function α we require a further independent relation between u, u', and w. This can be obtained using a Lorentz transformation to the other reference frame. Fig. 5.2 makes it clear that $\mathbf{u}'$ is related to $\mathbf{u}$ simply by a change of frame. This is a transverse velocity, so, using eqn (2.27),

$$u' = \frac{u}{\gamma_v}. \quad (5.3)$$

We write γ_v because there are several speeds in play, and we need to be clear which one we mean. Make sure you are convinced that the relative speed of the two reference frames here is v, not $2v$ or anything else.

We now have enough information to deduce the function α in eqn (5.2). Substituting eqn (5.3) in eqn (5.2), we have

$$\alpha(w) \equiv \gamma_v\, \alpha(u) \quad (5.4)$$

where the $\equiv$ symbol is to emphasize that this is an identity: i.e., valid for all values of u and v. To solve for the unknown function $\alpha(u)$, one can use a power series expansion, having first observed from Fig. 5.2 that $w^2 = v^2 + (u')^2$ and therefore, using eqn (5.3),

$$w^2 = v^2 + u^2 - u^2 v^2/c^2. \quad (5.5)$$

For $u \ll c$ we must have $\alpha(u) \to 1$ in order to produce the classical momentum formula mu. From (5.5), $w \to v$ as $u \to 0$ and then (5.4) becomes $\alpha(v) = \gamma_v$. With this hint, we guess that the general solution, for all speeds, is

$$\alpha(v) \overset{?}{=} \gamma_v = \frac{1}{\sqrt{1 - v^2/c^2}}. \quad (5.6)$$

Invariant, conserved

		Lorentz invariant	conserved
energy	E	×	✓
momentum	$\mathbf{p}$	×	✓
rest mass	m	✓	×
charge	q	✓	✓
charge density	ρ	×	×

If this is correct, then

$$\alpha(w) = \frac{1}{\sqrt{1 - w^2/c^2}}, \qquad \text{and} \qquad \alpha(u) = \frac{1}{\sqrt{1 - u^2/c^2}}.$$

Substitution in eqn (5.4) confirms the guess: the equation is satisfied for all u, v.

The conclusion is that if some function of particle velocity $\mathbf{v}$ is conserved in simple elastic (= non-rest-mass-changing) collisions between particles of the same rest mass, then that function must be $\gamma_v \mathbf{v}$.

We have shown that a conserved quantity related to rest mass and velocity must of necessity[1] have the form $\gamma m_0 \mathbf{v}$, but the argument has not addressed the question whether that is sufficient to satisfy the Principle of Relativity for all types of collision. By investigating further collisions, one can explore the more general question. One finds that the formula *is* sufficient to guarantee well-behaved (i.e. Lorentz covariant) physics in all collisions. We shall prove this more simply by using a 4-vector approach in the next section.

[1] See exercise 5.2.

Energy conservation and mass-energy equivalence

A further very interesting point emerges from a theoretical study of momentum conservation in collisions. We shall present this in the next section using 4-vectors, but it can also be deduced by applying the above type of reasoning to inelastic collisions. The argument is presented, for example, in *The Wonderful World*. One finds that, along with momentum, there is another conserved quantity that comes along 'for free': i.e., without the need for any further assumptions. This further property is a scalar, it increases monotonically with the speed of a particle of given rest mass, and it can be converted between stored and motion-related forms. In other words it has all the attributes we normally associate with 'energy'. Therefore that is what we call it. *The conservation of energy, and its connection to rest mass, is a necessary consequence of the requirement of momentum conservation combined with Lorentz invariance.*

This connection will be proved more generally in the next section. We have mentioned it here at the outset in order to make clear that it can also be found by careful reasoning using only simpler concepts such as 3-vectors.

5.2   Energy-momentum conservation using 4-vectors

The Lewis and Tolman argument has the merit of being unsophisticated for the simplest case, but it is not easy to generalize it to all collisions. The use of 4-vectors makes the general argument much more straightforward.

By considering the worldline of a particle, we showed in chapter 2.1 that various 4-vectors, such as spacetime position X, 4-velocity $U = dX/d\tau$, and 4-momentum $P = m_0 U$ could be associated with a single particle. In order to introduce a conservation law we need to define first of all what we mean by the 4-momentum of a *collection* of particles. The definition is the obvious one:

$$P_{tot} \equiv P_1 + P_2 + P_3 + \cdots + P_n. \tag{5.7}$$

That is, we define the total 4-momentum of a collection of n particles to be the sum of the individual 4-momenta. Now we can state what we mean by the conservation of energy and momentum:

Law of conservation of energy and momentum: the total energy-momentum 4-vector of an isolated system is independent of time. In particular, it is not changed by internal interactions among the parts of the system.

In order to apply the insights of Special Relativity to dynamics, we state this conservation law as an axiom. Before going further we must check that it is consistent with the other axioms. We shall find that it is. Then one can use the conservation law to make predictions which must be compared with experiment. Further insight will be provided in chapter 14, where this conservation law is related to invariance of the *Lagrangian* under translations in time and space.

Agreement with the Principle of Relativity

First we tackle the first stage, which is to show that energy-momentum conservation, as defined above, is consistent with the main Postulates (the Principle of Relativity and the Light Speed Postulate). To show this we write down the conservation law in one reference frame, and then use the Lorentz transformation to find out how the same situation is described in another reference frame.

Let $P_1, P_2, \ldots P_n$ be the 4-momenta of a set of particles, as observed in frame S. Then, by definition, the total 4-momentum is P_{tot}, given by eqn (5.7). By calling the result of this sum a '4-momentum' and giving it a symbol P_{tot} we are strongly implying that the sum total is itself a 4-vector. You might think that this is obvious, but in fact it requires further thought. After all, we have already noted that adding up 4-velocities does not turn out to be a sensible thing to do—so why is 4-momentum any different? When we carry out the mathematical sum, summing the 4-momentum of one particle and the 4-momentum of a

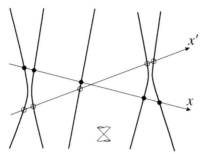

Fig. 5.3 A set of worldlines is shown on a spacetime diagram, with lines of simultaneity for two different reference frames. The energy-momenta at some instant in frame A are defined at a different set of events (shown dotted) from those obtaining at some instant in frame B (circled). Therefore each term in the sum defining the total energy-momentum at some instant in A is not necessarily the Lorentz-transform of the corresponding term in the sum defining the total energy-momentum at some instant in B. Nevertheless, when the terms are added together, as long as 4-momentum conservation holds and the total system is isolated, the totals P_{tot} and P'_{tot} are Lorentz-transforms of each other (see text).

different particle, we are adding up things that are specified at different events in spacetime. When the terms in the sum can themselves change with time, we need to clarify at what moment each individual P_i term is to be taken. Therefore a more careful statement of the definition P_{tot} would read:

$$P_{tot}(t = t_0) \equiv P_1(t = t_0) + P_2(t = t_0) + \cdots + P_n(t = t_0) \qquad (5.8)$$

where t_0 is the instant in some frame S at which the total 4-momentum is being specified.

Now, if we apply the definition to the same set of particles, but now at some instant t'_0 in a different reference frame S′, we find the total 4-momentum in S′ is

$$P'_{tot}(t' = t'_0) \equiv P'_1(t' = t'_0) + P'_2(t' = t'_0) + \cdots + P'_n(t' = t'_0). \qquad (5.9)$$

The problem is that the 4-momenta being summed in eqn (5.8) are taken at a set of events simultaneous in S, while the 4-momenta being summed in eqn (5.9) are being summed at a set of events simultaneous in S′. Owing to the relativity of simultaneity, these are two different sets of events (see Fig. 5.3). Therefore the individual terms are not necessarily Lorentz-transforms of each other:

$$P'_i(t' = t'_0) \neq \Lambda P_i(t = t_0). \qquad (5.10)$$

Therefore, when we take the Lorentz-transform of P_{tot} we will not obtain P'_{tot}, unless there is a physical constraint on the particles that makes their 4-momenta behave in such a way that ΛP_{tot} does equal P'_{tot}. Fortunately, the conservation law itself comes to the rescue, and provides precisely the constraint that is required! Proof: When forming the sum in one reference frame, one can always artificially choose a set of times t_i that lie in a plane of simultaneity for the other reference frame. Compared with the sum at t_0, the terms will either stay the same (for particles that move freely between t_0 and their t_i) or they will change (for particles that collide or interact in any way between t_0 and t_i), but if 4-momentum is conserved, such interactions do not change the total P_{tot}. Now the terms in one sum are Lorentz transforms of the terms in the other, hence so are the totals. QED.

4-momentum proof. Let A and B be two reference frames. Make the following definitions (cf. Fig. 5.3):

$P_{A,A}$ ≡ sum of 4-momenta evaluated in frame A at events simultaneous in A

$P_{A,B}$ ≡ sum of 4-momenta evaluated in frame A at events simultaneous in B

$P_{B,B}$ ≡ sum of 4-momenta evaluated in frame B at events simultaneous in B

The events at which $P_{A,A}$, $P_{A,B}$ are evaluated are shown dotted, circled, respectively in Fig. 5.3. By definition, $P_{A,A}$ is what is called 'the total 4-momentum' in frame A, and $P_{B,B}$ is what is called 'the total 4-momentum' in frame B. We would like to prove that $P_{B,B} = \Lambda P_{A,A}$. This is done in two steps.

(Step 1): $P_{A,B}$ is a sum of terms. By applying the Lorentz transformation to each term in the sum one obtains

$$P_{B,B} = \Lambda P_{A,B}.$$

(Step 2): Because of conservation of 4-momentum, the evolution in frame A only redistributes 4-momentum within the set of particles, without changing the total, therefore

$$P_{A,B} = P_{A,A}.$$

It follows that

$$P_{B,B} = \Lambda P_{A,A}. \qquad \text{QED.}$$

If the system of particles were not isolated, then its 4-momentum would not necessarily be conserved, and then we could not take step 2. Consequently, the sum of the 4-momenta of a non-isolated set of particles is not necessarily a 4-vector.

For the sake of clarity, the argument is repeated in the box above.

We originally introduced P in section 2.5.3 as a purely mathematical quantity: a 4-vector related to 4-velocity and rest mass. That did not in itself tell us that P is conserved. However, if the natural world is mathematically consistent and Special Relativity describes it, then only certain types of quantity can be universally conserved (i.e., conserved in all reference frames). It makes sense to postulate a conservation law for something like $\gamma m_0 \mathbf{u}$ (3-momentum) because this is part of a 4-vector. The formalism of Lorentz transformations and 4-vectors enables us to take three further steps:

(1) If a 4-vector is conserved in one reference frame then it is conserved in all reference frames.

(2) If one component of a 4-vector is conserved in all reference frames then the entire 4-vector is conserved.

(3) A sum of 4-vectors evaluated at spacelike separated events is itself a 4-vector if the sum is conserved.

Proof. We already dealt with item (3). For item (1), argue as follows. The word 'conserved' means 'constant in time' or 'the same before and after' any given process. For some chosen reference frame let P be the conserved quantity, with P_{before} signifying its value before some process, and P_{after}. The conservation of P is then expressed by

$$P_{before} = P_{after}. \tag{5.11}$$

Now consider the situation in another reference frame. Since P is a 4-vector, we know how it transforms: we shall find

$$P'_{before} = \Lambda P_{before} , \qquad P'_{after} = \Lambda P_{after}.$$

By applying a Lorentz transformation to both sides of eqn (5.11) we shall immediately find $P'_{before} = P'_{after}$: i.e., the quantity is also conserved in the new reference frame, QED. This illustrates how 4-vectors 'work': by expressing a physical law in 4-vector form we automatically take care of the requirements of the Principle of Relativity.

To prove item (2) above we make use of the following lemma:

Zero component lemma: If one component of a 4-vector is zero in all reference frames, then the entire 4-vector is zero.

Proof. Consider some 4-vector Q, pick a component such as the x—component, and suppose this component vanishes in all frames. If there is a frame in which the y or z component is non-zero, then we can rotate axes to make the x component non-zero, contrary to the claim that it is zero in all reference frames. Therefore the y and z components are zero also. If there is a reference frame in which the time-component Q^0 is non-zero, then we can apply a Lorentz transformation to make Q^1 non-zero, contrary to the claim. Therefore Q^0 is zero. A similar argument can be made starting from any of the components, which concludes the proof.

The proof of item (2) in our list now follows immediately, by applying the zero-component lemma to the 4-vector $Q \equiv P_{after} - P_{before}$.

5.2.1   Mass–energy equivalence

At first the zero component lemma might seem to be merely a piece of mathematics, but it is much more. It says that if we have conservation (in all reference frames) of a scalar quantity that is known to be one component of a 4-vector, then we have conservation of the whole 4-vector. This enables us to reduce the number of assumptions we need to make: instead of postulating conservation of 4-momentum, for example, we could postulate conservation of one of its components, say the x-component of momentum, in all reference frames, and we would immediately deduce not only conservation of 3-momentum but conservation of energy as well.

In classical physics the conservation laws of energy and momentum were separate: they do not necessarily imply one another. In Relativity they do. The conservation of the 3-vector quantity (momentum) is no longer separate from the conservation of the scalar quantity (energy). The unity of spacetime is here exhibited as a unity of energy and momentum. It is not that they are the same, but they are two parts of one thing.

Once we have found the formula relating the conserved 3-vector to velocity, i.e. $\mathbf{p} = \gamma m \mathbf{v}$ (the spatial part of $m\mathsf{U}$), we do not have any choice about the formula for the conserved scalar, up to a constant factor, it must be $E \propto \gamma m$ (the temporal part of $m\mathsf{U}$). Also, the constant factor must be c^2 in order to give the known formula for kinetic energy in the low-velocity limit, and thus match the classical definition of what we call energy:

$$\gamma = (1 - v^2/c^2)^{-1/2} \simeq 1 + \frac{1}{2}v^2/c^2 \qquad \Rightarrow \qquad \gamma mc^2 \simeq mc^2 + \frac{1}{2}mv^2.$$

Thus the important relation 'E = mc^2' follows from momentum conservation and the Main Postulates. This formula gives rise to a wonderful new insight—perhaps the most profound prediction of Special Relativity—namely the *equivalence of mass and energy*. By this we mean two things. First, in any process, kinetic energy of the reactants can contribute to rest mass of the products, and conversely. For example, in a collision where two particles approach and then stick together, there is a reference frame where the product is at rest. In that frame, we shall find $Mc^2 = \sum \gamma_i m_i c^2$ and therefore $M > \sum m_i$, where M is the rest mass of the product and m_i are the rest masses of the reactants.

The physical meaning of this rest mass M is *inertial*. It is 'that which increases the momentum': i.e., the capacity of a body to make other things move when it hits them. It does not immediately follow that it is the same thing as gravitational mass. One of the foundational assumptions of General Relativity is that this inertial mass is indeed the same thing as gravitational mass, for a body at rest with no internal pressure.

The second part of the meaning of 'equivalence of mass and energy' is that 'rest mass' and 'rest energy' are simply different words for the same thing (up to a multiplying constant: i.e., c^2). This is a strict equivalence. It is not that they are 'like' one another (as is sometimes asserted of space and time, where the likeness is incomplete), but they are strictly the same—just different words used by humans for the same underlying physical reality. In an exothermic reaction such as nuclear fission, therefore, rather than saying 'mass is converted into energy' it is arguably more correct to say simply that energy is converted from one form to another. We have only ourselves to blame if we gave it a different name when it was located in the nucleus. The point can be emphasized by considering a more everyday example such as compression of an ordinary metal spring. When under compression, energy has been supplied to the spring, and we are taught to call it 'potential energy'. We may equally call it 'mass-energy': it results in an increase in the rest mass of the spring (by the tiny amount of 10^{-17} kg per joule). When we enjoy the warmth from a wooden log fire, we are receiving benefit from a process of 'conversion of mass to energy' just as surely as when we draw on the electrical power provided by a nuclear power station. The 'binding energy' between the oxygen atoms and carbon atoms is another name for a rest mass deficit: each molecule has a smaller rest mass than the sum of the rest masses of the separate atoms. The tiny difference δm is enough to liberate noticeable amounts of energy $(\delta m c^2)$ in another form such as heat.

5.3   Collisions

We will now apply the conservation laws to a variety of collision-type processes, starting with the most simple and increasing in complexity as we proceed. We will make repeated use of the formula $E^2 - p^2 c^2 = m^2 c^4$ which we can now recognise both as a statement about mass and energy, and also as a Lorentz invariant quantity associated with the energy-momentum 4-vector.

The quantities E_i, p_i, m_i will usually refer to the energy, momentum and rest mass of the ith particle *after* the process. In particle physics experiments one typically gathers information on p and E (e.g., from curvature of particle tracks and from energy deposited in a detector, respectively), and some or all of the rest masses may be known. To extract a velocity one can use $\mathbf{v} = \mathbf{p}c^2/E$ (eqn (2.69)). However, not all the information is always available, and typically momenta can be obtained more precisely than energies. Even if one has a set of measurements that in principle gives complete information, it is still very useful to establish relations (constraints) that the data ought to obey, because this will allow the overall precision to be improved, consistency checks to be made, and systematic error uncovered. Also, it is crucial to have good systematic ways of looking for patterns in the data, because usually the interesting events are hidden in a great morass or background of more frequent but mundane processes.

1. *Spontaneous emission, radioactive decay.*
An atom at rest emits a photon and recoils. For a given energy level difference in the atom, what is the frequency of the emitted photon? A radioactive nucleus emits a single particle of given rest mass. For a given change in rest mass of the nucleus, what is the energy of the particle?

These are both examples of the same type of process. Before the process there is a single particle of rest mass M^* and zero momentum. The asterisk serves as a reminder that this is an excited particle that can decay. Afterwards there are two particles of rest mass m_1 and m_2. By conservation of momentum these move in opposite directions, so we only need to treat motion in one dimension. The conservation of energy and momentum gives

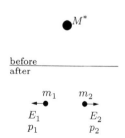

Fig. 5.4

$$M^* c^2 = E_1 + E_2, \tag{5.12}$$

$$p_1 = p_2. \tag{5.13}$$

The most important thing to notice is that, for given rest masses M^*, m_1, m_2, there is a *unique solution* for the energies and momenta (i.e., the sizes of the momenta; the directions must be opposed but otherwise they are unconstrained). This is because we have four unknowns, E_1, E_2, p_1, p_2, and four equations—the above and $E_i^2 - p_i^2 c^2 = m_i^2 c^4$ for $i = 1, 2$.

Taking the square of the momentum equation, we have $E_1^2 - m_1^2 c^4 = E_2^2 - m_2^2 c^4$. After substituting for E_2 using eqn (5.12), this is easily solved for E_1, giving

$$E_1 = \frac{M^{*2} + m_1^2 - m_2^2}{2M^*} c^2. \tag{5.14}$$

When the emitted particle is a photon, $m_1 = 0$ so this can be simplified. Let $E_0 = M^* c^2 - m_2 c^2$ be the gap between the energy levels of the decaying atom or nucleus in its rest frame. Then $M^{*2} - m_2^2 = (M^* + m_2)(M^* - m_2) = (2M^* - E_0/c^2)E_0/c^2$ so

$$E_1 = \left(1 - \frac{E_0}{2M^* c^2}\right) E_0. \tag{5.15}$$

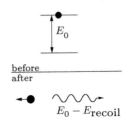

The energy of the emitted photon is slightly smaller than the rest energy change of the atom. The difference $E_0^2/(2M^* c^2)$ is called the **recoil energy**.

before
after

$E_0 - E_{\text{recoil}}$

Fig. 5.5

2. *Absorption.*

A moving particle collides with a stationary particle of rest mass m_2 and sticks to it or is absorbed. How does the change in rest mass relate to the incident energy?

This is like spontaneous emission 'run backwards', except that the final composite object of rest mass M^* is left with a non-zero recoil momentum p in the laboratory frame of reference. We adopt the notation 'incident (E_1, p_1) strikes stationary m_2 producing (E, p, M^*) final product'.

Energy-momentum conservation now gives

$$E_1 + m_2 c^2 = E, \qquad p_1 = p. \tag{5.16}$$

$m_1 \qquad m_2$

E_1, p_1

before
after

M^*

E, p

Fig. 5.6

Using the same method of solution as for spontaneous emission, one finds in general

$$E_1 = \frac{M^{*2} - m_1^2 - m_2^2}{2m_2} c^2, \tag{5.17}$$

and for the case of photon absorption

$$E_1 = \left(1 + \frac{E_0}{2m_2 c^2}\right) E_0. \tag{5.18}$$

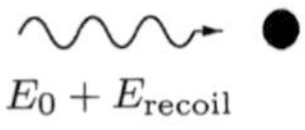

Now the recoil energy has to be provided by the incoming photon.

Note that if the atomic transition is narrow, then atoms at rest will not absorb photons of frequency tuned to match the internal resonance

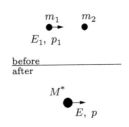

$E_0 + E_{\text{recoil}}$

Fig. 5.7

energy E_0. Or if there are two atoms of the same type at rest relative to one another, with one excited and one in the ground state, then if the excited atom decays, the photon emitted will not be at the right frequency to be absorbed by the other atom.

An important phenomenon related to this is the possibility of suppressing the recoil. If M^* or m_2 is large compared to E_0/c^2 then the recoil energy is negligible. This could in principle happen for a heavy atom or nucleus with closely spaced energy levels. However, a more interesting case is when the atom or nucleus is confined to a small region of space: for example, in a fabricated atom trap or as part of a solid material. When the region of confinement is small compared to the wavelength of the electromagnetic radiation, the momentum of the photon is taken up by the whole of the confining trap or solid. This is called the *Mössbauer effect*. The mass of the recoiling solid crystal can exceed that of an atom by a huge factor, so the recoil energy is essentially completely suppressed.

3. *In-flight decay.*
We already noted that absorption and emission are essentially the same process running in different directions, and therefore eqn (5.17) could be obtained from eqn (5.14) by a change of reference frame. To treat the general case of a particle moving with any speed decaying into two or more products, it is better to learn some more general techniques employing 4-vectors.

Suppose a particle with 4-momentum P decays into various products. The conservation of 4-momentum reads

$$P = \sum_i P_i. \tag{5.19}$$

Therefore

$$M^2 c^4 = E^2 - p^2 c^2 = \left(\sum E_i\right)^2 - \left(\sum \mathbf{p}_i\right) \cdot \left(\sum \mathbf{p}_i\right) c^2. \tag{5.20}$$

Thus if all the products are detected and measured, one can deduce the rest mass M of the original particle.

In the case of just two decay products (a so-called *two-body decay*), a useful simplification is available. We have

$$P = P_1 + P_2. \tag{5.21}$$

Take the scalar product of each side with itself:

$$P \cdot P = P^2 = P_1^2 + P_2^2 + 2P_1 \cdot P_2 \tag{5.22}$$

All these terms are Lorentz-invariant. By evaluating P^2 in any convenient reference frame, one finds $P^2 = -M^2 c^2$, and similarly $P_1^2 = -m_1^2 c^2$, $P_2^2 = -m_2^2 c^2$. Therefore

$$M^2 = m_1^2 + m_2^2 + \frac{2}{c^4}(E_1 E_2 - \mathbf{p}_1 \cdot \mathbf{p}_2 c^2) \tag{5.23}$$

(cf. eqn (2.77)). This shows that to find M it is sufficient to measure the sizes of the momenta and the angle between them, if m_1 and m_2 are known.

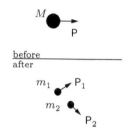

before
after

Fig. 5.8

Table 5.1 Some particles and their rest energies to six significant figures.

e	0.510999 MeV
p	938.272 MeV
π_0	134.977 MeV
$\pi_\pm$	139.570 MeV
Z	(91.1876 ± 0.0021) GeV

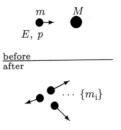

Fig. 5.9

The $P_1 \cdot P_2$ term in eqn (5.22) can also be interpreted using eqn (2.78), giving

$$M^2 = m_1^2 + m_2^2 + 2m_1 m_2 \gamma(u) \tag{5.24}$$

where u is the relative speed of the decay products.

Some further comments on the directions of the momenta are given in the discussion of elastic collisions below, in connection with Fig. 5.17 which applies to any 2-body process.

4. Particle formation and centre of momentum frame

A fast-moving particle of energy E, rest mass m, strikes a stationary one of rest mass M. One or more new particles are created. What are the energy requirements?

The most important idea in this type of collision is to consider the situation in the **centre of momentum frame**. This is the inertial frame of reference in which the total momentum is zero. The total energy of the system of particles in this reference frame is called the 'centre of momentum collision energy' E_{CM} or sometimes (by a loose use of language) the 'centre of mass energy'. The quickest way to calculate E_{CM} is to use the Lorentz-invariant '$E^2 - p^2 c^2$' applied to the total energy-momentum of the system. In the laboratory frame before the collision the total energy-momentum is $P = (E/c + Mc, \mathbf{p})$ where $\mathbf{p}$ is the momentum of the incoming particle. In the CM frame the total energy-momentum is simply $(E_{\mathrm{CM}}/c, 0)$. Therefore, by Lorentz invariance we have

$$\begin{aligned} E_{\mathrm{CM}}^2 &= (E + Mc^2)^2 - p^2 c^2 \\ &= m^2 c^4 + M^2 c^4 + 2Mc^2 E. \end{aligned} \tag{5.25}$$

If the intention is to create new particles by crunching existing ones together, then one needs to provide the incoming 'torpedo' particle with sufficient energy. In order to conserve momentum, the products of the collision must move in some way in the laboratory frame. This means that not all of the energy of the 'torpedo' can be devoted to providing rest mass for new particles. Some of it has to be used up furnishing the products with kinetic energy. The least kinetic energy in the CM frame is obviously obtained when all the products are motionless. This suggests that this is the optimal case: i.e., with the least kinetic energy in the laboratory frame also. To prove that this is so, apply eqn (5.25). This shows that the minimum E (hence the minimum laboratory frame energy) is attained at the minimum E_{CM} (i.e., CM energy). E_{CM}/c^2 can never be less than the sum of the post-collision rest masses, but it can attain that minimum if the products do not move in the CM frame. Therefore the threshold energy is when

$$E_{\mathrm{CM}} = \sum_i m_i c^2 \tag{5.26}$$

where m_i are the rest masses of the collision products. Substituting this into eqn (5.25) we obtain the general result:

Threshold energy

$$E_{\text{th}} = \frac{(\sum_i m_i)^2 - m^2 - M^2}{2M} c^2. \qquad (5.27)$$

This gives the threshold energy in the laboratory frame for a particle m hitting a free stationary target M, such that collision products of total rest mass $\sum_i m_i$ can be produced.

Let us consider a few examples. Suppose we would like to create antiprotons by colliding a moving proton with a stationary proton. The process $p + p \to \bar{p}$ does not exist in nature because it does not satisfy conservation laws associated with particle number, but the process $p + p \to p + p + p + \bar{p}$ is possible. Applying eqn (5.27) we find that the energy of the incident proton must be $7Mc^2$: i.e., 3.5 times larger than the minimum needed to create a proton/anti-proton pair.

In general, eqn (5.27) shows that there is an efficiency problem when the desired new particle is much heavier than the target particle. Suppose for example that we wanted to create Z bosons by smashing fast positrons into electrons at rest in the laboratory. Eqn (5.27) says the initial energy of the positrons must be approximately 90 000 times larger than the rest-energy of a Z boson! Almost all the precious energy, provided to the incident particle using expensive accelerators, is 'wasted' on kinetic energy of the products. In Rindler's memorable phrase, 'it is a little like trying to smash ping-pong balls floating in space with a hammer'. This is the reason why the highest-energy particle accelerators now adopt a different approach, where two beams of particles with equal and opposite momenta are collided in the laboratory. In such a case the laboratory frame is the CM frame, so all the energy of the incident particles can in principle be converted into rest mass energy of the products. Getting a pair of narrow intense beams to hit each other presents a great technical challenge, but formidable as the task is, it is preferable to attempting to produce a single beam of particles with energies thousands of times larger. This is the way the Z boson was experimentally discovered in the 'SPS' proton–antiproton collider at CERN, Geneva, in 1983, and subsequently produced in large numbers by that laboratory's large electron–positron collider ('LEP').

The process of creating particles through collisions is called *formation*. In practice the formed particle may be short-lived and never observed directly. The sequence of events may be, for example, $a + b \to X \to a + b$, or else X may be able to decay into other particles (in which case it is said to have more than one *decay channel*). The state consisting of X is a state of reasonably well-defined energy and momentum (broadened by the finite lifetime of the particle). It shows up in experiments as a large enhancement in the scattering cross-section when a and b scatter off one another; see Fig. 5.10. Such a signature is called a *resonance*.

Fig. 5.10 Data from several experiments operating at different energy regimes is here brought together, showing a resonance in the cross-section for electron–positron scattering at a centre-of-momentum energy of 91.1876 ± 0.0021 GeV. This resonance is interpreted as evidence that a particle with rest mass 91.1876 ± 0.0021 GeV/c^2 is formed in such collisions. The data is consistent with the Standard Model of the weak interaction; the particle is the Z boson, whose discovery and quantitative study was a major success for the Standard Model. (Figure from Physics Reports 427 (2006); the ALEPH, DELPHI, L3, OPAL and SLD Collaborations)

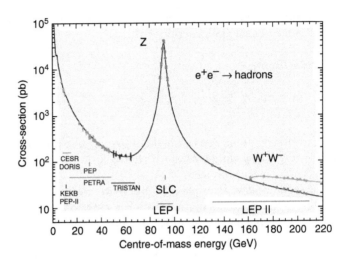

5. *CM frame properties*

The velocity of the CM frame relative to the laboratory frame is

$$\mathbf{v} = \mathbf{p}_{\text{tot}} c^2 / E_{\text{tot}} \tag{5.28}$$

where $\mathbf{p}_{\text{tot}}$ and E_{tot} are the total 3-momentum and energy in the laboratory frame. *Proof:* Without loss of generality, we can align the x axis of the laboratory frame with $\mathbf{p}_{\text{tot}}$. Applying the standard Lorentz transformation, we shall find that in another frame the momentum components are

$$p'_{\text{tot},x} = \gamma(-E_{\text{tot}} v/c^2 + p_{\text{tot},x}), \qquad p'_{\text{tot},y} = p'_{\text{tot},z} = 0$$

It follows that $\mathbf{p}'_{\text{tot}} = 0$ (i.e., the new frame is the CM frame) as long as $E_{\text{tot}} v/c^2 = p_{\text{tot}}$, which is eqn (5.28).

If an incoming particle of momentum p strikes a stationary particle of rest mass M, then the momentum of either particle in the CM frame is

$$p' = \frac{Mc^2}{E_{\text{CM}}} p. \tag{5.29}$$

Proof: Since the particle of rest mass M is stationary in the laboratory frame, it has speed v in the CM frame given by eqn (5.28). Hence its momentum is

$$p' = \gamma M v = \frac{Mpc^2/E_{\text{tot}}}{\sqrt{1 - p^2 c^2/E_{\text{tot}}^2}} = \frac{Mpc^2}{\sqrt{E_{\text{tot}}^2 - p^2 c^2}} = \frac{Mc^2 p}{E_{\text{CM}}}. \tag{5.30}$$

6. *3-body decay.* If a particle Y decays into three products $1, 2, 3$, then the conservation of 4-momentum reads

$$\mathsf{P}_Y = \mathsf{P}_1 + \mathsf{P}_2 + \mathsf{P}_3.$$

To find out about one of the products, say 3, bring it to the left and square:

$$-m_Y^2 c^2 - m_3^2 c^2 - 2\mathsf{P}_Y \cdot \mathsf{P}_3 = -m_1^2 c^2 - m_2^2 c^2 + 2(\mathsf{P}_1 \cdot \mathsf{P}_2) \tag{5.31}$$

Fig. 5.11

Now adopt the CM frame, then P_Y has zero spatial part, so $\mathsf{P}_Y \cdot \mathsf{P}_3 = -m_Y E_3$, hence

$$
\begin{aligned}
E_3 &= \frac{(m_Y^2 + m_3^2 - m_1^2 - m_2^2)c^4 + 2\mathsf{P}_1 \cdot \mathsf{P}_2 c^2}{2m_Y c^2} \\
&= \frac{(m_Y^2 + m_3^2 - m_1^2 - m_2^2)c^4 - 2E_1 E_2 + 2\mathbf{p}_1 \cdot \mathbf{p}_2 c^2}{2m_Y c^2}
\end{aligned} \tag{5.32}
$$

There is now a range of values of E_3, depending on the value of $\mathsf{P}_1 \cdot \mathsf{P}_2$. This is in contrast to the 2-body decay which gives a unique solution in the CM frame. Suppose not all the decay products are detected (for example because one of them is a neutrino), then a signature of 3-body decays compared to 2-body decays is the presence of a range of values of the energy of any one of the products, for a given direction of emission. This was used to deduce the presence of a further particle (the anti-neutrino) in radioactive β-decay, for example.

Now recall eqn (2.78), which we repeat here for convenience:

$$
\mathsf{P}_1 \cdot \mathsf{P}_2 = -m_1 m_2 c^2 \gamma_u \tag{5.33}
$$

where u is the relative speed of the particles. The maximum value of E_3 is when $\mathsf{P}_1 \cdot \mathsf{P}_2$ reaches its highest (i.e., least negative) value, which occurs when 1 and 2 have no relative velocity, and then $\mathsf{P}_1 \cdot \mathsf{P}_2 = -m_1 m_2 c^2$. This makes sense, because then the 1+2 system has the least internal energy. Hence the maximum possible value of E_3 is

$$
E_{3,\text{max}} = \frac{\left(m_Y^2 + m_3^2 - (m_1 + m_2)^2\right) c^2}{2m_Y}. \tag{5.34}
$$

Further quantities are explored in exercise 5.14.

2-stage 3-body decay. If a 3-body decay takes place in two stages: $Y \to 1 + X$ followed by $X \to 2 + 3$ then the end result is the same, but the presence of the intermediate particle X constrains the energies, since now we have only 2-body processes with unique solutions. In many cases the average lifetime of X is so short that it is never directly observed, but as long as its rest mass is reasonably well-defined (subject to the energy-time Heisenberg uncertainty limit) then its presence can be inferred. For example, in the final situation, particle 1 has a unique energy in the CM frame, and this shows up as a spike in the detected energy distribution.

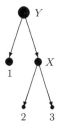

Fig. 5.12

5.3.1   'Isolate and square'

A method of algebraic manipulation that is often useful in collision problems may be called 'isolate and square'. The idea is useful when a rest mass is known but the energy and momentum are not. In order to simplify the equations, one wishes to focus on one unknown while discarding others. To this end, pick a 4-momentum term in the equation that you do not know and do not wish to know. Make this term the subject of the equation (i.e., 'isolate' it), then square both sides of the equation. The isolated term is thus converted into a squared rest mass.

To illustrate the method, consider once again the simple case of 2-body decay. The conservation of energy-momentum reads

$$\mathsf{P} = \mathsf{P}_1 + \mathsf{P}_2$$

where subscripts $1, 2$ label the products. We assume the rest masses are all known, and we would like to learn about particle number 1. In order to discard unknown information about particle 2, isolate P_2 and square the equation:

$$(\mathsf{P} - \mathsf{P}_1)^2 = \mathsf{P}_2^2$$

$$\Rightarrow \qquad \mathsf{P}^2 - 2\mathsf{P} \cdot \mathsf{P}_1 + \mathsf{P}_1^2 = -m_2^2 c^2$$

$$\Rightarrow \qquad M^2 + \frac{2}{c^2}\mathsf{P} \cdot \mathsf{P}_1 + m_1^2 = m_2^2$$

This can now be solved for P_1. For example, in the rest frame of the decaying particle (which is the CM frame) we shall find $\mathsf{P} = (Mc, 0)$ and $\mathsf{P}_1 = (E_1/c, p_1)$ so $\mathsf{P} \cdot \mathsf{P}_1 = -E_1 M$ and we find

$$E_1 = \frac{M^2 + m_1^2 - m_2^2}{2M}c^2 \qquad (5.35)$$

in agreement with eqn (5.14). For such a simple problem, the 'isolate and square' method is not especially advantageous, but for more sophisticated problems it is very useful.

5.4   Elastic collisions

We term a collision *elastic* when the rest masses of the colliding particles are all preserved. Such collisions form an important tool in particle physics for probing the structure of composite particles, and testing fundamental theories—for example, of the strong and weak interactions. They include particle formation processes of the form $a + b \to X \to a + b$, in which the formed particle X does not emerge but may be inferred from the presence of a resonance in the scattering cross-section. Even in the absence of a resonance, the experiment still tests whatever theoretical description exists for the scattering cross-section as a function of the 4-momenta.

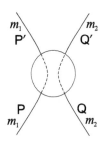

A generic 2-body elastic collision is shown in Fig. 5.13, in order to introduce notation. To conserve energy-momentum we have $\mathsf{P} + \mathsf{Q} = \mathsf{P}' + \mathsf{Q}'$. Squaring this gives $\mathsf{P}^2 + \mathsf{Q}^2 + 2\mathsf{P} \cdot \mathsf{Q} = \mathsf{P}'^2 + \mathsf{Q}'^2 + 2\mathsf{P}' \cdot \mathsf{Q}'$. But by hypothesis, $\mathsf{P}^2 = \mathsf{P}'^2$ and $\mathsf{Q}^2 = \mathsf{Q}'^2$. It follows that

$$\mathsf{P} \cdot \mathsf{Q} = \mathsf{P}' \cdot \mathsf{Q}'. \qquad (5.36)$$

Fig. 5.13 A generic elastic collision, in which the incoming 4-momenta are P, Q, the outgoing 4-momenta are P', Q'. The rest masses m_1, m_2 are unchanged.

Using eqn (5.33) it is seen that this implies the relative speed of the particles is the same before and after the collision, just as occurs in classical mechanics.

In the centre-of-momentum (CM) frame an elastic collision is so simple as to be almost trivial: the two particles approach one another along a line with equal and opposite momenta; after the collision they leave in opposite directions along another line, with the same relative speed and again equal and opposite momenta. The result in some other frame is easily obtained by Lorentz transformation from this one. The velocity of the CM frame relative to the laboratory is given by eqn (5.28).

5.4.1   Billiards

Consider the case of identical particles ('relativistic billiards'); see Fig. 5.14. We take an interest in the opening angle $\theta = \theta_1 + \theta_2$ between the final velocities $\mathbf{v}$, $\mathbf{w}$ after the collision, in the frame in which one of the colliding partners was initially at rest. This angle can be obtained from the dot product:

$$\cos\theta = \frac{\mathbf{v}\cdot\mathbf{w}}{vw}.$$

For this calculation, in contrast to all the collision problems we have considered up until now, we shall work in terms of velocity and Lorentz factor rather than energy and momentum. The conservation of energy and momentum yields (after cancelling common factors of m and c^2)

$$\gamma_u + 1 = \gamma_v + \gamma_w,$$

$$\gamma_u \mathbf{u} = \gamma_v \mathbf{v} + \gamma_w \mathbf{w}$$

Squaring the second equation, and employing eqn (2.10), we have

$$\gamma_u^2 u^2 = (\gamma_u^2 - 1)c^2 = \gamma_v^2 v^2 + \gamma_w^2 w^2 + 2\gamma_v\gamma_w \mathbf{v}\cdot\mathbf{w}.$$

Now substitute for γ_u using the first equation (energy), and we find

$$2\gamma_v\gamma_w \mathbf{v}\cdot\mathbf{w} = (\gamma_v + \gamma_w - 1)^2 c^2 - c^2 - \gamma_v^2 v^2 - \gamma_w^2 w^2$$

$$= 2c^2(\gamma_v - 1)(\gamma_w - 1).$$

Hence

$$\cos\theta = \frac{c^2(\gamma_v - 1)(\gamma_w - 1)}{vw\gamma_v\gamma_w} = \sqrt{\left(\frac{\gamma_v - 1}{\gamma_v + 1}\right)\left(\frac{\gamma_w - 1}{\gamma_w + 1}\right)}. \qquad (5.37)$$

At small speeds we obtain $\cos\theta \to 0$, which is the classical prediction (products emerging at right angles). At high speed we obtain $\cos\theta > 0$, so the opening angle is *reduced*. The opening angle is less than 90° because both particles are 'thrown forward' compared to the classical

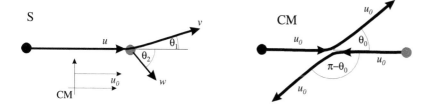

Fig. 5.14 An elastic collision between particles of equal rest mass. The 'lab frame' S is taken to be that in which one of the particles is initially at rest. The CM moves at speed u_0 relative to S, given by $u_0 = \gamma_u u/(\gamma_u + 1)$.

case; see Fig. 2.3. Elastic collisions with opening angles below 90° are frequently seen in particle accelerators and in cosmic-ray events in photographic emulsion detectors.

It is also useful to examine the result in terms of what went on in the CM frame. In that frame the initial and final speeds are all u_0, where u_0 is the speed of the CM frame relative to the laboratory frame, given by $u_0 = p_{tot}c^2/E_{tot} = \gamma_u u/(\gamma_u + 1)$. Choose the x axis along the incident direction of one of the particles. If the final velocity in the CM frame of one particle is directed at some angle θ_0 to the x axis in the anticlockwise direction, then the other is at $\theta_0 - \pi$: i.e., $\pi - \theta_0$ in the clockwise direction. The post-collision angles θ_1 (anti-clockwise) and θ_2 (clockwise) in the laboratory frame are related to θ_0 and $\pi - \theta_0$ by the angle transformation equation for particle velocities (2.70), with the substitutions $\theta \to (\theta_0 \text{ or } \pi - \theta_0)$, $\theta' \to (\theta_1 \text{ or } \theta_2)$, $u \to u_0$, $v \to -u_0$. Hence

$$\tan\theta_1 = \frac{\sin\theta_0}{\gamma(u_0)(\cos\theta_0 + 1)}, \quad \tan\theta_2 = \frac{\sin\theta_0}{\gamma(u_0)(-\cos\theta_0 + 1)},$$

Using these expressions we find, for $\theta_0 \neq 0$, that the opening angle $\theta_1 + \theta_2$ is given by

$$\tan(\theta_1 + \theta_2) = \frac{2\gamma_{u_0}}{(\gamma_{u_0}^2 - 1)\sin\theta_0}. \tag{5.38}$$

(The case $\theta_0 = 0$ has to be treated separately, but it has an obvious answer.) In terms of the relative speed u we have $\gamma_{u_0}^2 = \frac{1}{2}(\gamma_u + 1)$ by using the gamma relation (2.13). The relationship between θ_1 and θ_2 can also be written:

$$\tan\theta_1 \tan\theta_2 = \gamma_{u_0}^{-2}. \tag{5.39}$$

5.4.2 Compton scattering

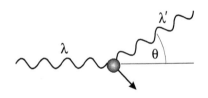

Fig. 5.15 Compton scattering.

'Compton scattering' is the scattering of light off particles, such that the recoil of the particles results in a change of wavelength of the light (Figs 5.15, 5.16). When Arthur Compton (1892–1962) and others discovered changes in the wavelength of X-rays and γ-rays scattered by electrons, and especially changes that depended on scattering angle, it was very puzzling, because it is hard to see how a wave of given frequency can cause any oscillation at some other frequency when it drives a free particle. Compton's careful experimental observations gave him sufficiently accurate data to lend focus to his attempts to model the phenomenon theoretically. He hit upon a stunningly simple answer by combining the quantum theory of light, still in its infancy, with Special Relativity.

Let the initial and final properties of the photon be $\mathsf{P} = (E/c, \mathbf{p})$ and $\mathsf{P}' = (E'/c, \mathbf{p}')$, and let Q, Q' be the initial and final properties of the target, of rest mass m. Supposing that the initial conditions P and Q are given, we would like to know the final properties of the photon: i.e.,

P'. To get rid of Q', isolate it and square:

$$(P + Q - P')^2 = Q'^2 \quad \Rightarrow \quad P^2 + P'^2 + 2(P \cdot Q - P \cdot P' - Q \cdot P') = 0$$

$$\Rightarrow \quad P \cdot P' = Q \cdot (P - P') \qquad (5.40)$$

where we used first $Q^2 = Q'^2$ and then $P^2 = P'^2 = 0$.

Assuming the target is initially at rest, we have $Q = (mc, 0)$ so we have

$$-EE'/c^2 + \mathbf{p} \cdot \mathbf{p}' = -m(E - E')$$

$$\Rightarrow \quad EE'(1 - \cos\theta) = mc^2(E - E')$$

$$\Rightarrow \quad \frac{1}{E'} - \frac{1}{E} = \frac{1}{mc^2}(1 - \cos\theta). \qquad (5.41)$$

So far the calculation has concerned particles and their energies and momenta. If we now turn to quantum theory then we can relate the energy of a photon to its frequency, according to Planck's famous relation $E = h\nu$. Then eqn (5.41) becomes

$$\lambda' - \lambda = \frac{h}{mc}(1 - \cos\theta). \qquad (5.42)$$

This is the Compton scattering formula.

A wave model of Compton scattering is not completely impossible to formulate, but the particle model presented above is much simpler. In a wave model, the change of wavelength arises from a Doppler effect owing to the motion of the target electron.

The quantity

$$\lambda_C \equiv \frac{h}{mc} \qquad (5.43)$$

is called the **Compton wavelength**. For the electron its value is $2.4263102175(33) \times 10^{-12}$ m. It is poorly named because, although it may be related to wavelengths of photons, it is best understood as the distance scale below which quantum field theory is required; both classical physics and non-relativistic quantum theory then break down. The non-relativistic Schrödinger equation for the hydrogen atom is

$$-\frac{1}{2}a_0 \nabla^2 \psi - \frac{1}{r}\psi = \frac{i}{\alpha c}\frac{\partial \psi}{\partial t}$$

where a_0 is the Bohr radius and α is the fine structure constant. The Bohr radius can be written as

$$a_0 = \frac{\lambda_C}{2\pi\alpha}.$$

Schrödinger's equation tells us that for a bound state of an electron in hydrogen, a_0 is the typical distance scale and αc the typical speed. Since $\alpha \ll 1$ we find that $v \ll c$ and $a_0 \gg \lambda_C$. Therefore relativistic quantum theory is not required to treat the structure of atoms, at least in first approximation: Schrödinger's equation will do.

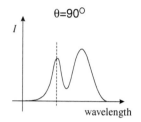

Fig. 5.16 Example spectrum in a Compton scattering experiment. Observations at $\theta \neq 0$ typically show two peaks—one at the incident wavelength (indicated by a dashed line) and one at a longer wavelength. The first peak is due to scattering by tightly bound electrons and nuclei, and the second peak is due to scattering by weakly bound electrons which behave to good approximation as if they were free.

Elastic terminology. Compton scattering appears here under the heading of 'elastic' processes because the rest masses do not change. However, the word 'elastic' can also be used to mean that the energies of the colliding parties are unchanged; Compton scattering is not elastic in that sense, except in the limit $m \to \infty$.

Inverse Compton scattering

Eqn (5.41) shows that a photon scattering off a stationary particle always loses energy. A photon scattering off a moving particle can either lose or gain energy; the latter case is sometimes called 'inverse Compton scattering'. It is of course just another name for Compton scattering viewed from a different reference frame. In astrophysics such inverse Compton scattering is more important (because a more useful source of observational information) than Compton scattering.

Eqn (5.40) is true for any initial conditions. We now assume the target particle may be moving, and for the sake of simplicity we specialize to the case of a head-on collision: i.e., $\mathsf{P} = E(1,1)$, $\mathsf{P}' = E'(1,-1)$, $\mathsf{Q} = \gamma m(1, -u)$ in one spatial dimension, and taking $c = 1$. We thus find

$$-2EE' = \gamma m[-E + E' - u(E + E')].$$

Solving for E' yields

$$E' = \frac{\gamma m(1 + u)}{2 + \gamma m(1 - u)/E}. \tag{5.44}$$

When $u \simeq 1$ (i.e., close to the speed of light) it is more useful to write $(1 + u) \simeq 2$ and $(1 - u) \simeq 1/2\gamma^2$, so

$$E' \simeq \frac{\gamma m}{1 + m/(4\gamma E)} \tag{5.45}$$

which further simplifies to $E' = 4\gamma^2 E$ (hence wave frequency $\nu' = 4\gamma^2\nu$) when $\gamma E \ll m$.

This process is relevant in various astrophysical phenomena, such as X-ray emission from active galactic nuclei, gamma-ray emission in some quasars, and X-ray emission in intergalactic space. For example, an electron with $\gamma \simeq 10^4$ colliding with a photon from the cosmic microwave background radiation (wavelength $\simeq 0.5$ cm) can result in a scattered X-ray photon. At higher energies the incident particle loses a large fraction of its energy in a single collision.

Compton and inverse Compton scattering are also related to *bremsstrahlung* or 'braking radiation', which is the radiation emitted when charged particles are slowed, for example, by elastic collisions with atomic nuclei.

5.4.3   More general treatment of elastic collisions*

Our treatment of 'relativistic billiards' above committed the treason of failing to use invariants when they are available. This was because we already had the angle transformation formula in hand. In this section we provide some guidance on the more general problem of elastic scattering using different particles. This is an important tool in high-energy physics. Typically one is interested in a case where the outcome (e.g., the distribution of scattering angles) is determined by a quantum mechanical process, resulting in a probability function.

Let the collision involve 4-momenta $\mathsf{P}, \mathsf{Q}, \mathsf{P}', \mathsf{Q}'$ satisfying

$$\mathsf{P} + \mathsf{Q} = \mathsf{P}' + \mathsf{Q}'. \tag{5.46}$$

with

$$\mathsf{P}^2 = \mathsf{P}'^2 = -m_1^2, \quad \mathsf{Q}^2 = \mathsf{Q}'^2 = -m_2^2, \tag{5.47}$$

where we have adopted units such that $c = 1$. The subscripts $1, 2$ refer to particles whose rest masses m_1, m_2 may differ, but P is the 4-vector of a particle with the same rest mass as P' (see Fig. 5.13). It may be the very same particle, but since we only assume the rest mass is the same, the treatment can apply to a variety of processes, such as

$$\pi + p \rightarrow \pi + p \qquad \text{pion–proton scattering}$$

$$p + \bar{p} \rightarrow p + \bar{p} \qquad \text{proton–antiproton scattering}$$

$$p + \bar{p} \rightarrow \pi + \bar{\pi} \qquad \text{ditto}$$

where to treat the last case we can use P, P' for the proton and antiproton, Q, Q' for the pion and antipion, and in order that eqn (5.46) still states the conservation of momentum, we must interpret $-\mathsf{P}'$ as the initial momentum of the antiproton, and $-\mathsf{Q}$ as the final momentum of the pion. More generally, by appropriately interpreting the signs of the momenta one can allow any pair of the 4-momenta to be incoming, then the other pair must be outgoing (for an elastic process).

The example proton–antiproton processes show that more than one type of process may happen in a given experiment. (We could also treat $p + p \rightarrow \pi + \pi$ but that process does not exist in Nature.)

For given rest masses, the probability amplitude of observing a given outcome is some function of the 4-momenta, $T = T(\mathsf{P}, \mathsf{Q}, \mathsf{P}', \mathsf{Q}')$. It appears from this that T may depend on sixteen variables. However, we can whittle that number down to just two. We argue that T must be Lorentz-invariant (for, in a fixed large number N of trials, if $N|T|^2$ are observed in some frame to give a particular outcome, for example given outgoing particles arriving in a given pair of buckets, then all frames must agree that that is where the particles went, and that it happened on $N|T|^2$ occasions; furthermore a process involving further particles would have interference terms whose probability depends on T itself, not just $|T|^2$, to which the same argument would apply). Therefore T depends only on the ten invariants that can be formed from the 4-momenta:

$$P^2 = -m_1^2, \quad P \cdot Q, \qquad P \cdot P', \qquad P \cdot Q'$$
$$Q^2 = -m_2^2, \quad Q \cdot P', \qquad Q \cdot Q'$$
$$P'^2 = -m_1^2, \quad P' \cdot Q'$$
$$Q'^2 = -m_2^2.$$

Of these, four are constants, reducing the number to six, and the conservation of energy-momentum equation (5.46) gives four constraints, so there are just two independent variables.

One cannot pick any pair, because some turn out to be equal to one another. We have already noted that $P \cdot Q = P' \cdot Q'$, (eqn (5.36)). By rearranging the conservation of 4-momentum formula and squaring one can obtain the other three constraints:

$$(P - Q')^2 = (P' - Q)^2 \quad \Rightarrow \quad P \cdot Q' = P' \cdot Q, \qquad (5.48)$$

$$(P - P')^2 = (Q' - Q)^2 \quad \Rightarrow \quad m_1^2 + P \cdot P' = m_2^2 + Q \cdot Q'. \qquad (5.49)$$

$$(P + Q - P')^2 = Q'^2 \quad \Rightarrow \quad m_1^2 = P \cdot Q - P \cdot P' - Q \cdot P'. \qquad (5.50)$$

A possible (not the only) choice of independent variables is $P \cdot Q$ and $P \cdot Q'$.

Some general considerations concerning the directions of the momenta are indicated in Fig. 5.17. For a 2-body process, the final momenta are equal and opposite in the CM frame, and of a fixed size. Therefore, they lie on a circle, when plotted on a diagram of the kind we introduced in Fig. 2.9. In the lab frame, therefore, the observed momenta will lie on an ellipse, as shown in Fig. 5.17. There may be more than one possible outcome of the experiment, and the data will be noisy. By looking for things like double peaks as a function of p at given θ (see the caption to Fig. 5.17) one can begin the process of interpreting

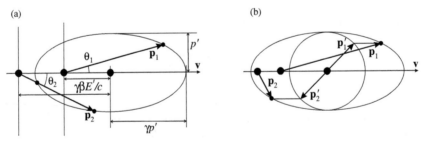

(a)   (b)

Fig. 5.17 Two uses of the ellipse construction that was introduced in Fig. 2.9. The ellipse permits the set of sizes and directions of momenta in the lab frame to be found, when the momentum in the primed frame (e.g., the CM frame) is of fixed size and any direction. Both diagrams show a case where two particles have the *same p'* (the size of the 3-momentum) in frame S', and therefore both give rise to the same ellipse, but their rest mass and therefore energy E' may differ. (a) The diagram at left shows that, depending on whether $\beta E'$ is smaller or larger than cp', the foot of the lab frame momentum vector lies inside or outside the ellipse. This means that for a given observed direction of flight θ in the lab frame, there is either a single size of momentum p_1, or a pair of sizes p_2 (corresponding to a single or a pair of directions in the CM frame). If one measures the number of detections as a function of p at a given angle θ, one sees either a single or a double peak. Also, in the latter case (large rest mass) there is a maximum angle $|\theta| < \theta_{\max}$ that occurs when $\mathbf{p}$ meets the ellipse at a tangent; in the former case (small rest mass) all angles are possible. (b) The diagram at right shows a case where the momenta in the CM frame are equal and opposite, as must be the case for a 2-body system.

the information and separating the data from the noise. By plotting
the lab frame momenta observed in many experiments one can find the
ellipse that best fits the observations. Similar considerations apply to
the interpretation of 2-body decay data, for example, if the identities of
the decay products are unknown, and their energies are hard to measure
accurately.

Now we consider a process where one particle (the 'target') is initially
at rest in the lab frame. Let P be the 4-momentum of the target. Then,
when written down in the lab frame, we will find that P has no spatial
(3-momentum) part. If follows that when dotted onto other 4-vectors, it
'picks out' their energy part. Therefore, we can write the energies in the
lab frame as

$$E_1 = m_1 \qquad\qquad \text{target rest energy}$$

$$E_2 = -\mathsf{P}\cdot\mathsf{Q}/m_1 \qquad \text{energy of incoming particle}$$

$$E_1' = -\mathsf{P}\cdot\mathsf{P}'/m_1 \qquad \text{energy of target after scattering}$$

$$E_2' = -\mathsf{P}\cdot\mathsf{Q}'/m_1 \qquad \text{outgoing energy of scattered particle}$$

The scattering angle θ is the angle between the 3-momenta $\mathbf{q}$ and $\mathbf{q}'$ in
the lab frame, which can be obtained from

$$\mathsf{Q}\cdot\mathsf{Q}' = -E_2 E_2' + qq'\cos\theta.$$

Using $q = (E_2^2 - m_2^2)^{1/2}$ one finds

$$\cos\theta = \frac{E_2 E_2' + \mathsf{Q}\cdot\mathsf{Q}'}{\sqrt{(E_2^2 - m_2^2)(E_2'^2 - m_2^2)}}$$

$$= \frac{(\mathsf{P}\cdot\mathsf{Q})(\mathsf{P}\cdot\mathsf{Q}') + m_1^2(\mathsf{Q}\cdot\mathsf{Q}')}{\sqrt{((\mathsf{P}\cdot\mathsf{Q})^2 - m_1^2 m_2^2)((\mathsf{P}\cdot\mathsf{Q}')^2 - m_1^2 m_2^2)}}. \qquad (5.51)$$

It is often helpful to introduce the *Mandelstam variables*

$$s \equiv -(\mathsf{P}+\mathsf{Q})^2 = -(\mathsf{P}'+\mathsf{Q}')^2,$$

$$t \equiv -(\mathsf{P}-\mathsf{P}')^2 = -(\mathsf{Q}-\mathsf{Q}')^2,$$

$$u \equiv -(\mathsf{P}-\mathsf{Q}')^2 = -(\mathsf{P}'-\mathsf{Q})^2.$$

s is the square of the CM energy if P and Q or P' and Q' are incoming;
t is the square of the CM energy if P and $(-\mathsf{P}')$ or Q and $(-\mathsf{Q}')$ are
incoming. The Mandelstam variables are not all independent, but (using
eqns (5.48)—(5.50)) satisfy

$$s + t + u = 2(m_1^2 + m_2^2).$$

In the CM frame[2] t can be interpreted as minus the square of the
momentum transfer (a positive value for t indicates that the scattering
is not elastic). One there has $(\mathsf{P}-\mathsf{P}') = (0, \mathbf{p}-\mathbf{p}')$ so

$$t = 2p^2(\cos\theta - 1) = -4p^2\sin^2\frac{\theta}{2} \qquad (5.52)$$

[2] To avoid clutter we do not
trouble to introduce a prime or superscript (CM)
here; the reader must understand that
eqns (5.52) through (5.54) deal exclu-
sively with quantities in the CM frame.

where we used the fact that $p = p'$ in the CM frame (the momenta change direction but not size in that frame). Also

$$s = m_1^2 + m_2^2 + 2(E_1 E_2 - \mathbf{p} \cdot \mathbf{q}).$$

When P, Q are incoming we have $\mathbf{p} = -\mathbf{q}$ in the CM frame, so this is

$$s = E_{CM}^2 = m_1^2 + m_2^2 + 2(E_1 E_2 + p^2). \tag{5.53}$$

We can convert this into a formula expressing p in terms of s and the rest masses. First make $E_1 E_2$ the subject of the formula, then square:

$$4(E_1 E_2)^2 = 4(m_1^2 + p^2)(m_2^2 + p^2) = (s - m_1^2 - m_2^2 - 2p^2)^2$$
$$\Rightarrow \qquad 4sp^2 = (s - m_1^2 - m_2^2)^2 - 4m_1^2 m_2^2. \tag{5.54}$$

Eqns (5.52) and (5.54) allow $\cos\theta$ to be expressed in terms of s, t, m_1, m_2.

5.5   Composite systems

In the discussion of Special Relativity in this book we have often referred to 'objects' or 'bodies' and not just to 'particles'. In other words, we have taken it for granted that one can talk of a composite entity such as a brick or a plank of wood as a single 'thing', possessing a position, velocity, and mass. The conservation laws are needed in order to make this logically coherent (the same is true in classical physics).

We use the word 'system' to refer to a collection of particles whose behaviour will be discussed. Such a system could consist of particles attached to one another, such as the atoms in a solid object, or it could be a loose collection of independent particles, such as the atoms in a low-density gas. In either case the particles do not 'know' that we have gathered them together into a 'system': the system is just our own selection, a notional 'bag' into which we have placed the particles, without actually doing anything to them. The idea of a system is usually invoked when the particles in question may interact with one another, but they are not interacting with anything else. Then we say we have an 'isolated system'. This terminology has already been invoked in the previous section, where we discussed the total energy and total 3-momentum of such a system. Now we would like to enquire what it might mean to talk about the velocity and rest mass of a composite system.

If a composite system can be discussed as a single object, then we should expect that its rest mass must be obtainable from its total energy momentum in the standard way: i.e.,

$$\mathsf{P}_{tot}^2 = -E_{tot}^2/c^2 + \mathbf{p}_{tot}^2 \equiv -m^2c^4. \qquad (5.55)$$

This serves as the definition of the rest mass m of the composite system. It makes sense because the conservation law guarantees that P_{tot} is constant if the system is not subject to external forces.

One convenient way to calculate m is to work it out in the CM frame, where $\mathbf{p}_{tot} = 0$. Thus we find

$$m = E_{CM}/c^2 \qquad (5.56)$$

where E_{CM} is the value of E_{tot} in the CM frame. Note that the *rest* mass of the composite system is equal to the *total energy* of the constituent particles (divided by c^2) in the CM frame, not the sum of their rest masses. For example, a system consisting of two photons propagating in different directions has a non-zero rest mass.[3] The photons propagating inside a hot oven or a bright star make a contribution to the rest mass of the respective system. The gluons (zero rest mass) propagating inside a proton contribute most of the mass of the proton.

Relative to any other reference frame, the CM frame has some well-defined 3-velocity $\mathbf{u}_{CM}$, and therefore a 4-velocity $\mathsf{U}_{CM} = \gamma(u_{CM})(c, \mathbf{u}_{CM})$. You can now prove that

$$\mathsf{P}_{tot} = m\mathsf{U}_{CM} \qquad (5.57)$$

(Method: both are 4-vectors and they agree in CM frame, hence in all frames.) This confirms that the composite system is behaving as we would expect for a single object of given rest mass and velocity.

5.6   Energy flux, momentum density, and force

There is an important general relationship between *flux* of energy $\mathbf{S}$ and *momentum per unit volume* $\mathbf{g}$. It is easily stated:

$$\mathbf{S} = \mathbf{g}c^2. \qquad (5.58)$$

$\mathbf{S}$ is the amount of energy crossing a surface (in the normal direction), per unit area per unit time, and $\mathbf{g}$ is the momentum per unit volume in the flow.

It would be natural to expect energy flux to be connected to *energy density*. For example, for a group of particles all having energy E and moving together at the same velocity $\mathbf{v}$, the energy density is $u = En$ where n is the number of particles per unit volume, and the number crossing a surface of area A in time t is $nAvt$, so $S = nvE = uv$: the energy flux is proportional to the energy density. However, if the particles are moving in some other way—for example, isotropically—then the

[3] For two or more photons all propagating in the same direction there is no CM frame, because reference frames cannot attain the speed of light.

relationship changes. For particles effusing from a hole in a chamber of gas, for example, we find $S = (1/4)uv$.

Eqn (5.58) is more general. For the case of particles all moving along together, it is easy to prove by using the fact that $\mathbf{p} = E\mathbf{v}/c^2$ for each particle. The momentum density is then $\mathbf{g} = n\mathbf{p}$, and the energy flux is $\mathbf{S} = nvE = npc^2 = \mathbf{g}c^2$. If we now consider more general scenarios, such as particles in a gas, we can apply this basic vector relationship to every small region and small range of velocities, to obtain

$$\mathbf{S}_{\text{tot}} = \sum_i n_i \mathbf{v}_i E_i = \sum_i n_i \mathbf{p}_i c^2 = \mathbf{g}_{\text{tot}} c^2.$$

Since the proportionality factor is c^2 for every term in the sum, it remains c^2 in the total.

The particles we considered may or may not have had rest mass: the relationship $\mathbf{p} = E\mathbf{v}/c^2$ is valid for either, so eqn (5.58) applies equally to light and to matter, and to the fields inside a material body. It is universal!

Another important idea is *momentum flow*.

We introduced force by *defining* it as the rate of change of momentum. We also established that momentum is conserved. These two facts, taken together, imply that another way to understand force is in terms of momentum flow. When more than one force acts we can have a balance of forces, so the definition in terms of rate of change of momentum is no longer useful: there is no such rate of change. In a case like that we know what we mean by the various forces in a given situation: we mean that we studied other cases and we claim that the momentum *would* change if the other forces were not present. In view of the primacy of conservation laws over the notion of force, it can sometimes be helpful to adopt another physical intuition of what a force represents. A force per unit area, in any situation, can be understood as an 'offered' momentum flux: i.e., an amount of momentum flowing across a surface, per unit area per unit time. When a field or a body offers a pressure force to its environment, it is as if it is continually bringing up momentum to the boundary, like the molecules in a gas hitting the chamber walls, and 'offering' the momentum to the neighbouring system. If the neighbour wants to refuse the offer of acquiring momentum, it has to push back with a force: it makes a counter-offer of just enough momentum flow to prevent itself from acquiring any net momentum. In the case of a gas such a picture of momentum flow is natural, but one could, if one chooses, claim that precisely the same flow is taking place in a solid, or anywhere that a force acts. The molecules do not have to move in order to transport momentum: they only need to push on their neighbours. It is a matter purely of taste whether one prefers the language of 'force' or 'momentum flow'.

These ideas are important when one considers energy and momentum exchange between continuous systems. This discussion is postponed until chapter 16, since it requires the introduction of the important but more difficult concept of the *stress-energy tensor*.

Exercises

(5.1) **Energy relation.** Show that, for a particle with constant rest mass, $dE/dp = v$. This is the same as the classical result. To get some insight into this formula, suppose we have the formula $\mathbf{p} = \gamma m_0 \mathbf{v}$ for momentum, but not yet a formula for kinetic energy. Define force by $\mathbf{f} = d\mathbf{p}/dt$, and suppose that it can be derived from a scalar function V called potential energy:

$$\mathbf{f} = -\boldsymbol{\nabla} V.$$

Suppose further that the motion is conservative, so that $V + K$ is constant, where K is kinetic energy. Then we can obtain K as follows. First simplify to a purely one-dimensional case, so

$$dK = -dV = f dx \qquad \text{and} \quad dp = f dt.$$

Therefore

$$\frac{dK}{dp} = \frac{dx}{dt} = v \quad \Rightarrow K = \int v \, dp$$

Using the known relationship between v and p, carry out the integration and hence derive $K = \gamma m_0 c^2 +$ const. Note, however, that this method is less general than those described in the main text.

(5.2) Confirm that $\alpha = \gamma$ is the unique solution to eqn (5.4) having $\alpha(0) = 1$, as follows. We have already shown that it is a solution; it remains to show there is no other choice. Using the power series $\alpha(x) = \sum_i a_i x^i$ with the coefficients a_i to be discovered, show that eqn (5.4) takes the form

$$\sum_i a_i v^i \left(1 - u^2/v^2\gamma_v^2\right)^{i/2} \equiv \gamma_v \sum_i a_i u^i$$

What can be learned from the coefficient of u^0 in this identity?

(5.3) The upper atmosphere of the Earth receives electromagnetic energy from the Sun at the rate 1400 Wm^{-2}. Find the rate of loss of mass of the Sun due to all its emitted radiation. (The Earth–Sun distance is 499 light-seconds.)

(5.4) Calculate the mass reduction owing to heat loss of a 100-kg bath of water (specific heat capacity

4.19 kJ/kg K) as it cools from 90°C to 20°C. How many additional water molecules would be needed to make up the loss? [*Ans.* 3.3×10^{-10} kg; 10^{16}]

(5.5) Find the energy, in joules, of a cosmic-ray proton having $\gamma = 10^{11}$.

(5.6) A particle of rest mass m and kinetic energy $3mc^2$ strikes a stationary particle of rest mass $2m$ and sticks to it. Find the rest mass and speed of the composite particle. [*Ans.* $\sqrt{21}$ m, $0.646c$]

(5.7) A system consists of two photons, each of energy E, propagating at right angles in the laboratory frame. Find the rest mass of the system and the velocity of its CM frame relative to the laboratory frame.

(5.8) A particle of rest mass m breaks up into two particles of equal rest mass αm. What are the largest and smallest possible values of α?

Particle formation

(5.9) (i) A proton beam strikes a target containing stationary protons. Calculate the minimum kinetic energy which must be supplied to an incident proton to allow pions to be formed by the process $\mathrm{p} + \mathrm{p} \to \mathrm{p} + \mathrm{p} + \pi_0$, and compare this to the rest energy of a pion.

(ii) An electron collides with another electron at rest to produce a pair of muons by the process $e + e \to e + e + \mu^+ + \mu^-$. Show that the threshold momentum of the incident electron for this process is

$$p_{\mathrm{th}} = 2Mc(1 + M/m)\sqrt{1 + 2m/M}$$

where m, M are the masses of the electron and muon respectively.

(iii) A photon is incident on a stationary proton. Find, in terms of the rest masses, the threshold energy of the photon if a neutron and a pion are to emerge.

(5.10) A particle formation experiment creates reactions of the form $A + B \to A + B + N$ where A is an incident particle of mass m, B is a target of mass M at rest in the laboratory frame, and N is a new particle. Define the 'efficiency' of the experiment as the ratio of the supplied kinetic energy to the rest energy of the new particle, $m_N c^2$. Show that, at threshold, the efficiency thus defined is equal to $M/(m + M + \frac{1}{2}m_N)$.

(5.11) Two photons of energies E_0, E may collide to produce an electron–positron pair. Find the threshold value of E for this reaction, in terms of E_0 and the electron rest mass m. Calculate this threshold for the case of high-energy galactic photons travelling through the cosmic microwave background radiation, which can be regarded as a gas of photons of energy 2.3×10^{-4} eV. [*Ans.* 1.1×10^{15} eV]

Decay

(5.12) Particle tracks are recorded in a bubble chamber subject to a uniform magnetic field of 2 tesla. A vertex consisting of no incoming and two outgoing tracks is observed. The tracks lie in the plane perpendicular to the magnetic field, with radii of curvature 1.67 m and 0.417 m, and separation angle 21°. It is believed that they belong to a proton and a pion respectively. Assuming this, and that the process at the vertex is decay of a neutral particle into two products, find the rest mass of the neutral particle. [*Ans.* 1103.8 MeV/c^2]

(5.13) §§A decay mode of the neutral Kaon is $K^0 \to \pi^+ + \pi^-$. The Kaon has momentum 300 MeV/c in the laboratory, and one of the pions is emitted, in the laboratory, in a direction perpendicular to the velocity of the Kaon. Find the momenta of both pions. [*Ans.* 166 MeV/c, 344 MeV/c at 29°]

(5.14) **Three-body decay** A particle Y decays into three other particles, with labels indicated by $Y \to 1 + 2 + 3$. Working throughout in the CM frame:

   (i) Show that the 3-momenta of the decay products are coplanar.

   (ii) Derive eqns (5.32) and (5.34).

   (iii) Show that, when particle 3 has its maximum possible energy, particle 1 has the energy

$$E_1 = \frac{m_1(m_Y c^2 - E_{3,\text{max}})}{m_1 + m_2}$$

   [Hint: first argue that 1 and 2 have the same speed in this situation]

   (iv) Let X be the system composed of particles 1 and 2. Show that its rest mass is given by

$$m_X^2 = m_Y^2 + m_3^2 - 2m_Y E_3/c^2$$

   (v) Write down an expression for the energy E^* of particle 2 in the rest frame of X, in terms of m_1, m_2 and m_X.

   (vi) Show that when particle 3 has an energy of intermediate size, $m_3 c^2 < E_3 < E_{3,\text{max}}$, the energy of particle 2 in the original frame (the rest frame of Y) is in the range

$$\gamma(E^* - \beta p^* c) \le E_2 \le \gamma(E^* + \beta p^* c)$$

where p^* is the momentum of particle 2 in the X frame, and γ, β refer to the speed of that frame relative to the rest frame of Y.

(5.15) This diagram illustrates a process in which an electron emits a photon:

Prove that the process is impossible. Prove also that a photon cannot transform into an electron–positron pair in free space. In the presence of a nucleus, however, it can. Find the threshold energy of the photon, if the nucleus of rest mass M is initially at rest. Verify that in the limit of large M the efficiency approaches 100% and therefore the nucleus acts as a perfect catalyst.

(5.16) §Prove that a photon in free space cannot decay, neither into a pair of photons with differing directions of propagation, nor into a pair of co-propagating photons with different frequencies.

(5.17) A 'photon rocket' propels itself by emitting photons in the rearwards direction. The rocket is initially at rest with mass m. Show that when the rest mass has fallen to αm the speed (as observed in the original rest frame) is given by

$$\frac{v}{c} = \frac{1 - \alpha^2}{1 + \alpha^2}$$

(Hint: conservation of momentum.)

It is desired to reach a speed giving a Lorentz factor of 10. What value of α is required? Supposing the rocket cannot pick up fuel *en route*, what proportion of its initial mass must be devoted to fuel if it is to make a journey in which it first accelerates to $\gamma = 10$, then decelerates to rest at the destination (the destination being a star with negligible speed relative to the Sun)?

(5.18) §A rocket propels itself by giving portions of its mass m a constant velocity **u** relative to its instantaneous rest frame. Let S$'$ be the frame in which the rocket is at rest at time t. Show that, if v' is the speed of the rocket in S$'$, then to first order in dv',

$$(-dm)u = mdv'.$$

Hence, prove that when the rocket attains a speed v relative to its initial rest frame, the ratio of final to initial rest mass of the rocket is

$$\frac{m_f}{m_i} = \left(\frac{1 - v/c}{1 + v/c}\right)^{c/2u}$$

Note that the least expenditure of mass occurs when $u = c$: i.e., the 'photon rocket'.

Prove that if the rocket moves with constant proper acceleration a_0 for a proper time τ, then $m_f/m_i = \exp(-a_0\tau/u)$.

(5.19) A collimated beam of X-rays of energy 17.52 keV is incident on an amorphous carbon target. Sketch the frequency spectrum you would expect to be observed at a scattering angle of 90°, including a quantitative indication of the frequency scale.

(5.20) §Consider a head-on elastic collision between a moving 'bullet' of rest mass m and a stationary target of rest mass M. Show that the post-collision Lorentz factor γ of the bullet cannot exceed $(m^2 + M^2)/(2mM)$. (This means that for large energies almost all the energy of the bullet is transferred to the target—very different from the classical result). (Hint: consider $P_t + Q'_b$ where P_t is the initial 4-momentum of the target and Q'_b is the final 4-momentum of the bullet.)

(5.21) Particles of mass m and kinetic energy T are incident on similar particles at rest in the laboratory. Show that if elastic scattering takes place, then the minimum angle between the final momenta in the laboratory is given by $\cos\theta_{min} = (1 + 4mc^2/T)^{-1}$.

6 Further kinematics

In this chapter we return to kinematics: that is, the study of generic properties of motion without regard to the forces which may be involved. Kinematics is mostly concerned with the structure of spacetime, rather than conservation laws. However, we will still feel free to bring in dynamical ideas when they provide insight.

It will be useful to bring in some more 4-vectors—especially the idea of a 4-gradient. This will enable us to discuss flow and density, and wave motion. After that we will discuss accelerated motion and what happens to the shape of bodies when they accelerate. It happens that acceleration is inextricably related to a bending or distortion of the shape of a body, which might otherwise have been considered to be rigid. Finally, we discuss Lorentz transformations for motion in any direction (not just along the coordinate axes), and we will discover a remarkable counter-intuitive rotational effect called 'Thomas–Wigner rotation'. We begin, however, with an important property of non-accelerated motion.

6.1 The Principle of Most Proper Time

Given two time-like-separated events in spacetime, what worldline between them has the most proper time? This is the kind of question which a travelling salesman might like to ask. He has an appointment at an agreed time and place in the future, and we suppose he wants to maximize the time he has available for preparing his notes, or for relaxation *en route*. In view of the twin paradox it should not take you long to guess that the salesman should arrange that his worldline is straight—in other words, he should travel at constant velocity from one meeting to another. By contrast, a salesman who wants to stay young between appointments should rush about making detours. Now, constant-velocity motion is also (Newton's First Law) inertial motion: i.e., motion in the absence of applied forces. Thus we have a connection between inertial motion and proper time. This connection is sufficiently important, especially in General Relativity, that we give it a name.

The Principle of Most Proper Time. *Given two time-like-separated events, of all worldlines connecting the events, that having the most proper time corresponds to inertial motion.*

The proper time of a worldline is, of course, the sum of all the $\mathrm{d}\tau$ contributions along it. In Special Relativity the Principle of Most Proper Time can be derived if we assume Newton's First Law. However, in

General Relativity it is better to regard Most Proper Time as axiomatic, and derive Newton's First Law from it. Therefore, we shall argue in that direction here.

We wish to identify which worldline (among all the possibly wiggly ones) has the most proper time between given time-like-separated events. For the derivation, it is convenient to pick the inertial frame in which the two events in question appear at the same place. Let t_1, t_2 be their times in this frame. Then the proper time along an arbitrary worldline $\mathcal{W}$ connecting the events is

$$\Delta \tau = \int_{\text{(event 1)}}^{\text{(event 2)}} \mathrm{d}\tau = \int_{t_1}^{t_2} \left(1 - \frac{v^2}{c^2}\right)^{1/2} \mathrm{d}t \qquad (6.1)$$

where we used $\mathrm{d}t/\mathrm{d}\tau = \gamma$, and in the integral v is some function of time determined by the worldline $\mathcal{W}$. Now, one possible worldline has $v = 0$ everywhere along it (for the frame we picked)—this is the straight worldline. It gives $\Delta \tau = t_2 - t_1$. It is obvious that any other function $v(t)$ can only ever give a smaller $\Delta \tau$ because $-v^2$ has to be negative. It follows that the constant-velocity worldline is the 'longest' (most proper time). QED.

Another way of looking at the same proof is to compare it to the twin paradox. The straight worldline is that of the 'stay-at-home' twin. As soon as the other twin ventures to move relative to home, her accumulated proper time, for a given amount $(t - t_1)$ of reference frame time, has fallen below $(t - t_1)$, and she can never make up the difference because the Lorentz factor γ is always greater than or equal to 1. This general idea may be called 'proof by twin paradox'.

6.2   Four-dimensional gradient

Now that we have got used to 4-vectors, it is natural to wonder whether we can develop 4-vector operators, the 'larger cousins', so to speak, of the gradient, divergence and curl. A first guess might be to propose a 4-gradient $((1/c)\partial/\partial t, \partial/\partial x, \partial/\partial y, \partial/\partial z)$. Although this quantity is clearly a sort of gradient operator, it is not the right choice because the gradient it produces is not a standard 4-vector. The reason is that it has a sign error. According to the Lorentz transformation, events at positive x and $t = 0$ occur at *negative* t'. However, a function V having positive $(\partial V/\partial x)$ and $(\partial V/\partial t) = 0$ ought to give *positive* $(\partial V/\partial t')$ (see Fig. 6.1).

The answer to this problem is that we must define the four-dimensional gradient operator as[1]

$$\Box \equiv \left(-\frac{1}{c}\frac{\partial}{\partial t}, \boldsymbol{\nabla}\right) = \left(-\frac{1}{c}\frac{\partial}{\partial t}, \frac{\partial}{\partial x}, \frac{\partial}{\partial y}, \frac{\partial}{\partial z}\right). \qquad (6.2)$$

The idea is that with this definition, $\Box V$ is a 4-vector, as we shall now prove.

We have in mind for V a scalar quantity that is itself Lorentz-invariant. This means, if we change reference frames, the value of V

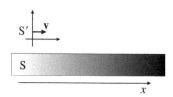

Fig. 6.1 The shading represents charge density ρ on a rigid glass bar fixed in S. As frame frame S$'$ sweeps from left to right, an observer there will observe an *increasing* $\rho(t')$ at any given position in S$'$.

[1] The symbol $\Box$ is commonly used for the d'Alembertian operator shown in eqn (6.22). In our notation that operator is written $\Box^2$. The student should beware of this issue when consulting other textbooks.

at any particular event in spacetime does not change. However, owing to time dilation and space contraction the rate of change of V with either of t' or x' is not necessarily the same as the rate of change with t or x.

Consider two neighbouring events. In some reference frame S their coordinates are t, x, y, z and $t + dt, x + dx, y + dy, z + dz$. The change in the function V between these events is

$$dV = \left(\frac{\partial V}{\partial t}\right)_x dt + \left(\frac{\partial V}{\partial x}\right)_t dx, \tag{6.3}$$

where for simplicity we have chosen a potential function that is independent of y and z. Therefore

$$\left(\frac{\partial V}{\partial t'}\right)_{x'} = \left(\frac{\partial V}{\partial t}\right)_x \left(\frac{\partial t}{\partial t'}\right)_{x'} + \left(\frac{\partial V}{\partial x}\right)_t \left(\frac{\partial x}{\partial t'}\right)_{x'}$$

$$\text{and} \quad \left(\frac{\partial V}{\partial x'}\right)_{t'} = \left(\frac{\partial V}{\partial t}\right)_x \left(\frac{\partial t}{\partial x'}\right)_{t'} + \left(\frac{\partial V}{\partial x}\right)_t \left(\frac{\partial x}{\partial x'}\right)_{t'} \tag{6.4}$$

where t', x' are coordinates in some other frame S′. The coordinate systems are related by the Lorentz transformation, so

$$t = \gamma(t' + (v/c^2)x'), \qquad x = \gamma(vt' + x')$$

from which

$$\left(\frac{\partial t}{\partial t'}\right)_{x'} = \gamma, \qquad \left(\frac{\partial t}{\partial x'}\right)_{t'} = \gamma v/c^2$$

$$\left(\frac{\partial x}{\partial t'}\right)_{x'} = \gamma v, \qquad \left(\frac{\partial x}{\partial x'}\right)_{t'} = \gamma.$$

Substituting these into eqn (6.4) we have

$$\left(\frac{\partial V}{\partial t'}\right)_{x'} = \gamma \left(\left(\frac{\partial V}{\partial t}\right)_x + v\left(\frac{\partial V}{\partial x}\right)_t\right),$$

$$\left(\frac{\partial V}{\partial x'}\right)_{t'} = \gamma \left(\frac{v}{c^2}\left(\frac{\partial V}{\partial t}\right)_x + \left(\frac{\partial V}{\partial x}\right)_t\right).$$

After multiplying the first equation by $(-1/c)$, this pair of equations can be written

$$\begin{pmatrix} \frac{-1}{c}\frac{\partial}{\partial t'} \\ \frac{\partial}{\partial x'} \end{pmatrix} V = \begin{pmatrix} \gamma & -\beta\gamma \\ -\beta\gamma & \gamma \end{pmatrix} \begin{pmatrix} \frac{-1}{c}\frac{\partial}{\partial t} \\ \frac{\partial}{\partial x} \end{pmatrix} V,$$

which is

$$\Box'V = \Lambda\Box V. \tag{6.5}$$

This proves that $\Box V$ is a 4-vector.

To gain some familiarity, let us examine what happens to the gradient of a scalar function $V(t, x) = \phi(x)$ that depends only on x in reference frame S. In this case the slope $(\partial V/\partial x)$ in S and the slope $(\partial V/\partial x')$ in

S′ are related by a factor γ:

$$\frac{\partial V}{\partial x'} = \gamma \frac{\partial V}{\partial x} \qquad [\text{ when } \tfrac{\partial V}{\partial t} = 0$$

This is a special relativistic effect, not predicted by the Galilean transformation. It can be understood in terms of space contraction. The observer S could pick two locations where the potential differs by some given amount $\Delta V = 1$ unit, say, and paint a red mark at each location, or place a stick extending from one location to the other. This is possible because V is independent of time in S. Suppose the marks are separated by 1 metre according to S (or the stick is 1 metre long in S). Any other observer S′ must agree that the potential at the first red mark differs from that at the other red mark by $\Delta V = 1$ unit, assuming that we are dealing with a Lorentz invariant scalar field. However, such an observer moving with respect to S must find that the two red marks are separated by a *smaller* distance (contracted by γ). He must conclude that the gradient is larger than 1 unit per metre by the Lorentz factor γ.

Similarly, when V depends on time but not position in S, then its rate of change in another reference frame is larger than $\partial V/\partial t$ owing to time dilation.

In classical mechanics we often take an interest in the gradient of potential energy or of electric potential. You should beware, however, that potential energy is not Lorentz-invariant, and neither is electric potential, so an attempt to calculate a 4-gradient of either of them on its own is misconceived.[2] Instead, they are each part of a 4-vector, and one may take an interest in the 4-divergence or 4-curl of the associated 4-vector. The definition of 4-divergence of a 4-vector field F is what one would expect:

$$\Box \cdot \mathsf{F} \equiv \Box^T (g\mathsf{F}) = \frac{1}{c}\frac{\partial \mathsf{F}^0}{\partial t} + \boldsymbol{\nabla} \cdot \mathbf{f} \qquad (6.6)$$

where $\mathbf{f}$ is the spatial part of F (i.e., $\mathsf{F} = (\mathsf{F}^0, \mathbf{f})$). Note that the minus sign in the definition of $\Box$ combines with the minus sign in the scalar product (from the metric g) to produce plus signs in eqn (6.6).

The four-dimensional equivalent of curl is more complicated, and will be discussed in chapter 12.

As an example, you should check that the 4-divergence of the spacetime displacement $\mathsf{X} = (ct, \mathbf{r})$ is simply

$$\Box \cdot \mathsf{X} = 4. \qquad (6.7)$$

Example (i) If ϕ and V are scalar fields (i.e., Lorentz scalar quantities that may depend on position and time), show that

$$\Box(\phi V) = V\Box\phi + \phi\Box V.$$

(ii) If ϕ is a scalar field and F is a 4-vector field (i.e., a 4-vector that may depend on position and time), prove that

$$\Box \cdot (\phi\mathsf{F}) = \mathsf{F}\cdot\Box\phi + \phi\Box\cdot\mathsf{F}.$$

[2] This does not rule out that one could introduce a Lorentz scalar field Φ with the dimensions of energy, as a theoretical device, for example to model a 4-force by $-\Box\Phi$; such a force would be impure. An example is the scalar meson theory of the atomic nucleus.

Solution

(i) Consider first of all the time component:

$$-\frac{1}{c}\frac{\partial}{\partial t}(\phi V) = -\frac{1}{c}\left(\frac{\partial \phi}{\partial t}V + \phi\frac{\partial V}{\partial t}\right)$$

which is the time component of $V\Box\phi + \phi\Box V$. Proceeding similarly with all the other components (paying attention to the signs), the result is soon proved.

(ii) This is just like the similar result for $\boldsymbol{\nabla}\cdot(\phi\mathbf{f})$ and may be proved similarly, by proceeding one partial derivative at a time.

6.3   Current density, continuity

The general pattern with 4-vectors is that a scalar quantity appears with a 'partner' vector quantity. So far, examples have included time with spatial displacement, speed of light with particle velocity, energy with momentum. Once one has noticed the pattern it becomes possible to guess at further such 'partnerships'. Our next example is density and flux.

The density ρ of some quantity is the amount per unit volume, and the *flux* or current density $\mathbf{j}$ is a measure of flow, defined as 'amount crossing a small area, per unit area per unit time.'

Suppose some fluid is distributed throughout a region of space. In general the fluid might move with different velocities at different places, but suppose the velocities are smoothly distributed, not jumping abruptly from one value to another for neighbouring places. Then, in any small enough region, the fluid in it all has the same velocity (Fig. 6.2). Then we can speak of a rest frame for that small region. We define the rest density ρ_0 to be the density of the local fluid in such a rest frame. ρ_0 can be a function of position and time, but note that by definition it is Lorentz invariant. It earns its Lorentz invariant status in just the same way that proper time does: it comes with reference frame 'pre-attached'. Now define

$$\mathsf{J} \equiv \rho_0 \mathsf{U} \tag{6.8}$$

where U is the 4-velocity of the fluid at the given time and position. Clearly J is a 4-vector because it is the product of an invariant and a 4-vector.

We shall now show that, when defined this way, J will turn out to be equal to $(\rho c, \mathbf{j})$, where ρ and $\mathbf{j}$ are the density and flux in whatever reference frame we choose to consider. In order to do this, it will be convenient to consider that the fluid is made of a large number of closely-spaced particles, so that we can keep track of a given amount of fluid by counting the particles. The particles could be water molecules, in the case of a flow of water, or charge carriers in the case of electric charge. We will take the limit where the flow is continuous, but using the word 'particles' helps to indicate that we are considering the flow of a

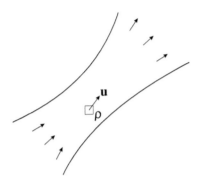

Fig. 6.2 A flowing fluid has, at each point in the flow, a local velocity $\mathbf{u}$ and density ρ.

3-surface. A flat section through spacetime, such as a time slice, is called a 'hyperplane' because it has one fewer dimensions than spacetime. Since spacetime has four dimensions, such a 'hyperplane' is three-dimensional. It could, for example, refer to the whole of space at some instant of time in some reference frame. More generally, a section through spacetime of some arbitrary shape (not necessarily flat) is called a '3-surface'. The word 'surface' here indicates that it has one fewer dimensions than spacetime; the '3' is a reminder that such a region is three-dimensional. It could, for example, be a three-dimensional volume of space at some instant of time, or a two-dimensional spatial surface persisting through an extended duration of time.

Lorentz-invariant quantity. For particles one can simply count the number of worldlines crossing some given 3-surface in spacetime (see box above). Since this is merely a matter of counting, it is obviously Lorentz-invariant. We do not need to assume that the particle number is a conserved quantity. Non-conservation would mean that particles can appear or disappear, which means that particle worldlines can begin or end—for example, this would happen for water molecules flowing around a lump of sodium metal, or for positrons flowing through ordinary matter. In such cases the particle number is not a conserved quantity, but it is Lorentz invariant because the number of worldlines crossing a given 3-surface is Lorentz invariant whether or not the worldlines are infinitely long.

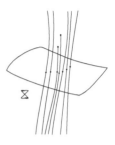

Fig. 6.3 Worldlines crossing a 3-surface. (Since this spacetime diagram has one spatial dimension suppressed, the 3-surface appears two-dimensional.)

In the local rest frame there is density ρ_0 and zero flux, so $\mathsf{J} = (\rho_0 c, \mathbf{0})$. If we pass from the rest frame to any other frame, then, by the Lorentz transformation, the zeroth component of J changes from $\rho_0 c$ to $\gamma \rho_0 c$. This is equal to ρc where ρ is the density in the new frame, because any given region of the rest frame (containing a fixed number of particles) will be Lorentz-contracted in the new frame, so that its volume is reduced by a factor γ. Therefore the number per unit volume in the new frame is higher by that factor. Let $\mathbf{u}$ be the local flow velocity in the new frame. Then the flux is given by $\mathbf{j} = \rho \mathbf{u}$. It is obvious that this u is also the relative speed of the new frame and the local rest frame, so

$$\mathbf{j} = \rho \mathbf{u} = \gamma_u \rho_0 \mathbf{u}. \qquad (6.9)$$

But this is just the spatial part of $\rho_0 \mathsf{U}$. Since we can use such a Lorentz transformation from the rest frame to connect ρ_0 and U to ρ and $\mathbf{j}$ for any part of the fluid, we have proved in complete generality that

$$\rho_0 \mathsf{U} = (\rho c, \mathbf{j}). \qquad (6.10)$$

Hence $\mathsf{J} = (\rho c, \mathbf{j})$, as we suspected.

In the case of an electric current, ρ would be the charge density; in the case of a flow of mass, ρ would be the density of *rest mass*—not the density of some other quantity such as E/c^2, where E is the energy.

This is because J is, by definition, a measure of the density and flux of a *Lorentz invariant* scalar quantity. Energy density can be defined as well, but it is not a component of any 4-vector; it is part of a higher-order quantity called a tensor, to be discussed in chapter 16.

Next, let us consider a special case: the flow of a quantity that is not only Lorentz-invariant but also conserved. This could be a flow of water if there are no chemical reactions or phase changes, or a flow of electric charge carriers, or the flow of the charge itself if the carriers are not conserved but the charge is (the question of two different signs for charge is easily kept in the account and will not be explicitly indicated in the following). We will continue to use the generic word 'particles' to track whatever is flowing.

If the particles are conserved, then the number of particles present in some closed region of space can only grow or shrink if there is a corresponding net flow in or out across the boundary of the region. The mathematical expression of this is

$$\frac{\partial}{\partial t} \int_R \rho \, dV = - \int_{(R)} \mathbf{j} \cdot d\mathbf{S} \tag{6.11}$$

where R signifies some closed region of space; the integral on the left is over the volume of the region, and the integral on the right is over the surface of the region. The minus sign is needed because by definition, in the surface integral, $d\mathbf{S}$ is taken to be an outward-pointing vector so the surface integral represents the net flow *out* of R. By applying Gauss's divergence theorem, and arguing that the relation holds for all regions R, one obtains the *continuity equation*

$$\frac{\partial \rho}{\partial t} + \boldsymbol{\nabla} \cdot \mathbf{j} = 0. \tag{6.12}$$

This equation is reminiscent of the 4-divergence equation (6.6). Indeed, by combining the definition of the 4-gradient operator with our 4-vector equation (6.10), we can immediately see that the continuity equation can be written

$$\Box \cdot \mathsf{J} = 0. \qquad [\text{Continuity equation} \tag{6.13}$$

What we have gained from all this is some practice at identifying 4-vectors, and a useful insight into the continuity equation (6.13). Because the left-hand side can be written as a scalar product of a 4-vector-operator and a 4-vector, it must be Lorentz-invariant. So the whole equation relates one invariant to another (zero). Therefore, if the continuity equation is obeyed in one reference frame, then it is obeyed in all. The equation is said to be *Lorentz-covariant*.

The continuity equation is a statement about conservation of particle number (or electric charge etc.). The 4-flux J is not itself conserved, but its null 4-divergence shows the conservation of the quantity whose flow it expresses. The conserved quantity is here a Lorentz scalar. This is in contrast to energy-momentum where the conserved quantity was the set of all components of a 4-vector. The latter can be treated by writing the divergence of a higher-order quantity called the stress-energy tensor—something we will do in chapter 16.

6.4   Wave motion

A plane wave (whether of light or of anything else, such as sound, or oscillations of a string, or waves at sea) has the general form

$$h = h_0 \cos(\mathbf{k} \cdot \mathbf{r} - \omega t) \tag{6.14}$$

where h is the oscillating quantity (electric field component; pressure; height of a water wave; etc.), h_0 is the amplitude, ω the angular frequency and $\mathbf{k}$ the wave vector. As good relativists, we suspect that we may be dealing with a scalar product of two 4-vectors:

$$\mathsf{K} \cdot \mathsf{X} = (\omega/c, \mathbf{k}) \cdot (ct, \mathbf{r}) = \mathbf{k} \cdot \mathbf{r} - \omega t. \tag{6.15}$$

Let's see if this is right. That is, does the combination $(\omega/c, \mathbf{k})$ transform as a 4-vector under a change of reference frame?

A nice way to see that it does is simply to think about the phase of the wave,

$$\phi = \mathbf{k} \cdot \mathbf{r} - \omega t. \tag{6.16}$$

To this end we plot the wavefronts on a spacetime diagram. Figure 6.4 shows a set of wavefronts of a wave propagating along the positive x axis of some frame S. Be careful to read the diagram correctly: the whole wave appears 'static' on a spacetime diagram, and the lines represent the locus of a mathematically defined quantity. For example, if we plot the wave crests then we are plotting those events where the displacement h is at a maximum. For plane waves in one spatial dimension, each such locus is a line in spacetime. Note also that because the phase velocity ω/k can be either smaller, equal to, or greater than the speed of light, a wavecrest locus (='ray') in spacetime can be either time-like, null, or spacelike.

One may plot the wavecrests in the first instance from the point of view of one particular reference frame (each line then has the equation $\omega t = kx - \phi$). However, a maximum excursion is a maximum excursion: all reference frames will agree on those events where the displacement is maximal, even though the amplitude (h_0 or h_0') may be frame-dependent. It follows that the wavecrest locations are Lorentz invariant, and more generally so is the phase ϕ, because the Lorentz transformation is linear, so all frames agree on how far through the cycle the oscillation is between wavecrests: see Fig. 6.5.

We can now obtain K as the gradient of the phase:

$$\mathsf{K} = \Box\phi = \left(-\frac{1}{c}\frac{\partial}{\partial t}, \boldsymbol{\nabla} \right)\phi$$

$$= (\omega/c, \mathbf{k}), \tag{6.17}$$

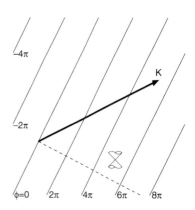

Fig. 6.4 Wavefronts (surfaces of constant phase) in spacetime. It is easy to get confused by this picture, and imagine that it shows a snapshot of wavefronts in space. It does not. It shows the complete propagation history of a plane wave moving to the right in one spatial dimension. By sliding a space-like slot up the diagram you can 'watch' the wavefronts march to the right as time goes on in your chosen reference frame (each wavefront will look like a dot in your slot). The direction of the wave 4-vector K may be constructed by drawing a vector in the direction up the phase gradient (shown dotted), and then changing the sign of the time component. The waves shown here have a phase velocity less than c. (For light-waves in vacuum the wavefronts and the wave 4-vector are both null: i.e., sloping at 45° on such a diagram.) The wavelength λ in any given reference frame F is the distance between events where successive wavecrest lines cross a line of simultaneity (=position axis) of F. The period T is the time between events where successive wavecrest lines cross the time axis of F.

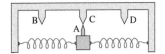

Fig. 6.5 Phase of an oscillation. If the bob A oscillates sinusoidally in one frame, then it oscillates sinusoidally in all frames since the Lorentz transformation is linear. B,C,D are pointers attached to a rigid frame, with B and D at the maximum excursion and C in the middle. The distance CD depends on reference frame, but all frames agree that A reaches D and does not pass it. Therefore in all frames the event 'A meets D' has phase $\pi/2$ (plus a multiple of 2π). Similar arguments apply to the events 'A meets B', 'A meets C', etc. Therefore the phase at all events in the cycle is Lorentz-invariant.

using eqn (6.16). Since this is a 4-gradient of a Lorentz scalar, it is a 4-vector.

Writing v_p for the phase velocity ω/k, we find the associated invariant

$$\mathsf{K}^2 = \omega^2 \left(\frac{1}{v_\mathrm{p}^2} - \frac{1}{c^2} \right). \tag{6.18}$$

Therefore when $v_p < c$ the 4-wave-vector is spacelike, and when $v_p > c$ the 4-wave-vector is time-like. For light-waves in vacuum the 4-wave-vector is null. The invariant also shows that a wave of any kind whose phase velocity is c in some reference frame will have that same phase velocity in all reference frames.

6.4.1   Wave equation

Wave motion such as that expressed in eqn (6.14) is a solution of the wave equation

$$\frac{\partial^2 h}{\partial t^2} = v_\mathrm{p}^2 \nabla^2 h. \tag{6.19}$$

Writing this

$$-\frac{1}{c^2} \frac{\partial^2 h}{\partial t^2} + \frac{v_\mathrm{p}^2}{c^2} \nabla^2 h = 0 \tag{6.20}$$

we observe that for the special case $v_p = c$ the wave equation takes the Lorentz covariant form

$$\Box^2 h = 0. \qquad [\text{ Wave equation!}] \tag{6.21}$$

The operator is called the *d'Alembertian*[3]:

$$\Box^2 \equiv \Box \cdot \Box = -\frac{1}{c^2} \frac{\partial^2}{\partial t^2} + \nabla^2 \tag{6.22}$$

(a product of three minus signs made the minus sign here!). Hence the general idea of wave propagation can be very conveniently treated in Special Relativity when the waves have phase velocity c. This will be used to great effect in the treatment of electromagnetism in chapter 8.

6.4.2   Particles and waves

While we are considering wave motion let us briefly look at a related issue: the wave–particle duality. We will not try to introduce that idea with any great depth, that would be the job of another textbook, but it is worth noticing that the introduction of the photon model for light can be guided by Special Relativity, and de Broglie's introduction of a wave model for particles was guided by Special Relativity.

Max Planck is associated with the concept of the photon, owing to his work on black-body radiation. However, when he introduced the idea of energy quantization he did not, in fact, have in mind that this should serve as a new model for the electromagnetic field. It was sufficient for his purpose merely to assert that energy was absorbed by matter

[3] Beware: as noted previously, it has become common practice to use the symbol $\Box$ (without the ^{2}) for the d'Alembertian, even though ∇^2 is used for the Laplacian. Confusing! This practice arose in the context of index notation, which we will introduce in chapter 12, where it makes sense as long as some other symbol is used for the 4-gradient, the standard choice being ∂^α. However, in vector/matrix notation, $\Box$ is a natural choice for the generalization of ∇, and I believe this choice to be the least confusing for learning purposes (no one ever mistakes $\Box^2$ for a y-gradient, for example, and the 2 reminds us that it is a second derivative). Finally, the d'Alembertian may also be defined as $c^{-2}\partial^2/\partial t^2 - \nabla^2$ (the negative of our $\Box^2$).

in quantized 'lumps'. It was Einstein who extended the notion to the electromagnetic field itself, through his March 1905 paper. This paper is often mentioned in regard to the photoelectric effect, but this does not do justice to its full significance. It was a revolutionary rethinking of the nature of electromagnetic radiation.

When teaching students about the photoelectric effect and its impact on the development of quantum theory, it makes sense, and it is the usual practice, to emphasize that the *energy* of the emitted electrons has no dependence on the *intensity* of the incident light. Rather, the energy depends linearly on the frequency of the light, while the light-intensity influences the rate at which photoelectrons are generated. This leads one to propose the model $E = h\nu$ relating the energy of the light-particles to the frequency of the waves.

However, this information was not available in 1905. There was evidence that the electron energy did not depend on the intensity of the light, and for the existence of a threshold frequency, but the linear relation between photoelectron energy and light-frequency was *predicted* in Einstein's paper, not extracted from experimental data. Einstein reasoned from thermodynamics and what we now call statistical mechanics: he calculated the entropy per unit volume of thermal radiation, and showed that the thermodynamic behaviour of the radiation at a given frequency ν was the same as that of a gas of particles each carrying energy $h\nu$. The relationship $E = h\nu$ as applied to what we now call photons was thus first proposed by Einstein. However, his 1905 paper was still far short of a full model; it was not until Compton's experiments (1923) that the photon idea began to gain wide acceptance, and a thorough model required the development of quantum field theory, the work of many authors, with Dirac (1927) playing a prominent role.

In this section we shall merely point out one feature (which is not the one historically emphasized in 1905): *if one is going to attempt a particle model for electromagnetic waves, then Special Relativity can guide you on how to do it.* That is, we shall play the role of theoretical physicist, and assume merely that we know about classical electromagnetism and we would like to investigate what kind of photon model might be consistent with it.

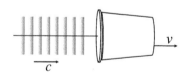

Consider a parallel beam of light falling on a moving bucket (Fig. 6.6). We shall use this situation to learn about the way the energy and intensity of light transform between reference frames. In fact we have already made a general observation about this in the discussion of the headlight effect in section 3.2, in connection with eqn (3.8). The present discussion will proceed more simply, restricting the motion to one spatial dimension.

Fig. 6.6 A parallel beam of light falls into a moving bucket.

Suppose that in frame S the light and the bucket move in the same direction, with speeds c and v respectively. Let u be the energy per unit volume in the light-beam. The amount of energy flowing across a plane fixed in S of cross-section A during time t is then $uA(ct)$. The 'intensity' (or energy flux) I is defined to be the power per unit area, so

$$I = uc. \tag{6.23}$$

We would like to calculate the amount of energy entering the bucket, and compare this between reference frames. To this end it is convenient to use the Lorentz invariance of the phase of the wave. We consider the energy and momentum that enters the bucket during a period when N wavefronts move into the bucket. In frame S these waves fill a total length $L = N\lambda$ where λ is the wavelength, so the energy entering the bucket is $E = N\lambda Au$. The portion or 'lump' of the light-field now in the bucket (we can suppose the bucket is deep so the light has not been absorbed yet) possesses energy E and propagates at speed c. It follows that its momentum must be $p = E/c$. Note that we have not invoked a particle model in order to assert this; we have merely claimed that the relation $p/E = v/c^2$, which we know to be valid for $v < c$, is also valid in the limit $v = c$. (In chapter 13 we will show that electromagnetic field theory also confirms $p = E/c$ for light-waves.) Applying a Lorentz transformation to the energy-momentum of the light, we obtain for the energy part:

$$E' = \gamma(1 - \beta)E = \sqrt{\frac{1 - \beta}{1 + \beta}}E. \qquad (6.24)$$

By Lorentz transforming the 4-wave-vector $(\omega/c, \mathbf{k})$, or by using the Doppler effect equation (A.10), we obtain for the wavelength

$$\lambda' = \sqrt{\frac{1 + \beta}{1 - \beta}}\lambda.$$

Substituting these results in $E' = N\lambda'Au'$ (using that the area A is transverse so uncontracted) we find

$$u' = \frac{E'}{N\lambda'A} = \frac{1 - \beta}{1 + \beta}\frac{E}{N\lambda A}$$

$$\Rightarrow \qquad I' = \frac{1 - \beta}{1 + \beta}I \qquad (6.25)$$

where the last step uses eqn (6.23).

Two things are striking in this argument. First, the energy of the light entering the bucket does *not* transform in the same way as its intensity. Second, the energy does transform in the same way as the frequency. When making an approach to a particle model, therefore, although one might naively have guessed that the particle energy should be connected to the intensity of the light, we see immediately that this will not work: it cannot be true in all reference frames for a given set of events. For, just as the number of wavefronts entering the bucket is a Lorentz invariant, so must the number of particles be: those particles could be detected and counted, after all, and the count displayed on the side of the bucket. Therefore the energies E and E' that we calculated must correspond to the *same* number of particles, so they are telling us about the energy per particle.

One will soon run into other difficulties with a guess that the particle energy is proportional to $\sqrt{I}$ or to $I\lambda$. It seems most natural to try

$E \propto \nu$, the frequency. Indeed, with the further consideration that we need a complete energy-momentum 4-vector for our particle, not just a scalar energy, and we have to hand the 4-wave-vector of the light with just the right direction in spacetime (i.e., the null direction), it is completely natural to guess the right model, $E = h\nu$ and $\mathsf{P} = \hbar \mathsf{K}$.

6.4.3   Group velocity and particle velocity

Recall eqns (2.70) and (3.3) for the angle change of the velocity of a particle and the wave vector of a plane wave, respectively. We reproduce these here for convenience:

$$\tan \theta = \frac{\sin \theta_0}{\gamma(\cos \theta_0 + v/u_0)}, \tag{6.26}$$

$$\tan \theta = \frac{\sin \theta_0}{\gamma(\cos \theta_0 + v v_{\mathrm{p}}/c^2)}, \tag{6.27}$$

where the frames are labelled S and S_0, $v_{\mathrm{p}} \equiv \omega_0/k_0$ is the phase velocity of the waves in the frame S_0, and u_0 is the speed of the particle in the frame S_0. These are both examples of a direction-change of a 4-vector, so they amount to the same formula: the first can be obtained from the second by the replacement $u_0 \to (k_0/\omega_0)c^2$. However, the result is that a particle travelling along at the phase velocity of the waves (i.e., having the same speed and direction) in frame S_0 does not in general have the same speed or direction as the phase velocity in frame S (if it is riding the crest of the wave, it still does so in the new frame but not in the normal direction).

Something interesting emerges if we look at *group velocity*. The group velocity of a set of waves is defined

$$v_{\mathrm{g}} \equiv \frac{\mathrm{d}\omega}{\mathrm{d}k}. \tag{6.28}$$

Thus the group velocity depends on the way the frequency of the waves is related to their wavevector. There is no general formula for this, because it depends on the physical conditions, such as the behaviour of the refractive index for light-waves in a transparent medium, or the dispersion relation for sound waves, etc. However, an interesting case to consider is waves that have the property that $\mathsf{K} \cdot \mathsf{K}$ is independent of k. Note that this does not necessarily have to happen: $\mathsf{K} \cdot \mathsf{K}$ is guaranteed to be Lorentz-invariant, but its value might in general be a function of frequency. However, if it does not depend on frequency then we have

$$-\omega^2/c^2 + k^2 = \text{const.}$$

After multiplying by c^2 and taking the derivative with respect to k, we obtain

$$v_{\mathrm{g}} = \frac{\mathrm{d}\omega}{\mathrm{d}k} = \frac{kc^2}{\omega} = \frac{c^2}{v_{\mathrm{p}}}. \tag{6.29}$$

(Note that $v_{\mathrm{g}} < c$ if $v_{\mathrm{p}} > c$).

An instance of this case is de Broglie waves. Those waves satisfy the condition $K \cdot K = \text{const}$, the constant in question being $-(mc/\hbar)^2$, where m is the rest mass of the particle. The idea of de Broglie waves proceeded historically in two stages. The suggestion to treat light-waves in terms of particles—photons—came first. The suggestion that all particles had associated waves—de Broglie waves—was a profound further step and came considerably later. The equation de Broglie proposed for his waves was strongly motivated by Special Relativity.

Consider a classical particle whose speed and direction u_0, θ_0 in frame S_0 matches that of the *group* velocity of a set of waves, as given by (6.29). Then we have $u = c^2/v_\text{p}$. Substituting this into eqn (6.26) we find now that the change in direction of the particle motion matches that of the wave motion. Also, by using eqn (2.69), the size of the speed continues to match the group velocity in the new frame. In short, a group of waves at nearby frequencies (a 'wavepacket') behaves like a particle. The de Broglie formula relating wavelength of a quantum mechanical *wavefunction* to momentum of the associated particle, $\lambda = h/p$, comes essentially from the 4-vector relationship $K = P/\hbar$ and is fully consistent with Special Relativity. It is more general than Schrödinger's equation, for example.

We have just seen that the group velocity of de Broglie waves behaves like a particle velocity. The phase velocity can also be given a physical interpretation by appealing to Special Relativity. Imagine a group of particles sharing the same velocity, and furnish each one with a little pointer that rotates. Let these pointers rotate in synchrony in the rest frame. Then, in other frames, the events 'pointer is vertical' for the group of particles are not simultaneous, but occur at times $t = \beta\gamma x_0/c = vx/c^2$ where x_0 is the position in the rest frame and we considered $t_0 = 0$ for convenience. Hence the sequence of events at which successive pointers reach the vertical is a sequence that sweeps down the group of particles at the velocity $x/t = c^2/v$. This velocity is the phase velocity of de Broglie waves. Thus the de Broglie wave can be regarded as a *wave of simultaneity in the rest frame*. The little pointers we imagined in this argument correspond, in quantum theory, to the complex numbers describing the phase of the wavefunction.

6.5   Acceleration and rigidity

Consider a stick that accelerates as it falls. For example, suppose that in some reference frame $S(x, y, z)$ a stick is extended along the x direction, and remains straight at all times. It accelerates in the y direction all as a piece (without bending) at constant acceleration a in S. The worldline of any particle of the stick is then given by

$$x = x_0 \tag{6.30}$$

$$y = \frac{1}{2}at^2 \tag{6.31}$$

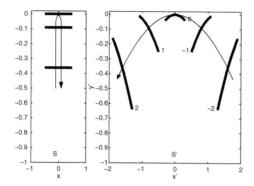

Fig. 6.7 A rigid stick that remains straight and parallel to the x axis in frame S (left diagram), is here shown at five successive instants in frame S' (right diagram). The stick has an initial velocity in the upwards (+ve y) direction and accelerates in the $-$ve y direction; frame S' moves to the right (+ve x direction) at speed v relative to frame S. (In the example shown the stick has proper length 1, $v = 0.8$, and $a = -2$, all in units where $c = 1$.).

during some interval for which $t < c/a$, where x_0 takes values in the range $-L_0/2$ to $L_0/2$. L_0 is the rest length of the stick.

Now consider this stick from the point of view of a reference frame moving in the x direction (relative to S) at speed v. In the new frame the coordinates of a particle on the stick are given by

$$\begin{pmatrix} ct' \\ x' \\ y' \\ z' \end{pmatrix} = \begin{pmatrix} \gamma & -\gamma\beta & 0 & 0 \\ -\gamma\beta & \gamma & 0 & 0 \\ 0 & 0 & 1 & 0 \\ 0 & 0 & 0 & 1 \end{pmatrix} \begin{pmatrix} ct \\ x_0 \\ \frac{1}{2}at^2 \\ 0 \end{pmatrix} = \begin{pmatrix} \gamma(ct - \beta x_0) \\ \gamma(x_0 - vt) \\ \frac{1}{2}at^2 \\ 0 \end{pmatrix} \quad (6.32)$$

Use the first line to express t in terms of t', obtaining $t = (t'/\gamma + \beta x_0/c)$, and substitute this into the rest, to find

$$x' = \frac{x_0}{\gamma} - vt', \qquad y' = \frac{1}{2}a\left(\frac{t'}{\gamma} + \frac{\beta x_0}{c}\right)^2. \quad (6.33)$$

When we allow x_0 to take values between $-L_0/2$ and $L_0/2$, these equations tells us the location in S' of the all the particles of the stick, at any given t'. It is seen that they lie along a parabola. Fig. 6.7b shows the stick in frame S' at five successive values of t', and Fig. 6.8 shows the spacetime diagram.

This example shows that accelerated motion while maintaining a fixed shape in one reference frame will result in a changing shape for the object in other reference frames. This is because the worldlines of the particles of the object are curved, and the planes of simultaneity for most reference frames must intersect such a set of worldlines along a curve. This means that, for accelerating objects, the concept of 'rigid' behaviour is not Lorentz-invariant. The notion of 'remaining undeformed' cannot apply in all reference frames when a body is accelerating (see the exercises for further examples).

A related issue is the concept of a 'rigid body'. In classical physics this is a body which does not deform when a force is applied to it; it accelerates all of a piece. In Special Relativity this concept has to be abandoned. There is no such thing as a rigid body, if by 'rigid' we mean a body that does not deform when struck. This is because when a force

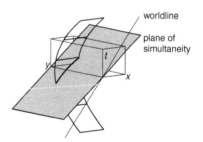

Fig. 6.8 Spacetime diagram showing the worldsheet of the accelerating stick shown in Fig. 6.7, and a plane of simultaneity for frame S' at $t' = 0$ (shaded). The stick is straight in the reference frame whose axes are indicated by the rectangular box (x, y, t); it is curved in the other reference frame because the plane of simultaneity intersects its worldsheet along a curve.

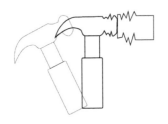

Fig. 6.9 There is no such thing as a body that does not deform when struck.

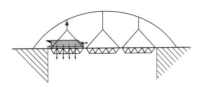

Fig. 6.10 A fast train moves over a bridge across a chasm. The rest length of the train is equal to that of the chasm. The picture shows the situation in the rest frame of the bridge, in which the train is Lorentz-contracted by a factor 3, and therefore the whole of the train has to be supported by one section of the bridge.

[4] The following discussion is based loosely on a treatment by Fayngold.

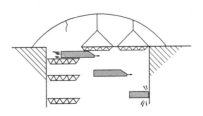

Fig. 6.11 The cable holding up the bridge section cannot support the train. It breaks, the section falls, and the train drops into the chasm, eventually crashing into the far wall.

is applied to one part of a body, only that part of the body is causally influenced by the force. Other parts, outside the future light-cone of the event at which the force began to be applied, cannot possibly be influenced, whether to change their motion or whatever. It follows that the application of a force to one part of body *must* result in deformation of the body. Another way of stating this is to say that a rigid body is one for which the group velocity of sound goes to infinity, but this is ruled out by the Light Speed Postulate.

There can exist accelerated motion of a special kind, such that the different parts of a body move in synchrony so that proper distances are maintained. Such a body can be said to be 'rigid' while it accelerates. This is described in section 9.2.1 of chapter 9.

6.5.1 The great train disaster

Full fathom five thy father lies,
Of his bones are coral made:
Those are pearls that were his eyes,
Nothing of him that doth fade,
But doth suffer a sea-change
Into something rich and strange.
Sea-nymphs hourly ring his knell:
Hark! now I hear them, ding-dong, bell.

(Ariels's song from *The Tempest*
by William Shakespeare)

The relativity of the shape of accelerated objects is nicely illustrated by a paradox in the general family of the contracted stick gliding through a hole (see, for example, *The Wonderful World*). Or perhaps, now that we understand Relativity moderately well (let us hope), it is not a paradox so much as another fascinating example of the relativity of simultaneity and the transformation of force.[4]

So, imagine a super-train, 300 metres long (rest length), that can travel at about 600 million miles per hour, or, to be precise, $\sqrt{8}c/3$. The train approaches a chasm of width 300 metres (rest length) which is spanned by a bridge made of three suspended sections, each of rest length 100 metres. Owing to its Lorentz contraction by a factor $\gamma = 3$, the whole weight of the train has to be supported by just one section of the bridge; see Fig. 6.10. Unfortunately, the architect has forgotten to take this into account: the cable snaps, the bridge section falls, and the train drops into the chasm: Fig. 6.11.

At this point the architect arrives, both shocked and perplexed.

'But I did take Lorentz contraction into account', he says. 'In fact, in the rest frame of the train, the chasm is contracted to 100 metres, so the train easily extends right over it [as in Fig. 6.12]. Each section of the bridge only ever has to support one ninth of the weight of the train. I cannot understand why it failed, and I certainly cannot understand how the train could fall down, because it will not fit into such a short chasm.'

Before we resolve the architect's questions, it is only fair to point out that the example is somewhat unrealistic in that, on all but the largest planets, the high speed of the train will exceed the escape velocity. Instead of pulling the train down any chasm, an ordinary planet's gravity would not even suffice to keep such a fast train on the planet surface: the train would continue in an almost straight line, moving off into space as the planet's surface curved away beneath it. A gravity strength such as that near the event horizon of a black hole would be needed to cause the crash. However, we could imagine that the train became electrically charged by rubbing against the planet's (thin) atmosphere and the force on it is electromagnetic in origin, then a more modest planet could suffice.

In any case, it is easy to see that whatever vertical force f' acted on each particle of the train in the rest frame of the planet, the force per particle in the rest frame of the train is considerably larger: $f = \gamma f' = 3f'$ (see eqn (4.6)). The proper breaking strength of the cable would appear to be something less than Nf', where N is the number of particles in the whole of the train. In the train frame, this breaking strength will be reduced (says eqn (4.6)) to something less than $Nf'/\gamma = Nf'/3$. Hence, in the train frame, the number of particles n that the bridge section can support is given by

$$n\gamma f' < Nf'/\gamma \quad \Rightarrow \quad n < N/\gamma^2. \tag{6.34}$$

It is not surprising, therefore, that the cable should break when only one ninth of the length of the train is on the bridge section (in the train's rest frame).[5]

The architect's second comment is that the train in its rest frame will not fit horizontally into the chasm. This is of course true. However, by using the Lorentz transformation it is easy to construct the trajectories of all the particles of the train, and they are as shown in Fig. 6.13. The vertical acceleration of the falling train is Lorentz-transformed into

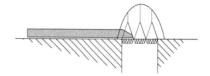

Fig. 6.12 The situation in the rest frame of the train, when the train has run on to the first bridge section. In this reference frame the train has its 300-metre rest length, while the bridge and chasm are Lorentz-contracted. The architect is perplexed because it is clear that the bridge section only needs to support one ninth of the train, and in any case the train cannot possibly fit into the chasm.

[5] We are treating a simplified model in which the force is evenly distributed along the train, and we do not examine the details of the time taken for the extension force to propagate along the cables of the bridge. We wish merely to illustrate that the whole scenario can be consistent between two different reference frames.

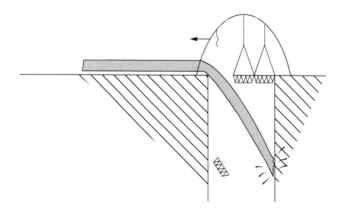

Fig. 6.13 The architect failed to account for the Lorentz transformation of force—in the train's rest frame the transverse force on it is larger, and the tensile strength of the fast-moving bridge cable is smaller. The cable cannot support even a small part of the train. The architect also failed to realise that the vertical acceleration of the falling train corresponds to a change of shape of the train in this frame, so that it does fit into the chasm after all, and strikes the chasm wall at the same point as is observed in the other reference frame.

a bending downwards. The train which appeared 'rigid' in the planet frame is revealed in the horizontally moving frame to be as floppy as a snake as it plunges headlong through the narrow gap in the bridge.

6.5.2   Lorentz contraction and internal stress

The Lorentz contraction results in distortion of an object. The contraction is purely that: a contraction, not a rotation, but a contraction can change angles as well as distances in solid objects. For example, a picture frame that is square in its rest frame will be a parallelogram at any instant of time in most other inertial reference frames. The legs of a given ordinary table are not at right angles to its surface in most inertial reference frames.

For accelerating bodies, the change of shape associated with a change of inertial reference frame is more extreme. Examples are the twisted cylinder (see exercise 6.4), and the falling train of the previous section. Things that accelerate can suffer a Lorentz-change into something rich and strange.

These observations invite the question: are these objects still in internal equilibrium, or are they subject to internal stresses? What is the difference between Lorentz contraction and the distortion that can be brought about by external forces?

John Bell proposed the following puzzle. Suppose that two short identical rockets are at rest relative to a space station S, one behind the other, separated by a gap $L = 100$ metres. They are programmed to blast off simultaneously in reference frame S, and thereafter to burn fuel at the same rate. It is clear that the trajectory of either rocket will be identical in S, apart from the 100-metre gap. If the front rocket moves as $x(t)$, then the back rocket moves as $x(t) - L$. Therefore their separation remains 100 metres in reference frame S.

Now suppose that before they blast off, a string of rest length $L_0 = 100$ metres is connected between the rockets, and suppose any forces exerted by the string are negligible compared to those provided by the rocket engines. Then, in frame S the string will suffer a Lorentz contraction to less than 100 metres, but the rockets are still separated by 100 metres. So what will happen? Does the string break?

I hope it is clear to you that the string will eventually break. It undergoes acceleration owing to the forces placed on it by the rockets. It will in turn exert a force on the rockets, and its Lorentz contraction means that that force will tend to pull the rockets together to a separation smaller than 100 metres in frame S. This means that it begins to act as a tow rope. The fact that its length remains (very nearly) constant at 100 metres in S, whereas it 'ought to' be L_0/γ, shows us that the engine of the rear rocket is not doing enough to leave the tow rope nothing to do: the tow rope is being stretched by the external forces. The combination of this stretching and the Lorentz contraction results in the observed constant string length in frame S.

100 m

Fig. 6.14

Such a string is not in internal equilibrium. It will only be in internal equilibrium, exerting no outside forces, if it attains the length L_0/γ. As the rockets reach higher and higher speed relative to S, γ gets larger and larger, so the string is stretched more and more relative to its equilibrium length. If you need to be further convinced of this, then jump aboard the rest frame of the front rocket at some instant of time, and you will find the back rocket is trailing behind by considerably more than 100 metres. At some point the material of the string cannot withstand further stretching, and the string breaks.

In the study of springs and Hooke's law, the length of a spring when it exerts zero force is called its 'natural' length. In Special Relativity we call the length of a body in the rest frame of the body its 'proper' length: you might say this is the length that it 'thinks' it has. The proper length is, by definition, a Lorentz invariant. The natural length depends on reference frame, however, and the proper length does not have to be equal to the natural length.

A spring with no external forces acting on it, and for which any oscillations have damped away, will have its natural length. Suppose that length is $L_n(0)$ in the rest frame of the spring. In inertial reference frames moving relative to the spring in a direction along its length, the natural length will be $L_n(v) = L_n(0)/\gamma$.

We now have three lengths to worry about: the length L that a body actually has in any given reference frame, its natural length $L_n(v)$ in that reference frame, and its proper length L_0. The Lorentz contraction affects the length and the natural length. A stretched or compressed spring has a length in any given reference frame different from its natural length in that reference frame. Its proper length is $L_0 = \gamma L$. If $L \neq L_n(v)$ then $L_0 \neq L_n(0)$: i.e., a stretched or compressed spring has a proper length different from what the natural length would be in its rest frame.

In the example of the rockets joined by a string, in reference frame S the natural length of the string shortens, but the string does not, owing to the forces on it. In the sequence of rest frames of the centre of the string the natural length is constant but the actual length grows, owing again to the forces which stretch it.

If a moving object is abruptly stopped, so that all of its parts stop at the same time in a reference frame other than the rest frame, then the length in that frame remains constant but the proper length gets shorter (it was γL, now it is L). If the object was previously moving freely with no internal stresses, then now it will try to expand to its new natural length, but it has been prevented from doing so. Therefore, it now has internal stresses: it is under compression.

Similarly, if an object having no internal stresses is set in motion so that all parts of the object get the same velocity increase at the same time in the initial rest frame S, then the length of the object in S stays constant while the proper length gets longer (it was L, now it is γL). Since the proper length now exceeds the proper natural length, such a procedure results in internal stresses such that the object is now under tension.

More generally, to discover whether internal stresses are present, it suffices to discover whether the distance between neighbouring particles of a body is different from the natural distance. In the example of the great train disaster, the train is without internal stress as it bends during the free fall. The passengers too are without internal stress—except the pyschological kind, of course. If the natural (unstressed) shape of an accelerating object remains straight in some inertial reference frame, then in most other reference frames the natural (unstressed) shape will be bent and/or twisted.

6.6   General Lorentz boost

So far we have considered the Lorentz transformation only for a pair of reference frames in the standard configuration, where it has the simple form presented in eqn (2.32). More generally, inertial reference frames can have relative motion in a direction not aligned with their axes, and they can be rotated or suffer reflections with respect to one another. To distinguish these possibilities, the transformation for the case where the axes of two reference frames are mutually aligned, but they have a non-zero relative velocity, is called a *Lorentz boost*. A more general transformation, involving a rotation of coordinate axes as well as a relative velocity, is called a Lorentz transformation but not a boost.

The most general Lorentz boost, therefore, is for the case of two reference frames of aligned axes, whose relative velocity $\mathbf{v}$ is in some arbitrary direction relative to those axes. In order to obtain the matrix representing such a general boost, it is instructive to write the simpler case given in eqn (2.32) in the vector form:

$$ct' = \gamma(ct - \boldsymbol{\beta} \cdot \mathbf{x})$$

$$\mathbf{x}' = \mathbf{x} + \left(-\gamma ct + \frac{\gamma^2}{1+\gamma} \boldsymbol{\beta} \cdot \mathbf{x} \right) \boldsymbol{\beta} \tag{6.35}$$

This gives a strong hint that the general Lorentz boost is

$$\Lambda(\mathbf{v}) = \begin{pmatrix} \gamma & -\gamma\beta_x & -\gamma\beta_y & -\gamma\beta_z \\ . & 1+\alpha\beta_x^2 & \alpha\beta_x\beta_y & \alpha\beta_x\beta_z \\ . & . & 1+\alpha\beta_y^2 & \alpha\beta_y\beta_z \\ . & . & . & 1+\alpha\beta_z^2 \end{pmatrix} \tag{6.36}$$

where $\alpha = \gamma^2/(1+\gamma)$ and the lower left part of the matrix can be filled in by using the fact that the whole matrix is symmetric. One can prove that this matrix is indeed the right one by a variety of dull but thorough methods; see the exercises.

6.7   Lorentz boosts and rotations

Suppose a large regular polygon (e.g., 1 km to the side) is constructed out of wood and laid on the ground. A pilot then flies an aircraft around this polygon (at some fixed distance above it); see Fig. 6.15a.

Let N be the number of sides of the polygon. As the pilot approaches any given corner of the polygon, he observes that the polygon is Lorentz-contracted along his flight direction. If, for example, he considers the right-angled triangle formed by a continuation of the side he is on and a hypotenuse given by the next side of the polygon, then he will find the lengths of its sides to be $(L_0 \cos \theta)/\gamma$ and $L_0 \sin \theta$, where $\theta = 2\pi/N$; see Fig. 6.15b. He deduces that the angle he will have to turn through, in order to fly parallel to the next side, is θ' given by

$$\tan \theta' = \gamma \tan \theta. \tag{6.37}$$

Having made the turn, he can also consider the side receding from him and confirm that it makes this same angle θ' with the side he is now on.

For large N we have small angles, so

$$\theta' \simeq \gamma \theta. \tag{6.38}$$

After performing the manoeuvre N times, the aircraft has completed one circuit and is flying parallel to its original direction, and yet the pilot considers that he has steered through a total angle of

$$N\theta' = \gamma 2\pi. \tag{6.39}$$

Since $\gamma > 1$ we have a total steer by more than two pi radians, in order to go once around a circuit! The extra angle is given by

$$\Delta \theta = N\theta' - 2\pi = (\gamma - 1)2\pi. \tag{6.40}$$

This is a striking result. What is going on? Are the pilot's deductions faulty in some way? Perhaps something about the acceleration needed to change direction renders his argument invalid?

It will turn out that the pilot's reasoning is quite correct, but some care is required in the interpretation. The extra rotation angle is an example of a phenomenon called *Thomas–Wigner rotation*.[6] It is also

[6] The effect was discovered by Thomas in the context of atomic physics, and subsequently elucidated more generally by Wigner.

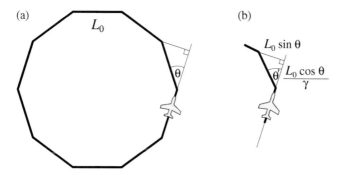

(a)   (b)

L_0

$L_0 \sin \theta$

$\dfrac{L_0 \cos \theta}{\gamma}$

θ

Fig. 6.15 (a) An aircraft flies around a regular polygon. The polygon has N sides, each of rest length L_0. The angle between one side and the next, in the polygon rest frame, is $\theta = 2\pi/N$. (b) shows the local situation in the rest frame of the pilot as he approaches a corner and is about the make a turn through θ'. Since $\theta' > \theta$, the pilot considers that the sequence of angle turns he makes, in order to complete one circuit of the polygon, amounts to more than 360°.

often called *Thomas precession*, because it was first discovered as a changing direction of an angular momentum vector. We will provide the interpretation and some more details in section 6.7.2. First we need a result concerning a simple family of three inertial reference frames.

6.7.1 Two boosts at right angles

Fig. 6.16 shows a set of three reference frame axes, all aligned with one another at any instant of time in frame S'. Frame S'' is moving horizontally with respect to S' at speed v. Frame S is moving vertically with respect to S' at speed u. Let A be a particle at the origin of S, and B be a particle at the origin of S''.

We will calculate the angle between the line AB and the x-axis of S, and then the angle between AB and the x''-axis of S''. This will reveal an interesting phenomenon.

First consider the situation in S. Here A stays fixed at the origin, and B moves. We use the velocity transformation equations (2.27), noting that we have the simple case where the pair of velocities to be 'added' are mutually orthogonal. In S' the velocity components of B are (horizontal, vertical)$=(v, 0)$, therefore in S they are $(v/\gamma(u), -u)$. Therefore the angle θ between AB and the $x-$axis of frame S is given by

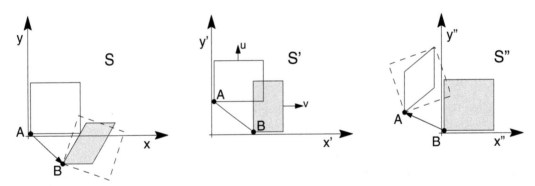

Fig. 6.16 Two squares (i.e., each is a solid object that is square in its rest frame: it helps to think of them as physical bodies, not just abstract lines) of the same proper dimensions are in relative motion. In frame S' the white square moves upwards at speed u, and the grey square moves to the right at speed v. The central diagram shows the situation at some instant of time in S': each square is contracted along its direction of motion. Frame S is the rest frame of the white square; S'' is the rest frame of the grey square. A and B are particles at the origins of S and S'' respectively. The left and right diagrams show the situation at some instant of time in S and S'' respectively. The reference frame axes of S and S'' have been chosen parallel to the sides of the fixed square in each case; those of S' have been chosen parallel to the sides of both objects as they are observed in that frame. Note that there are three diagrams here, not one! The diagrams have been oriented so as to bring out the fact that S and S' are mutually aligned, and S' and S'' are mutually aligned. However the fact that S and S'' are *not* mutually aligned is not directly indicated, it has to be inferred. The arrow AB on the left diagram indicates the velocity of S'' relative to S. The arrow BA on the right diagram indicates the velocity of S relative to S''. These two velocities are colinear (they are equal and opposite). The dashed squares in S and S'' show a shape that, if Lorentz contracted along the relative velocity AB, would give the observed parallelogram shape of the moving object in that reference frame. It is clear from this that the relationship between S and S'' is a boost combined with a rotation, not a boost alone. This rotation is the kinematic effect that gives rise to the Thomas precession. Take a long look at this figure: there is a lot here—it shows possibly the most mind-bending aspect of Special Relativity!

$$\tan\theta = \frac{u\gamma_u}{v}. \tag{6.41}$$

Now consider the situation in S''. Here B is fixed at the origin, and A moves. In S' the velocity components of A are $(0, u)$ so in S'' they are $(-v, u/\gamma(v))$. Therefore the angle θ'' between AB and the x''−axis of frame S'' is given by

$$\tan\theta'' = \frac{u}{v\gamma_v}. \tag{6.42}$$

Thus we find that $\theta \neq \theta''$. Since the three origins all coincide at time zero, the line AB is at all times parallel to the relative velocity of S and S''. This velocity is constant, and it must be the same (equal and opposite) when calculated in the two reference frames whose relative motion it describes. Therefore the interpretation of $\theta \neq \theta''$ must be that *the coordinate axes of S are* not *parallel to the coordinate axes of S''* (when examined either in reference frame S or in S'').

This is a remarkable result, because we started by stating that the axes of S and S'' are mutually aligned in reference frame S'. It is as if we attempted to line up three soldiers, with Private Smith aligned with Sergeant Smithers, and Sergeant Smithers aligned with Captain Smitherson, though somehow Private Smith is not aligned with Captain Smitherson. With soldiers, or lines purely in *space*, this would not be possible. What we have found is a property of constant-velocity motion in *spacetime*, owing to the relativity of simultaneity.

The sequence of passing from frame S to S' to S'' consists of two Lorentz boosts, but the overall result is not merely a Lorentz boost to the final velocity $\mathbf{w}$, but a boost combined with a rotation. Mathematically, this is

$$\Lambda(\mathbf{v})\Lambda(-\mathbf{u}) = R(\Delta\theta)\Lambda(\mathbf{w}) \tag{6.43}$$

where $\Delta\theta = \theta - \theta''$ and $\mathbf{w} = (v/\gamma_u, -u, 0)$ is the velocity of S'' relative to S. We have proved the case where $\mathbf{u}$ and $\mathbf{v}$ are orthogonal. We will show in sections 6.8 and 6.9 that the pattern of this result holds more generally: a sequence of Lorentz boosts in different directions gives a net result that involves a rotation, even though each boost on its own produces no rotation. The rotation angle for orthogonal $\mathbf{u}$ and $\mathbf{v}$ can be obtained from eqns (6.41) and (6.42) using the standard trigonometric formula, $\tan(\theta - \theta'') = (\tan\theta - \tan\theta'')/(1 + \tan\theta\tan\theta'')$:

$$\tan\Delta\theta = \frac{uv(\gamma_u\gamma_v - 1)}{u^2\gamma_u + v^2\gamma_v}. \tag{6.44}$$

Note that the rotation effect is a purely kinematic result: it results purely from the geometry of spacetime. That is to say, the amount and sense of rotation is determined purely by the velocity changes involved, not by some further property of the forces which cause the velocity changes in any particular case. It is at the heart of the Thomas precession, which we will now discuss.

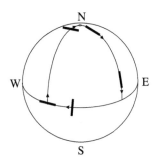

Fig. 6.17 Parallel transport on the surface of a sphere (see text).

6.7.2 The Thomas precession

Let us return to the thought experiment with which we began section 6.7: the aircraft flying around the polygon. This thought experiment can be understood in terms of the rotation effect that results from a sequence of changes of inertial reference frame, as discussed in the previous section. The pilot's reasoning is valid, and it implies that a vector carried through a sequence of velocity changes by *parallel transport in velocity-space* will undergo a net rotation: it will finish pointing in a direction different to the one in which it started.

Parallel transport is the type of transport when an object is translated as a whole, in some given direction, without rotating it. For example, if you pass someone a book, you will normally find that your action will rotate the book as your arm swings. However, with care you could adjust the angle between your hand and your arm, as your hand moves, so as to maintain the orientation of the book fixed. That would be a parallel transport.

Parallel transport in everyday situations (in technical language, in flat Euclidean geometry) never results in a change of orientation of an object. However, it is possible to define parallel transport in more general scenarios, and then a net rotation can be obtained. To get a flavour of this idea, consider motion in two dimensions, but allow the 'two dimensional' surface to be curved in some way, such as the surface of a sphere (Fig. 6.17). Define 'parallel transport' in this surface to mean the object has to lie in the surface, but it is not allowed to rotate relative to the nearby surface as it moves. For a specific example, think of carrying a metal bar over the surface of a non-rotating spherical planet. Hold the bar always horizontal (i.e., parallel to the ground at your location), and when you walk make sure the two ends of the bar move through the same distance relative to the ground: that is what we mean by parallel transport in this example. Start at the equator, facing north, so that the bar is oriented east–west. Walk due north to the north pole. Now, without rotating yourself or the bar, step to your right, and continue until you reach the equator again. You will find on reaching the equator that you are facing around the equator, and the bar is now oriented north–south. Next, again without turning, walk around the equator back to your starting point. You can take either the long route by walking forwards, or the short route by stepping backwards. In either case, when you reach your starting point, the bar, and your body, will have undergone a net rotation through 90°.

In the case of the aircraft flying around a polygon, the transport of the rod is not a parallel transport in spacetime (spacetime is *not* curved in Special Relativity), but the rotation can be conveniently understood by relating it to a parallel transport in a certain abstract space. This is a *4-velocity space* which we define by setting up axes in the U^t, U^x, U^y, U^z 'directions'. Then the set of allowed 4-velocities does not fill the space, but lies on the region defined by the constraint $U \cdot U = -c^2$. This is the equation of a hyperboloid of revolution:

$$U_t^2 - U_x^2 - U_y^2 - U_z^2 = c^2. \tag{6.45}$$

This hyperboloid of revolution is a curved 3-surface (Fig. 6.18); let us call it $\mathcal{H}$. Now use the fact that, in any frame with 4-velocity U, the hyperplane of simultaneity for that frame is orthogonal to U, and therefore it is parallel to (i.e., tangential to) the surface $\mathcal{H}$. As a physical object, such as a straight rod, is accelerated from one velocity to another, its 4-velocity moves around $\mathcal{H}$, and we can imagine that the object also casts a short straight 'shadow' onto $\mathcal{H}$, signifying its spatial orientation. A Lorentz boost through a small velocity change will move the 4-velocity to a nearby point in $\mathcal{H}$, in such a way that the shadow moves by a parallel transport: this signifies that the object was accelerated without rotating it relative to the instantaneous rest frame. Nonetheless, after a series of such boosts, finishing back at the initial velocity, the shadow has rotated owing to the fact that $\mathcal{H}$ is not flat, and this implies that the physical object (the straight rod, or whatever) must have undergone a net rotation in space, relative to any unaccelerated object.[7]

In the aircraft example, the aircraft did not undergo such a transport, but if the pilot kept next to him a rod, initially parallel to the axis of the aircraft, and made it undergo acceleration by a sequence of Lorentz boosts without rotation, then after flying around the polygon he would find the rod was no longer parallel to the axis of the aircraft. He could ensure that the rod had a parallel transport in velocity space by applying each required velocity change to all particles of the rod simultaneously in the instantaneous rest frame. His observations of his journey convince him that the angle between himself and such a rod increases by more than $360°$, and he is right. On completing the circuit in an anticlockwise direction, the aircraft is on a final flight path parallel to its initial one, but the rod has undergone a net rotation clockwise; see fig. 6.19. If instead of a rigid rod we consider a gyroscope, with its axis aligned with the rod, then this axis also will rotate, by exactly the same amount as the rod, and such a rotation is called 'Thomas precession'.

More generally one may speak of Thomas- or Wigner- rotation of a 'vector'. This is simply to liberate the definition from the need to talk about any particular physical object, but note that such a vector ultimately has to be defined in physical terms. It is a mathematical quantity behaving in the same way as a spatial displacement in the instantaneous rest frame, where, as always, spatial displacement is displacement relative to a reference body in uniform motion.

Is there a torque?

Students (and more experienced workers) are sometimes confused about the distinction between kinematic and dynamic effects. For example, the Lorentz contraction is a kinematic effect because it is the result of examining the same set of worldlines (those of the particles of a body) from the perspective of two different reference frames. Nonetheless, if a given object starts at rest and then is made to accelerate, then any

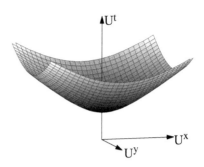

Fig. 6.18 The 4-velocity surface $\mathcal{H}$.

[7] For further details, see S. Ben-Menahem, J. Math. Phys. **27**, 1284 (1986).

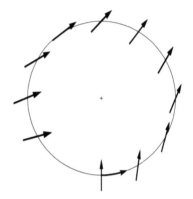

Fig. 6.19 The evolution, in ordinary space, of an object (e.g., a wooden arrow) when it undergoes a parallel transport in velocity-space, such that it is carried around a circle in some inertial reference frame. 'Parallel transport in velocity-space' means that at each moment, the evolution in the next small time interval can be described by a Lorentz boost—that is, an acceleration without rotation.

change in its shape in a given frame (such as the initial rest frame) is caused by the forces acting on it—a dynamic effect. The concept of Lorentz contraction enables us to see what kind of dynamical contraction is the one that preserves the proper length. To be specific, if a rod starts at rest in S and then is accelerated to speed v in S by giving the same velocity-change to all the particles in the rod, then if the proper length is to remain unchanged, the particles must not be pushed simultaneously in S. Rather, the new velocity has to be acquired by the back of the rod first. No wonder, then, that it contracts.

In the case of Thomas rotation, similar considerations apply. Recall the example of the aircraft, and suppose that the aircraft first approached the polygon in straight line flight along a tangent, and then flew around it. From the perspective of a reference frame fixed on the ground, the rod initially has a constant orientation (until the aircraft reaches the polygon), and then it begins to rotate. It must therefore be subject to a torque to set it rotating. It is not hard to see how the torque arises. Transverse forces on the rod are needed to make it accelerate with the aircraft around the polygon. If the application of these forces is simultaneous in the rest frame of the aircraft, then in the rest frame of the polygon the force at the back of the rod happens first, so there is a momentary torque about the centre of mass.

In the case of a gyroscope there must exist a torque in any inertial frame relative to which the gyroscope precesses, because its angular momentum vector changes. This is explored further in chapter 15.

The two effects (Thomas rotation and Lorentz contraction) are close companions. They both arise from the way planes of simultaneity associated with one inertial reference frame or another 'slice up' spacetime differently.

6.7.3   Analysis of circular motion

We shall now analyze the case of motion around a circular trajectory. We already know the answer because the simple argument given at the start of this section for the aircraft flying above the polygon is completely valid, but to get a more complete picture it is useful to think about the sequence of rest frames of an object following a curved trajectory.

Fig. 6.20 shows the case of a particle following a circular orbit. The axes xy are those of the 'laboratory' frame S in which the circle is at rest. The particle is momentarily at rest in frame S′ at proper time τ and in frame S″ at the slightly later proper time $\tau + d\tau$. The axes of both S and of S″ are constructed to be parallel to those of S′ for an observer at rest in S′. Nevertheless, as we have already shown in section 6.7, the axes of S and S″ are not parallel in S or S″.

Let $\mathbf{a}_0$ be the proper acceleration of the particle at proper time τ. This is the acceleration it has in the instantaneous rest frame S′. During the next small time interval, the velocity change, as observed in frame S′, is

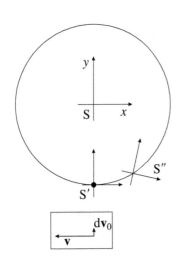

Fig. 6.20 Analysis of motion around a circle. The 'lab' frame S is that of the fixed circle. Frames S′ and S″ are successive rest frames of an object moving around the circle. The axes of S″ and S′ are arranged to be parallel in either of those frames. (Therefore they are not parallel in frame S which has been used to draw the diagram). The box shows the velocities of S and S″ relative to S′.

$$d\mathbf{v}_0 = \mathbf{a}_0 d\tau. \tag{6.46}$$

This gives the velocity of S'' relative to S'. The acceleration $\mathbf{a}_0$ is directed towards the centre of the circle, which at the instant indicated in the diagram is in the vertical direction, so the velocity of S'' relative to S' has components $(0, dv_0)$.

Let $\mathbf{v}$ be the velocity of S' relative to S. This is horizontal at the chosen instant. $\mathbf{v}$ and $d\mathbf{v}_0$ are mutually perpendicular, so we have a situation exactly as was discussed in section 6.7, with the speeds u and v now replaced by dv_0 and v. Angles θ and θ'' (eqns (6.41) and (6.42)) are both small, so we have

$$\theta = \frac{dv_0 \gamma (dv_0)}{v}, \qquad \theta'' = \frac{dv_0}{v\gamma(v)}. \tag{6.47}$$

Using $\gamma(dv_0) \simeq 1$, we find that the rotation of the rest frame axes is by

$$d\theta = \theta - \theta'' = \frac{dv_0}{v}\left(1 - \frac{1}{\gamma}\right). \tag{6.48}$$

In this equation, dv_0 is a velocity *in* the instantaneous rest frame, whereas v is a velocity *of* that frame relative to the centre of the circle. It is more convenient to express the result in terms of quantities all in the latter frame. The size of the change in velocity observed in S is $|d\mathbf{v}| = dv_0/\gamma$ (by using the velocity addition equations, (2.27)). Hence

$$d\theta = \frac{|d\mathbf{v}|}{v}(\gamma - 1). \tag{6.49}$$

The motion completes one circuit of the circle when $\int |d\mathbf{v}| = 2\pi v$, at which point the net rotation angle of the axes is $2\pi(\gamma - 1)$, in agreement with eqn (6.40).

We conclude that the axes in which the particle is momentarily at rest, when chosen such that each set is parallel to the previous set for an observer on the particle, are found to rotate in the reference frame of the centre of the circle (and therefore in any inertial reference frame) at the rate

$$\frac{d\theta}{dt} = \frac{a}{v}(\gamma - 1), \tag{6.50}$$

and it is easy to check that the directions are such that this can be written in vector notation

$$\boldsymbol{\omega}_T = \frac{\mathbf{a} \wedge \mathbf{v}}{c^2}\frac{\gamma^2}{1+\gamma}, \tag{6.51}$$

where we made use of eqn (2.10). In fact, the derivation did not need to assume that the motion was circular, and we can always choose to align the axes with the local velocity, so we have proved the vector result (6.51) for any motion where the acceleration is perpendicular to the velocity. By analysing a product of two Lorentz boosts, it can be shown that the result is valid in general.

Application

In order to apply eqn (6.51) to dynamical problems, one uses a standard kinematic result for rotating frames (whether classical or relativistic): namely, that if some vector **s** has a rate of change $(ds/dt)_{\rm rot}$ relative to axes that are themselves rotating with angular velocity $\boldsymbol{\omega}_T$, then its rate of change relative to non-rotating axes is

$$\left(\frac{d\mathbf{s}}{dt}\right)_{\rm nonrot} = \left(\frac{d\mathbf{s}}{dt}\right)_{\rm rot} + \boldsymbol{\omega}_T \wedge \mathbf{s}. \tag{6.52}$$

The dynamical equations applying in the instantaneous rest frame will dictate the proper rate of change $d\mathbf{s}/d\tau$ of any given vector **s** describing some property of the particle. For a particle describing circular motion the sequence of instantaneous rest frames supplies axes rotating at $\boldsymbol{\omega}_T$, to which eqn (6.52) applies, with the substitution

$$\left(\frac{d\mathbf{s}}{dt}\right)_{\rm rot} = \frac{1}{\gamma}\left(\frac{d\mathbf{s}}{d\tau}\right)_{\rm rest\ frame}. \tag{6.53}$$

For motion which is curved but not circular, the equation applies to each short segment of the trajectory.

For electrons in atoms there is a centripetal acceleration given by the Coulomb attraction to the nucleus, $\mathbf{a} = -e\mathbf{E}/m$ where **E** is the electric field at the electron, calculated in the rest frame of the nucleus, and $-e$ is the charge on an electron. For atoms such as hydrogen, the velocity $v \ll c$ so we can use $\gamma \simeq 1$ and we obtain to good approximation,

$$\boldsymbol{\omega}_T = \frac{e\mathbf{v} \wedge \mathbf{E}}{2mc^2}. \tag{6.54}$$

The spin-orbit interaction calculated in the instantaneous rest frame of the *electron* gives a Larmor precession frequency

$$\boldsymbol{\omega}_L = \frac{-g_{\rm s}\mu_B}{\hbar}\frac{\mathbf{v} \wedge \mathbf{E}}{c^2}, \tag{6.55}$$

where $g_{\rm s}$ is the gyromagnetic ratio of the spin of the electron and the Bohr magneton is $\mu_B = e\hbar/2m$. To find what is observed in an inertial frame, such as the rest frame of the nucleus, we must add the Thomas precession to the Larmor precession,

$$\boldsymbol{\omega} = \boldsymbol{\omega}_L + \boldsymbol{\omega}_T = \left(\frac{-g_{\rm s}\mu_B}{\hbar} + \frac{e}{2m}\right)\frac{\mathbf{v} \wedge \mathbf{E}}{c^2} = -\frac{e}{2m}(g_{\rm s} - 1)\frac{\mathbf{v} \wedge \mathbf{E}}{c^2} \tag{6.56}$$

If we now substitute the approximate value $g_{\rm s} = 2$, we find that the Thomas precession frequency for this case has the opposite sign and half the magnitude of the rest frame Larmor frequency. This means that the precession frequency observed in the rest frame of the nucleus will be half that in the electron rest frame. More precisely, the impact is to replace $g_{\rm s}$ by $g_{\rm s} - 1$ (not $g_{\rm s}/2$): it is an additive, not a multiplicative correction (see exercise 6.12).

The above argument treated the motion as if classical rather than quantum mechanics was adequate. This is wrong. However, upon re-examining the argument starting from Schrödinger's equation, one finds that the spin-orbit interaction gives a contribution to the potential energy of the system, and the precession of the spin of the electron may still be observed. For example, when the electron is in a non-stationary state (a superposition of states of different orientation), the spin direction precesses at ω_L in the rest frame of the electron, and at $\omega_L + \omega_T$ in the rest frame of the nucleus. This precession must be related to the gap between energy levels by the universal factor $\hbar$, so it follows that eqn (6.56), after multiplying by $\hbar$, must describe the observed energy level splittings.

6.8   Generators of boosts and rotations

For frames in the standard configuration, the Lorentz transformation matrix can be written, to first order in β, as

$$
\Lambda \simeq I - \beta \begin{pmatrix} 0 & 1 & 0 & 0 \\ 1 & 0 & 0 & 0 \\ 0 & 0 & 0 & 0 \\ 0 & 0 & 0 & 0 \end{pmatrix}
$$

where I is the identity matrix. Since $\gamma(v)$ is an even function, the lowest order neglected terms are $O(\beta^2)$ on the diagonal and $O(\beta^3)$ elsewhere.

It is easy to see that boosts in the other two coordinate directions can be written in matrix terms in a similar way, and so can pure rotations. We introduce the sets of matrices

$$
S_x = \begin{pmatrix} 0 & 0 & 0 & 0 \\ 0 & 0 & 0 & 0 \\ 0 & 0 & 0 & -1 \\ 0 & 0 & 1 & 0 \end{pmatrix}, \quad K_x = \begin{pmatrix} 0 & 1 & 0 & 0 \\ 1 & 0 & 0 & 0 \\ 0 & 0 & 0 & 0 \\ 0 & 0 & 0 & 0 \end{pmatrix}, \tag{6.57}
$$

$$
S_y = \begin{pmatrix} 0 & 0 & 0 & 0 \\ 0 & 0 & 0 & 1 \\ 0 & 0 & 0 & 0 \\ 0 & -1 & 0 & 0 \end{pmatrix}, \quad K_y = \begin{pmatrix} 0 & 0 & 1 & 0 \\ 0 & 0 & 0 & 0 \\ 1 & 0 & 0 & 0 \\ 0 & 0 & 0 & 0 \end{pmatrix}, \tag{6.58}
$$

$$
S_z = \begin{pmatrix} 0 & 0 & 0 & 0 \\ 0 & 0 & -1 & 0 \\ 0 & 1 & 0 & 0 \\ 0 & 0 & 0 & 0 \end{pmatrix}, \quad K_z = \begin{pmatrix} 0 & 0 & 0 & 1 \\ 0 & 0 & 0 & 0 \\ 0 & 0 & 0 & 0 \\ 1 & 0 & 0 & 0 \end{pmatrix}. \tag{6.59}
$$

Then a small boost in an arbitrary direction can be written

$$
I - (\beta_x K_x + \beta_y K_y + \beta_z K_z)
$$

and a small rotation can be written

$$
I - (\theta_x S_x + \theta_y S_y + \theta_z S_z)
$$

where the direction of $\boldsymbol{\theta} = (\theta_x, \theta_y, \theta_z)$ gives the rotation axis, and the size θ (which is here small) gives the amount of rotation.

The matrices K_x, K_y, K_z do not commute:

$$[K_x, K_y] \equiv K_x K_y - K_y K_x = -S_z \tag{6.60}$$

which the reader is invited to verify. It follows that Lorentz boosts in different directions do not commute, and the Thomas-Wigner rotation also follows. Consider, for example, a sequence of four boosts around a square—that is, the sequence

$$\Lambda = (I + \beta K_y)(I + \beta K_x)(I - \beta K_y)(I - \beta K_x) \tag{6.61}$$

(plus terms of $O(\beta^2)$). This corresponds to a boost by βc in the x-direction, then the y-direction, then the negative x-direction, then the negative y-direction. The term of $O(\beta)$ in this expression is

$$\beta K_y + \beta K_x - \beta K_y - \beta K_x = 0$$

therefore the lowest order term involving β is the quadratic term, and in order to calculate it fully we should include the $O(\beta^2)$ terms in the expression for each boost, replacing eqn (6.61) by

$$\Lambda = \left(I + \beta K_y + \frac{\beta^2}{2}K_y^2\right)\left(I + \beta K_x + \frac{\beta^2}{2}K_x^2\right)\left(I - \beta K_y + \frac{\beta^2}{2}K_y^2\right)$$
$$\left(I - \beta K_x + \frac{\beta^2}{2}K_x^2\right)$$

By expanding the brackets and neglecting terms of $O(\beta^3)$ one finds

$$\Lambda = I - \beta^2 [K_x, K_y]. \tag{6.62}$$

Using eqn (6.60), this is a rotation about the z axis by an angle $-\beta^2$, in agreement with two applications of eqn (6.44).

The matrices K_x, K_y, K_z are said to *generate* Lorentz boosts, and the matrices S_x, S_y, S_z are said to *generate* rotations. All possible boosts and rotations can be formed by combining them (e.g., a large rotation can always be decomposed into many small rotations).

It can be shown that a general boost and rotation can be written[8]

$$\Lambda = e^{-\boldsymbol{\rho}\cdot\mathbf{K} - \boldsymbol{\theta}\cdot\mathbf{S}} \tag{6.63}$$

where $\boldsymbol{\rho}$ is a rapidity vector, $\boldsymbol{\theta}$ is a rotation angle (the direction of the vector specifying the axis of rotation). This result is easy to verify for simple cases. For example, a boost in the x direction would be given by $\boldsymbol{\theta} = 0$ and $\boldsymbol{\rho} = (\rho, 0, 0)$ with $\tanh\rho = v/c$.

[8] The exponential of a matrix M is defined $\exp(M) \equiv 1 + M + M^2/2! + M^3/3! + \cdots$. It can also be calculated from $\exp(M) = U\exp(M_D)U^\dagger$, where M_D is a diagonalized form of M: i.e., $M_D = U^\dagger M U$ where U is the (unitary) matrix whose columns are the normalized eigenvectors of M.

6.9   The Lorentz group*

A product of two rotations is a rotation, but a product of two Lorentz boosts is not always a Lorentz boost (cf. eqn (6.43)). This invites one to look into the question: to what general class of transformations does the Lorentz transformation belong?

We define the Lorentz transformation as that general type of transformation of coordinates that preserves the interval $-(ct)^2 + x^2 + y^2 + z^2$ unchanged. Using eqn (2.48) this definition is conveniently written

$$L \equiv \{\Lambda : \Lambda^T g \Lambda = g\}. \tag{6.64}$$

where L is the set of all Lorentz transformations, Λ is a general Lorentz transformation, and g is the Minkowski metric defined in eqn (2.46).

We will now prove that the set L is in fact a *group*, and furthermore it can be divided into 4 distinct parts, one of which is a sub-group. Here a mathematical *group* is a set of entities that can be combined in pairs, such that the combination rule is associative (i.e., $(ab)c = a(bc)$), the set is closed under the combination rule, there is an identity element and every element has an inverse. *Closure* here means that for every pair of elements in the set, their combination is also in the set. We can prove all these properties for the Lorentz group by using matrices that satisfy eqn (6.64). The operation or 'combination rule' of the group will be matrix multiplication. The matrices are said to be a *representation* of the group.

(1) *Associativity.* This follows from the fact that matrix multiplication is associative.

(2) *Closure.* The net effect of two successive Lorentz transformations $\mathsf{X} \to \mathsf{X}' \to \mathsf{X}''$ can be written $\mathsf{X}'' = \Lambda_2 \Lambda_1 \mathsf{X}$. The combination $\Lambda_2 \Lambda_1$ is a Lorentz transformation, since it satisfies eqn (6.64):

$$(\Lambda_2 \Lambda_1)^T g \Lambda_2 \Lambda_1 = \Lambda_1^T \Lambda_2^T g \Lambda_2 \Lambda_1 = \Lambda_1^T g \Lambda_1 = g.$$

(3) *Inverses.* We have to show that the inverse matrix Λ^{-1} exists and is itself a Lorentz transformation. To prove its existence, take determinants of both sides of $\Lambda^T g \Lambda = g$ to obtain

$$|\Lambda|^2 |g| = |g|$$

but $|g| = -1$ so

$$|\Lambda|^2 = 1, \qquad |\Lambda| = \pm 1. \tag{6.65}$$

Since $|\Lambda| \neq 0$ we deduce that the matrix Λ does have an inverse. To show that Λ^{-1} satisfies eqn (6.64) we need a related formula. First consider

$$(\Lambda g)(\Lambda^T g \Lambda)(g \Lambda^T) = \Lambda g^3 \Lambda^T = \Lambda g \Lambda^T$$

Now (following Taylor) pre-multiply by $(\Lambda g \Lambda^T g)^{-1}$:

$$\Lambda g \Lambda^T = g^{-1} \tag{6.66}$$

where we used $(AB)^{-1} = B^{-1} A^{-1}$ (eqn (1.13)), and we can be sure that $(\Lambda g \Lambda^T g)^{-1}$ exists because $|\Lambda g \Lambda^T g| = |\Lambda|^2 |g|^2 = 1$. Now, to show that Λ^{-1} is a Lorentz transformation, take the inverse of both sides of (6.66):

$$(\Lambda^T)^{-1} g^{-1} \Lambda^{-1} = g \qquad \Rightarrow \qquad (\Lambda^{-1})^T g \Lambda^{-1} = g \tag{6.67}$$

which shows Λ^{-1} satisfies the condition eqn (6.64).

(4) *Identity element.* The identity matrix satisfies eqn (6.64), and so can serve as the identity element of the Lorentz group.

Since the complete set of 4×4 real matrices can themselves be considered as a representation of a sixteen-dimensional real space, we can think of the Lorentz group as a subset of sixteen-dimensional real space. The defining condition (6.64) might appear to set sixteen separate conditions, which would reduce the space to a single point; but there is some repetition since g is symmetric, so there is a continuous 'space' of solutions. There are ten linearly independent conditions (a symmetric 4×4 matrix has ten independent elements); it follows that L is a six-dimensional subset of R^{16}. That is, a general member of the set can be specified by six real parameters; you can think of these as three to specify a rotation and three to specify a velocity.

We can move among some members of the Lorentz group by continuous changes, such as by a change in relative velocity between reference frames or a change in rotation angle. However, we can show that not all parts of the group are continuously connected in this way. The condition (6.65) is interesting because it is not possible to change the determinant of a matrix discontinuously by a continuous change in its elements. This means that we can identify two subsets:

$$L_\uparrow \equiv \{\Lambda \in L : |\Lambda| = +1\}$$
$$L_\downarrow \equiv \{\Lambda \in L : |\Lambda| = -1\} \tag{6.68}$$

and one cannot move between $L_\uparrow$ and $L_\downarrow$ by a continuous change of matrix elements. The subsets are said to be *disconnected*. One can see that the subset $L_\downarrow$ is not a group because it is not closed (the product of any two of its members lies in $L_\uparrow$), but it is not hard to prove that $L_\uparrow$ is a group, and therefore a sub-group of L. An important member of $L_\downarrow$ is the spatial inversion through the origin, also called the *parity* operator:

$$P \equiv (t \to t, \; \mathbf{r} \to -\mathbf{r}).$$

Its matrix representation in rectangular coordinates is

$$P = \begin{pmatrix} 1 & 0 & 0 & 0 \\ 0 & -1 & 0 & 0 \\ 0 & 0 & -1 & 0 \\ 0 & 0 & 0 & -1 \end{pmatrix}. \tag{6.69}$$

What is interesting is that if $\Lambda \in L_\uparrow$ then $P\Lambda \in L_\downarrow$. Thus to understand the whole group it suffices to understand the sub-group $L_\uparrow$ and the effect of P. The action of P is to reverse the direction of vector quantities such as the position vector or momentum vector; the subscript arrow notation $L_\uparrow, L_\downarrow$ is a reminder of this. Members of $L_\uparrow$ are said to be *proper* and members of $L_\downarrow$ *improper*. Rotations are in $L_\uparrow$, reflections are in $L_\downarrow$.

We can divide the Lorentz group a second time by further use of eqn (6.64). We adopt the notation Λ^μ_ν for the (μ, ν) component of Λ. Examine the $(0,0)$ component of eqn (6.64). If we had the matrix product $\Lambda^T \Lambda$

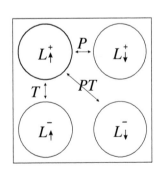

Fig. 6.21 The structure of the Lorentz group. The proper orthochronous set $L_\uparrow^+$ is a subgroup. It is continuous and six-dimensional. The other subsets can be obtained from it.

this would be $\sum_\mu (\Lambda_0^\mu)^2$, but the g matrix in the middle introduces a sign change, so we obtain

$$-(\Lambda_0^0)^2 + \sum_{i=1}^3 (\Lambda_0^i)^2 = g_{00} = -1$$

$$\Rightarrow \quad \Lambda_0^0 = \pm \left(1 + \sum_{i=1}^3 (\Lambda_0^i)^2 \right)^{1/2}. \tag{6.70}$$

The sum inside the square root is always positive since we are dealing with real matrices, and we deduce that

$$\text{either} \quad \Lambda_0^0 \geq 1 \quad \text{or} \quad \Lambda_0^0 \leq -1.$$

That is, the time-time component of a Lorentz transformation can either be greater than or equal to 1, or less than or equal to -1, but there is a region in the middle, from -1 to 1, that is forbidden. It follows that the transformations with $\Lambda_0^0 \geq 1$ form a set disconnected from those with $\Lambda_0^0 \leq 1$. We define

$$L^+ \equiv \{ \Lambda \in L : \Lambda_0^0 \geq 1 \} \tag{6.71}$$

$$L^- \equiv \{ \Lambda \in L : \Lambda_0^0 \leq 1 \}. \tag{6.72}$$

An important member of L^- is the time-reversal operator

$$T \equiv (t \to -t, \, \mathbf{r} \to \mathbf{r})$$

whose matrix representation is[9]

$$T = \begin{pmatrix} -1 & 0 & 0 & 0 \\ 0 & 1 & 0 & 0 \\ 0 & 0 & 1 & 0 \\ 0 & 0 & 0 & 1 \end{pmatrix}. \tag{6.73}$$

[9] The time-reversal operator is not the same as the Minkowski metric, although they may look the same in a particular coordinate system such as rectangular coordinates. Their difference is obvious as soon as one adopts another coordinate system such as polar coordinates.

It is now straightforward to define the sub-sets $L_\uparrow^+$ $L_\downarrow^+$ $L_\uparrow^-$ $L_\downarrow^-$ as intersections of the above. It is easy to show furthermore (left as an exercise for the reader) that $L_\uparrow^+$ is a group and the operators P, T and PT allow one-to-one mappings between $L_\uparrow^+$ and the other distinct sets, as shown in Fig. 6.21.

A member of $L_\uparrow^+$ is called a 'proper orthochronous' Lorentz transformation. All transformations in this group can be written as shown in eqn (6.63).

All known fundamental physics is invariant under proper orthochronous Lorentz transformations, but examples of both parity violation and time reversal violation are known in weak radioactive processes. Thus one cannot always ask for Lorentz invariance under the whole Lorentz group, but as far as we know it is legitimate to require invariance under transformations in $L_\uparrow^+$. This group is also called the 'restricted' Lorentz group.

6.9.1   Further group terminology

The Lorentz group consists of boosts, rotations, and combinations of those. One can therefore regard it as the group of symmetry operations on spacetime that leave the origin fixed. It is a subgroup of the Poincaré group, which consists of the Lorentz group plus translations. Operations such as rotation, boost and translation, which can act on elements in a space with a metric (here, spacetime), while preserving the metric, are called *isometries*.

The general theory of groups can be used to find connections between mathematical or physical entities whose close relationship might not otherwise be obvious. This can give some very useful insights. We have already exhibited one such connection: on the one hand we have 4×4 real matrices obeying eqn (6.64), and on the other we have the operations of boost and rotation, regarded as abstract mathematical operations in spacetime. We have found that the former can be used to mathematically model the complete behaviour of the latter. To be precise, we have found that the Lorentz group (the group of boosts and rotations) is the same as the 'indefinite orthogonal group $O(1,3)$'. The latter is defined as a matrix group whose elements are matrices of real numbers, whose operation is matrix multiplication, and whose members have the property that when acting on a 4×1 vector the combination $x_0^2 - x_1^2 - x_2^2 - x_3^2$ is preserved. The term 'indefinite' in the group name alludes to the fact that this invariant is not positive definite but may have either sign or none. The notation $O(1,3)$ arises because this can be generalised to more dimensions with more terms in the metric. The technically correct way to say the groups are 'the same' is to say they are *isomorphic*. This means there exists a one-to-one mapping between elements of the first group and elements of the second, such that the structure of the group is preserved: i.e., if $M_i M_j = M_k$ for elements M_α in the first group, and the mapping to elements N_α of the second group is $M_\alpha \leftrightarrow N_\alpha$, then $N_i N_j = N_k$.

The groups we are discussing are also called *Lie groups*. A Lie group is a group where the members can be described in terms of smoothly varying parameters such as real or complex numbers. These parameters, as they vary in combination with one another, can be regarded as mapping out a smooth space. If the space is indeed smooth, according to a technical definition, then it is said to be a 'differentiable manifold' (often abbreviated to 'manifold') and we have a Lie group (as long as the members also obey the four defining properties of a group, of course: closure, inverses, identity element, associative).

Now we can have some fun with group theory.

The four parts of $O(1,3)$ are called its four *connected components*: each is a component that is smoothly connected within itself. The two components giving all the proper Lorentz transformations form a subgroup called $SO(1,3)$ (the 'special indefinite orthogonal group of order 1,3'). The two components giving all the orthochronous Lorentz transformations form a subgroup called $O^+(1,3)$. The single component that includes the identity operation is called the 'identity component'; this is the proper, orthochronous Lorentz group, also called the 'restricted Lorentz group' $\mathrm{SO}^+(1,3)$.

For matrix groups, it often happens that the restriction to matrices of determinant $+1$ is itself a group, a subgroup of the original group. It is called the 'special' subgroup: hence '$\mathrm{SO}(1,3)$'. (Another example, $\mathrm{SU}(2)$, is described in volume 2.)

The discrete set

$$\{I, P, T, PT\}$$

itself forms a group! This is called the Klein group V (for 'Vierergruppe'). Because all of $\mathrm{O}(1,3)$ can be reconstructed precisely by a combination of $\mathrm{SO}^+(1,3)$ and V, we say that $\mathrm{O}(1,3)$ contains a 'quotient group' $\mathrm{O}(1,3)/\mathrm{SO}^+(1,3)$, and this 'quotient group' is isomorphic to the Klein group.

The restricted Lorentz group is closely related to another group: the 'special linear group' $\mathrm{SL}(2,\mathrm{C})$. This groups consists of all 2×2 matrices of complex numbers, with determinant 1:

$$M = \begin{pmatrix} a & b \\ c & d \end{pmatrix}, \qquad ab - cd = 1.$$

To make the connection, consider Hermitian matrices (those such that $H^\dagger = H$, where $H^\dagger = (H^T)^*$: transpose and complex conjugate). Any 2×2 Hermitian matrix X can be described by four real numbers, and we can make the clever choice

$$X = \begin{pmatrix} t+z & x-iy \\ x+iy & t-z \end{pmatrix}, \tag{6.74}$$

then the determinant of X is

$$|X| = t^2 - x^2 - y^2 - z^2.$$

Now, if we allow members of the group $\mathrm{SL}(2,\mathrm{C})$ to act on such X by

$$X \to MXM^\dagger \tag{6.75}$$

then the determinant of X is preserved. Looking into this a little further, one finds that the Lie group $\mathrm{SL}(2,\mathrm{C})$ can provide an exact copy of the Lie group $\mathrm{SO}^+(1,3)$. In fact, $\mathrm{SL}(2,\mathrm{C})$ provides not just one copy of $\mathrm{SO}^+(1,3)$, but two, because replacing M by $-M$ will result in the same transformation of X. That is, for every member N_i of the restricted Lorentz group $\mathrm{SO}^+(1,3)$, there are two members M_i and $-M_i$ of the special linear group that map onto it. Such a '2 to 1' mapping is called a *double cover*. If one takes just one of the two parts of $\mathrm{SL}(2,\mathrm{C})$, one has the 'projective special linear group' $\mathrm{PSL}(2,\mathrm{C})$ and then one has an isomorphism (i.e., one-to-one mapping) with $\mathrm{SO}^+(1,3)$.

It is also known that $\mathrm{PSL}(2,\mathrm{C})$ is isomorphic to yet another important group: the *Möbius group*. This opens up further insights into the Lorentz group, many of which are important in particle physics.

The connection between the Lorentz group and the $\mathrm{SL}(2,\mathrm{C})$ group leads to the concept of *spinors*, discussed in volume 2.

Exercises

(6.1) An oscillator undergoes periodic motion. State, with reasons, which of the following are Lorentz-invariant:

    (i) the amplitude of the oscillation,

    (ii) the phase of the oscillation,

    (iii) the maximum velocity.

(6.2) The 4-vector field F is given by $\mathsf{F} = 2\mathsf{X} + \mathsf{K}(\mathsf{X} \cdot \mathsf{X})$ where K is a constant 4-vector and $\mathsf{X} = (ct, x, y, z)$ is the 4-vector displacement in spacetime. Evaluate the following:

    (i) $\Box \cdot \mathsf{X}$

    (ii) $\Box(\mathsf{X} \cdot \mathsf{X})$

    (iii) $\Box^2(\mathsf{X} \cdot \mathsf{X})$

    (iv) $\Box \cdot \mathsf{F}$

    (v) $\Box(\Box \cdot \mathsf{F})$

    (vi) $\Box^2 \sin(\mathsf{K} \cdot \mathsf{X})$

    [*Ans.* 4, $2\mathsf{X}$, 8, $8 + 2\mathsf{K} \cdot \mathsf{X}$, $2\mathsf{K}$, $-\mathsf{K}^2 \sin(\mathsf{K} \cdot \mathsf{X})$]

(6.3) Prove that if a distribution of current density $\mathbf{j}$ is confined to a finite region, then

$$\int x \boldsymbol{\nabla} \cdot \mathbf{j} \, dV = - \int j_x \, dV$$

where $dV = dx dy dz$ and the integral is over all space. (Hint: write $\boldsymbol{\nabla} \cdot \mathbf{j} = (\partial j_x / \partial x) + (\partial j_y / \partial y) + (\partial j_z / \partial z)$ and treat the three integrals separately.) Hence, show that if ρ and $\mathbf{j}$ are the density and flux of a conserved quantity, then

$$\frac{d}{dt} \int \rho \mathbf{x} \, dV = \int \mathbf{j} \, dV. \qquad (6.76)$$

Note the physical interpretation: the LHS is the rate of change of dipole moment, and the RHS is the total current. A non-zero current means charge is being carried from one place to another, hence changing the dipole moment; if the dipole moment is constant then the currents must be flowing in loops.

(6.4) A cylinder has no translational motion in some reference frame S, but rotates rigidly: i.e., all its particles move around the axis of the cylinder with the same angular velocity ω in S. Show that in other reference frames moving in the direction of the cylinder's axis, the cylinder is twisted. Find the speed of a reference frame in which the total twist from one end of the cylinder to the other is by 2π radians. (Hint: write down the trajectory of a particle, or, as suggested by Rindler, refer to exercise 2.3 of chapter 2.)

(6.5) Explain qualitatively what sort of motion should be given to a banana, relative to a given inertial reference frame, in order that its natural shape would be straight in that reference frame.

(6.6) *Rotating disk paradox.* A disk of radius a sits on top of a piece of paper, both initially at rest in frame S. A line is painted on the paper around the edge of the disk; such a line must have length $2\pi a$, equal to the circumference of the disk. A device now puts the disk into rigid rotation at angular frequency ω, such that all parts of the disk acquire the appropriate initial velocity simultaneously in S. It is observed that the edge of the rotating disk still lies directly above the circle painted on the paper—whose circumference remains $2\pi a$. However, any small region of the disk's edge is moving at speed ωa relative to S and so suffers a Lorentz contraction by a factor $\gamma = (1 - \omega^2 a^2 / c^2)^{-1/2}$. The circumference of the disk, observed in S, is therefore Lorentz-contracted by this amount. Yet the disk still matches with the painted line. How is this to be explained? (Hint: section 6.5.2). If $a = 10^6$ m, $\omega = 100 \, \text{s}^{-1}$, what is the circumference of the disk as observed in frame S? How many sticks of rest length 1 metre can be attached to the disk, laid end to end around its circumference?

(6.7) Here are two ways to find the general Lorentz boost, eqn (6.36). (i) First rotate the axes of S so that the relative velocity $\mathbf{v}$ lies along the rotated x-axis, then apply the simple boost for frames in the standard configuration, then rotate axes back again. (ii) Confirm that a particle at rest in S′ has velocity $\mathbf{v} = \boldsymbol{\beta} c$ in S, that the coordinate axes of S and S′ are aligned, and that $\Lambda^T g \Lambda = g$. Pick one or both of these methods, and carry it through to check eqn (6.36).

(6.8) In S′ a rod parallel to the x' axis moves in the y' direction with velocity u. Show that in S the rod is inclined to the x-axis at an angle $-\tan^{-1}(\gamma uv/c^2)$. (Hint: let one end of the rod pass through the origin at $t = 0$ and locate the other end at $t = 0$.)

(6.9) A 10 foot pole (proper length) remains parallel to the x axis of frame S while it glides at velocity $(v, -w, 0)$ towards a hole of diameter 5 feet in a steel plate lying in the plane $y = 0$. Describe how the pole passes through the hole, in three frames: the rest frame of the pole, the frame S, and the frame S′ in standard configuration with S (i.e. S′ moves relative to S at speed v in the x direction).

(6.10) §A certain elastic band will break when it is stretched to twice its natural length. Such a band has initially its natural length and lies at rest on the x axis of some frame S. The ends are then made to move in the x direction with constant proper acceleration a_0, starting simultaneously in S. Show that the elastic breaks at time $t = \sqrt{3}\,a_0/c$.

(6.11) Prove eqn (6.43) by calculating $\Lambda(\mathbf{v})\Lambda(-\mathbf{u})\Lambda^{-1}(\mathbf{w})$ explicitly.

(6.12) According to Quantum Electrodynamics, the g-factor (gyromagnetic ratio) of the electron is given by $g_s = 2 + (\alpha/\pi) - 0.657(\alpha/\pi)^2 + \cdots$, and therefore $g_s - 1 \simeq 1 + \alpha/\pi$ whereas $g_s/2 \simeq 1 + \alpha/2\pi$. Calculate the fractional error introduced by using $g_s/2$ instead of the correct value $g_s - 1$ in a theory of the spin-orbit interaction in hydrogen. Compare this with the precision of the most accurate measurements of the spin-orbit splittings in hydrogen.

(6.13) Prove (e.g., by evaluating the power series expansion) that $\exp(-\rho K_x)$ gives the Lorentz boost matrix shown in eqn (2.39), where K_x is given in eqn (6.57).

(6.14) Confirm that eqns (6.62) and (6.44) are mutually consistent. (Check both the size and sense of rotation.)

(6.15) For any two future-pointing time-like vectors $\mathbf{V}_1$, $\mathbf{V}_2$, prove that $\mathbf{V}_1 \cdot \mathbf{V}_2 = -V_1 V_2 \cosh\rho$ where ρ is the relative rapidity of frames in which $\mathbf{V}_1$ and $\mathbf{V}_2$ are purely temporal.

7 Relativity and electromagnetism

Relativity and electromagnetism go hand in hand. Historically, the basic equations of the theory of electromagnetism were discovered first, but its application to complicated problems such as the structure of materials was handicapped by a lack of insight into fundamental principles, such as the invariance of the speed of light in vacuum. On the other hand, it opened the way to the discovery of Special Relativity. Einstein himself remarked that one of his guiding principles, leading him to the discovery of Relativity, was the hunch that magnetic effects were none other than a manifestation of electric effects under another guise.

We can now recognize that the essential idea of Relativity is a kind of symmetry principle: both the Principle of Relativity and the Light Speed Postulate speak of symmetry: i.e., the idea that a system or a dynamics should stay unchanged when an operation such as a change of reference frame is performed. It is natural to regard this as the more basic insight into the Laws of Nature, so that whereas it would interest us but not unsettle us too much to find that Maxwell's equations were wrong, it would be very disturbing if their replacement did not obey the Main Postulates of Relativity.

With the hindsight, or insight, of Relativity theory, we can now return to electromagnetism and use it as an object lesson in how to discover physics by using fundamental symmetry principles. That is, supposing we did not already know Maxwell's equations, could we discover them from a few simple observations by applying the powerful machinery of Lorentz transformations? Also, was Einstein's hunch true: can the magnetic force be regarded as none other than the electric force seen from another point of view? What else can Relativity teach us about electromagnetism?

We have already partially answered the last question: all the examples of physical behaviour, such as contracting bodies, headlight effect, Doppler shifts, collisions, time dilation, could be regarded as predictions from Maxwell's equations, many of which we would be hard pressed to derive with confidence directly from the latter. The answer to the second question—are magnetic forces just electric forces in another reference frame?—will turn out to be 'sometimes'. The full answer is that electric and magnetic fields should be regarded as two parts of a single thing: the electromagnetic field. They are like two components of a single vector (actually, two parts of a single matrix-like quantity called a *tensor*— but more of this later).

In this chapter we shall start by considering electric and magnetic fields as essentially *fields of force*. That is, we say there is a 'something' which exerts forces on charged matter, and our starting point is an equation saying how the force depends on the charge and motion of the matter (the Lorentz force equation). We then get to work with Relativity and discover what we can about the 'something', such as how it changes from one inertial frame of reference to another. After that we shall introduce Maxwell's equations without attempting to derive them, and work our way towards showing their full consistency with Relativity ('Lorentz covariance') and finding solutions, such as the fields due to a moving particle.

In chapter 13 we shall adopt some more powerful mathematical methods, and this will allow us to take a different perspective on the whole subject. There we shall take the view that the electromagnetic field is itself the primary thing, whose existence can be postulated. By claiming that it is a tensor field we will automatically know how it behaves under Lorentz transformations, and one can argue that the Maxwell equations and the Lorentz force equation are to be expected, since they amount to one of the most simple and natural field theories that Special Relativity allows.

Both approaches yield important insights. It is not a case of one being 'better' than the other. The tensor methods are needed to get certain insights, especially into the energy and momentum of the field itself, but the vector methods are more straightforward for obtaining solutions to many types of problem. First of all, then, let us take the more 'humble' approach of 3-vectors and 4-vectors.

7.1   Definition of electric and magnetic fields

Who has seen the wind?
Neither you nor I.
But when the trees bow down their heads,
The wind is passing by.

Christina Rossetti

The very concept of a 'field' is unsettling when one first meets it as a student of physics. We naturally wonder what a magnetic field or an electric field 'is', when it seems that we cannot smell it or touch it. Is it there at all? The current-carrying wire is visible enough, and the compass needle certainly swings, and we can believe that the electrons in the wire somehow pushed on the electrons in the needle, but what is this 'field' our teachers are telling us about? We suspect that it is just some sort of mathematical method, such as integration. Integration is fine: we know what that is—it is just adding up lots of little bits. So what is the field really? What are its bits?

These same questions were much discussed in the nineteenth century, and perturbed some of the greatest minds in science. We should not dismiss them or imagine that scientists such as Maxwell, Lorentz, and Poincaré were somehow less intelligent than modern physicists. It is just that we have learned that the fields do not have any 'bits': they are themselves fundamental, and we just have to get used to them. However, we have learned that it does make a lot of sense to regard these fields as real physical things. They store energy, they carry momentum, and they move energy and momentum around from one place to another. This does not rule out that new insights in the future may tell us more about fields, but we are pretty sure that any such insights will not make fields seem more everyday.

The best way to define an electric field seems to be to define it in terms of the effect it has on charged particles. So our first attempt is:

Throughout all of spacetime there is a vector field called an electric field (whose size tends to zero a long way from matter). Its value at any given position and time is equal to the force per unit charge exerted on a small fixed test charge at that position and time:

$$\mathbf{E}(t, x, y, z) \equiv \lim_{q \to 0} \frac{\mathbf{f}(t, x, y, z)}{q} \tag{7.1}$$

It has to be understood that the force talked of here is the one that cannot be accounted for in other terms. For example, it does not include any gravitational force there may be. We could handle that by adding a condition that we take a limit of zero mass for the test charge, or else insist that we are referring only to that part of the force which changes with the amount of electric charge. (The definition of electric charge can be handled separately: the main point is that it is something that particles can possess and it is conserved.)

The trouble with the above definition is that there exist physical scenarios where an electric field is present in a region of spacetime when we apply the definition in one inertial reference frame, but it vanishes when we pick another inertial reference frame! (For an example, put a permanent magnet on board a train, and consult eqns (7.13)). So what does it mean to say 'there is an electric field', when a mere change of inertial reference frame can make it vanish?

To do better we have to define electric and magnetic fields together, as follows:

Throughout all of spacetime there is a tensor field called an electromagnetic field (whose size tends to zero a long way from matter). Its value at any given position and time can be expressed in terms of two vector fields $\mathbf{E}$ and $\mathbf{B}$, through the force per unit charge exerted on a small test charge at that position and time. The 'electric' part is defined as above, and the 'magnetic' part is defined through

$$\mathbf{v} \wedge \mathbf{B}(t,x,y,z) \equiv \left(\lim_{q \to 0} \frac{\mathbf{f}(t,x,y,z)}{q} \right) - \mathbf{E}(t,x,y,z) \qquad (7.2)$$

where $\mathbf{v}$ is the velocity of the test charge in the reference frame in which the fields are being evaluated.

This definition survives a change of reference frame, because under such changes electric fields and magnetic fields can come and go, but the electromagnetic field survives (this is like the way the components of a vector can vary under rotations, but the vector is still there). Note that the force can itself change under a change of inertial reference frame—in fact, we already know how: eqn (4.6). It will turn out that this 3-force is γ^{-1} times the spatial part of a 4-vector 4-force.

The main point is to arrive at the

Lorentz force equation

$$\mathbf{f} = q \, (\mathbf{E} + \mathbf{v} \wedge \mathbf{B}) \qquad (7.3)$$

and remark that this can be regarded as the defining equation of the electric and magnetic fields. In the Lorentz force equation, $\mathbf{f}$ is the force on a particle located at some position, $\mathbf{E}$ and $\mathbf{B}$ are the electric and magnetic fields at that position, produced there by everything other than the particle in question, and q and $\mathbf{v}$ are the charge and velocity of the particle.

7.1.1   Transformation of the fields (first look)

The Lorentz force equation (7.3) together with the equations (4.6) describing the transformation of force between one reference frame and another can teach us some of the properties of electric and magnetic fields. In fact, if you think about it, it ought to be possible to discover from them how electric and magnetic fields transform under a change of reference frame. In order to do this, we need to assume that we have a sensible theory: i.e. one that respects the Postulates of Relativity, and in particular the Principle of Relativity. This means that we shall assume the Lorentz force equation is valid in all reference frames. We shall further assume that it describes a pure force (one that does not change the rest mass of particles on which it acts): this is not required by the Principle of Relativity, but it will turn out to be right for electromagnetic forces. It means that we can use the simpler form (4.7) for the transformation of $\mathbf{f}_{\parallel}$. Finally, we assume the charge on a particle is Lorentz invariant. If it all works out, then these assumptions will have been shown to be consistent. After that we can march in hope to an experimental laboratory, to determine whether the theory matches experimental observations.

The idea is to start with a reference frame S in which there are fields $\mathbf{E}$ and $\mathbf{B}$, then pick another frame S' moving at velocity $\mathbf{v}$ relative to the first. We can probe the field by putting a test particle of charge q

in it. We use the transformation of force to tell us what the force on this test particle is in the new frame, and we can use its dependence on velocity to unpick which part is electric and which part magnetic. Thus we deduce the fields $\mathbf{E}'$ and $\mathbf{B}'$ in the new frame.

It is convenient to treat the fields in S one at a time, and then add them together at the end. So, first suppose there is an electric field $\mathbf{E}$ but no magnetic field in S. Let our test particle have velocity $\mathbf{u}$ in S. The force on it, in frame S, is

$$\mathbf{f} = q\mathbf{E}.$$

In S' the force on the test particle is given by eqn (4.6):

$$\mathbf{f}'_\parallel = \frac{q(\mathbf{E}_\parallel - \mathbf{v}(\mathbf{E} \cdot \mathbf{u})/c^2)}{1 - \mathbf{u} \cdot \mathbf{v}/c^2}, \qquad \mathbf{f}'_\perp = \frac{q\mathbf{E}_\perp}{\gamma(1 - \mathbf{u} \cdot \mathbf{v}/c^2)}, \qquad (7.4)$$

where the subscripts $\parallel$ and $\perp$ refer to parallel and perpendicular to $\mathbf{v}$, the relative velocity of the reference frames, and γ is γ_v.

For any $\mathbf{v}$, we can pick $\mathbf{u} = \mathbf{v}$, then S' is the rest frame of the test particle, and we expect the force on it to be wholly electric in nature. Eqns (7.4) give for this case $\mathbf{f}'_\parallel = q\mathbf{E}_\parallel$ and $\mathbf{f}'_\perp = \gamma q\mathbf{E}_\perp$, so we deduce

$$\mathbf{E}'_\parallel = \mathbf{E}_\parallel, \qquad \mathbf{E}'_\perp = \gamma \mathbf{E}_\perp. \qquad (7.5)$$

This is the complete behaviour of the electric field when there is no magnetic field in the first frame.

Now to find the magnetic field in the second frame.

Do not forget that we can choose any velocity $\mathbf{u}$ we like for our probe: we are using it to explore the fields. So, next let us choose $\mathbf{u}$ parallel to $\mathbf{v}$ but not equal to it. Then eqns (7.4) give the same result as before for $\mathbf{f}'_\parallel$, and, after using eqn (4.8),

$$\mathbf{f}'_\perp = q\mathbf{E}_\perp \gamma(1 + \mathbf{u}' \cdot \mathbf{v}/c^2). \qquad (7.6)$$

When $\mathbf{u}$ is parallel to $\mathbf{v}$, so is $\mathbf{u}'$, so $\mathbf{u}' \wedge \mathbf{B}'$ is perpendicular to $\mathbf{v}$. Therefore the magnetic contribution to the Lorentz force in S' goes completely into $\mathbf{f}'_\perp$ (not $\mathbf{f}'_\parallel$), and we must have

$$\mathbf{f}'_\perp = q(\mathbf{E}'_\perp + \mathbf{u}' \wedge \mathbf{B}')$$

Substituting this into eqn (7.6) and using eqn (7.5), we have

$$\mathbf{u}' \wedge \mathbf{B}' = \mathbf{E}_\perp \gamma u'v/c^2. \qquad (7.7)$$

After thinking about the directions you should be able to see that the solution for $\mathbf{B}'$ is

$$\mathbf{B}'_\perp = \gamma \frac{-\mathbf{v} \wedge \mathbf{E}}{c^2}. \qquad (7.8)$$

We cannot learn anything about $\mathbf{B}'_\parallel$ from this case because we launched our probe in such a way that its velocity in S' is along $\mathbf{v}$, so even if there were a non-zero $\mathbf{B}'_\parallel$ it would not exert a force on the test particle. To get further information we would need to launch the probe in other directions. The general case (an arbitrary $\mathbf{u}$) involves a lot of rather unenlightening algebra, but it is not necessary to treat it. It suffices to examine the case $\mathbf{u}$ perpendicular to $\mathbf{v}$; this is not too hard and is left as an exercise for the reader. One finds that, when the field in S is purely electric, the magnetic field in S′ is perpendicular to $\mathbf{v}$: i.e., $\mathbf{B}'_\parallel = 0$ (and $\mathbf{B}'_\perp$ is still given by eqn (7.8), of course). This completes the calculation of the field transformation for this case.

Next let us perform a similar analysis for the case where the field in S is purely magnetic:

$$\mathbf{f} = q\mathbf{u} \wedge \mathbf{B}.$$

To keep things simple we shall assume from the outset that $\mathbf{u}$ is parallel to $\mathbf{v}$, so we have $\mathbf{f}'_\parallel = \mathbf{f}_\parallel = 0$ and

$$\mathbf{f}' = \mathbf{f}'_\perp = \frac{q\mathbf{u} \wedge \mathbf{B}}{\gamma(1 - uv/c^2)}. \tag{7.9}$$

As before, first take $\mathbf{u} = \mathbf{v}$ to get the electric field:

$$\mathbf{E}' = \gamma\mathbf{v} \wedge \mathbf{B}. \tag{7.10}$$

Then substitute this in to the Lorentz force equation in reference frame S′:

$$q(\gamma\mathbf{v} \wedge \mathbf{B} + \mathbf{u}' \wedge \mathbf{B}') = \frac{q\mathbf{u} \wedge \mathbf{B}}{\gamma(1 - uv/c^2)}. \tag{7.11}$$

After some algebra similar to that we employed for eqn (7.7) you can confirm that the solution for $\mathbf{B}'$ is

$$\mathbf{B}'_\perp = \gamma\mathbf{B}_\perp. \tag{7.12}$$

As before, we gain no knowledge of $\mathbf{B}'_\parallel$ from a test particle launched along it, but by considering other directions for the test particle it is not hard to prove that $\mathbf{B}'_\parallel = \mathbf{B}_\parallel$.

Finally we can gather all the results together, using linearity of the Lorentz transformation. The effect of a combination of both an electric and a magnetic field in S is to produce a sum of forces, and the Lorentz transformation results in a sum of corresponding terms in the new frame. Therefore the complete set of equations for the transformation of the electromagnetic field between one reference frame and another is

Transformation of electromagnetic field

$$\mathbf{E}'_\parallel = \mathbf{E}_\parallel$$

$$\mathbf{E}'_\perp = \gamma\left(\mathbf{E}_\perp + \mathbf{v} \wedge \mathbf{B}\right),$$

$$\mathbf{B}'_\parallel = \mathbf{B}_\parallel$$

$$\mathbf{B}'_\perp = \gamma\left(\mathbf{B}_\perp - \mathbf{v} \wedge \mathbf{E}/c^2\right). \tag{7.13}$$

As usual, the subscripts $\parallel$ and $\perp$ refer to the components parallel and perpendicular to the relative velocity $\mathbf{v}$ of the reference frames. By using $\mathbf{E}_\parallel = (\mathbf{E} \cdot \mathbf{v})\mathbf{v}/v^2$ and $\mathbf{E}_\perp = \mathbf{E} - \mathbf{E}_\parallel$ the results can also be written as

$$\mathbf{E}' = \gamma(\mathbf{E} + \mathbf{v} \wedge \mathbf{B}) - \frac{\gamma^2}{1+\gamma} \frac{(\mathbf{E} \cdot \mathbf{v})\mathbf{v}}{c^2}$$

$$\mathbf{B}' = \gamma(\mathbf{B} - \mathbf{v} \wedge \mathbf{E}/c^2) - \frac{\gamma^2}{1+\gamma} \frac{(\mathbf{B} \cdot \mathbf{v})\mathbf{v}}{c^2}$$

(and see exercise 13.4 of chapter 13 for yet another form).

The derivation of eqn (7.13) that we have given is perfectly correct and complete, and at the same time we have completed another task: namely, to prove that the Lorentz force equation is valid in all reference frames if it is valid in one (under the assumption that we are dealing with a pure force). We did not explicitly show the case of a test particle moving in an arbitrary direction, but any velocity can be written $\mathbf{u} = \mathbf{u}_\parallel + \mathbf{u}_\perp$ and the force is linear in $\mathbf{u}$ so can be obtained by summing the cases we did treat. The main lesson is that this method of proof is quite correct and complete, should you wish to check it further.

We shall present alternative derivations of eqn (7.13) using two different 4-vector methods at the end of this chapter, one of which is much more direct. We shall also exhibit the Lorentz force as part of a 4-vector 4-force. Indeed, at this point I must 'come clean'. A modern theoretical physicist reading the above treatment, eqns (7.4) to (7.12), would feel a certain impatience. 'Why is he offering this laborious 3-vector treatment?', an expert would surely ask, 'when it can be done so much more easily and elegantly using the proper language of Relativity, which is 4-vectors and tensors?' The answer is that I think you can learn something useful from looking at the 3-vectors and thinking it through. However, my hope is that by the time you finish this book you will be that impatient expert!

Generalizing from Coulomb's law

The above argument invoked the transformation of *force*. Suppose that to start with we only knew Coulomb's law for the force between static point charges. Then by invoking a change of reference frame we would immediately deduce that there must be further contributions to the force when charges are in motion—what we call the magnetic contribution. It is interesting that quite a lot of electromagnetism can thus be discovered by building on Coulomb's law and insisting on consistency with Special Relativity. One cannot discover the whole subject by this method, because it does not reveal the effect of accelerations, for example, but if we had to discover electromagnetism without the benefit of Ampère's law or Faraday's law, then Relativity would provide some strong pointers. It would show us, for example, that forces of the form $q\mathbf{v} \wedge \mathbf{B}$ must exist when charged particles are in motion, which would direct us to the magnetic field definition (7.2). A similar discussion of the force due to gravity suggests that it too should have a contribution that depends on motion of the source, as indeed it does, according to General Relativity.

7.2   Maxwell's equations

The theory of electromagnetism discovered by Faraday, Ampère, Maxwell and others is encapsulated by the Lorentz force equation (7.3) and the

Maxwell equations:

$$\boldsymbol{\nabla} \cdot \mathbf{E} = \frac{\rho}{\epsilon_0}$$

$$\boldsymbol{\nabla} \cdot \mathbf{B} = 0$$

$$\boldsymbol{\nabla} \wedge \mathbf{E} = -\frac{\partial \mathbf{B}}{\partial t}$$

$$c^2 \boldsymbol{\nabla} \wedge \mathbf{B} = \frac{\mathbf{j}}{\epsilon_0} + \frac{\partial \mathbf{E}}{\partial t}, \tag{7.14}$$

where ρ is the charge per unit volume in some region of space, $\mathbf{j}$ is the current density[1] and ϵ_0 is a fundamental constant called the permittivity of free space.

In relativity theory the issue immediately arises: are these equations satisfactory? Can we proceed and apply them in any reference frame we might choose, or do they include a hidden assumption that one reference frame is preferred above others?

The answer turns out to be that the equations are fine as they are: they do not prefer one reference frame to another. To prove this, we can consider a change of reference frame, and work out how Maxwell's equations are affected. We already know how the position and time coordinates will change, and we know how the charge density and current density will change (because together they form a 4-vector, eqn (6.10)), so we can work out how Maxwell's equations and the Lorentz equation will look in the new coordinate system. After a lot of algebra, the answer turns out to be

$$\boldsymbol{\nabla}' \cdot \mathbf{E}' = \frac{\rho'}{\epsilon_0}$$

$$\boldsymbol{\nabla}' \cdot \mathbf{B}' = 0$$

$$\boldsymbol{\nabla}' \wedge \mathbf{E}' = -\frac{\partial \mathbf{B}'}{\partial t'}$$

$$c^2 \boldsymbol{\nabla}' \wedge \mathbf{B}' = \frac{\mathbf{j}'}{\epsilon_0} + \frac{\partial \mathbf{E}'}{\partial t'}, \tag{7.15}$$

$$\mathbf{f}' = q\left(\mathbf{E}' + \mathbf{u}' \wedge \mathbf{B}'\right) \tag{7.16}$$

[1] In SI units the last equation is often written $\boldsymbol{\nabla} \wedge \mathbf{B} = \mu_0 \mathbf{j} + \epsilon_0 \mu_0 (\partial \mathbf{E}/\partial t)$, where μ_0 is another constant called the permeability of free space, defined by $\mu_0 \equiv 1/(\epsilon_0 c^2)$. In the SI system c, ϵ_0 and μ_0 all have exactly defined values; these are given in the inside cover.

where $\boldsymbol{\nabla}'\cdot$ and $\boldsymbol{\nabla}'\wedge$ are the div and curl operators in the primed coordinate system (i.e. $\boldsymbol{\nabla}' = (\partial/\partial x', \partial/\partial y', \partial/\partial z')$), and $\mathbf{E}', \mathbf{B}'$ are given by eqns (7.13).

Eqn (7.16) confirms that the symbols $\mathbf{E}'$ and $\mathbf{B}'$ refer to vector fields which fit the definition of electric and magnetic fields in reference frame S$'$. Eqns (7.15) then confirm that the Maxwell equations are the same in the second reference frame as they were in the first one. Therefore every physical phenomenon they describe will show no preference for one reference frame above another, so the Principle of Relativity is upheld. The set of equations is said to be *Lorentz covariant*. The word 'covariant' rather than 'invariant' is used for technical and historical reasons. One can think of it as expressing the idea that whereas all the bits and pieces in the equations ($\mathbf{E}, \mathbf{B}, \mathbf{j}, \rho, t, x, y, z$) do change from one reference frame to another, they all conspire together, or co-vary, in such a way that the form of the equations does not change.

The lengthy algebra we mentioned (but did not go into), to derive eqns (7.15) and (7.13), can be considered a 'brute force' method to show that Maxwell's equations are Lorentz covariant and to find out how the fields transform. One of the aims of this chapter is to introduce some powerful concepts and tools that will enable us to prove the former and to derive the latter in a slicker way. We will re-express the equations using 4-vectors and the $\square$ operator, so that their Lorentz covariance is obvious. This will make the result seem less like a 'conspiracy' and more like an elegant symmetry.

7.2.1   Moving capacitor plates

To get some insight into eqns (7.13) let us consider some simple cases. Consider, for example, a parallel plate capacitor, carrying charges $Q, -Q$ on two parallel plates of area A and separation d, at rest in reference frame S (see Fig. 7.1). The electric field between the plates of such a capacitor is uniform, directed perpendicular to the plates, and of size $E = Q/\epsilon_0 A$.

Now consider a reference frame S$'$ moving parallel to $\mathbf{E}$. The charges on the plates are invariant, the area is unchanged since it is transverse to the motion, while the plate separation is Lorentz-contracted to $d' = d/\gamma$. However, the electric field is independent of d'. One finds, therefore, $E' = E$, in agreement with eqn (7.13i).

Next suppose that instead of moving parallel to $\mathbf{E}$, S$'$ moves relative to S in a direction perpendicular to $\mathbf{E}$ (i.e., it moves parallel to the plates). Now $d' = d$ but the Lorentz contraction leads to $A' = A/\gamma$, therefore the charge per unit area on the plates is larger in S$'$, and we have $E' = \gamma E$, in agreement with eqn (7.13ii).

In fact, this simple argument from the capacitor plates is sufficient to prove (7.13i) and (7.13ii) in general when the relative velocity is either parallel to or perpendicular to $\mathbf{E}$, and there is no magnetic field in the first (unprimed) reference frame. This is because the field at a given

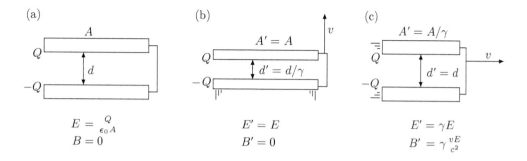

(a)

$$E = \frac{Q}{\epsilon_0 A}$$
$$B = 0$$

(b)

$$E' = E$$
$$B' = 0$$

(c)

$$E' = \gamma E$$
$$B' = \gamma \frac{vE}{c^2}$$

Fig. 7.1 The moving capacitor. (a) The situation in frame S where the capacitor is at rest. (b) The situation in a frame moving along the field direction, normal to the plates. (c) The situation in a frame moving perpendicular to the field direction, parallel to the surface of the plates. The charges are invariant; the capacitor dimensions change as shown. (These simple cases are easily remembered and cover much of what one needs to know about field transformation.)

point must transform in the same way, independent of what charges or movement of charge gave rise to it.

The capacitor example also illustrates the second term in eqn (7.13iv). A flat sheet of charge moving parallel to its own plane represents a sheet of current (Fig. 7.2). It gives rise to a magnetic field above and below it, in a direction parallel to the sheet and perpendicular to the current, of size $\mu_0 I / 2w$ where I is the current flowing through a width w of the sheet (this is easily proved from Ampère's Law or by integrating the field due to a wire). Applying this result to the case of a capacitor, we have two oppositely charged sheets moving at speed v in reference frame S'. For $\mathbf{v}$ perpendicular to $\mathbf{E}$ the magnetic fields of the two sheets add (in the region in between the capacitor plates) to give $B' = \mu_0 I'/w'$, where $I' = Qv/L'$ and L', w' are the dimensions of the plates in S'. Using $w'L' = A' = A/\gamma$ owing to Lorentz contraction of L, we have

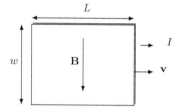

Fig. 7.2

$$B' = \frac{\mu_0 Qv}{A'} = \frac{\gamma Qv}{c^2 \epsilon_0 A} = \gamma \frac{vE}{c^2}$$

in agreement with eqn (7.13iv).

Charge from nowhere?

Similar arguments can be made concerning the transformation of magnetic fields, but one needs to be more careful because there are more movements of charge to keep track of. Consider the following, which seems paradoxical at first. An ordinary current-carrying wire is electrically neutral, but has a current I in it. Therefore the 4-vector current density is $\mathbf{J} = (\rho c, \mathbf{j}) = (0, I/A)$, where A is the cross-sectional area of the wire. Now adopt some other reference frame, moving parallel to the wire, which we shall take to define the x axis. The Lorentz transformation gives the charge density in the new reference frame: it is $\rho' = \gamma(\rho - vj/c^2) = -\gamma vI/(Ac^2)$. This charge density is non-zero! So where did the charge come from? It was not there in the first reference frame; now it has 'magically' appeared.

Before we resolve this, consider another puzzle. A stationary electron in the vicinity of the wire, say 1 metre from it, experiences no force in the first reference frame, since its velocity is zero and a neutral wire does not produce an electric field. Therefore it does not accelerate. But now consider a reference frame moving at the drift velocity v of the electrons in the wire. This drift velocity is small. It is related to the current by $I = Anqv$, where n is the number density of electrons in the wire and q is the charge of an electron. For a typical metal such as copper, $n \simeq 8 \times 10^{28}$ m^{-3}, so for a 10-amp current in a wire of diameter 1 mm, we find $v \simeq 1$ mm/s. In the new reference frame the electron flow is zero, but now all the other parts of the wire (the nuclei and bound electrons) are in motion. They carry a net positive charge, so their motion constitutes a current $I' = \gamma I$, where the γ comes from the Lorentz transformation of J. (We could neglect γ here because it is extremely close to 1, but let us keep it anyway.) In the new frame, therefore, there is a magnetic field around the wire $B' = \mu_0 \gamma I /(2\pi r)$: this is an example of eqn (7.13iv). Now, the interesting part is that in the new reference frame, the electron situated near the wire is in motion, so it experiences a magnetic force! The force is

$$f' = qvB' = \frac{qv\mu_0\gamma I}{2\pi r}. \tag{7.17}$$

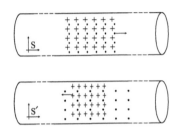

Fig. 7.3 A neutral current-carrying wire consists of positive and negative charges of equal number density in frame S (upper diagram). The negative charges are shown as dots; the arrow indicates their drift velocity. In frame S′ moving at the drift velocity the current is caused by the positive charges moving to the left. Compared with frame S, the lattice of positive charges suffers a Lorentz contraction, while the opposite happens to the negative charges (since in S they were moving and in S′ they are not). Therefore in S′ the wire is not neutral: it carries a net positive charge density. Charge is still conserved (count the dots and crosses!); the extra density has come at the expense of the charge distribution elsewhere, where the current flow must be in the opposite direction to complete the electrical circuit.

We find the B field is about 2 micro-tesla, and the force is $f' \simeq 3 \times 10^{-28}$ newton, leading to an acceleration approximately 350 ms^{-2} away from the wire. So, according to this argument, the wire will very quickly accelerate electrons in a large volume around it ... whereas in the first reference frame we found no such acceleration.

These two paradoxes are, of course, related. The non-zero charge density in the new reference frame is correct. It creates an electric field in the second frame and thus a further contribution to the force on any particle near the wire: this exactly balances the magnetic force we have just calculated.

Fig. 7.3 explains what is going on. An object that is overall electrically neutral but which carries a current must have two sets of charged particles in it: one positive and one negative. The overall neutrality, in a given reference frame S, means these sets have equal densities, $n_+ = n_- = n$, in S. The non-zero current means that one set of particles is moving and the other is not, or else they both move with different velocities. When we change to another reference frame, the Lorentz contraction is by a different amount for one set of particles than for the other, because of their different velocities. Indeed, in going from the frame where the copper nuclei are at rest to the frame where the conduction electrons are at rest, the nuclei get closer together while the conduction electrons *spread out* because we are transforming to their rest frame. So $n'_+ = \gamma n_+$ and $n'_- = n_-/\gamma$. The charge density in S′ is then

$$\rho' = q_+ \left(\gamma n_+ - \frac{n_-}{\gamma} \right) = \gamma n q_+ \left(1 - \frac{1}{\gamma^2} \right) = \gamma n q_+ v^2/c^2 = -\gamma j v/c^2$$

where we used $j = nqv = n(-q_+)v$. j is the current density in S, and q_+ is the charge on a proton. (Here j and v are in opposite directions so ρ' is positive.) This result agrees with the one we obtained by transforming J.

To complete the analysis let us check the electric field produced by this non-zero charge density. We have a line of charge, with charge per unit length $\lambda' = \rho' A$. The electric field at distance r from such a line charge is

$$E' = \frac{\lambda'}{2\pi\epsilon_0 r} = \frac{\gamma n q_+ v^2 A}{2\pi\epsilon_0 c^2 r} = \frac{v\mu_0 \gamma I}{2\pi r}.$$

Compare this with eqn (7.17). You can see that the electric and magnetic forces in S' are everywhere balanced.

Such a perfect balance of forces that, if they were not balanced, would have substantial effects, should arouse our suspicion. It looks like a conspiracy, but we do not like conspiracies in Nature. We think they are a sign that we do not have the right perspective on something. In this case the answer is that the two forces are not two but one: we must regard the electric and magnetic parts as two parts of one thing. If the 'one force' is zero, then we have only ourselves to 'blame' for supposedly 'marvellous' effects if we start interpreting it as two forces. Of course, we will find that they are balanced.

The strength of materials

Let us examine another issue nicely illustrated by the parallel-plate capacitor. In section 4.1.1 we noted that a moving body loses its strength in the direction transverse to its motion. Now, most ordinary bodies are made of atoms, and the forces inside them, when they are stretched or compressed away from their natural length, are almost entirely electromagnetic in origin: a complicated combination of the electrostatic attractions between the unlike charges (nuclei and electrons), repulsions between the like charges, and the magnetic forces. It requires a quantum-mechanical treatment to treat materials correctly, but to get a simple insight, suppose we argue that an attempt to break an ordinary object by pulling on it is somewhat like pulling apart a pair of capacitor plates. You should not treat this simple idea as anything like a quantitative model of the structure of materials, but it does illustrate the kind of thing that happens to electromagnetic forces inside an object when it is set in motion.

For a stationary capacitor, the force on any given charge q in one of the plates is equal to q times the electric field due to the *other* plate (you can soon convince yourself that the forces from other charges within the same plate will cancel to very good approximation near the middle of a large enough plate). Therefore the force on such a charge is

$$f = qE_1 = \frac{qQ}{2\epsilon_0 A},$$

where E_1 is the field due to the charges on one plate (this is half the total field between the plates). Now consider a reference frame in which the capacitor is moving in a direction parallel to the plates: i.e., perpendicular to **E**. According to eqn (7.13) the electric field between the plates is now *larger*, but according to eqn (4.6ii) the force on the particle we picked is now *smaller*. What is going on?

In the new reference frame there is a magnetic as well as an electric contribution to the force. The magnetic field due to either one of the plates on its own is

$$B_1' = \frac{\mu_0 I'}{2w'} = \frac{\gamma v E_1}{c^2}$$

and the charged particle now has speed v, in a direction perpendicular to $\mathbf{B}_1'$. The magnetic force in this example has a direction opposite to the electric force. It follows that the total force on the particle in the new reference frame is

$$
\begin{aligned}
f' &= q(E_1' - vB_1') = qE_1'(1 - v^2/c^2) & (7.18) \\
&= q\frac{\gamma E_1}{\gamma^2} = \frac{qE_1}{\gamma}. & (7.19)
\end{aligned}
$$

Thus the argument from Maxwell's equations does agree with the prediction from the Lorentz transformation of forces: physical objects become weaker in the transverse direction when they are in motion (see the box above, however, for a comment on all this).

At speeds small compared to c, the magnetic contribution to the force is very much smaller than the electric contribution. Some people, on observing the factor v^2/c^2 in eqn (7.18), like to say that it is as if magnetic effects are a 'relativistic correction' to electric effects. When we put a current in a wire, and observe the magnetic field through its effect on a nearby compass needle, for example, one might say that we are observing at first hand the influence of a tiny relativistic correction! In practice, magnetic effects can very often be traced to a moving electric charge.[2] Since no magnetic monopoles have ever been discovered, and since motion is relative while charge is not, one may well feel that the electric field is the 'senior partner'. I would prefer to say that magnetic and electric fields are two parts of a single thing, as I already mentioned, but it is good to be aware of the relative sizes of the effects. In the case of a current-carrying wire, the electrostatic effects have been cancelled extremely well by the presence of equal amounts of positive and negative charge in the wire, to a precision of order $v^2/c^2 \simeq 10^{-23}$, which allows us to see the tiny magnetic contribution.

At speeds approaching c, on the other hand, the electric and magnetic contributions have similar sizes.

[2] ...but not always: it is found that magnetic dipoles are associated with the intrinsic spin angular momentum of charged particles; this spin cannot be associated with a movement of matter.

Ask a silly question ... 'Who cares about the 3-force? It is just part of a 4-vector, and it is not really fundamental: it is a way of keeping track of momentum changes. If the spatial part of a 4-vector changes in some way, it is simply a hang-over from pre-spacetime thinking to agonise about this. We need to think in terms of the whole 4-vector, including the temporal part. The 4-vector F is what it is, independent of reference frame.'

Answer. I agree with this position, up to a point. It is true that spacetime physics should be discussed with the right language: i.e., 4-vectors. However, in the application to physical examples we have to pick a reference frame. The fact that at high speeds the electric and magnetic contributions tend to cancel for transverse forces is memorable, and worth noticing. Also, we find that to treat the motion of particles subject to forces, the 3-force can sometimes provide the most direct route to the result.

7.3   The fields due to a moving point charge

A point charge at rest produces an electric field in the radial direction and no magnetic field. By using the field transformation equations, it is straightforward to find the fields produced by a point charge in uniform motion.

Place a point charge Q is at the origin of frame S', in standard configuration with S. The fields in S' are then

$$\mathbf{E}' = \frac{Q}{4\pi\epsilon_0 r'^3} \begin{pmatrix} x' \\ y' \\ z' \end{pmatrix}, \qquad \mathbf{B}' = 0$$

where we wrote the components of the vector $\mathbf{r}' = (x', y', z')$ explicitly. Consider the event at $\{ct', x', y', z'\}$. The fields at this same event, but evaluated in frame S, are, by using eqn (7.13),

$$E_x = E'_{x'} = \frac{Q}{4\pi\epsilon_0} \frac{x'}{r'^3}$$

$$E_y = \gamma E'_{y'} = \frac{Q}{4\pi\epsilon_0} \frac{\gamma y'}{r'^3}$$

$$E_z = \gamma E'_{z'} = \frac{Q}{4\pi\epsilon_0} \frac{\gamma z'}{r'^3}.$$

This is the complete and correct expression for the electric field in S, but we would prefer to have it in terms of the position and time coordinates of S. To this end, we use the Lorentz transformation for the coordinates, which gives

$$x' = \gamma(x - vt), \qquad y' = y, \qquad z' = z.$$

The field at any given place in S is time-dependent. The most useful way to understand the result is to pick the moment $t = 0$, because at this moment the moving charge is at the origin of S. At any time, the coordinates (x, y, z) give the vector from the origin, but at $t = 0$ this is also a vector from the point charge to the place where we are calculating the field. At this moment we have $x' = \gamma x$ and we obtain the result:

Electric field of point charge moving with constant velocity

$$\mathbf{E} = \frac{\gamma Q \mathbf{r}}{4\pi\epsilon_0 (\gamma^2 x^2 + y^2 + z^2)^{3/2}}. \tag{7.20}$$

where the charge is at the origin and moving in the x-direction, and $\mathbf{r}$ is the vector (x, y, z).

Using eqns (7.13) we obtain for the magnetic field,

$$\left. \begin{array}{l} \mathbf{B}_\parallel = 0 \\ \mathbf{B}_\perp = \gamma \frac{\mathbf{v} \wedge \mathbf{E}'}{c^2} \end{array} \right\} \qquad \Rightarrow \mathbf{B} = \frac{\mathbf{v} \wedge \mathbf{E}}{c^2} \tag{7.21}$$

(the final expression for $\mathbf{B}$ correctly matches both $\mathbf{B}_\parallel$ and $\mathbf{B}_\perp$ because the cross product only involves $\mathbf{E}'_\perp$). In the limit of low velocities, eqns (7.20) and (7.21) lead to the Biot–Savart law.

Transverse and longitudinal directions

Eqn (7.20) states that compared to a charge at rest, the field is reduced in the longitudinal direction and enhanced in the transverse direction; see Fig. 7.4. In view of the fact that Gauss's law holds in any frame, and the charge Q is a Lorentz invariant, the flux of $\mathbf{E}$ out through a closed surface around the source particle must be the same in the two frames. Therefore we must expect that an enhancement in one direction must come at the expense of a reduction in another.

It is useful to ask whether we can understand the result for these cases by simple arguments.

For points on the line of motion of the charge—i.e., directly in front or behind—the field is parallel to the motion, so naively one might expect, from the equation $\mathbf{E}'_\parallel = \mathbf{E}_\parallel$, that the field at such points is the same for the moving charge as for a stationary one. However, the distance of the

Fig. 7.4 Electric field lines due to a stationary charge (left) and a moving charge (right). The lines are along the field direction; their density (per unit area in three dimensions) represents the field strength. A remarkable property is that the right diagram (moving charge) could be obtained by applying a Lorentz contraction to the left diagram (stationary charge).

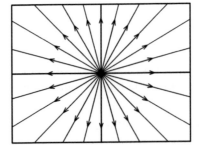
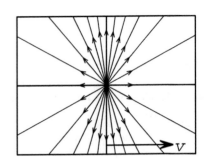

point from the charge is contracted by the Lorentz factor γ. Therefore we have the same field at a closer distance, hence at any given distance the field is smaller, and from the inverse-square law in the rest frame it must be smaller by a factor γ^2, in agreement with eqn (7.20). To be clear that we have accounted correctly for Lorentz contraction in this argument, imagine that the charge is situated at one end of a rigid rod, and let the rod lie along the x axis. Owing to Lorentz contraction of this rod, any *given event* at the far end of the rod is further from the charge in the rest frame S′ than in the lab frame S; cf. Fig. 7.6.

Now consider a point in the transverse direction, as in Fig. 7.5(b). The relative motion of the frames is perpendicular to the field, so we expect an enhanced field, $\mathbf{E} = \gamma \mathbf{E}'$. The distance being a transverse distance, it is uncontracted. However, this is not completely obvious, because the source is moving, and we need to keep in mind the relativity of simultaneity. To see it, imagine again a rigid rod attached to the particle. For example, suppose the arrow used to indicate the vector $\mathbf{r}$ in Fig. 7.5b were a solid material object attached to the source particle Q. This arrow will be found to have the same length in either frame, because there is no contraction in a transverse direction, and the event when q reaches the end of the arrow is one and the same event no matter what frame is adopted. The conclusion is, then, that at any given distance the field at points in the transverse direction is enhanced.

Both these results follow equally from a consideration of force per unit charge. For a given situation, the force in the transverse direction is higher in the frame where the particle *on which the force acts* is at rest, says eqn (4.6). A test particle at rest in S experiences only the electric, not magnetic, contribution to the force in either frame (in S because it is not moving, in S′ because there is no magnetic field).

Discussion of the result

Eqn (7.20) has some remarkable properties. For one thing, it says that the electric field due to a moving source particle is in a direction radially outward from the particle; see Fig. 7.4. This seems sensible at first, but on reflection one realizes that the field has no business pointing outwards from the *present* location of the particle! The field at x, y, z at time $t = 0$ can only 'know about' or be caused by what the source particle was doing earlier on, in the past light-cone. If one had to guess, one might guess that the field at any event t, x, y, z would point in the direction away from the source's *earlier* position, not from where it is now. But instead the field seems to 'know' where the moving source is *now*. Of course, we are discussing a *uniformly* moving source, so the information on where the source is going to be is contained in its past history, assuming the uniform motion continues. That the result should turn out so simple is, however, important. If the field were not radial from the present position, then a system of two particles moving uniformly abreast would exert a non-zero net total force on itself, leading to a self-acceleration in the absence of external forces. This would violate momentum conservation.

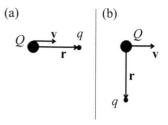

Fig. 7.5 Two simple cases of forces between point charges. (a) Source particle moves directly towards the test particle, $\mathbf{r}$ and $\mathbf{v}$ are parallel. (b) Source particle moves past a test particle, we consider the force on the latter at the event when $\mathbf{r}$ and $\mathbf{v}$ are perpendicular.

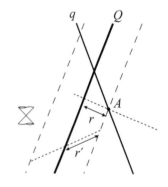

Fig. 7.6 Spacetime diagram for the situation (a) in Fig. 7.5. The full lines show the worldlines of the source Q and test particle q. The dashed lines show a set of events at given distance r' from Q in its rest frame S′. The dotted lines are lines of simultaneity; the one through A is a line of simultaneity for q's rest frame S. The distance from Q to q at event A is smaller in frame S than in frame S′.

Fig. 7.7 'B of the Bang': a sculpture designed by Thomas Heatherwick, and erected in Manchester, England. The sculpture draws its inspiration from the explosive start of a sprint race at the 'bang' of the starting pistol; but to a physicist it is also reminiscent of the electric field due to a fast-moving charged particle—perhaps a muon arriving in Manchester from a cosmic-ray event. (Photograph by Nick Smale.)

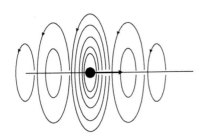

Fig. 7.8 Magnetic field due to a uniformly moving point charge. The field lines loop around the line of motion of the charge. There is no magnetic field directly in front of or behind the charge.

The equations succeed in avoiding that situation. It is as if the source gives its 'marching orders' to the field in the form 'line yourself up on my future position, assuming that I will continue at constant velocity'. We shall re-examine this point in section 8.2.3.

Eqn (7.21) says that the magnetic field has a similar forward-back symmetry. It loops around the direction of motion of the charge, with a maximum strength at positions to the side, falling to zero in front and behind (Fig. 7.8).

We have already noticed that the electric field is diminished in front and behind the moving particle, and enhanced at the sides. The next remarkable feature is that the size of these changes is just as if the field lines of a stationary particle had been 'squeezed' by a Lorentz contraction. This is not self-evident, but is suggested by Fig. 7.4. The field lines from a point source transform like rigid spikes attached to the source. You should not deduce that this is a universal feature of electric field lines: just add a magnetic field in the first reference frame and this behaviour is lost. However, the picture does give a good insight into the way the Lorentz contraction of moving objects is brought about and embodied by the fields inside them.

In the 'relativistic limit'—i.e., as the speed approaches c—a charged particle such as an electron appears like a stealthy pancake with a mighty force field around it. There is little sign of its approach, but as it whizzes by it exerts, for a moment, a powerful lateral force, like a shock wave. However, because this force appears in a short burst, the net impulse delivered is not enhanced, but varies in proportion to $1/v$ (exercise 7.4).

7.4 Covariance of Maxwell's equations

We have already stated that Maxwell's equations are *Lorentz covariant*: they take the same form in one reference frame as they do in another. However, when written down in the standard way, eqns (7.14), this covariance is far from obvious. Now we shall develop some concepts that allow the covariance to be easily seen.

Any textbook on electromagnetism will tell you that the electric and magnetic fields can be obtained from two potentials ϕ and $\mathbf{A}$, called the scalar and vector potential, through

$$
\begin{aligned}
\mathbf{E} &= -\boldsymbol{\nabla}\phi - \frac{\partial \mathbf{A}}{\partial t} \\
\mathbf{B} &= \boldsymbol{\nabla} \wedge \mathbf{A}.
\end{aligned}
\tag{7.22}
$$

It is not hard to see where this idea comes from. If you look at M2 (the second Maxwell equation, (7.14ii)) you will see that $\mathbf{B}$ has zero divergence. This implies that $\mathbf{B}$ can be written as the curl of something, so we write it that way and call the 'something' a 'vector potential' $\mathbf{A}$. You should also see that another vector $\tilde{\mathbf{A}} = \mathbf{A} + \boldsymbol{\nabla}\chi$—for any scalar field χ—would be just as good, because it has the same curl: more on that in a moment. Next turn to Faraday's law M3. Now it looks like

$$
\boldsymbol{\nabla} \wedge \mathbf{E} = -\frac{\partial}{\partial t}\boldsymbol{\nabla} \wedge \mathbf{A}.
$$

The order of differentiation with respect to time and space can be reversed, so this can be written

$$
\boldsymbol{\nabla} \wedge \left(\mathbf{E} + \frac{\partial \mathbf{A}}{\partial t} \right) = 0.
$$

The combination in the bracket has zero curl, therefore it can be written as the gradient of something. We write the something $-\phi$ with ϕ called the 'scalar potential' (the minus sign comes in for convenience: it means that this potential behaves like a potential energy per unit charge in electrostatics).

By using the potentials $\mathbf{A}$ and ϕ, and eqns (7.22) we guarantee that no matter what functional form we put into $\mathbf{A}$ and ϕ, two of the Maxwell equations will be automatically satisfied! Our work is reduced because now we only have to find four potential functions (ϕ and the three components of $\mathbf{A}$) instead of six field components.

When looking for solutions for $\mathbf{A}$ and ϕ it proves to be very useful to keep in mind that we have some flexibility, as we already noted. We can add to $\mathbf{A}$ any field with zero curl, without in the least affecting the $\mathbf{B}$ field that is obtained from it, eqn (7.22ii). However, since $\mathbf{A}$ influences $\mathbf{E}$ as well we need to check what goes on there. You can easily confirm that we can keep the flexibility if both potentials are changed together, as

$$
\tilde{\mathbf{A}} = \mathbf{A} + \boldsymbol{\nabla}\chi, \quad \tilde{\phi} = \phi - \frac{\partial \chi}{\partial t}
\tag{7.23}
$$

where χ is an arbitrary function. If the potentials are changed in this way, the derived fields are not changed at all. This is no more mysterious than the well-known fact that the gradient of a function does not change if you add a constant to the function; it is just that in three dimensions the possibilities are more rich. The change from $\mathbf{A}, \phi$ to $\tilde{\mathbf{A}}, \tilde{\phi}$ given in eqn (7.23) goes by the fancy name of a 'gauge transformation'. We say that the electric and magnetic fields are 'invariant under gauge transformations'. A simple example is to shift the scalar potential by a constant: $\tilde{\phi} = \phi + V_0$. This is a gauge transformation with $\chi = -V_0 t$.

Now, anyone studying Relativity who comes across a vector paired with a scalar, and who sees eqn (7.23), begins to suspect that we have a 4-vector in play. Let us see if it works. We form the '4-vector potential'

$$\mathsf{A} \equiv (\phi/c, \, \mathbf{A}) \qquad (7.24)$$

and note that the gauge transformation equation (7.23) can be written

$$\tilde{\mathsf{A}} = \mathsf{A} + \Box\chi. \qquad (7.25)$$

We have not yet proved that A is a four-vector, but the fact that we can write the gauge transformation in four-vector notation is promising.

Next we shall plug the forms (7.22) into Maxwell's equations M1 and M4 (eqns 7.14i and iv). One obtains

$$-\nabla^2\phi - \frac{\partial}{\partial t}\nabla \cdot \mathbf{A} \;=\; \frac{\rho}{\epsilon_0}, \qquad (7.26)$$

$$c^2\nabla(\nabla \cdot \mathbf{A}) + \frac{\partial}{\partial t}\nabla\phi + \frac{\partial^2 \mathbf{A}}{\partial t^2} - c^2\nabla^2\mathbf{A} \;=\; \frac{\mathbf{j}}{\epsilon_0}. \qquad (7.27)$$

As things stand this does not look very simple! However, the second equation is suggestive. The last two terms look like $-c^2\Box^2$ acting on $\mathbf{A}$ (recall that the d'Alembertian $\Box^2$ was defined in eqn (6.22)). The trouble is that we also have the first two terms, which together form the 3-gradient of $(c^2\nabla \cdot \mathbf{A} + \partial\phi/\partial t)$. Now we take a clever step. We are going to take advantage of the idea of gauge transformation. We recall that we have some flexibility in picking the potential functions, and we propose that by taking advantage of this flexibility it is always possible to arrange that

$$\nabla \cdot \mathbf{A} + \frac{1}{c^2}\frac{\partial\phi}{\partial t} = 0. \qquad [\text{ Lorenz gauge} \qquad (7.28)$$

When we impose this condition, the first two terms in eqn (7.27) cancel and the equation reduces to the simple form

$$\Box^2\mathbf{A} = \frac{-\mathbf{j}}{c^2\epsilon_0}. \qquad (7.29)$$

You can also confirm that eqn (7.26) becomes

$$\Box^2\phi = \frac{-\rho}{\epsilon_0}. \qquad (7.30)$$

Eqn (7.28) is called the *Lorenz gauge condition*,[3] and imposing it is called 'choosing the Lorenz gauge'. One needs to be aware that once such a gauge choice has been made, results based on it no longer have the full flexibility offered by eqns (7.23). However, that is merely a statement about the potentials. The fields that are obtained through any given choice of gauge are completely valid and 'care nothing' about how they were calculated.

Before commenting on the beautifully simple eqns (7.29) and (7.30) we need to check that it is always possible to impose the Lorenz gauge condition. To this end, first suppose we have some arbitrary $\mathbf{A}$ and ϕ not necessarily in the Lorenz gauge. They have

$$\nabla \cdot \mathbf{A} + \frac{1}{c^2}\frac{\partial \phi}{\partial t} = f(\mathbf{r}, t)$$

for some function f. Let us try a gauge transformation and see what happens:

$$\nabla \cdot \tilde{\mathbf{A}} + \frac{1}{c^2}\frac{\partial \tilde{\phi}}{\partial t} = f(\mathbf{r}, t) + \nabla^2\chi - \frac{1}{c^2}\frac{\partial^2 \chi}{\partial t^2}.$$

If follows that we can achieve the Lorenz condition as long as χ can be chosen such that it satisfies the equation

$$\frac{1}{c^2}\frac{\partial^2 \chi}{\partial t^2} - \nabla^2\chi = f.$$

This is a wave equation with f as source. The important point is that it is known that there always exist solutions to this equation, no matter what form the source function f takes. The method of solution is explained in section 8.2.2. If follows that we can always adjust the potentials so that they satisfy the Lorenz gauge condition.

Eqns (7.29) and (7.30) are important because they are uncoupled (you can solve them for ϕ on its own, and then for $\mathbf{A}$ on its own) and because they are both wave equations with a source term, for which powerful methods of solution exist. Furthermore, they open the way to writing down Maxwell's equations in a 4-vector notation that makes their Lorentz covariance explicit and obvious.

We have already learned in chapter 6 that for the flow of a quantity such as electric charge, the combination $(\rho c, \mathbf{j})$ is a 4-vector. We can write all the formulae leading up to eqns (7.29) and (7.30) in 4-vector notation. We have

$$\mathsf{J} = (\rho c, \mathbf{j}), \quad \mathsf{A} = (\phi/c, \mathbf{A}).$$

The Lorenz gauge condition is $\Box \cdot \mathsf{A} = 0$, and the final result is

Maxwell's equations

$$\Box^2 \mathsf{A} = \frac{-1}{c^2\epsilon_0}\mathsf{J}, \qquad \text{with } \Box \cdot \mathsf{A} = 0. \tag{7.31}$$

This equation does two jobs at once. First it shows that A is indeed a 4-vector as we suspected (because we already know that J is a 4-vector, c^2 and ϵ_0 are constants, and $\Box^2$ is a Lorentz scalar operator). Secondly, it expresses *all* of Maxwell's equations in one go, in explicitly Lorentz covariant form! I say 'all' because we have already noted that two of the equations were already taken care of when adopting the potentials, so there are only two left to worry about. The point is that we can see immediately that a change of reference frame will give the equation $\Box'^2 A' = -J'/(c^2 \epsilon_0)$: i.e., the same equation with primed symbols, and therefore, by reversing the argument, we would obtain Maxwell equations in their 3-vector form just as we claimed in eqns (7.15).

Coulomb gauge

We chose the Lorenz gauge above because it leads to a simple statement of Maxwell's equations. For some calculations, another choice of gauge (i.e., choice of constraint to impose on A) can be more convenient. There is an infinite variety of constraints one could choose. One that has proved sufficiently useful to earn a name is the **Coulomb gauge**, also called *radiation gauge*, where the constraint is

$$\mathbf{\nabla} \cdot \mathbf{A} = 0, \qquad [\text{Coulomb gauge} \qquad (7.32)$$

i.e., the divergence of the 3-vector potential is zero. Note that this is a three-vector equation. Therefore, if the potentials are in Coulomb gauge in one inertial frame, they are not guaranteed to be in Coulomb gauge in all inertial frames. This does not make the calculations invalid: the fields are obtained correctly, no matter what gauge is adopted.

If the scalar potential is independent of time then the potentials can satisfy both Lorenz and Coulomb gauge conditions.

The proof that it is always possible to find a gauge transformation so as to satisfy the Coulomb gauge condition is left as an exercise for the reader. In the Coulomb gauge, the first Maxwell equation (7.26) becomes Poisson's equation

$$\nabla^2 \phi = -\rho/\epsilon_0.$$

This is the same equation as one would obtain in electrostatics, but now we are treating general situations! If ρ changes with time, the influence on ϕ happens instantaneously in the Coulomb gauge. However, the influence on the *fields* is not instantaneous: once the contribution of both the scalar and the vector potential is taken into account, one gets the same result as one would in any other gauge: i.e., light-speed-limited cause and effect.

7.4.1   Transformation of the fields: 4-vector method*

Now that we have established that A is a four-vector, we know how it transforms for a change of reference frame: $A' = \Lambda A$. Hence for two reference frames in standard configuration,

$$\phi'/c = \gamma(\phi/c - \beta A_x)$$
$$A'_x = \gamma(-\beta\phi/c + A_x)$$
$$A'_y = A_y$$
$$A'_z = A_z. \tag{7.33}$$

One can now plug these into eqn (7.22) and thus find the fields in the primed frame in terms of the potentials in the unprimed frame. With a little care one can then derive eqn (7.13). It is not particularly quick, but at least it is thorough and automatic. The only way to get a full appreciation of the virtues and pitfalls of this method is to try it oneself, but to help you I shall show how it works for $\mathbf{E}'$.

To reduce clutter we will drop c from all the equations, and then put it back in at the end using dimensional analysis.

First we would like to find $\mathbf{E}'_\| = E'_x$, given by

$$E'_x = -\frac{\partial\phi'}{\partial x'} - \frac{\partial A'_x}{\partial t'}. \tag{7.34}$$

We have ϕ' and A'_x in terms of ϕ and A_x, eqn (7.33), but the problem is that the derivatives in eqn (7.34) are with respect to x' and t', not x and t as we would like (since we are trying to relate the field to E_x). We shall have to make use of the standard result for partial derivatives,

$$d\phi' = \frac{\partial\phi'}{\partial t}dt + \frac{\partial\phi'}{\partial x}dx + \frac{\partial\phi'}{\partial y}dy + \frac{\partial\phi'}{\partial z}dz$$

$$\Rightarrow \quad \frac{\partial\phi'}{\partial x'} = \frac{\partial\phi'}{\partial t}\frac{\partial t}{\partial x'} + \frac{\partial\phi'}{\partial x}\frac{\partial x}{\partial x'} + \frac{\partial\phi'}{\partial y}\frac{\partial y}{\partial x'} + \frac{\partial\phi'}{\partial z}\frac{\partial z}{\partial x'}.$$

Note that this is nothing especially to do with 4-vectors or the $\square$ operator; it is just what happens when you express a change in a function in terms of its changes with respect to different sets of coordinate variables. We can find out what $\partial t/\partial x'$ etc. are by using the inverse Lorentz transformation of coordinates, $t = \gamma(t' + vx')$ and $x = \gamma(vt' + x')$, $y = y'$, $z = z'$, so

$$\left(\frac{\partial t}{\partial t'}\right)_{x',y',z'} = \gamma, \quad \left(\frac{\partial t}{\partial x'}\right)_{t',y',z'} = \gamma v,$$

$$\left(\frac{\partial x}{\partial t'}\right)_{x',y',z'} = \gamma v, \quad \left(\frac{\partial x}{\partial x'}\right)_{t',y',z'} = \gamma,$$

$$\left(\frac{\partial y}{\partial y'}\right)_{t',x',z'} = 1,$$

$$\left(\frac{\partial z}{\partial z'}\right)_{t',x',y'} = 1,$$

and all the others are zero (all these partial derivatives are simply the elements of the matrix Λ^{-1}). Using these results along with the potential transformation equation (7.33), we find

$$\frac{\partial \phi'}{\partial x'} = \gamma \left(v \frac{\partial \phi'}{\partial t} + \frac{\partial \phi'}{\partial x} \right)$$

$$= \gamma^2 \left(v \frac{\partial \phi}{\partial t} - v^2 \frac{\partial A_x}{\partial t} + \frac{\partial \phi}{\partial x} - v \frac{\partial A_x}{\partial x} \right), \qquad (7.35)$$

$$\frac{\partial A'_x}{\partial t'} = \gamma \left(\frac{\partial A'_x}{\partial t} + v \frac{\partial A'_x}{\partial x} \right)$$

$$= \gamma^2 \left(-v \frac{\partial \phi}{\partial t} + \frac{\partial A_x}{\partial t} - v^2 \frac{\partial \phi}{\partial x} + v \frac{\partial A_x}{\partial x} \right). \qquad (7.36)$$

[4] Do not forget that we have dropped c, so $\gamma^2(1 - v^2) = 1$ here.

When these are added to make E'_x, four terms cancel, leaving[4]

$$E'_x = \gamma^2 \left(-(1 - v^2) \frac{\partial \phi}{\partial x} - (1 - v^2) \frac{\partial A_x}{\partial t} \right) = -\frac{\partial \phi}{\partial x} - \frac{\partial A_x}{\partial t} = E_x.$$

It seems like a lot of trouble just to get the simplest of the results, but the other terms now go smoothly, because $\partial \phi'/\partial y' = \partial \phi'/\partial y$ and $\partial \phi'/\partial z' = \partial \phi'/\partial z$, and $A'_y = A_y$ leads to

$$\frac{\partial A'_y}{\partial t'} = \frac{\partial A_y}{\partial t} \gamma + \frac{\partial A_y}{\partial x} \gamma v.$$

Hence

$$E'_y = \gamma \left(-\frac{\partial \phi}{\partial y} - \frac{\partial A_y}{\partial y} - v \left(\frac{\partial A_y}{\partial x} - \frac{\partial A_x}{\partial y} \right) \right) = \gamma \left(E_y + (\mathbf{v} \wedge \mathbf{B})_y \right)$$

and the calculation of E'_z is similar. We have now derived the full transformation equation for $\mathbf{E}$, without restriction. The calculation for $\mathbf{B}$ is left as an exercise for the reader. If the reader would rather avoid it, then read on!

7.5   Introducing the Faraday tensor

We would now like to introduce a new mathematical tool that, among other things, can greatly simplify the calculation presented in the previous section. The idea is to extend the 'apparatus' of 4-vector analysis by introducing a matrix-like quantity called a *tensor*. In fact, this is part of a more extensive apparatus called *tensor analysis* that is introduced in chapter 12, but here we shall not need the whole apparatus, so to keep things simple we will concentrate on the minimum we need in order to gain some useful insights into electromagnetism.

7.5.1   Tensors

Take two arbitrary 4-vectors A and B and form the product

$$AB^T$$

where, as usual, we have in mind that the 4-vectors are considered to be column vectors. By the rules of matrix multiplication, this is a valid combination because it is the product of a 4×1 'matrix' with a 1×4 'matrix'. The outcome is a 4×4 matrix, and the operation is called 'taking the **outer product**'. You can confirm that the resulting matrix is

$$
\begin{pmatrix}
A^0B^0 & A^0B^1 & A^0B^2 & A^0B^3 \\
A^1B^0 & A^1B^1 & A^1B^2 & A^1B^3 \\
A^2B^0 & A^2B^1 & A^2B^2 & A^2B^3 \\
A^3B^0 & A^3B^1 & A^3B^2 & A^3B^3
\end{pmatrix},
$$

In other words, we just write out every possible combination of elements of A and B and arrange them in a matrix. For example:

$$
\begin{pmatrix} 2 \\ 3 \\ 1 \\ 5 \end{pmatrix} \begin{pmatrix} 1 \\ 10 \\ 0 \\ -2 \end{pmatrix}^T = \begin{pmatrix} 2 \\ 3 \\ 1 \\ 5 \end{pmatrix} (1,\, 10,\, 0,\, -2) = \begin{pmatrix} 2 & 20 & 0 & -4 \\ 3 & 30 & 0 & -6 \\ 1 & 10 & 0 & -2 \\ 5 & 50 & 0 & -10 \end{pmatrix}
$$

The result can also be expressed conveniently by writing down the expression for an arbitrary element of the matrix. If $M = AB^T$, then

$$
M^{mn} = A^m B^n \tag{7.37}
$$

where the indices m, n run over the values $(0, 1, 2, 3)$.

Note the contrast with the inner product $A^T g B$ which leads to a scalar. Both inner and outer product are much used in quantum theory, where in Dirac notation they are expressed $\langle \phi | \psi \rangle$ and $| \phi \rangle \langle \psi |$. They are different again from the 'tensor product' (written $| \phi \rangle \otimes | \psi \rangle$), which we shall not need[5].

Clearly, since A and B are 4-vectors, their outer product $M = AB^T$ cannot be Lorentz invariant, but must transform as

$$
\left. \begin{array}{ccc} A & \rightarrow & \Lambda A \\ B & \rightarrow & \Lambda B \end{array} \right\} \Rightarrow AB^T \rightarrow \Lambda A (\Lambda B)^T = \Lambda A B^T \Lambda^T.
$$

In other words, under a change of reference frame the matrix M transforms into M' given by

$$
M' = \Lambda M \Lambda^T. \tag{7.38}
$$

We now have two different types of matrix in play. Λ is a matrix describing the transformation from one frame to another, whereas M is a matrix that can be written down in any one frame, and which transforms as shown in eqn (7.38) under a change of reference frame. To distinguish them, M is called a *tensor* whereas Λ is not.

More generally, we *define* a tensor (or, to give the full name, a 'contravariant second-rank tensor') to be any 4×4 matrix that transforms, under a change of reference frame, as given in eqn (7.38), whether or not the matrix can be written as an outer product of 4-vectors. This makes perfect sense, because one can show that any such tensor can be written as a sum of outer products.

Having introduced the tensor, it is natural to ask whether or not it can multiply a 4-vector, as in the product MU. Why should it not? It is a perfectly well-defined mathematical operation, obeying the rules

[5] *Alternative notations.* Sometimes the outer product is written $A \otimes B$ and sometimes you see simply AB. In the latter form it is to be understood that the outer product is intended. The outer product is also called 'dyadic product'. The symbol $\otimes$ is also used, in other contexts, for a tensor product, and sometimes you will find the dyadic product called a 'tensor product', but strictly that is an abuse of terminology.

of matrix multiplication. However, beware! The outcome is a column vector, but it is not a 4-vector, because it does not transform the right way:

$$\mathsf{M}\mathsf{U} = \mathsf{A}\mathsf{B}^T\mathsf{U} = \mathsf{A}(\mathsf{B}^T\mathsf{U})$$

This is the product of a 4-vector A with a scalar quantity $(\mathsf{B}^T\mathsf{U})$ that is mathematically well-defined but is not a Lorentz-invariant scalar, because it is missing the crucial metric g in the middle. Therefore the combination $\mathsf{M}\mathsf{U}$ is not a 4-vector. It is *well-defined but not useful*. To get a more useful quantity, it is obvious what we need to do: put the metric into the calculation. That is, we consider the product

$$\mathsf{M}g\mathsf{U} = \mathsf{A}\mathsf{B}^T g\mathsf{U} = \mathsf{A}(\mathsf{B}\cdot\mathsf{U}). \tag{7.39}$$

This is a 4-vector because it is the product of an invariant and a 4-vector.

More generally, we can now show that any tensor (that is, any entity transforming as eqn (7.38)), when multiplying the combination $g\mathsf{U}$ for an arbitrary 4-vector U, yields a 4-vector, whether or not the tensor can be written as an outer product. The proof is easy:

$$\mathsf{M}g\mathsf{U} \to (\Lambda\mathsf{M}\Lambda^T)g\Lambda\mathsf{U} = \Lambda\mathsf{M}(\Lambda^T g\Lambda)\mathsf{U} = \Lambda(\mathsf{M}g\mathsf{U}) \tag{7.40}$$

by using the definition of Λ, eqn (2.48). In view of this, we extend the dot notation already employed for inner products, such that, for any tensor M and 4-vector U,

$$\mathsf{M}\cdot\mathsf{U} \equiv \mathsf{M}g\mathsf{U} \qquad \text{and} \qquad \mathsf{U}\cdot\mathsf{M} \equiv \mathsf{U}^T g\mathsf{M}. \tag{7.41}$$

Thus the use of a dot takes care of the presence of g, and also makes sense, because these types of matrix multiplication are closely related to inner products.

We will show in chapter 12 that g is itself a tensor, though of a different type, that transforms as $g \to (\Lambda^{-1})^T g\Lambda^{-1}$, but since this equals g, we do not need to worry about it. We shall take it for granted, just as we have done all along in the discussion of 4-vectors. It is only in General Relativity that this property is no longer guaranteed.

The tensors that are relevant to physics tend to be either symmetric $(\mathsf{M} = \mathsf{M}^T)$ or antisymmetric $(\mathsf{M} = -\mathsf{M}^T)$. The antisymmetric type are the simplest, because they have only six independent elements: the diagonal elements must be zero, and elements in the lower left triangle must be the negative of those in the upper right triangle. It is useful to write a generic antisymmetric tensor in the form

$$\mathsf{M} = \begin{pmatrix} 0 & a^1 & a^2 & a^3 \\ -a^1 & 0 & b^3 & -b^2 \\ -a^2 & -b^3 & 0 & b^1 \\ -a^3 & b^2 & -b^1 & 0 \end{pmatrix} = \begin{pmatrix} 0 & a_x & a_y & a_z \\ -a_x & 0 & b_z & -b_y \\ -a_y & -b_z & 0 & b_x \\ -a_z & b_y & -b_x & 0 \end{pmatrix} \tag{7.42}$$

where (a^1, a^2, a^3) and (b^1, b^2, b^3) are the six independent numbers, and the second version implies (correctly, as we shall see) that these numbers in fact form the elements of two 3-vectors. Note carefully the placement of the signs in the **b** part. The logic is that b_x goes into the (y, z) slot of the tensor (third row, fourth column), and the other components go as given by cyclic permutation: i.e., b_y goes into the (z, x) slot and b_z into the (x, y) slot. This assignment is like the rules for a vector product of two 3-vectors; we shall display that connection in full in chapter 12.

7.5.2   Application to electromagnetism

We shall now make a claim: the Lorentz force equation (7.3) can be written

$$\mathsf{F} = q\mathbb{F} \cdot \mathsf{U} \tag{7.43}$$

where F is the 4-force on a particle of charge q, U is the 4-velocity of the particle, and $\mathbb{F}$ is a tensor. To be precise, the Lorentz force equation emerges as the spatial part of this 4-vector equation.

We shall prove the claim by finding the tensor $\mathbb{F}$. As a first step, consider

$$\mathsf{F} \cdot \mathsf{U} = q(\mathbb{F} \cdot \mathsf{U}) \cdot \mathsf{U} = q(\mathbb{F}g\mathsf{U})^T g\mathsf{U} = q\mathsf{U}^T g(\mathbb{F}^T g\mathsf{U}) = q\mathsf{U} \cdot (\mathbb{F}^T \cdot \mathsf{U}). \tag{7.44}$$

On the right-hand side we have the scalar product of U with the 4-vector $q(\mathbb{F}^T \cdot \mathsf{U})$, which is almost the same as F. If $\mathbb{F}$ were symmetric then this combination would be exactly equal to F and we would have $\mathsf{F} \cdot \mathsf{U} = \mathsf{U} \cdot \mathsf{F}$, which is true, but it is not the only possible solution. If $\mathbb{F}$ were antisymmetric then we would have $q(\mathbb{F}^T \cdot \mathsf{U}) = -q(\mathbb{F} \cdot \mathsf{U}) = -\mathsf{F}$, and then eqn (7.44) reads

$$\mathsf{F} \cdot \mathsf{U} = -\mathsf{U} \cdot \mathsf{F} = -\mathsf{F} \cdot \mathsf{U}.$$

This is more interesting, because it implies $\mathsf{F} \cdot \mathsf{U} = 0$. This means that the force is a pure force (cf. eqn (4.3)). Therefore, if our claim (7.43) is valid, then the use of an antisymmetric tensor will guarantee that the resulting force is pure. This is the very property we require for electromagnetism, so we propose that $\mathbb{F}$ is antisymmetric.

Any antisymmetric tensor can be written as in eqn (7.42). We have, therefore,

$$\frac{\mathsf{F}}{q} = \begin{pmatrix} 0 & a_x & a_y & a_z \\ -a_x & 0 & b_z & -b_y \\ -a_y & -b_z & 0 & b_x \\ -a_z & b_y & -b_x & 0 \end{pmatrix} \begin{pmatrix} -\gamma c \\ \gamma v_x \\ \gamma v_y \\ \gamma v_z \end{pmatrix} = \gamma \begin{pmatrix} \mathbf{a} \cdot \mathbf{v} \\ a_x c + b_z v_y - b_y v_z \\ a_y c + b_x v_z - b_z v_x \\ a_z c + b_y v_x - b_x v_y \end{pmatrix}$$

$$= \gamma \begin{pmatrix} \mathbf{a} \cdot \mathbf{v} \\ \mathbf{a}c + \mathbf{v} \wedge \mathbf{b} \end{pmatrix}$$

where we first multiplied U by g and then completed the calculation. If this is to give the Lorentz force equation, then the spatial part must be equal to $\gamma(\mathbf{E} + \mathbf{v} \wedge \mathbf{B})$ (do not forget the factor γ in eqn (2.75)). Therefore we have the correct force as long as

$$\mathbf{a} = \mathbf{E}/c, \qquad \mathbf{b} = \mathbf{B}. \tag{7.45}$$

The conclusion is that the electric and magnetic field vectors can now be regarded as two parts of a single entity: the **electromagnetic field tensor**, often called the *Faraday tensor*. Here it is written out in full:

$$\mathbb{F} = \begin{pmatrix} 0 & E_x/c & E_y/c & E_z/c \\ -E_x/c & 0 & B_z & -B_y \\ -E_y/c & -B_z & 0 & B_x \\ -E_z/c & B_y & -B_x & 0 \end{pmatrix}. \tag{7.46}$$

For example, a uniform electric field pointing in the x-direction would be expressed by the field tensor:

$$\mathbb{F} = \frac{E}{c} \begin{pmatrix} 0 & 1 & 0 & 0 \\ -1 & 0 & 0 & 0 \\ 0 & 0 & 0 & 0 \\ 0 & 0 & 0 & 0 \end{pmatrix}.$$

Using this in eqn (7.43) gives the explanation for eqns (4.35) and (4.36).

All the equations of electromagnetism can now be written in terms of the field tensor $\mathbb{F}$ instead of the 4-vector potential A. For example, consider the combination[6]

$$\Box\mathsf{A}^T = \Box^a \mathsf{A}^b = \begin{pmatrix} -\frac{1}{c}\frac{\partial A^t}{\partial t} & -\frac{1}{c}\frac{\partial A^x}{\partial t} & -\frac{1}{c}\frac{\partial A^y}{\partial t} & -\frac{1}{c}\frac{\partial A^z}{\partial t} \\ \frac{\partial A^t}{\partial x} & \frac{\partial A^x}{\partial x} & \frac{\partial A^y}{\partial x} & \frac{\partial A^z}{\partial x} \\ \frac{\partial A^t}{\partial y} & \frac{\partial A^x}{\partial y} & \frac{\partial A^y}{\partial y} & \frac{\partial A^z}{\partial y} \\ \frac{\partial A^t}{\partial z} & \frac{\partial A^x}{\partial z} & \frac{\partial A^y}{\partial z} & \frac{\partial A^z}{\partial z} \end{pmatrix}. \tag{7.47}$$

This is a kind of 'gradient of a vector', saying how every component of the vector changes in every direction. You may recognize it as a Jacobian matrix. You can now use this to show that the field tensor is related to the 4-vector potential by

$$\mathbb{F}^{mn} = \Box^m \mathsf{A}^n - \Box^n \mathsf{A}^m \tag{7.48}$$

where the equation gives the matrix element-by-element, with indices m and n running over $(0, 1, 2, 3)$. For example, $(m = 0, n = 1)$ gives

$$\mathbb{F}^{01} = -\frac{1}{c}\frac{\partial}{\partial t}A_x - \frac{\partial}{\partial x}\frac{\phi}{c}$$

which is equal to E_x/c, in agreement with eqn (7.46).

Recalling the version of Maxwell's equations that we obtained before, eqn (7.31), it should now not surprise us too much to learn that those equations can also be expressed in terms of a first derivative of $\mathbb{F}$. We have

$$\Box \cdot \mathbb{F} = \left(\frac{-1}{c}\frac{\partial}{\partial t}, \nabla \cdot\right) \begin{pmatrix} 0 & & -\mathbf{E}/c & \\ \hline & 0 & B_z & -B_y \\ -\mathbf{E}/c & -B_z & 0 & B_x \\ & B_y & -B_x & 0 \end{pmatrix}$$

$$= \left(\frac{-\nabla \cdot \mathbf{E}}{c}, \frac{1}{c^2}\frac{\partial \mathbf{E}}{\partial t} - \nabla \wedge \mathbf{B}\right)$$

[6] The symbol $\Box^m$ is often written ∂^m, in view of the fact that

$$g\Box^m = \frac{\partial}{\partial x^m}$$

where $x^m = (ct, x, y, z)$; note that g is needed here in order to get the signs right.

(where the matrix is $g\mathbb{F}$). Therefore Maxwell's equations M1, M4 can be written

$$\Box \cdot \mathbb{F} = -\mu_0 \mathsf{J}^T. \tag{7.49}$$

We will explore this approach further in chapter 13. The main point to register at this stage is that we now have in clear view the idea that has been a repeated theme of this chapter: namely, that *the electric and magnetic fields are two parts of one thing.* That 'thing' is a *tensor field.* It is a set of values (six of them altogether) that is associated with every event in spacetime, and that captures something of the physical nature of, or situation at, every event in spacetime. It results in a four-force on any charged particle that happens to be at the event in question. This tensor is of a type (the antisymmetric type) that can also be interpreted as a pair of vectors. It is not that the magnetic field is derived from the electric field, nor vice versa, but that they each furnish part of the larger thing (the tensor field). The relative contributions they make at any given event can vary from one reference frame to another.

The fields due to a uniformly moving point charge are given by

$$\mathbb{F}^{mn} = \frac{qc}{4\pi\epsilon_0} \frac{\mathsf{U}^m \mathsf{R}^n - \mathsf{R}^m \mathsf{U}^n}{(-\mathsf{R} \cdot \mathsf{U})^3} \tag{7.50}$$

where U is the 4-velocity of the charge and R is a null vector from an event on the worldline of the charge to the event at which the fields are being calculated. These events are called the *source event* and the *field event.* They will be studied in detail in the next chapter, which will elucidate the expression in full.

Finally, we are now in a position to do the calculation promised at the end of section 7.1.1: namely, to obtain the field transformation equations by an algebraically easier method. All we need to do is some matrix multiplication:

$$\mathbb{F}' = \Lambda \mathbb{F} \Lambda^T = \begin{pmatrix} 0 & \mathbb{F}^{tx} & \gamma(\mathbb{F}^{ty} - \beta\mathbb{F}^{xy}) & \gamma(\mathbb{F}^{tz} - \beta\mathbb{F}^{xz}) \\ \cdot & 0 & \gamma(\mathbb{F}^{xy} - \beta\mathbb{F}^{ty}) & \gamma(\mathbb{F}^{xz} - \beta\mathbb{F}^{tz}) \\ \cdot & \cdot & 0 & \mathbb{F}^{yz} \\ \cdot & \cdot & \cdot & 0 \end{pmatrix}, \tag{7.51}$$

where we wrote down the result for two frames in standard configuration, and the dots indicate that the lower elements are to be assigned in an antisymmetric fashion. By extracting the two vectors, and recalling that the direction of relative motion is along x, one finds our old friend eqn (7.13). This is undoubtedly the most direct route to that result (and we shall present in section 12.2.3 a method to obtain $\mathbb{F}'$ that even avoids the need to perform the matrix product).

Exercises

(7.1) Examine the effects of Lorentz contraction and time dilation on a long straight solenoid in motion along the direction of its axis, and hence prove that the magnetic field inside the solenoid is unaffected by such motion (an example of $\mathbf{B}'_\parallel = \mathbf{B}_\parallel$.)

(7.2) Find the magnetic field due to a long straight current by Lorentz transform from the electric field due to a line charge.

(7.3) A pair of parallel particle beams separated by distance d have the same uniform charge per unit length λ. They propagate in a region where a magnetic field is applied with a direction and strength just sufficient to overcome the repulsion between the beams, so that they both propagate in a straight line at constant speed v. Find the size B of this applied magnetic field, and comment on the limit $v \to c$.

(7.4) Suppose a particle of charge q moves at constant velocity $\mathbf{v}$ at distance d above a surface. Let the xy plane be parallel to the surface, and let the x axis be along the path of the particle. Show that the fields at a point on the surface are

$$\mathbf{E} = \frac{\gamma q}{(d^2 + (\gamma vt)^2)^{3/2}} (-vt, \ d, \ 0)$$

$$\mathbf{B} = \frac{\gamma q}{(d^2 + (\gamma vt)^2)^{3/2}} (0, \ 0, \ dv/c).$$

Show that the time integral of the field is proportional to $1/v$.

(7.5) A circular ring of proper radius a carries current I and lies in the $x'y'$ plane at the origin of frame S′, in standard configuration with S. Find the electric field in S at points on the z axis when $t = 0$.

(7.6) Electrostatic problems are usually treated using $\phi = \phi(\mathbf{r})$, $\mathbf{A} = 0$. Show that the same physical results can be obtained using potentials $\phi = 0$, $\mathbf{A} = t\nabla\phi$.

(7.7) Show that the vector potentials $\mathbf{A} = (-y, x, 0)$ and $\tilde{\mathbf{A}} = (-2y, 0, 0)$ both give rise to the same uniform B field, and find a scalar function χ such that $\tilde{\mathbf{A}} = \mathbf{A} + \nabla\chi$.

(7.8) Confirm that eqn (7.48) gives eqn (7.46).

(7.9) A rectangular loop carries current I and lies in the xy plane with sides a, b parallel to x, y axes respectively. Its magnetic dipole moment is $\mathbf{m} = Iab\hat{\mathbf{z}}$ in its rest frame. Show that, in a frame where the loop moves in the x direction at speed v, there is a charge $\pm aIv/c^2$ on the two sides separated by b, and thus an electric dipole moment $\mathbf{p} = \mathbf{v} \wedge \mathbf{m}/c^2$.

Electromagnetic radiation

In this chapter we examine the solution of Maxwell's equations in general, and in particular the phenomenon of electromagnetic radiation. We start with the fact that even in the absence of any charge or current, Maxwell's equations have a rich variety of solutions (in addition to the solution $\mathbf{E} = 0$, $\mathbf{B} = 0$): namely, the plane wave solutions and their superpositions. This fact comes first because it is needed in the analysis of what happens when there is an accelerating charge or a changing current. We then consider the general problem of calculating the fields for any arbitrary situation, when the distribution of charge, and how it is moving, have been given and it is desired to find the fields. This seems to be an ambitious calculation, but by using the scalar and vector potential it becomes tractable. We then go on to consider electromagnetic radiation in more general terms, and especially the power radiated by simple sources such as moving point charges and oscillating dipoles. The chapter contains much that might be found in a moderately advanced textbook on electromagnetism, but we will focus our interest on areas where Lorentz covariance has something to teach us, or where charges are moving fast.

8.1 Plane waves in vacuum

First we shall derive the possibility of electromagnetic plane waves in vacuum, assuming the Maxwell equations as a starting point. The quickest way is simply to present them as trial functions and prove that they are solutions.

It is convenient to write a general electromagnetic plane wave using the complex number notation:

$$\mathbf{E} = \mathbf{E}_0 \, e^{i(\mathbf{k}\cdot\mathbf{r}-\omega t)}, \qquad \mathbf{B} = \mathbf{B}_0 \, e^{i(\mathbf{k}\cdot\mathbf{r}-\omega t)}, \qquad (8.1)$$

where $\mathbf{E}_0$ and $\mathbf{B}_0$ are constant vectors, independent of both time and space, as is $\mathbf{k}$, the wave vector. It is understood that the physical fields are given by the real part of this solution, $\mathbf{E}_{\text{observed}} = \Re[\mathbf{E}]$, $\mathbf{B}_{\text{observed}} = \Re[\mathbf{B}]$. If the constant vectors $\mathbf{E}_0$ and $\mathbf{B}_0$ are real then the plane waves are linearly polarized; if one allows $\mathbf{E}_0$ and $\mathbf{B}_0$ to be complex then one can treat any type of polarization. The waves are plane because we are assuming $\mathbf{k}$ is constant, so the wavefronts are flat and the direction of propagation is everywhere the same.

It is very easy to 'plug' this trial solution into Maxwell's equations if one once learns (e.g., by exhaustive coordinate analysis) that for vectors **a**, **k** that are independent of time and position, and constant ω:

$$\frac{\partial}{\partial t}\left(\mathbf{a}\,e^{i(\mathbf{k}\cdot\mathbf{r}-\omega t)}\right) = -i\omega\mathbf{a}\,e^{i(\mathbf{k}\cdot\mathbf{r}-\omega t)}, \tag{8.2}$$

$$\boldsymbol{\nabla}\cdot\left(\mathbf{a}\,e^{i(\mathbf{k}\cdot\mathbf{r}-\omega t)}\right) = i\mathbf{k}\cdot\mathbf{a}\,e^{i(\mathbf{k}\cdot\mathbf{r}-\omega t)}, \tag{8.3}$$

$$\boldsymbol{\nabla}\wedge\left(\mathbf{a}\,e^{i(\mathbf{k}\cdot\mathbf{r}-\omega t)}\right) = i\mathbf{k}\wedge\mathbf{a}\,e^{i(\mathbf{k}\cdot\mathbf{r}-\omega t)}. \tag{8.4}$$

It is useful to learn these, and they are easy to remember. They are saying that in the case of the function 'position-independent vector times $\exp(i\mathbf{k}\cdot\mathbf{r})$' the $\boldsymbol{\nabla}$ operator performing a div or curl acts just like the vector **k** producing a scalar or vector product. This makes the process of putting our trial solution in to Maxwell's equations in free space extremely easy. In the case of waves in free space (zero charge and current density), we find by using the above and dividing out the exp function:

Fig. 8.1 Field directions in a linearly polarized plane wave.

M1: $i\mathbf{k}\cdot\mathbf{E}_0 = 0.$ **E** is orthogonal to the wave vector.
M2: $i\mathbf{k}\cdot\mathbf{B}_0 = 0.$ **B** is orthogonal to the wave vector.
M3: $i\mathbf{k}\wedge\mathbf{E}_0 = i\omega\mathbf{B}_0$ **E** is $\perp$ to **B**; $E_0 = (\omega/k)B_0$
M4: $ic^2\mathbf{k}\wedge\mathbf{B}_0 = -i\omega\mathbf{E}_0$ $\omega = kc,\ E_0 = cB_0$

The last equation (M4) on its own gives a statement about the mutual directions, and it says the sizes are related by $c^2kB = \omega E$. The directions are consistent with M3, and the sizes agree with M3 as long as $c^2k = \omega c$, leading to the conclusion $\omega = kc$ and $E_0 = cB_0$ that has been given on the last line of the table.

Since the above are all mutually consistent, they confirm that the trial solution is indeed a solution, and we find the constraints on the plane waves: they must be transverse (with **E**, **B**, **k** forming a right-handed set), the sizes of the fields must be 'equal'—i.e., related by $|E_0| = c|B_0|$—and the phase velocity ω/k must be equal to c.

In terms of the 4-vector potential, the Maxwell equations in Lorenz gauge (7.31) in free space ($\mathsf{J} = 0$) give the wave equation, so there are plane wave solutions

$$\mathsf{A} = \mathsf{A}_0 e^{i\mathsf{K}\cdot\mathsf{X}}$$

where A_0 is a constant 4-vector amplitude. The choice of Lorenz gauge $\Box\cdot\mathsf{A} = 0$ (eqn (7.28)) is required in order to get the wave equation, so we have the constraint

$$\Box\cdot\mathsf{A} = i\mathsf{K}\cdot\mathsf{A} = 0 \qquad \Rightarrow \mathsf{K}\cdot\mathsf{A}_0 = 0. \tag{8.5}$$

Therefore the waves of A are 'transverse' in spacetime. Often, a polarization 4-vector ε is introduced, such that

$$\mathsf{A} = A\varepsilon, \tag{8.6}$$

where $A = (A_x^2 + A_y^2 + A_z^2)^{1/2}$ and then the Lorenz condition is

$$\varepsilon \cdot \mathsf{K} = 0 \quad \Rightarrow \quad \varepsilon^0 = \boldsymbol{\varepsilon} \cdot \mathbf{k}\frac{c}{\omega} = \boldsymbol{\varepsilon} \cdot \hat{\mathbf{k}}. \tag{8.7}$$

Note that $\boldsymbol{\varepsilon}$ can have a component along $\mathbf{k}$. This possibility is called *longitudinal polarization*. It does not mean that the fields have longitudinal polarization: they remain transverse.

In free space we can always choose that the scalar potential is zero, $\phi = 0$ (in addition to the Lorenz gauge condition), since there exists a gauge transformation within the Lorenz gauge that accomplishes this (see below). Then $\Box \cdot \mathsf{A} = \boldsymbol{\nabla} \cdot \mathbf{A}$ so the Coulomb gauge condition is satisfied as well. In this case the polarization vector has $\varepsilon^0 = 0$, and then eqn (8.7) implies that $\boldsymbol{\varepsilon}$ is transverse (i.e. $\mathbf{A} \cdot \mathbf{k} = 0$).

Example A plane wave in free space is described by a 4-vector potential $\mathsf{A} = \mathsf{A}_0 \exp(i\mathsf{K} \cdot \mathsf{X})$ satisfying the Lorenz gauge condition, with $\mathsf{A}^0 = \phi/c \neq 0$. Find a gauge change $\mathsf{A} \to \tilde{\mathsf{A}}$ that results in a 4-potential still in Lorenz gauge, but with $\tilde{\phi} = 0$.

Solution
Since we want to get rid of ϕ, we suggest the gauge function $\chi = \int \phi \mathrm{d}t$, so that $\partial\chi/\partial t = \phi$. In order to stay in the Lorenz gauge we need this χ to satisfy the wave equation. It does, because $\Box^2 \chi = \Box^2 \int \phi \mathrm{d}t = \int \Box^2 \phi \mathrm{d}t$ which is zero because here ϕ satisfies the wave equation.

More generally, a change of 4-polarization by

$$\varepsilon \to \varepsilon + a\mathsf{K}, \tag{8.8}$$

where a is an arbitrary constant, amounts to a gauge change and therefore does not affect the fields. Since K is null, eqn (8.7) is still satisfied so the 4-potential remains within the Lorenz gauge. In this way one can always arrange that one of the components of ε is zero. The Lorenz condition gives a further constraint, and therefore there remain just two independent components of the polarization 4-vector.

We have already discovered some of the kinematics of these plane wave solutions, through our study of the headlight effect and the Doppler effect, and the energy falling into a bucket. A Lorentz transformation applied to the 4-wave-vector, and eqns (7.13) to transform the fields, must reproduce all those effects. For example, suppose a linearly polarized plane wave has its electric field along the y direction, its magnetic field along the z direction, and propagates along the x direction. In another reference frame S' in standard configuration with the first, one finds

$$E_x' = E_z' = 0, \qquad E_y' = \gamma(E_0 - vB_0)\,e^{i\varphi} = \gamma(1 - \beta)E_0\,e^{i\varphi}$$

$$B_x' = B_y' = 0, \qquad B_z' = \gamma(B_0 - vE_0/c^2)\,e^{i\varphi} = \gamma(1 - \beta)B_0\,e^{i\varphi}$$

where the phase $\varphi = kx - \omega t = k'x' - \omega't'$ is an invariant. Notice the similarity with the longitudinal Doppler effect: the field amplitudes transform in the same way as frequency.

We shall show in section 16.4 that the intensity (power per unit area) is proportional to $\mathbf{E} \wedge \mathbf{B}$, so we have $I' = \gamma^2(1 - \beta)^2 I$, in agreement with eqn (6.25).

8.2   Solution of Maxwell's equations for a given charge distribution

We shall now use the potentials to acquire some more information about electromagnetic fields. A common type of problem would be of the form, 'given that there are charges here and here, moving thus, what can you tell me about the fields?' That is, we would like to solve the equations in such a way that we can obtain the fields from given information about the charges and currents.

An important example is the case of no charge and no current. One possible solution for this case is zero field everywhere, but that is not the only solution: in vacuum the fields also can have forms that propagate as waves at the speed of light, as we saw in the previous section.

Another simple case is that of a single point charge in uniform motion. We studied this in section 7.3. It will serve as a useful introduction to methods based on potentials.

8.2.1   The 4-vector potential of a uniformly moving point charge

As in section 7.3 we suppose a point charge is at rest in one reference frame and therefore moving in another. As before we will choose the primed frame S' to be the one in which the source particle is at rest, and S to be the frame for which we want to write down the result. We are preparing now for a more general treatment in which we want to learn the potentials in a given reference frame in terms of the charge and current distribution in that frame. It will save a lot of clutter if we adopt unprimed symbols for the reference frame that is the 'final destination' of our calculation.

So, suppose a charge q is at rest in frame S', and this frame is in standard configuration with S. Then the charge is moving along the x-axis of S with speed v. The potentials for the case of a point charge at rest are

$$\phi' = \frac{q}{4\pi\epsilon_0 r'}, \qquad \mathbf{A}' = 0. \tag{8.9}$$

By applying an inverse Lorentz transformation to the 4-vector $\mathbf{A}'$ we obtain

$$
\begin{aligned}
\phi &= \gamma(\phi' + vA'_x) = \gamma\frac{q}{4\pi\epsilon_0 r'} \\
A_x &= \gamma(v\phi'/c^2 + A'_x) = v\phi/c^2, \\
A_y &= A_z = 0.
\end{aligned}
\tag{8.10}
$$

Now

$$r' = ((x')^2 + (y')^2 + (z')^2)^{1/2} = (\gamma^2(x - vt)^2 + y^2 + z^2)^{1/2}$$

(by Lorentz transformation of the coordinates) so

$$\phi = \frac{q}{4\pi\epsilon_0} \frac{\gamma}{(\gamma^2(x - vt)^2 + y^2 + z^2)^{1/2}},$$

$$\mathbf{A} = \mathbf{v}\phi/c^2. \tag{8.11}$$

The source particle is located at $\mathbf{r}_p = (vt, 0, 0)$ at any given time t in S.

Now we apply eqns (7.22) to find the fields. One obtains

$$\mathbf{E} = \frac{q}{4\pi\epsilon_0} \frac{\gamma(\mathbf{r} - \mathbf{r}_p)}{(\gamma^2(x - vt)^2 + y^2 + z^2)^{3/2}} \tag{8.12}$$

in agreement with eqn (7.4), and[1]

$$\mathbf{B} = \frac{q}{4\pi\epsilon_0 c^2} \frac{\gamma \mathbf{v} \wedge (\mathbf{r} - \mathbf{r}_p)}{(\gamma^2(x - vt)^2 + y^2 + z^2)^{3/2}}. \tag{8.13}$$

One can notice that $\mathbf{B} = \mathbf{v} \wedge \mathbf{E}/c^2$, as previously remarked.

So what have we learned from this? We knew the fields already (section 7.3), though perhaps the new method of calculation is simpler because it does not need to assume the field transformation equations. The more important point is that we have the potentials, eqns (8.11). They will prove to be very useful in what follows.

[1] The vector in the numerator of $\mathbf{B}$ is found to be $(0, -z, y)$ multiplied by v; here, owing to the fact that the source travels through the origin, $\mathbf{r}_p$ and $\mathbf{v}$ are parallel so one can write this either as $\mathbf{v} \wedge \mathbf{r}$ or as $\mathbf{v} \wedge (\mathbf{r} - \mathbf{r}_p)$. A shift of origin must not affect the result, however, so the latter form is more general.

8.2.2   The general solution

So far we have mentioned two types of solution to the Maxwell equations: the waves in free space, and the field due to a uniformly moving point charge. Next we shall consider the general solution for the type of problem where the distribution of charge and current is known.

Our aim is to solve eqns (7.29) and (7.30), which we shall rewrite here for convenience:

$$\Box^2 \phi = \frac{-\rho}{\epsilon_0}, \qquad \Box^2 \mathbf{A} = \frac{-\mathbf{j}}{c^2 \epsilon_0}. \tag{8.14}$$

There are four equations (three for the components of $\mathbf{A}$, and 1 for ϕ) but they are all of the same form,

$$\frac{1}{c^2} \frac{\partial^2 f}{\partial t^2} - \nabla^2 f = s(\mathbf{r}, t). \tag{8.15}$$

This equation is called the *inhomogeneous wave equation* or *wave equation with a source term*. We want to solve such equations for the unknown function $f(\mathbf{r}, t)$ when the source function s has been given.

The Poisson equation

To get the general idea, first consider the situation of electrostatics: i.e., there are just fixed charges and no currents, with no time-dependence.

The Dirac δ-function. The Dirac δ-function is a mathematical tool that can be employed to simplify some calculations involving integration. It is a function of x that is defined by taking the limit of a narrow tall function of fixed area, as the width tends to zero and the height tends to infinity. The result is $\delta(x) = 0$ everywhere except at one point, the point $x = 0$, and its value there is infinite in such a way that $\int_{-\epsilon}^{\epsilon} \delta(x)\mathrm{d}x = 1$. A δ-function in three dimensions takes the form $\delta(x)\delta(y)\delta(z)$, which may be abbreviated to $\delta^{(3)}(\mathbf{r})$.

In this case the vector potential is zero, and eqn (8.14i) for the scalar potential becomes the Poisson equation

$$\nabla^2 \phi = \frac{-\rho}{\epsilon_0} \qquad (8.16)$$

since $\partial \phi / \partial t = 0$. We know that the potential due to a fixed point charge is $\phi = q/4\pi\epsilon_0 r$ where r is the distance from the charge to the point where the potential is to be evaluated. We say r is the distance from the *source point* to the *field point*. The potential due to a set of charges can be obtained simply by adding the contributions from each charge. This follows from the fact that the Poisson equation is linear. We can consider any charge distribution ρ to be made of many tiny elements, each containing an amount of charge $\mathrm{d}q = \rho \mathrm{d}V_s$ where $\mathrm{d}V_s$ is a volume element at the source point. Therefore the solution for the potential can be written

$$\phi(\mathbf{r}) = \int \frac{\rho(\mathbf{r}_s)}{4\pi\epsilon_0 |\mathbf{r} - \mathbf{r}_s|} \mathrm{d}V_s. \qquad (8.17)$$

This method of solution, by dividing up the source function ρ into many tiny pieces, is called Green's method, and one can see that it will work whenever the differential equation is linear. The function

$$\frac{-1}{4\pi|\mathbf{r} - \mathbf{r}_s|}$$

is called the Green function (or Green's function) for Poisson's equation. It is the solution of eqn (8.16) when the right-hand side takes the form of a sharp spike having unit integral over volume: i.e. a δ-function (see box above).

Cautionary note. The solution we have just presented is perfectly valid, but by quoting the known answer for a point charge we avoided a mathematical issue that needs to be examined for a thorough understanding of the method. For the case of a point charge at the origin we are considering $\nabla^2 \phi$ with $\phi \propto 1/r$. But $\nabla^2(1/r)$ is a strange function. We will show that $\nabla^2(1/r) = 0$ for $r \neq 0$, but $\nabla^2(1/r) = -\infty$ for $r = 0$. It is not that the function tends gradually to infinity as r becomes smaller, but rather, $\nabla^2(1/r)$ is zero, and zero, and still zero, as r becomes smaller, and then it suddenly shoots off to infinity when r reaches zero! To understand this, consider

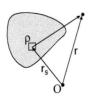

Fig. 8.2

$$\nabla \frac{1}{r} = \frac{-1}{r^2}\hat{\mathbf{r}}. \tag{8.18}$$

This gradient tends to infinity smoothly as $r \to 0$. When we take its divergence, for $r \neq 0$ there is a cancellation of terms that may be very large but of opposite sign. It is only at $r = 0$ that the cancellation fails, and then the terms are infinite and they add up. Here is the proof:

$$\nabla^2 \frac{1}{r} = \nabla \cdot \left(\frac{-\mathbf{r}}{r^3}\right) = -\frac{\nabla \cdot \mathbf{r}}{r^3} - \mathbf{r} \cdot \nabla \left(\frac{1}{r^3}\right) = -\frac{3}{r^3} + \frac{3r}{r^4}$$

$$= \frac{3}{r^3}\left(\frac{r}{r} - 1\right). \tag{8.19}$$

It is extremely tempting to evaluate the bracket as equal to zero, and hence obtain $\nabla^2(1/r) = 0$. This is correct almost everywhere. However, at $r = 0$ it is not legitimate and our expression is there ill-defined.

To understand the behaviour at $r = 0$ we can use a nice trick. Rather than working with the troublesome $\nabla^2(1/r)$ directly, we integrate it over a volume of space and then apply Gauss's divergence theorem:

$$\int \nabla^2 \left(\frac{1}{r}\right) dV = \int \nabla \cdot \left(-\frac{\hat{\mathbf{r}}}{r^2}\right) dV = \oint \frac{-1}{r^2}\hat{\mathbf{r}} \cdot d\mathbf{S}.$$

Now choose the region integrated over to be a sphere centred at the origin. The surface integral then evaluates to the surface area of the sphere, and we find

$$\int \nabla^2 \left(\frac{1}{r}\right) dV = -4\pi. \tag{8.20}$$

The volume integral thus 'tames' the function, and we conclude that $\nabla^2(1/r)$ is not zero at the origin, but takes such a value there that its volume integral is finite and equal to -4π. Using δ-function notation, the result is expressed as

$$\nabla^2 \frac{1}{r} = -4\pi\delta(x)\delta(y)\delta(z) = -4\pi\delta^{(3)}(\mathbf{r}). \tag{8.21}$$

The wave equation

Now we are ready to tackle the inhomogeneous wave equation. The equation is linear, so it can be treated by Green's method. To use the full method we would start by finding the solution of the wave equation when the source term is concentrated in a tiny region of both space and time. However, it saves a little working if we use some general knowledge of waves to jump straight to a solution where the source is unrestricted in time. That is, we suppose the function s on the right-hand side of (8.15) can have any time-dependence, but it is zero everywhere except near one spatial point, which we may as well take to be the origin. This means that elsewhere, away from the origin, the differential equation is just the

Spherical waves

We seek a spherically symmetric solution to the wave equation $\Box^2 f = 0$. For spherical symmetry, the function f does not depend on angles, so the Laplacian reduces to

$$\nabla^2 f \to \frac{1}{r^2}\frac{\partial}{\partial r}\left(r^2\frac{\partial f}{\partial r}\right) = \frac{2}{r}\frac{\partial f}{\partial r} + \frac{\partial^2 f}{\partial r^2} = \frac{1}{r}\frac{\partial^2}{\partial r^2}(rf). \qquad (8.22)$$

Now let $u = rf$ and substitute into $\Box^2 f = 0$. For $r \neq 0$ we can multiply both sides of the resulting equation by r, and we obtain

$$\frac{1}{c^2}\frac{\partial^2 u}{\partial t^2} - \frac{\partial^2 u}{\partial r^2} = 0.$$

This is the one-dimensional wave equation. Its general solution is $u(r,t) = g(t - r/c) + h(t + r/c)$ where g, h are arbitrary functions. The general spherically-symmetric solution of the three-dimensional problem is therefore

$$f = \frac{g(t - r/c)}{r} + \frac{h(t + r/c)}{r}.$$

This solves the homogeneous wave equation everywhere except at the origin ($r = 0$), which requires special consideration: see main text. The $t - r/c$ dependence means that g gives waves propagating towards positive r: i.e., outwards from the origin; h gives waves propagating inwards towards the origin. These are also called the *retarded* and *advanced* parts of the solution, respectively, see Fig. 8.4. For a situation in which the waves are caused by a source in the past, the h function is zero: the solution is purely retarded.

wave equation in free space. We already know that this has plane wave solutions, but they are not the solutions we need here because they will not have the right behaviour near our source at $r = 0$. However, another type of wave is the spherical wave, which has the general form

$$f = \frac{g(t - r/c)}{r} \qquad (8.23)$$

and this does have a non-trivial behaviour near $r = 0$. You can check that this is a solution of $\Box^2 f = 0$ for any function g, except at the origin (see box above).

Physically this corresponds to waves excited by a point source that oscillates with some time-dependence described by the function g. The waves travel outwards from the source, with speed c and spherical wavefronts. The $1/r$ factor means they diminish in amplitude as they go, thus ensuring energy conservation. Another solution is $h(t + r/c)/r$ for any function h: this corresponds to waves collapsing in towards the origin.

Fig. 8.3 A spherical wave. In general, the time-dependence need not be sinusoidal, but the amplitude always falls as $1/r$ as the waveform propagates out.

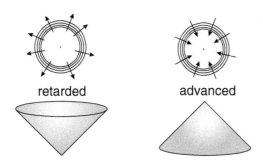

retarded   advanced

Fig. 8.4 Retarded and advanced solutions of the wave equation. A retarded solution propagates outwards from the source, forming a light-cone in the future of the source. An advanced solution propagates inwards towards the central point; it is possible for such waves to occur, but the point towards which they propagate is not their source. When calculating the field due to sources, we only need retarded solutions because the sources influencing any given field event lie in its past.

By comparing the situation with the one we have already treated (Poisson's equation), we can now guess the answer. The general solution to the inhomogeneous wave equation (8.15), using retarded contributions only, is (see Fig. 8.5)

$$f(\mathbf{r},t) = \int \frac{s(\mathbf{r_s}, t - |\mathbf{r} - \mathbf{r_s}|/c)}{4\pi|\mathbf{r} - \mathbf{r_s}|} \, dV_s. \qquad (8.24)$$

That is, we add up all the spherical waves produced by the sources, where each source has a strength $s\,dV_s$. We have shown most of the proof of this. In order to complete the proof, we need to show that eqn (8.24) accounts correctly for the time-dependence of the source for points near to and right at a given source. To do this we return to the case of a single point source and consider again the function $f = g(t-r/c)/r$. We already know that this function has $\Box^2 f = 0$ except at $r = 0$ (see box above, or exercise 8.2). To find the behaviour at $r = 0$, first consider the situation at small but non-zero values of r. The first derivatives with respect to r and t are

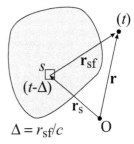

Fig. 8.5 Calculating the potentials due an arbitrary distribution of an arbitrarily moving charge. For a given field event $(\mathbf{r}, t)$ we sum over source events. The source events occur at the positions $\mathbf{r_s}$ and times $t - r_{sf}/c$.

$$\frac{\partial}{\partial r}\left(\frac{g(t-r/c)}{r}\right) = -\frac{g(t-r/c)}{r^2} - \frac{1}{c}\frac{g'(t-r/c)}{r}, \qquad (8.25)$$

$$\frac{\partial}{\partial t}\left(\frac{g(t-r/c)}{r}\right) = \frac{g'(t-r/c)}{r}, \qquad (8.26)$$

where g' refers to the first derivative of the function g. In the limit $r \to 0$, the g/r^2 term dominates all the others, unless g' tends to infinity. We will assume g varies smoothly, never having an abrupt change, and therefore g' is finite and hence we only need to keep the g/r^2 term. Applying this argument again to the second derivative, we have

$$\lim_{r\to0}\Box^2\left(\frac{g(t-r/c)}{r}\right) = \lim_{r\to0}\Box^2\left(\frac{g(t)}{r}\right) = g(t)\lim_{r\to0}\nabla^2\left(\frac{1}{r}\right) \qquad (8.27)$$

where we can replace $g(t-r/c)$ by $g(t)$ because the spatial dependence of $g(t-r/c)$ only introduces terms like the second term on the right of eqn (8.25), which we have just shown are negligible, and then we can ignore the time-derivatives because we have just shown that they too are negligible. Therefore, for the solution under consideration, the wave equation reduces to the Poisson equation for locations near to or at the point-like source.

We deduce that we have a solution to eqn (8.15) as long as

$$g(t) \lim_{r \to 0} \nabla^2 \left(\frac{1}{r} \right) = -s(t).$$ (8.28)

Therefore, using eqn (8.21), we have

$$4\pi g(t) = \int s(t) \mathrm{d}V.$$ (8.29)

This relates the function g appearing in our solution ($f = g(t - r/c)/r$) to the source term $s(t)$ in the equation that we are trying to solve.

It is convenient to absorb the 4π factor by defining $q \equiv 4\pi g$, then we have $f = q(t - r/c)/4\pi r$. The overall conclusion is as follows.

If the source in the inhomogeneous wave equation is concentrated at a point in space but has an arbitrary time-dependence $s(t)$ of total strength

$$q(t) = \int s(t) \mathrm{d}V,$$

then a solution of eqn (8.15) is

$$f(\mathbf{r}, t) = \frac{q(t - r/c)}{4\pi r}.$$ (8.30)

This solution looks just like the Coulomb potential, except instead of evaluating the 'charge' q at the time t, it is evaluated at the 'retarded' time $t - r/c$. The interpretation is that the potential at a given position receives waves from the source, and they take time to get there. This makes sense: it is the mathematical expression of the cause–effect relationship between the source and the potential, with a finite speed for signals.

Another solution exists, with 'advanced' time $t + r/c$, but this corresponds to waves moving in towards the source, so it does not correspond to the physical situation we are treating.

We can now complete the Green method and deduce that for any source function (now spread out in space and time), the solution to the wave equation (8.15), with retarded potentials, is eqn (8.24).

Application to Maxwell's equations

Using eqn (8.24), we are now in a position to write down the solutions we wanted, for given charge and current distributions in Maxwell's equations. The complete story is given in the box below.

Maxwell's equations:

$$\mathbf{\nabla} \cdot \mathbf{E} = \frac{\rho}{\epsilon_0}, \qquad \mathbf{\nabla} \cdot \mathbf{B} = 0,$$

$$\mathbf{\nabla} \wedge \mathbf{E} = -\frac{\partial \mathbf{B}}{\partial t}, \qquad c^2 \mathbf{\nabla} \wedge \mathbf{B} = \frac{\mathbf{j}}{\epsilon_0} + \frac{\partial \mathbf{E}}{\partial t}.$$

Their solution:

$$\mathbf{E} = -\mathbf{\nabla}\phi - \frac{\partial \mathbf{A}}{\partial t},$$

$$\mathbf{B} = \mathbf{\nabla} \wedge \mathbf{A},$$

$$\phi(\mathbf{r}, t) = \frac{1}{4\pi\epsilon_0} \int \frac{\rho(\mathbf{r}_\text{s}, t - r_\text{sf}/c)}{r_\text{sf}} \mathrm{d}^3 \mathbf{r}_\text{s}$$

$$\mathbf{A}(\mathbf{r}, t) = \frac{1}{4\pi\epsilon_0 c^2} \int \frac{\mathbf{j}(\mathbf{r}_\text{s}, t - r_\text{sf}/c)}{r_\text{sf}} \mathrm{d}^3 \mathbf{r}_\text{s} \qquad (8.31)$$

where $r_\text{sf} = |\mathbf{r} - \mathbf{r}_\text{s}|$ and $\mathrm{d}^3 \mathbf{r}_\text{s} \equiv \mathrm{d}x_\text{s} \mathrm{d}y_\text{s} \mathrm{d}z_\text{s}$.

One can verify that the potentials written here do satisfy the Lorenz gauge condition (7.28).

It might seem to be unwarranted to call eqn (8.31) 'the solution' of Maxwell's equations, because it still leaves some work to do: we have to carry out the integrals, and having done that we have to differentiate to get the fields. However, in principle an integral is nothing more nor less than adding up lots of tiny bits, and the equation tells us precisely what has to be added up: the amount of charge (for ϕ), or current (for $\mathbf{A}$) at the event $(\mathbf{r}_\text{s}, t - r_\text{sf}/c)$, divided by r_sf, and we have to sum over all source points $\mathbf{r}_\text{s}$. Differentiation is even more straightforward. This is an explicit set of instructions, as opposed to the very different sort of demand 'solve this partial differential equation'.

To write down the integral, we had to pick a reference frame in order to allow us to talk about things like distance, volume, and charge density. Obviously the integral is designed to tell you what the potentials are *in that reference frame*, but it does not matter what reference frame you choose. This fact can be made self-evident by writing the whole problem, and its solution, in 4-vector notation. The box below, eqns (8.32), shows this. The relation between the field tensor and the potential takes care of the second and the third Maxwell equations; the other two are given by the $\square^2 A$ equation (recall section 7.4). The integral used to calculate the 4-vector potential is now written in a form designed to bring out its Lorentz covariance.

Maxwell's equations:[2]

$$\mathbb{F} = \square \wedge \mathsf{A},$$

$$\square^2 \mathsf{A} = -\mu_0 \mathsf{J} \quad (\text{for } \square \cdot \mathsf{A} = 0).$$

Their solution:

$$\mathsf{A} = \frac{\mu_0}{4\pi} \int_{(\text{plc})} \mathsf{J} \, \mathrm{d}\sigma \qquad \text{where} \qquad \mathrm{d}\sigma = \frac{\mathrm{d}^3 \mathbf{r}_{\mathrm{s}}}{r_{\mathrm{sf}}}, \qquad (8.32)$$

in which $r_{\mathrm{sf}} = |\mathbf{r} - \mathbf{r}_{\mathrm{s}}|$ and (plc) is the past light-cone of the field event.

⇔ For an arbitrarily moving point charge:

$$\mathsf{A} = \frac{q}{4\pi\epsilon_0 c} \frac{\mathsf{U}}{(-\mathsf{R} \cdot \mathsf{U})} \qquad (8.33)$$

where U is its 4-velocity at the source event, and R is the (null) 4-vector from the source event to the field event.

The idea behind version (8.32) is that since the numerators in the integrands giving ϕ and $\mathbf{A}$ can be gathered into a 4-vector J, and the result of the integration is also a 4-vector, it must be that the combination $\mathrm{d}^3 \mathbf{r}_s / r_{\mathrm{sf}}$ is a Lorentz invariant. This is a subtle point, because this would not be true for any arbitrary region of integration. It is owing to the fact that while we allow x_s, y_s and z_s to explore all values, for any given field event, the integrand in eqn (8.31) forces t_s to vary as well, in such a way that the events contributing to the integral all lie on the past light-cone of the field event. The proof that the 'light-cone volume element' $\mathrm{d}^3 \mathbf{r}_s / r_{\mathrm{sf}}$ is an invariant is given in appendix D (we shall not need it again in this book).

The last equation (8.33) illustrates the method by supplying the result of the integral when the source is a single point charge. This will be derived in the next section.

8.2.3   The Liénard–Wiechert potentials

We are now in a position to find the potential and field of an *arbitrarily* moving point charge: i.e., one that may accelerate, and change its acceleration, and so on, and not just maintain a constant velocity. This is a wonderful possibility, because *all* fields come from point charges moving somehow or other (or at least we can model them that way), so we can encapsulate a great deal of insight into electromagnetism into one small but powerful result. We can get it because we have in eqn (8.32) all the information we need.

First consider a fixed point charge. For this case the integrals in eqn (8.32) are easy, and we already know the answer:

$$A_0 = \frac{q}{4\pi\epsilon_0 r_{sf}} \begin{pmatrix} 1/c \\ 0 \\ 0 \\ 0 \end{pmatrix}. \qquad (8.34)$$

The new information is that this solution is still valid even if the charge was accelerating at the source event! For, according to eqn (8.32), the acceleration of the source at the source event does not affect the result—only the position and velocity matter. Nonetheless, the influence of acceleration is there in the answer, because the distance r_{sf} is the distance between source event and field event. As we explore different field events, the source events also change. It is important to keep in mind that r_{sf} is here not the distance from any fixed point in the reference frame under consideration, such as the origin.

For an accelerating source, the solution (8.34) as it stands is of very limited usefulness, because it only applies at those field events where the charge happens to be at rest at the corresponding source event. However, we can use it to find the answer in other frames, by Lorentz transformation. In some other frame S we shall find

$$A = \Lambda A_0 \qquad (8.35)$$

To calculate this transformation for an arbitrary direction of motion, by far the best way is to express eqn (8.34) in terms of 4-vectors, and then the transformation is obvious. This approach is like the method of invariants, except that rather than combining suitable 4-vectors to form an invariant, we are trying to combine them to form a 4-vector. Eqn (8.34) is suggestive, because the column vector on the right has the form of a 4-velocity U evaluated in the rest frame (up to a factor c^2), and we already know what 4-velocity this must be: it is that of the point charge *at the source event* (see Fig. 8.6). In view of all that we discovered in the previous section, the factor r_{sf} in the denominator must be something to do with the 4-vector R from the source event to the field event. In fact, the 4-vectors U and R are the only ones that can possibly be relevant. Let us take a look at their scalar product:

$$R \cdot U = (r_{sf}, \mathbf{r}_{sf}) \cdot (\gamma c, \gamma \mathbf{v}) = \gamma(-r_{sf}c + \mathbf{r}_{sf} \cdot \mathbf{v}). \qquad (8.36)$$

This is promising, because it evaluates to $-r_{sf}c$ in the rest frame, so it will give the correct $1/r_{sf}$ Coulomb potential if it is in the denominator. Therefore, we propose the solution

$$A = \frac{q}{4\pi\epsilon_0} \frac{U/c}{(-R \cdot U)}. \qquad (8.37)$$

We can assert that this is what an evaluation of the right-hand side of eqn (8.35) must give, because (1) it is a 4-vector, (2) it reproduces the known result (8.34) in the rest frame, and (3) it does not introduce any extraneous quantities. We shall comment a little further on this method of derivation below. Before we do that, let us complete the derivation of

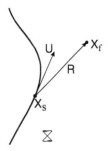

Fig. 8.6 Definition of 4-vectors R and U for the calculation of the 4-potential of an arbitrarily moving charge.

eqn (8.33). All we need do is to claim that eqn (8.37) is the complete solution for an *arbitrarily* moving charge, not just a constant-velocity one, because we knew from eqn (8.32) that the answer in the general case was going to depend only on the position and velocity of the charge at the source event, not its acceleration or rate of change of acceleration etc. In other words, we have found the completely general solution

4-vector potential of a charge in arbitrary motion

$$\mathsf{A} = \frac{q}{4\pi\epsilon_0} \frac{\mathsf{U}/c}{(-\mathsf{R} \cdot \mathsf{U})}. \tag{8.38}$$

The pair of potentials (scalar potential, vector potential) given by eqn (8.38) are called the *Liénard–Wiechert potentials*. Writing them out separately, we have

$$\phi = \frac{q}{4\pi\epsilon_0 \left[r_{sf} - \mathbf{v} \cdot \mathbf{r}_{sf}/c \right]}. \tag{8.39}$$

and

$$\mathbf{A} = \frac{q}{4\pi\epsilon_0 c^2} \left[\frac{\mathbf{v}}{r_{sf} - \mathbf{v} \cdot \mathbf{r}_{sf}/c} \right]. \tag{8.40}$$

The square brackets serve as a reminder that whereas we are evaluating the potential at the field point at some time t, the $\mathbf{r}_{sf}$ and $\mathbf{v}$ appearing in the formula are understood to mean $\mathbf{r}_{sf}(t_s)$ and $\mathbf{v}(t_s)$: i.e., their values at the source event which occurred at time $t_s = t - r_{sf}/c$.

We shall now gain further confidence by commenting on the method of derivation, and illustrating it for a uniformly moving point charge.

Method of 4-vectors

It is common when beginning the study of Relativity to write down what one knows to be the case in one frame, and then apply a Lorentz transformation. However, where possible one should use another type of reasoning that can save a lot of trouble. Rather than laboriously transforming from one frame to another, we simply express the result in terms of 4-vectors that correctly produce it in the starting frame, and then we use physical reasoning to show that no further terms could appear in other frames: i.e., terms that just happened to cancel or vanish in the starting frame. This is the generalization of the 'method of invariants' (section 2.6). It is now a 'method of 4-vectors'.

We are familiar with this type of reasoning in the case of 3-vectors. To take an example, consider the expressions (8.11) for the potentials of a uniformly moving point charge. In the denominator we have a term

$$(\gamma^2(x - vt)^2 + y^2 + z^2).$$

This expression clearly depends on the choice of coordinate system. However, by inspection of Fig. 8.7 you can see that the same result can

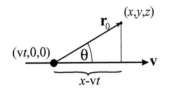

Fig. 8.7 Vectors and angle used to express the potential due to a uniformly moving charge.

be written down by substituting $(x - vt) = r_0 \cos\theta$ and $(y^2 + z^2)^{1/2} = r_0 \sin\theta$ where $\mathbf{r}_0$ is the vector from the charge at time t to the field point at time t, and θ is the angle between this vector and the velocity $\mathbf{v}$ of the charge. Thus the expression is

$$(\gamma^2(x - vt)^2 + y^2 + z^2) = r_0^2(\gamma^2 \cos^2\theta + \sin^2\theta),$$
$$\text{with} \quad \mathbf{r}_0 \cdot \mathbf{v} = r_0 v \cos\theta. \tag{8.41}$$

(The second equation serves to define θ in terms of the vectors.) We know for sure that the vector form of the expression is valid in the coordinate system from which we began, and we can see that there is no reason for things to stray from this form in other coordinate systems. Therefore, we now have the general formula, and eqn (8.11i) can be written as

$$\phi = \frac{q}{4\pi\epsilon_0} \frac{\gamma}{r_0(\gamma^2 \cos^2\theta + \sin^2\theta)^{1/2}}. \qquad [\text{ constant velocity } \tag{8.42}$$

The use of vectors saves us the trouble of applying rotation matrices to the original formula. If you are happy with the 3-vector example leading to eqn (8.42), then you should be similarly convinced of eqn (8.37).

We are now in a position to understand how the wonderful 'magic' of the electric field pointing away from the uniformly moving charge (Fig. 7.4) comes about. For a charge in an arbitrary state of motion, we focused attention on two positions: that of the source event and that of the field event. We can also take an interest in another position: the 'projected position'. This is the position the particle would have 'now' (i.e., at the time of the field event, in our chosen reference frame) *if* it were to continue on from the source event at the velocity it then had. The 'projected position' is not usually on the particle's trajectory: the particle does not go there (unless of course its velocity happens to be constant), but it is a well-defined place that we can take an interest in if we like. So, define the vector $\mathbf{r}_0$ to be the vector from the projected position to the field event. It is the vector that appeared in our formula (8.42) for the uniformly moving case, but now we are considering the general case. Using $\mathbf{r} = \mathbf{v}(r/c) + \mathbf{r}_0$ (Fig. 8.8) we obtain

$$\mathbf{r}_0 = \mathbf{r} - \mathbf{v}r/c \tag{8.43}$$

where we are dropping the subscript on $\mathbf{r}_{sf}$ because we hope that it is now obvious that this is the crucially important vector in terms of which the field is calculated.

We shall now write the general potential again, but expressing $\mathbf{r}$ in terms of $\mathbf{r}_0$ and $\mathbf{v}$. We have $\mathbf{r} \cdot \mathbf{v} = rv \cos\alpha$ and using Fig. 8.8 you can see that $r \sin\alpha = r_0 \sin\theta$. So after using $\cos^2\alpha = 1 - \sin^2\alpha$ we have

$$(\mathbf{r} \cdot \mathbf{v})^2 = r^2 v^2 (1 - \sin^2\alpha) = r^2 v^2 - v^2 r_0^2 \sin^2\theta.$$

Using this result you can easily confirm that

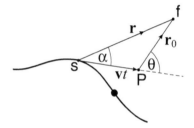

Fig. 8.8 Defining the *projected position* P. At the moment when the field is to be calculated at the field point f, the particle (large blob) has moved to some position of no interest. The field at f is caused by what occurred at the source point s. We can express it in a useful way in terms of the vector r_0 between the projected position and the field point. The time $t = r/c$ is the time taken for the influence from s to reach f.

$$\left(r - \frac{\mathbf{r} \cdot \mathbf{v}}{c}\right)^2 = r_0^2 \left(1 - \frac{v^2}{c^2} \sin^2 \theta\right).$$

Next replace 1 by $\cos^2 \theta + \sin^2 \theta$ on the right hand side and multiply by γ^2 to obtain

$$\gamma(r - \mathbf{r} \cdot \mathbf{v}/c) = r_0(\gamma^2 \cos^2 \theta + \sin^2 \theta)^{1/2}. \tag{8.44}$$

Substituting this into the Liénard–Wiechert potentials (8.39) and (8.40), we have

$$(\phi/c, \mathbf{A}) = \frac{q}{4\pi\epsilon_0 c^2} \frac{\gamma\,(c, \mathbf{v})}{r_0(\gamma^2 \cos^2 \theta + \sin^2 \theta)^{1/2}}. \tag{8.45}$$

What is this? It is the same expression we obtained for the uniformly moving charge, of course (cf. eqn (8.42)). We have confirmed that all our derivations are mutually consistent, and although the field for the case of uniform motion has the interesting form we noticed, we have confirmed that it is caused to assume that pattern by means of light-speed-limited communication.

With hindsight, one could now reason backwards from the potential of a particle at constant velocity (which is very easily derived by using the knowledge that $\mathbf{A}$ is a 4-vector) to the Liénard–Wiechert potentials, by introducing a change of 'position of interest' from the projected position back to the source event. Since the fields can then be derived from the potentials, even for an arbitrarily moving charge, people sometimes claim that all of electromagnetism can be derived from Coulomb's law and Lorentz transformations. Such a claim is wrong, however, because much more is needed. For example, we need to know that the potentials form a 4-vector, and how the fields relate to the potentials, and we need to know the non-trivial fact that the potentials depend only on the position and velocity of the charge at the source event, not on its acceleration. This is far from obvious: after all, the *fields* do depend on the acceleration. We also need to know that only properties at the source event are important, not some kind of integral over the history of the particle up to the source event.

The attempt to derive electromagnetism from Coulomb's law and Lorentz covariance therefore fails. However, the goal of developing fundamental theories from a minimal set of assumptions is valid and important. In chapter 13 we shall shall exhibit a construction of electromagnetic theory—i.e., Maxwell's equations and the Lorentz equation—based on a set of assumptions that we state explicitly, and that we try to make as small and simple as possible. This theme will also re-emerge when we consider field theory more generally in volume 2.

Integrating for a point-like source

We derived the Liénard–Wiechert potentials above by starting in the rest frame of the source event, and using it to help construct the 4-vector answer given in eqn (8.38). It should be possible to obtain this same answer by direct evaluation of the integrals given in eqn (8.31).

Consider first the zeroth element of A, i.e. ϕ/c, and look at its formula in eqns (8.31). Faced with the integral in eqn (8.31) and the desire to evaluate it in the case of a point charge, most of us would note that since ρ is then a sharply peaked function, the $1/r_{sf}$ can be brought outside the integral, and then we would take the volume integral of ρ to be the charge q, thus obtaining

$$\phi \overset{?}{=} \frac{q}{4\pi\epsilon_0 [r_{sf}]} \qquad \text{(wrong)}$$

It is what one might think, but it is wrong (compare with eqn (8.39)). The reason is because this does not correctly treat the time-dependent nature of the integrand when the charge is moving. Fig. 8.9 explains the problem and its solution. The correct answer is

$$\phi = \frac{q}{4\pi\epsilon_0 [r_{sf}(1 - v_r/c)]} = \frac{q}{4\pi\epsilon_0 [r_{sf} - \mathbf{v} \cdot \mathbf{r}_{sf}/c]} \qquad (8.46)$$

which agrees with eqn (8.39). The tricky integration here is a lesson in the care that is needed when dealing with δ-functions.

8.2.4   The field of an arbitrarily moving charge

The electric and magnetic fields of an arbitrarily moving charge can be obtained directly from the Liénard-Wiechert potentials, by applying the relations $\mathbf{E} = -\nabla\phi - \partial\mathbf{A}/\partial t$, $\mathbf{B} = \nabla \wedge \mathbf{A}$ (eqn (7.22)). Carrying out the differentiations with respect to time and space is a lot of work, however. The effort is reduced (though not to nothing) by some modest use of tensor methods, starting from eqn (7.48). The steps are shown in appendix D, which you should consult after reading chapter 12.

For a uniformly moving point charge one has that U is constant and the calculation via eqn (7.48) is consequently somewhat easier, though not as easy as the methods we employed in section 7.3. Its main use is to provide some practice and to provide the manifestly covariant expression (7.50). The overall form of this result should now be reasonably intuitive. To extract the electric field, consider the elements ($m = 0$, $n = 1, 2, 3$) in the expression $U^m R^n - R^m U^n$. They yield the vector

$$U^0 \mathbf{r} - ct U^i = \gamma c\mathbf{r} - ct\gamma\mathbf{u} = \gamma c(\mathbf{r} - \mathbf{u}t) \qquad (8.47)$$

where t is the time between source event and field event, so $(\mathbf{r} - \mathbf{u}t) = \mathbf{r}_0$, which leads to an electric field radially outwards from the projected position, as already noted. The denominator $(\mathbf{R} \cdot \mathbf{U})^3$ can then be expressed using eqn (8.44), and we find the same expression we found before, eqn (7.20).

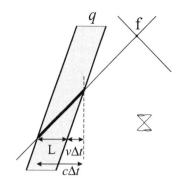

Fig. 8.9 Spacetime diagram to help calculate the potential at the field event f due to the charged particle q. We must allow the particle a finite spatial extent and take the limit as this becomes small compared to all other distances. The diagonal lines show the past light-cone of f. The events contributing to the integral are those shown bold. Suppose we want to calculate ϕ in the reference frame whose lines of simultaneity are horizontal in the diagram. Then the (spatial) length of the contributing line of events is $s = c\Delta t$, where Δt is the time taken for a light-pulse to travel $s = L + v\Delta t$ while the lump of charge travels $v\Delta t$, where L is equal to the length of the lump. Eliminating Δt we find $s = L/(1 - v/c)$. Thus the moving charge contributes as much to the integral as a non-moving charge of the same density but longer length would contribute. This leads to the 'enhancement' factor $1/(1 - v/c)$, where v is the component of velocity towards the field point.

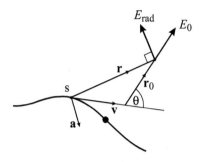

Fig. 8.10 The electric field due to an arbitrarily moving charge, illustrating the directions of the non-radiative and radiative parts of the field.

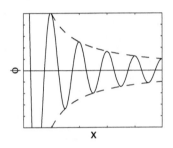

Fig. 8.11 The scalar potential $\phi(t, x, y, z)$ plotted as a function of distance at some instant of time in an inertial frame, for a case where the source charge has been undergoing oscillatory motion about the origin. The dashed lines show the characteristic $1/r$ decay of the potential of a stationary charge. At large r the gradient of the potential at the zero crossings is approximately proportional to this envelope, hence $\nabla\phi$ varies as $1/r$ not $1/r^2$.

[3] R. P. Feynman, R. B. Leighton, M. Sands, *The Feynman Lectures on Physics*, Vol. 1, eqn (28.3). A related expression for the magnetic field was found by Heaviside; see O. Heaviside *Electromagnetic Theory*, Vol. 3, eqn (214).

For an arbitrarily moving point charge, one finds

Field of a moving charge:

$$\mathbf{E} = \frac{q}{4\pi\epsilon_0(r - \mathbf{r}\cdot\mathbf{v}/c)^3}\left(\frac{(\mathbf{r} - \mathbf{v}r/c)}{\gamma^2} + \frac{\mathbf{r}\wedge[(\mathbf{r} - \mathbf{v}r/c)\wedge\mathbf{a}]}{c^2}\right)$$

$$= \frac{q}{4\pi\epsilon_0\kappa^3}\left(\frac{\mathbf{n} - \mathbf{v}/c}{\gamma^2 r^2} + \frac{\mathbf{n}\wedge[(\mathbf{n} - \mathbf{v}/c)\wedge\mathbf{a}]}{c^2 r}\right) \tag{8.48}$$

where $\mathbf{n} = \mathbf{r}/r$, $\kappa = 1 - v_r/c = 1 - \mathbf{n}\cdot\mathbf{v}/c$

$$\mathbf{B} = \mathbf{n}\wedge\mathbf{E}/c \tag{8.49}$$

$$= \frac{\mathbf{v}\wedge\mathbf{E}}{c^2} \text{ when } \mathbf{a} = 0 \tag{8.50}$$

where $\mathbf{r} = \mathbf{r}_f - \mathbf{r}_s$; the source event is at $(t_s = t - r/c, \mathbf{r}_s)$; $\mathbf{v}$, $\mathbf{a}$ are velocity and acceleration of the charge at the source event.

In terms of the displacement $\mathbf{r}_0 = \mathbf{r} - \mathbf{v}r/c$ from the projected position,

$$\mathbf{E} = \frac{q}{4\pi\epsilon_0 r_0^3(\gamma^2\cos^2\theta + \sin^2\theta)^{3/2}}\left(\gamma\mathbf{r}_0 + \frac{\gamma^3}{c^2}\mathbf{r}\wedge[\mathbf{r}_0\wedge\mathbf{a}]\right) \tag{8.51}$$

where θ is the angle between $\mathbf{r}_0$ and $\mathbf{v}$.

Alternative form (Feynman):

$$\mathbf{E} = \frac{q}{4\pi\epsilon_0}\left(\frac{\mathbf{n}}{r^2} + \frac{r}{c}\frac{\mathrm{d}}{\mathrm{d}t}\left(\frac{\mathbf{n}}{r^2}\right) + \frac{1}{c^2}\frac{\mathrm{d}^2}{\mathrm{d}t^2}\mathbf{n}\right) \tag{8.52}$$

Examining eqn (8.48), we see two terms. The first term is independent of the acceleration, and can be recognized as the field due to a uniformly moving charge. Its form is brought out by version (8.51). The second term is proportional to the acceleration. It varies as $1/r$ not $1/r^2$, so it dominates at large r. This is the radiation field. Its electric field vector is at right angles to $\mathbf{r}$ and in the plane containing the vectors $\mathbf{r}_0$ and $\mathbf{a}$ (by using the triple vector product rule, $\mathbf{r}\wedge(\mathbf{r}_0\wedge\mathbf{a}) = (\mathbf{r}\cdot\mathbf{a})\mathbf{r}_0 - (\mathbf{r}\cdot\mathbf{r}_0)\mathbf{a}$).

We started with a $1/r$ potential, eqn (8.39), so how can it come about that the radiation field varies as $1/r$, when differentiation of $1/r$ ought to give $1/r^2$? The answer to this is illustrated by Fig. 8.11. Owing to the propagation of the waves, a time-dependence at the source is converted into a spatial dependence in the potential around it. For a sinusoidally oscillating source, for example, over any given wavelength in space, the potential varies up or down by an amount of order $1/r$. The wavelength is independent of r, so the *slope* of the potential, at the positions of maximum slope, must be falling off approximately as $1/r$ not $1/r^2$.

The alternative form (8.52), due to Feynman,[3] brings out some further features. It has three terms. The first is the familiar Coulomb field, but

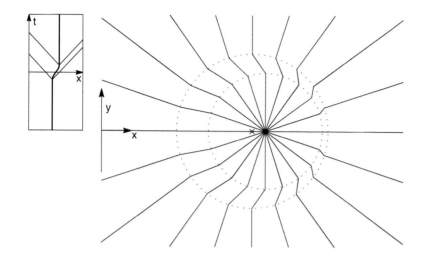

Fig. 8.12 A point charge is 'nudged' to the right. That is, the charge moves uniformly, undergoes a brief period of acceleration and deceleration, then moves uniformly again at the original velocity. The inset shows the world-line; the main figure shows the lines of electric field in a plane containing the acceleration vector, in the initial (and final) rest frame, at some moment shortly after the acceleration ceased. The dotted circles show the current position of two light-spheres that propagate outwards from source events at the beginning and end of the nudge (radii $c(t - t_A)$ and $c(t - t_B)$ respectively). Near the charge the field is that of a uniformly moving charge, which points radially outwards from the current position of the charge (eqn (7.20)). Beyond the second light-sphere the field is again that of a uniformly moving charge, but now pointing outwards from the projected position (the position the charge would now have, had it not accelerated), shown by a cross. Between the light-spheres the field has a bound part and a radiative part. The radiative part at any point is transverse to the light-sphere passing through that point.

evaluated—note—at the retarded position and time. The second term says we need to correct the retarded Coulomb field. We multiply the rate of change of that field by r/c, which is just the retardation time; this is like a linear extrapolation from the retarded time to the present time. For a slowly changing field this extrapolation turns out to be a very good approximation, but it is clear that it cannot be exactly right. The last term corrects it. This term varies as $1/r$; it contains all the radiation effects.

A simple but useful and correct insight into the connection between radiation and acceleration is contained in the following argument (see Fig. 8.12). Suppose a particle moves at constant velocity for a while, then at event A it starts to accelerate and shortly after, at event B, it assumes a constant velocity again. Then the electromagnetic field for field points whose source event is either before A or after B is easy to write down: it is just the one associated with constant-velocity motion (eqn (7.20)). This provides the information about the field throughout most of spacetime. The part between light-cones through events A and B can be obtained exactly from eqn (8.48), or approximately by simply joining up the field lines already obtained, since the total field is divergenceless (in vacuum).[4] Figure (8.12) shows the result for an example case in which the charge begins and ends at the same velocity, so the accelerated motion includes both a speeding-up and a slowing-down part.

One can immediately see from this simple (and correct) argument that there is a 'kink' in the field lines, that this kink propagates outwards at the speed of light, that the propagating part of the field is transverse (so as to introduce the observed change in direction of the field lines), and that it falls to zero along the line of the acceleration. This propagating pulse is the part of the total field that we call electromagnetic radiation.

[4] 'Field lines' are continuous for a field of zero divergence.

Identifying the radiation

The claim that the term proportional to $\mathbf{a}$ signifies 'radiation', while the term without $\mathbf{a}$ does not merits some attention. By 'radiation' or a 'radiative field' we mean a field that, once it is produced, can be regarded as a separate entity independent of the source. It propagates outwards at the speed of light, carrying a well-defined amount of energy and momentum with it, and it can be assigned its own energy-momentum 4-vector. The latter point is not self-evident because we are talking about an extended entity.

According to eqn (8.48) we can always separate an electromagnetic field at any given event into two parts:

$$\mathbf{E}_{\mathrm{I}} \equiv \frac{q}{4\pi\epsilon_0 \kappa^3} \frac{\mathbf{n} - \mathbf{v}/c}{\gamma^2 r^2}, \qquad \mathbf{E}_{\mathrm{II}} \equiv \frac{q}{4\pi\epsilon_0 \kappa^3} \frac{\mathbf{n} \wedge ((\mathbf{n} - \mathbf{v}/c) \wedge \mathbf{a})}{c^2 r},$$

(8.53)

$$\mathbf{B}_{\mathrm{I}} \equiv \mathbf{n} \wedge \mathbf{E}_{\mathrm{I}}/c, \qquad\qquad \mathbf{B}_{\mathrm{II}} \equiv \mathbf{n} \wedge \mathbf{E}_{\mathrm{II}}/c. \qquad\qquad (8.54)$$

In order to make this separation we would have to identify the source event and thus $\mathbf{v}$, $\mathbf{a}$ and all the other parts of these formulae. This may not always be easy (perhaps we have a field in our lab but do not know what produced it in the past), but in principle it could be done by an all-knowing investigator. We should like to propose that $\mathbf{E}_{\mathrm{I}}$, $\mathbf{B}_{\mathrm{I}}$ (hereafter called EM$_{\mathrm{I}}$) should be identified as a 'bound', non-radiative field, which may be regarded as a field owned by or in permanent interaction with the source, while $\mathbf{E}_{\mathrm{II}}$, $\mathbf{B}_{\mathrm{II}}$ (hereafter called EM$_{\mathrm{II}}$) is a radiative field having an independent existence, possessing a well-defined energy-momentum. Can we prove such a statement?

First note that $\mathbf{B}_{\mathrm{II}}$ is perpendicular to $\mathbf{E}_{\mathrm{II}}$, and since $\mathbf{E}_{\mathrm{II}}$ is perpendicular to $\mathbf{n}$, their sizes are related by $B = E/c$. A field with these properties is called **light-like**.

The formula for $\mathbf{E}_{\mathrm{I}}$ looks just like the formula for the field of a non-accelerating charge. In fact, it does not just look like it; it is precisely the formula for the field of a non-accelerating charge (eqn (8.51) makes this clear). However, the 'position vector' $\mathbf{r}$ is not here the position in space at some given time in a reference frame; it is a position vector *on a light-cone* from the source event. In some reference frame at a given time, for fixed values of the rest of the parts of the formula, $\mathbf{r}$ picks out positions on the surface of a light-sphere centred on the source event. The bound field at other positions is given by a different source event, where the charge may have had a different velocity. Therefore the whole bound field at any given reference frame time is *not* simply the field of a charge in uniform motion. In fact, one may show that it is not even a solution of Maxwell's equations! For example, for $a \neq 0$ one finds $\nabla \cdot \mathbf{E}_{\mathrm{II}} \neq 0$ and therefore $\nabla \cdot \mathbf{E}_{\mathrm{I}} \neq 0$ in empty space (but then we have $\nabla \cdot \mathbf{E}_{\mathrm{I}} = -\nabla \cdot \mathbf{E}_{\mathrm{II}}$ of course, since the total field *is* a solution of Maxwell's equations). For this reason the separation of the field into type I and type II has to be interpreted with care. It turns out to be a useful way to consider energy movements in the field.

As we follow $\mathbf{E}_{II}, \mathbf{B}_{II}$ out along the light-cone of a given source event, we see their sizes diminishing as $1/r$, whereas $\mathbf{E}_I, \mathbf{B}_I$ diminish as $1/r^2$. These statements are not about the dependence on position at any given time; they describe the dependence on the radii of a succession of light-spheres all centred at the same source event. Clearly, except in the direction along $\mathbf{a}$ (where EM_{II} vanishes but EM_I does not), the EM_{II} field dominates at large r, and furthermore if the energy content of the field goes as the square of the field amplitudes (as we shall show in chapter 13), the total amount of energy in the EM_{II} field is *undiminished* as it propagates out, while the energy in the EM_I field, in a spherical shell of fixed thickness, falls to zero. This enables one to identify the energy content of the EM_{II} field purely from the behaviour of the total field on a huge light-sphere in the distant future. Therefore a large enough light-sphere offers information about the division of the field into two parts without requiring knowledge of the sources. The far field is sometimes called the 'radiation zone' or 'wave zone'.

Note that the total energy movement in the field is caused by both contributions. For example, if the net energy flow is zero it does not necessarily imply there is no radiative part; rather it implies that the various contributions to the total energy flow are balanced. (This point was widely misunderstood in the first half of the twentieth century, and is still a possible area of confusion for students.) There are three contributions to $\mathbf{E} \wedge \mathbf{B}$:

$$\mathbf{E} \wedge \mathbf{B} = \mathbf{E}_I \wedge \mathbf{B}_I + \mathbf{E}_{II} \wedge \mathbf{B}_{II} + (\mathbf{E}_I \wedge \mathbf{B}_{II} + \mathbf{E}_{II} \wedge \mathbf{B}_I).$$

An example where the $\mathbf{E}_{II} \wedge \mathbf{B}_{II}$ term is equal and opposite to the rest occurs in the case of a charge in hyperbolic motion.

Another important property of the EM_{II} field of a given charge is that it can be zero. It is zero for all field events for which there is no acceleration at the source event. Therefore, if we assume the particle has not been undergoing permanent acceleration from the distant past until now, then at any given instant in a given frame, the non-zero part of EM_{II} is completely contained in a finite region of space.

Thus EM_{II} has the following properties:

- At any moment it is completely contained in a finite region of space, not necessarily including the point where the particle is located.

- Its total energy content is constant when the particle is not accelerating.

We shall discuss the energy flow in more detail in section 16.4, and show that the total energy and momentum of EM_{II} transform in the right way to form a 4-vector. This allows us to conclude that it is legitimate to call EM_{II} the *radiative field*.[5] It also follows that, when observed in an inertial reference frame, accelerated charges always radiate, and radiation fields always have their source in accelerated (not constant velocity) motion.

[5] In the far field—i.e., far from the source event—one may say the field is 'only' the EM_{II} part since it dominates, and this is sufficient for examining the interaction of the field with other things such as detectors. However, even though EM_I is small its divergence is not small compared to that of EM_{II} (they are equal and opposite); this is because the divergence of EM_{II} involves a cancellation of terms of opposite sign: they almost balance but not quite. The weaker EM_I field has a larger divergence relative to its size, and can supply a matching contribution.

8.2.5   Two example fields

The far field of a slowly oscillating dipole

The most important type of light-source, or source of electromagnetic radiation in general, is the oscillating dipole. Most of the light we see around us is sourced by oscillating electric dipoles in atoms and molecules. Radio waves are produced by antennae that may be treated as dipoles to first approximation.

We shall obtain the form of the electromagnetic field of an oscillating dipole as simply as possible, by assuming that the speed of motion of the charge is small compared to c, and that the dipole is itself small compared to the distance to the field point. This covers most cases of practical importance, and is the first step to treating more general cases.

Consider a dipole made of two charges $\pm q$ separated by a displacement $\mathbf{x}_q$, so the dipole moment is

$$\mathbf{d} = q\mathbf{x}_q. \tag{8.55}$$

We suppose the $-q$ charge is fixed and the q charge moves with velocity $\mathbf{v} = \dot{\mathbf{x}}_q$. We shall obtain the fields from the 4-vector potential. We could start with the electric field, but it turns out that the calculation is easier if we first obtain the magnetic field, which only depends on the 3-vector potential $\mathbf{A}$.

To calculate the magnetic field we only need to consider the contribution to $\mathbf{A}$ due to the moving charge. Starting from eqn (8.38), using $\mathsf{U} = \gamma(c, \mathbf{v})$ and $\mathsf{R} = (c(t - t_s), \mathbf{r}_{sf})$ we obtain for the moving charge

$$\mathsf{A} = \frac{q}{4\pi\epsilon_0} \frac{(c, \mathbf{v})}{c(r_{sf}c - \mathbf{r}_{sf} \cdot \mathbf{v})}. \tag{8.56}$$

This is true in general.

Now we make an approximation: we treat a 'slowly' oscillating dipole, meaning the speed of movement of the charge is small compared to c: i.e., $v \ll c$. For sinusoidal oscillation, this implies that the wavelength of the emitted radiation is large compared to size of the dipole. For example, for a dipole of atomic dimensions we are restricted to treating radiation in the electromagnetic spectrum from radio waves to soft X-rays. With this approximation we have

$$\mathsf{A} \simeq \frac{1}{4\pi\epsilon_0 c^2} \frac{\left(qc, \dot{\mathbf{d}}[t - r_{sf}/c]\right)}{r_{sf}} \tag{8.57}$$

where we used $q\mathbf{v} = \dot{\mathbf{d}}$ and we have explicitly indicated the fact that this has to be evaluated at the source time $t_s = t - r_{sf}/c$. For example, for a sinusoidally oscillating source,

$$\mathbf{d} = q\mathbf{x}_0 \sin \omega t, \tag{8.58}$$

$$\Rightarrow \quad \dot{\mathbf{d}}[t - r_{sf}/c] = \omega q\mathbf{x}_0 \cos(\omega t - kr_{sf})$$

where k is the wave vector.

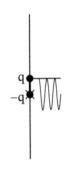

Fig. 8.13 Oscillating dipole.

We calculate the $\mathbf{B}$ field from

$$\mathbf{B} = \nabla \wedge \mathbf{A} \simeq \frac{1}{4\pi\epsilon_0 c^2} \nabla \wedge \left(\frac{\dot{\mathbf{d}}[t - r_{sf}/c]}{r_{sf}} \right). \tag{8.59}$$

At this stage it is helpful to introduce a further approximation: namely, to set $r_{sf} \simeq r$ where r is the distance from the origin to the field point (for a dipole oscillating at the origin). This is allowable as long as two conditions hold. The field point must be far from the dipole, $r \gg x_q$, and also the speed of movement of the charge must be small, $v \ll c$ (as we already assumed). The second condition arises because the derivative of r_{sf} with respect to (for example) x involves $\partial r_s/\partial x$: i.e., the change in the source point position when we ask about a change in field point by dx. If you think about the light-cones you should see that this amounts to asking how far the source moves during a time dx/c. Clearly, the movement of the source is small compared to dx when $v \ll c$, so under this condition we may take $dr_{sf} \simeq dr$.

Having made both the low speed and the far field approximations, we now have

$$\mathbf{B} \simeq \frac{1}{4\pi\epsilon_0 c^2} \nabla \wedge \left(\frac{\dot{\mathbf{d}}[t - r/c]}{r} \right) \tag{8.60}$$

which is reasonably straightforward to evaluate. Assume the oscillation of the dipole is along the z axis. Then $\mathbf{A}$ is along z, so for B_x we only need to evaluate

$$B_x = \frac{\partial A_z}{\partial y} = \frac{1}{4\pi\epsilon_0 c^2} \left(\left(\frac{-y}{r^3} \right) \dot{d} - \frac{y}{cr^2} \ddot{d} \right) \tag{8.61}$$

using $\partial r/\partial y = y/r$ twice (and dropping the $\simeq$). The expression for B_y can be calculated similarly (it is given by the same formula with the substitution $-y \to x$). Bringing together all three components (see also exercise 8.6), we have

$$\mathbf{B} = \frac{1}{4\pi\epsilon_0 c^3} \left(\frac{c\dot{\mathbf{d}} \wedge \mathbf{r}}{r^3} + \frac{\ddot{\mathbf{d}}[t - r/c] \wedge \mathbf{r}}{r^2} \right) \tag{8.62}$$

The first term is the same as the result of the Biot–Savart law, except $\dot{\mathbf{d}} = q\mathbf{v}$ is to be evaluated at the source time (=retarded time) not the field time t. It falls off as $1/r^2$, whereas the second term varies as $1/r$. Since we are here interested in the far field, we drop the first term, and the result is

$$\mathbf{B} = \frac{1}{4\pi\epsilon_0 c^3} \frac{\ddot{\mathbf{d}}[t - r/c] \wedge \mathbf{r}}{r^2} = \frac{\omega^2 d_0}{4\pi\epsilon_0 c^3} \frac{\sin\theta}{r} \sin(kr - \omega t)\hat{\boldsymbol{\phi}} \tag{8.63}$$

where the first version treats a general time-dependence of the source, and the second version gives the result for a sinusoidally oscillating dipole on the z axis (using $\hat{\mathbf{z}} \wedge \hat{\mathbf{r}} = \sin(\theta)\,\hat{\boldsymbol{\phi}}$). Note that the field is directed in the $\hat{\boldsymbol{\phi}}$ direction: i.e., in loops around the z axis.

[6] In a calculation of the electric field from the potentials, one finds that the approximation $r_{\text{sf}} \simeq r$ is inadequate for the scalar potential: higher-order terms are needed. However, one can avoid this difficulty by adopting the Lorenz gauge and obtaining ϕ from $\partial\phi/\partial t = -c^2 \boldsymbol{\nabla} \cdot \mathbf{A}$.

To obtain the electric field one can go via the potentials again,[6] but for the far field we do not need to. We have found previously that the far field is purely a radiation field (it falls as $1/r$ and is proportional to $\ddot{\mathbf{d}}$, hence to the acceleration of the source). Therefore we know it is light-like: i.e., $E = cB$ and $\mathbf{E}$, $\mathbf{B}$ and $\mathbf{r}$ form a right-handed set. Hence

$$\mathbf{E} = c\mathbf{B} \wedge \mathbf{r}/r = \frac{1}{4\pi\epsilon_0 c^2} \frac{(\ddot{\mathbf{d}}[t - r/c] \wedge \mathbf{r}) \wedge \mathbf{r}}{r^3}. \tag{8.64}$$

The vector product implies that the sizes of B and E vary with direction as

$$E = cB \propto \frac{\sin\theta}{r}, \tag{8.65}$$

where θ is the angle between $\mathbf{d}$ and $\mathbf{r}$. This pattern of the strength of the radiation field is called a 'dipole pattern'. For example, the sinusoidally oscillating dipole (8.58) gives

$$E = cB = \frac{\omega^2 d_0}{4\pi\epsilon_0 c^2} \frac{\sin\theta}{r} \sin(kr - \omega t) \tag{8.66}$$

with $\mathbf{E}$ and $\mathbf{B}$ directed around the surface of the light-sphere, $\mathbf{E}$ in the $\boldsymbol{\theta}$ direction, $\mathbf{B}$ in the $\boldsymbol{\phi}$ direction.

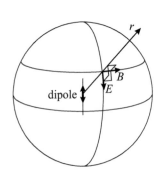

Fig. 8.14 Directions in the far field (or 'wave zone').

Antenna

The combination $\omega d_0 = \omega q l$ can be recognized as Il where I is the size of the current oscillations in a short segment of wire of length l. Therefore we can write eqn (8.66) as

$$E = cB = \frac{-iI}{2\epsilon_0 c} \frac{l}{\lambda} \frac{\sin\theta}{r} e^{i(kr - \omega t)}$$

where the complex notation is convenient in order to signal that the fields are a quarter cycle out of phase with the current.

An *antenna* is a short length of wire carrying an oscillating current and intended for use in either broadcasting or receiving electromagnetic waves. In that application we are interested in maximizing the transmitted or received power. Consider, for example, an antenna that is fed in the middle. Then the current oscillations are maximal at the centre of the antenna and zero at the ends. For a short antenna of length L we can approximate the current distribution as roughly linear: $I = I_0(1 - 2|z|/L)$. Integrating this along the antenna gives $I_0 L/2$, so the emitted power varies as L^2. This suggests that there is interest in using longer antennae. However, to calculate the field correctly we should allow for the phase lag: i.e., the fact that the distance from a current element on the antenna to the field point is also a function of z. This is very much like a diffraction calculation in optics. The essential point is that once the antenna is longer than about $\lambda/2$, further increases in length alter the directional distribution of the radiated field significantly, rather than the total emitted power.

Far field of an antenna. The approximation $I = I_0 \cos kz$ is quite good for a half-wave antenna, but not exact, because the radiation itself extracts power from the antenna. Within this approximation an accurate expression for the far field can be obtained by integrating along the antenna, allowing for the phase (just as in a Fraunhofer diffraction calculation). Each element on the antenna contributes $d\mathbf{E}$ to the field. First we make the approximation that the field point P is sufficiently far away that the *directions* of all these contributions agree, so

$$\mathbf{E} = \frac{-iI_0 \sin\theta}{2\epsilon_0 c\lambda} \int \frac{1}{r'} \cos(kz) e^{i(kr'-\omega t)} dz\, \hat{\boldsymbol{\theta}}$$

where $r'(z)$ is the distance from each current element to P (we wish to reserve the symbol r for the distance from the centre of the antenna to P). The assumption of far field allows us to use the Fraunhofer approximation

$$r' = r - z\cos\theta$$

so the integral in the expression above is

$$\frac{1}{r} e^{i(kr-\omega t)} \int_{-\lambda/4}^{\lambda/4} \cos(kz) e^{-ikz\cos\theta} dz$$

where we brought $1/r'$ outside the integral, since the variation of r' is negligible except through its effect on the phase. The integration can now be carried out easily by writing $\cos kz = (e^{ikz} + e^{-ikz})/2$. One finds

$$\frac{\sin(\frac{\pi}{2}(\cos\theta + 1))}{k(\cos\theta + 1)} + \frac{\sin(\frac{\pi}{2}(\cos\theta - 1))}{k(\cos\theta - 1)} = \frac{2\cos(\frac{\pi}{2}\cos\theta)}{k\sin^2\theta}$$

where in the last step we used $\sin(A + B) = \sin A \cos B + \cos A \sin B$ and added the two terms. Upon multiplying the various factors together, one power of $\sin\theta$ cancels and we have $k\lambda = 2\pi$, so the field is

$$E = \frac{-iI_0}{2\pi\epsilon_0 c} \frac{\cos(\frac{\pi}{2}\cos\theta)}{r\sin\theta} e^{i(kr-\omega t)}. \tag{8.67}$$

This expression is the more accurate (but still not exact) replacement for eqn (8.68).

A centre-fed antenna of length $L = \lambda/2$ is called a *half-wave dipole antenna*. We can model the current distribution roughly as $I = I_0 \cos kz$ (this falls to zero at $z = \pm\lambda/4$: i.e., the ends of the antenna). Then $\int I dz = I_0 \lambda/\pi$, and therefore (ignoring the diffraction effects) the fields are given approximately by

$$E = cB \simeq \frac{-iI_0}{2\pi\epsilon_0 c} \frac{\sin\theta}{r} e^{i(kr-\omega t)}. \tag{8.68}$$

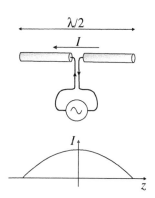

Fig. 8.15 The half-wave dipole antenna.

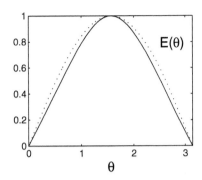

Fig. 8.16 Electric field far from a half-wave dipole antenna, as a function of angle. The full curve is $\cos((\pi/2) \cos\theta)/\sin\theta$ (eqn 8.67); the dashed curve is $\sin\theta$.

[7] Following E. Eriksen and Ø. Grøn, *Annals of Physics* **286**, 320–42 (2000).

(For a more accurate result, see the box above.) The dependence on wavelength has now dropped out. The constant $(2\pi\epsilon_0 c)^{-1}$ has the value 59.96 ohms; it is equal to $Z_0/2\pi$, where $Z_0 \equiv \mu_0 c$ is the characteristic impedance of free space.

A charge in hyperbolic motion*

The field due to a charge in hyperbolic motion is also most easily obtained from the 4-vector potential, eqn (8.38). We shall calculate it without approximation.[7]

For convenience, adopt units such that $c = 1$, and let

$$Q \equiv \frac{q}{4\pi\epsilon_0}.$$

Then $\mathsf{A} = Q\mathsf{U}/(-\mathsf{R}\cdot\mathsf{U})$. Let the position 4-vector of the charge be $\mathsf{R}_s = (t_s, x_s, y_s, z_s)$. We suppose the charge moves along the x direction, with the origin placed so that the equation of motion is $y_s = 0, z = 0$ and

$$x_s = \sqrt{L^2 + t_s^2}.$$

L is the distance of the charge from the origin at $t_s = 0$, it is related to the proper acceleration by $L = c^2/a_0$ (see table 4.1). Let $\mathsf{R}_f = (t, x, y, z)$ be the field event. For a given field event the source event is identified by solving $\mathsf{R}\cdot\mathsf{R} = 0$ for t_s, where $\mathsf{R} = \mathsf{R}_f - \mathsf{R}_s$:

$$(t - t_s)^2 - \left((x - \sqrt{L^2 + t_s^2})^2 + y^2 + z^2\right) = 0. \tag{8.69}$$

This yields a quadratic equation for t_s, whose solution is

$$t_s = \frac{t\delta - x\zeta}{2(x^2 - t^2)} \qquad \Rightarrow x_s = \frac{x\delta - t\zeta}{2(x^2 - t^2)}. \tag{8.70}$$

where we picked a sign corresponding to the retarded (not advanced) solution, and

$$\delta \equiv L^2 + \rho^2 + x^2 - t^2, \qquad \zeta \equiv \sqrt{\delta^2 - 4L^2(x^2 - t^2)}, \tag{8.71}$$

with $\rho = (y^2 + z^2)^{1/2}$. The 4-velocity at the source point is

$$\mathsf{U} = \gamma(1, \mathbf{v}) = (x_s/L, t_s/L, 0, 0) \tag{8.72}$$

where we used $x_s = \gamma L$ and $v_s = c^2 t_s/x_s$ from table 4.1. It follows that $\mathsf{R}\cdot\mathsf{U} = -\zeta/2L$, and one obtains

$$\phi = Q\frac{x\delta - t\zeta}{\zeta(x^2 - t^2)}, \qquad A_x = Q\frac{t\delta - x\zeta}{\zeta(x^2 - t^2)}, \qquad A_y = A_z = 0. \tag{8.73}$$

The electric and magnetic fields are found by differentiation. Adopting now cylindrical coordinates to express the result, one finds

$$E_x = -4QL^2(L^2 + \rho^2 - x^2 + t^2)/\zeta^3, \qquad E_\rho = 8QL^2\rho x/\zeta^3, \qquad E_\phi = 0$$

$$B_x = B_\rho = 0, \qquad\qquad\qquad\qquad\qquad B_\phi = E_\rho t/x. \tag{8.74}$$

In applying these results one must keep in mind that these equations are valid only in regions of spacetime for which the charge can source the

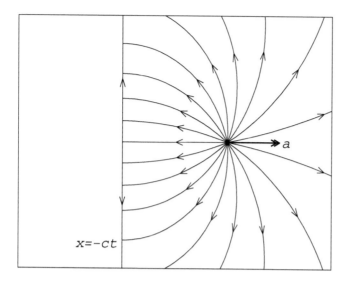

Fig. 8.17 Electric field of a charge undergoing hyperbolic motion. The field lines are shown at the moment when the charge is at rest in the chosen frame. Each field line is an arc of a circle, until it hits the plane $x = -ct$, where it is directed outwards.

field: i.e., for field events in the future light-cone of some event on the worldline of the charge. For the (somewhat artificial) case of a charge in permanent hyperbolic motion, this is the region $x + t > 0$. The lines of electric field are plotted in Fig. 8.17. This field has a number of interesting properties, explored in the exercises and in chapter 13. It contains both bound (type EM_I) and radiation (type EM_{II}) contributions.

8.3 Radiated power

It is very useful to have a formula for the power in the radiation part of the field. Such a formula can be obtained for a charge in an arbitrary state of motion.

To calculate the power in the emitted radiation, for convenience choose the frame that is the instantaneous rest frame of the particle at the source event, so $\mathbf{v} = 0$. Then the radiation field in eqn (8.48) reduces to

$$\mathbf{E}_{\text{rad}} = \frac{q}{4\pi\epsilon_0 c^2} \left[\frac{\mathbf{n} \wedge (\mathbf{n} \wedge \mathbf{a})}{r} \right].$$

The energy flux is given by the Poynting vector $\mathbf{N} \equiv \epsilon_0 c^2 \mathbf{E} \wedge \mathbf{B}$ (see chapter 13). It is allowable to calculate the Poynting vector of the radiative field alone (rather than the total field), since, in any case, for large enough r this part of the field contains all the energy crossing the light-sphere of radius r. We obtain:

$$\mathbf{N} = \epsilon_0 c \mathbf{E}_{\text{rad}} \wedge (\mathbf{n} \wedge \mathbf{E}_{\text{rad}}) = \epsilon_0 c E_{\text{rad}}^2 \mathbf{n}. \tag{8.75}$$

A solid angle $d\Omega$ on a sphere around the source event receives this flux onto an area $r^2 d\Omega$ at normal incidence, so the power radiated per unit solid angle is[8]

[8] We continue to use SI units throughout this section. The corresponding expressions in Gaussian units can be obtained by replacing $(q^2/4\pi\epsilon_0)$ by q^2.

$$\frac{d\mathcal{P}}{d\Omega} = Nr^2 = \frac{q^2}{4\pi\epsilon_0}\frac{a^2\sin^2\theta}{4\pi c^3} \qquad (8.76)$$

where θ is the angle between $\mathbf{n}$ and $\mathbf{a}$. This exhibits the $\sin^2\theta$ dependence characteristic of dipole radiation. The radiation is emitted primarily to the sides: i.e., in directions orthogonal to the acceleration.

The total power emitted is obtained by integrating eqn (8.76) over all solid angle, giving

$$\mathcal{P}_L = \frac{2}{3}\frac{q^2}{4\pi\epsilon_0}\frac{a^2}{c^3}. \qquad (8.77)$$

This is Larmor's formula for the power emitted by a non-relativistic accelerating charge.

We should like to generalize this to all velocities. It is not necessary to re-do the calculation, because we can argue that $\mathcal{P}_L$ is a Lorentz-invariant quantity. The argument hinges on the idea that we can regard the radiated part of the field as an 'isolated system' whose total energy and momentum form the components of a 4-vector. This is not obvious (it is not true of the non-radiative part of the field, for example), but it is valid because after the charge stops accelerating the radiation field continues to propagate outwards, so that it can be completely contained in a region of space where there are no charged particles in interaction with it. A more thorough discussion involving a consideration of the momentum flow in the field is provided in chapter 13.

Let $d\mathcal{E}_0$ be the total energy emitted into the radiation field in the instantaneous rest frame during some small time $d\tau$ (this is a proper time), then $\mathcal{P}_L = d\mathcal{E}_0/d\tau$. Since the radiation is emitted equally in opposite directions in the rest frame, the total momentum of the radiation field is zero in that frame, and since this energy and momentum form a 4-vector, we know how they transform. Clearly, the energy will be $d\mathcal{E} = \gamma d\mathcal{E}_0$ in some other frame, and the time interval $dt_s = \gamma d\tau$, therefore $d\mathcal{E}/dt_s = d\mathcal{E}_0/d\tau$, so the power (i.e., energy per unit time taken to emit it) is Lorentz-invariant. Note that to obtain an invariant quantity we choose to *define* 'power radiated' to mean energy per unit time taken to emit it, not receive it. dt_s is not a proper time; it is a reference frame time between events at the source (called a retarded time).

We can now find the general formula for the power by writing down a Lorentz scalar quantity that depends only on velocity, acceleration, and proper time, and that reduces to eqn (8.77) in the rest frame. The unique answer (Heaviside 1902) is

---

Power emitted by an accelerating charge

$$\mathcal{P}_L = \frac{2}{3}\frac{q^2}{4\pi\epsilon_0 c^3}\dot{\mathsf{U}}\cdot\dot{\mathsf{U}} \quad = \frac{2}{3}\frac{q^2 a_0^2}{4\pi\epsilon_0 c^3} \qquad (8.78)$$

---

where the dot signifies $d\mathsf{U}/d\tau$ (we have not used A for the 4-acceleration here in order to avoid confusion with the 4-vector potential; the mention of proper time in $\dot{\mathsf{U}}$ does not change the fact that $\mathcal{P}_L$ is an energy per unit reference frame time). In order to use the formula in practice it can be helpful to have it expressed in terms of 3-velocity and 3-acceleration at the source event, using eqn (2.61):

$$\mathcal{P}_L = \frac{2}{3}\frac{q^2}{4\pi\epsilon_0 c^3}\gamma^6\left(a^2 - \frac{(\mathbf{v}\wedge\mathbf{a})^2}{c^2}\right). \tag{8.79}$$

This version is associated with Liénard (1898).

To prepare for discussions of momentum in chapter 13 we shall quote also the 4-vector giving the rate at which 4-momentum is carried away by the radiation (Abraham 1903):

$$\frac{d\mathsf{P}}{d\tau} = \frac{\mathcal{P}_L\mathsf{U}}{c^2} = \frac{2}{3}\frac{q^2}{4\pi\epsilon_0 c^5}(\dot{\mathsf{U}}\cdot\dot{\mathsf{U}})\mathsf{U}. \tag{8.80}$$

This is obtained by arguing that the energy and momentum of the radiation field form a 4-vector (see above), and the radiation pattern is symmetric in the rest frame of the particle at the source event, so that it has no 3-momentum in that frame. Hence we seek a 4-vector whose zeroth component gives the power that we have already calculated, and which is parallel to U in the rest frame. Eqn (8.80) gives the only such 4-vector.

8.3.1 Linear and circular motion

For linear acceleration—i.e., $\mathbf{a}$ parallel to $\mathbf{v}$—we have from eqn (8.79):

$$\mathcal{P}_L = \frac{2}{3}\frac{q^2}{4\pi\epsilon_0 c^3}(\gamma^3 a)^2 = \frac{2}{3}\frac{q^2}{4\pi\epsilon_0 m^2 c^3}\left(\frac{dp}{dt}\right)^2$$

For fixed rest mass, the rate of change of momentum is equal to the change of energy per unit distance (exercise), $dp/dt = dE/dx$, so for linear motion the power radiated depends only on the externally provided force (potential energy gradient), not on the actual energy or momentum of the particle.

Consider the cases of a linear accelerator and a dipole oscillator. Writing $dE/dx = (dE/dt)(dt/dx)$ we find the ratio of radiated power to supplied power is

$$\frac{\mathcal{P}_L}{dE/dt} = \frac{2}{3}\frac{q^2}{4\pi\epsilon_0 m^2 c^3}\frac{1}{v}\frac{dE}{dx}. \tag{8.81}$$

The infinity for $v \to 0$ here is quite interesting: it says that if a particle accelerates through $v = 0$ then there is a moment at which it continues to emit radiation even though the externally applied forces are not providing any energy! We shall investigate this in volume 2, and argue that the bound field provides the energy. For high-velocity particles $(v \to c)$, the result shows that energy losses by radiation are negligible unless an energy equal to the rest energy of the particle is provided in a distance $q^2/(4\pi\epsilon_0 mc^2)$. For an electron this distance is 2.8×10^{-15} m;

the acceleration would have to reach 10^{14} MeV/m before losses become significant. Radiation loss in linear particle accelerators (on Earth) is utterly insignificant.

For an electric dipole oscillator of dipole moment $d(t) = qz = d_0 \cos \omega t$ we have $a = d^2 z/dt^2 = -\omega^2 (d_0/q) \cos \omega t$, so the instantaneous emitted power is

$$\mathcal{P}_L = \frac{2/3}{4\pi\epsilon_0 c^3} (\gamma^3 \omega^2 d_0 \cos \omega t)^2 .$$

Taking the non-relativistic limit $\gamma \simeq 1$, and taking the average over a cycle (the average value of the $\cos^2$ function is $1/2$), we find that the average power emitted is

$$\bar{\mathcal{P}}_L = \frac{1}{3} \frac{\omega^4 d_0^2}{4\pi\epsilon_0 c^3} = \frac{2\pi^2}{3} \frac{\omega d_0^2}{\epsilon_0 \lambda^3} . \tag{8.82}$$

This provides an important general insight into power radiation by small oscillators: the ω^4 dependence shows that for an oscillator of given size, energy is much more rapidly emitted via high-frequency than low-frequency oscillation. This explains why mobile phones have to use microwave not radio-wave technology. It also explains why the ultraviolet transitions in atoms and molecules are typically much stronger than the visible or infrared ones. This general insight lies behind the much-beloved problem of explaining why the sky is blue. Molecules and dust particles in the atmosphere scatter light from the Sun; owing mainly to eqn (8.82) they do so more efficiently for blue than for red light; we receive the scattered light—except during a sunrise or sunset, when we see primarily the remaining non-scattered part.

The d_0^2 term in eqn (8.82) is also significant. It shows why radio masts are tall. Its cousin in gravitational-wave physics is the reason why no gravitational waves have ever been detected by detectors of modest (a few metres) size.

For circular motion the acceleration is perpendicular to the velocity, and in synchrotrons it is typically much larger than in linear accelerators, since a given force can cause a much larger transverse than longitudinal acceleration (by a factor γ^2; see eqn (4.13)). Using $|\mathbf{v} \wedge \mathbf{a}| = va$ and $a = v^2/r$ for motion around a circle of radius r, eqn (8.79) gives

$$\mathcal{P}_L = \frac{2}{3} \frac{q^2}{4\pi\epsilon_0 c^3} \frac{\gamma^4 v^4}{r^2} . \tag{8.83}$$

The radiative loss per revolution is therefore

$$\Delta E = \frac{q^2}{3\epsilon_0 r} \gamma^4 (v/c)^3 .$$

For electrons the quantity $e^2/(3\epsilon_0 r)$ is 6×10^{-9} eV when $r = 1$ metre. When one wants a bright source of X-rays, the synchrotron radiation is welcome. When one wants to accelerate particles to high velocities, on the other hand, the radiation is a problem. It represents a continuous energy loss that must be compensated by the accelerator. This limits the velocity that can be achieved in circular particle accelerators, and is a major reason why these accelerators have had to be made larger and larger: by increasing the radius of curvature, the acceleration and thus synchrotron radiation is reduced for any given particle energy. A 10 GeV electron synchrotron has $\gamma = E/(mc^2) \simeq 2 \times 10^4$, so $\Delta E \simeq 880$ MeV if the radius is 1 metre; such a high loss would be prohibitive. At Cornell such a synchrotron was built with $r = 100$ m, producing a loss per turn of 8.8 MeV.

8.3.2   Angular distribution

From eqns (8.48) and (8.75) the Poynting vector of the radiative part of the field is given by

$$\mathbf{N} \cdot \mathbf{n} = \frac{1}{4\pi c^3} \frac{q^2}{4\pi\epsilon_0 r^2} \frac{|\mathbf{n} \wedge ((\mathbf{n} - \mathbf{v}/c) \wedge \mathbf{a})|^2}{(1 - \mathbf{v} \cdot \mathbf{n}/c)^6}. \tag{8.84}$$

This is the rate per unit area at which energy is detected at some time t, having been emitted at the prior time $t_s = t - r/c$. The total energy per unit area radiated in direction $\mathbf{n}$ during some period of acceleration is given by

$$E = \int_{t_1}^{t_2} \mathbf{N} \cdot \mathbf{n} \, dt = \int_{t_{s,1}}^{t_{s,2}} \mathbf{N} \cdot \mathbf{n} \frac{dt}{dt_s} dt_s.$$

This shows that if one is interested in the emission (as opposed to the detection) then the more useful quantity to consider is $(\mathbf{N} \cdot \mathbf{n})(dt/dt_s)$. We have (exercise 8.7)

$$\frac{dt}{dt_s} = 1 - \mathbf{v} \cdot \mathbf{n}/c. \tag{8.85}$$

We choose to *define* the 'power radiated' to mean energy per unit *source time* (= 'retarded time'), so the 'flux radiated' is

$$\mathbf{N} \cdot \mathbf{n} \frac{dt}{dt_s} = \mathbf{N} \cdot \mathbf{n}(1 - \mathbf{v} \cdot \mathbf{n}/c)$$

and therefore the power radiated per unit solid angle is

$$\left. \frac{d\mathcal{P}}{d\Omega} \right|_{t_s} = \frac{1}{4\pi c^3} \frac{q^2}{4\pi\epsilon_0} \frac{|\mathbf{n} \wedge ((\mathbf{n} - \mathbf{v}/c) \wedge \mathbf{a})|^2}{(1 - \mathbf{v} \cdot \mathbf{n}/c)^5}. \tag{8.86}$$

This formula is an example of the headlight effect; cf. eqn (3.14). Previously we treated a pattern isotropic in the source frame; now we are treating a dipole pattern. Where previously we had the 4th power of $(1 - (v/c)\cos\theta)$ in the denominator, now we have the 5th power, and a more complicated numerator. Going to energy per unit observer time instead of source time introduces a further factor of $(1 - (v/c)\cos\theta)$, making a 6th power, but after taking the numerator into account, the overall result is roughly a 4th power, as before.

Fig. 8.18 Angular distribution of radiation from an accelerating point charge. Each polar plot shows a curve whose distance from the origin is proportional to $\mathrm{d}\mathcal{P}/\mathrm{d}\Omega$. The lower left curve is for a charge at rest, giving the dipole pattern. The upper and right curves show the pattern at $\gamma = 2$ for a charge moving vertically and horizontally, respectively, both with vertical acceleration—eqns (8.87) and (8.89). The curves are not to scale; if the charge at rest has a maximum power per unit solid angle equal to 1, then the maximum values for the other two cases are 576 and 416, respectively. The radiation emitted in the backward direction in the last case is so comparatively weak that it does not register on the plot. See also Fig. 8.19.

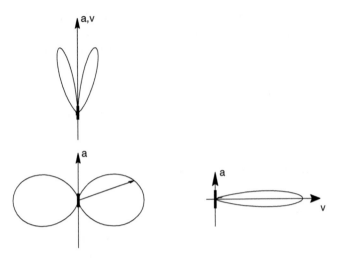

For linear motion eqn (8.86) can be expressed

$$\left.\frac{\mathrm{d}\mathcal{P}}{\mathrm{d}\Omega}\right|_{t_s} = \frac{q^2 a^2}{4\pi\epsilon_0 4\pi c^3} \frac{\sin^2\theta}{(1 - (v/c)\cos\theta)^5}, \tag{8.87}$$

where θ is the angle between the direction of emission and the acceleration (or velocity) vector. At low speed this is the dipole pattern. As $v \to c$ the power grows and the distribution is tipped more and more towards the forwards direction. In the limit it becomes a function of $\gamma\theta$ alone (exercise 8.8), with a maximum at the angle $\theta_{\max} = 1/2\gamma$. The spread of angles can be expressed by the r.m.s. value

$$\sqrt{\langle\theta^2\rangle} \to \frac{1}{\gamma}. \tag{8.88}$$

Figs 8.18 and 8.19 show the angular distribution.

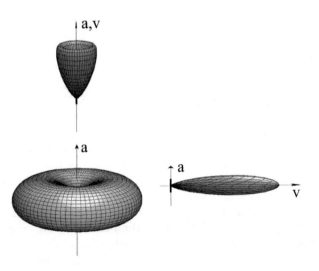

Fig. 8.19 Like Fig. 8.18, but giving a sense of the distribution in three dimensions.

For a charge undergoing circular motion, one finds

$$\frac{\mathrm{d}\mathcal{P}}{\mathrm{d}\Omega}\bigg|_{t_s} = \frac{q^2}{4\pi\epsilon_0}\frac{a^2/(4\pi c^3)}{(1-(v/c)\cos\theta)^3}\left[1 - \frac{\sin^2\theta\cos^2\phi}{\gamma^2(1-(v/c)\cos\theta)^2}\right] \quad (8.89)$$

where θ, ϕ are the usual polar angles if we align the z axis with $\mathbf{v}$ and the x axis with $\mathbf{a}$.

Approximate width of the frequency distribution

A charge moving around a circle emits in the angular distribution given above. For high speeds this is a narrow 'searchlight' beam in the direction of the velocity, which sweeps around as the particle moves around the circle. Consequently, a user situated at a fixed position receives short pulses of electromagnetic waves, as the beam sweeps across him. The frequency spectrum of the received radiation is a (complicated) universal function of ω/ω_c, where the 'critical frequency' ω_c is conventionally defined as

$$\omega_c \equiv \frac{3}{2}\frac{c}{r}\gamma^3$$

for a synchrotron of radius r. This is $\omega_c = (3/2)\gamma^3\omega_0$, where $\omega_0 = c/r$ is the angular frequency of the particle orbit for a fast particle (one with $v \simeq c$). The spectrum is plotted in Fig. 8.20. The main properties of this spectrum can be estimated from Fourier analysis of the received pulse. A pulse of duration Δt must have a frequency width $\Delta\omega$ satisfying

$$\Delta\omega\Delta t \gtrsim 1.$$

In the case of synchrotron radiation in the limit of high particle velocity (short pulses) there is no reason for the spectral width to greatly exceed the minimum possible, so we may estimate

$$\Delta\omega \simeq \frac{1}{\Delta t}.$$

Now consider the forward lobe in the emitted radiation pattern. It has an angular half-width θ given by the headlight effect, hence approximately satisfying $\cos\theta = v/c$, from which

$$\sin\theta = \sqrt{1-\cos^2\theta} = \frac{1}{\gamma}.$$

Therefore, for $\gamma \gg 1$,

$$\theta \simeq 1/\gamma$$

by using the small angle approximation. The beam sweeps across the user when the particle moves around its circular orbit by this same angle θ, hence a distance $r\theta$ which takes a time $\delta t = r\theta/v = r/\gamma v$. However, the pulse of energy received by the user is not of duration δt. It is of shorter duration because the leading edge of the pulse was emitted before the trailing edge, and has further to travel. During the time δt the leading edge travels a distance $c\delta t$ while the particle travels $v\delta t$ in the same

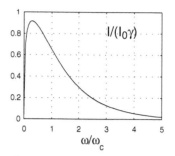

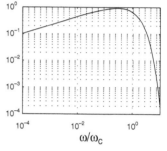

Fig. 8.20 Spectrum of synchrotron radiation, on linear (top) and logarithmic (bottom) scales. The spectrum is $I = I_0\gamma x \int_1^\infty K_{5/3}(x)\mathrm{d}x$, where $x = \omega/\omega_c$; $I_0 = \sqrt{3}e^2/4\pi\epsilon_0 c$ and K is a modified Bessel function.

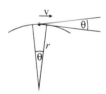

Fig. 8.21

Fig. 8.22

direction. Consequently, the trailing edge of the pulse lags behind the leading edge by

$$\Delta t = \frac{c\delta t - v\delta t}{c} = \frac{r}{\gamma v}(1 - v/c).$$

To convert this into an expression in terms of γ alone, multiply top and bottom by $(1 + v/c)$ and argue that for high speed, $v \simeq c$. Therefore

$$\Delta t \simeq \frac{r}{2c\gamma^3} \qquad \Rightarrow \qquad \Delta\omega \simeq \frac{4}{3}\omega_c. \tag{8.90}$$

This explains how the critical frequency arises.

One may expect that the spectrum extends to low frequency (including, for example, the comparatively low frequency ω_0), therefore the width $\Delta\omega$ also indicates the typical frequency near the peak of the spectrum. For example, a modern synchrotron might operate at $\omega_0 = 3 \times 10^6$ s^{-1}, $\gamma = 10^4$, yielding a typical wavelength $\lambda = 6 \times 10^{-10}$ m: i.e., 'hard' X-rays.

Radiation having the characteristic spectrum and polarization of synchrotron radiation is observed in many astrophysical sources, such as the Crab Nebula. It is believed to result from electrons spiralling in the interstellar magnetic field. Artificial sources using synchrotron radiation for experimental facilities on Earth normally use further techniques such as *undulators* that result in much higher X-ray brightness for a given electron energy.

Exercises

(8.1) Prove that if the 4-polarization of a plane wave in Lorenz gauge is changed as in eqn (8.8) then the fields are unaffected.

(8.2) Evaluate $\Box^2 g(t - r/c)/r$. (Hint: you may find it helpful to introduce a new variable $u = t - r/c$, and use $\partial g/\partial r = (\mathrm{d}g/\mathrm{d}u)(\partial u/\partial r)$.)

(8.3) Find the vector potential due to a long straight current-carrying wire, in Lorenz gauge. Hence find the magnetic field around such a wire.

(8.4) In a frame S a point charge first moves uniformly along the negative x-axis in the positive x direction, reaching the point $(-d, 0, 0)$ at $t = -\Delta t$, and then is brought to rest at the origin at $t = 0$. Sketch the lines of electric field in S at $t = 0$.

(8.5) A charged particle moves along the x axis with constant proper acceleration ('hyperbolic motion'), its worldline being given by

$$x^2 - t^2 = \alpha^2$$

in units where $c = 1$. Find the electric field at $t = 0$ at points in the plane $x = \alpha$, as follows.

(i) Consider the field event $(t, x, y, z) = (0, \alpha, y, 0)$. Show that the source event is at

$$x_s = \alpha + \frac{y^2}{2\alpha}$$

(ii) Show that the velocity and acceleration at the source event are

$$v_s = -\frac{\sqrt{x_s^2 - \alpha^2}}{x_s}, \qquad a_s = \frac{\alpha^2}{x_s^3}.$$

(iii) Consider the case $\alpha = 1$, and the field point $y = 2$. Write down the values of x_s, v_s, a_s. Draw on a diagram the field point, the source point, and the location of the charge at $t = 0$. Mark at the field point on the diagram the directions of the vectors $\mathbf{n}$, $\mathbf{v}$, $\mathbf{a}$, $\mathbf{n} \wedge$

$(\mathbf{n} \wedge \mathbf{a})$. Hence, by applying the formula above, establish the direction of the electric field at $(t, x, y, z) = (0, 1, 2, 0)$.

(iv) If two such particles travel abreast, undergoing the same motion, but fixed to a rod perpendicular to the x axis so that their separation is constant, comment on the forces they exert on one another. This is an example of a *self force*, also called *radiation reaction*.

(8.6) Obtain eqn (8.62) from (8.60) using vector methods, making use of the general result

$$\nabla \wedge (u\mathbf{f}) = (\nabla u) \wedge \mathbf{f} + u\nabla \wedge \mathbf{f}$$

where u and $\mathbf{f}$ are scalar and vector fields. (Hint: first you need to substitute $u = (1/r)$ and $\mathbf{f} = \dot{\mathbf{d}}$, and then to evaluate $\nabla \wedge \dot{\mathbf{d}}$ use the fact that the dipole $\mathbf{d}$ has a fixed direction, so $\dot{\mathbf{d}} = \dot{d}\mathbf{e}$ where $\mathbf{e}$

is a constant unit vector along the direction of the dipole.)

(8.7) Derive eqn (8.85).

(8.8) Show that in the limit $v \to c$, eqn (8.87) takes the form

$$\left. \frac{\mathrm{d}\mathcal{P}}{\mathrm{d}\Omega} \right|_{t_s} = \frac{q^2}{4\pi\epsilon_0} \frac{8a^2}{\pi c^3} \frac{\gamma^8(\gamma\theta)^2}{(1 + \gamma^2\theta^2)^5}$$

and that this has a maximum at $\gamma\theta = 1/2$.

(8.9) According to a classical model, an electron orbiting a proton should emit synchrotron radiation. Consider a classical model of a hydrogen atom in which an electron initially follows a circular orbit at the Bohr radius, at a speed $c/137$. Estimate how long it would take the electron to radiate enough energy to move to an orbit of substantially smaller radius, according to classical physics.

Part II

An Introduction to General Relativity

The Principle
of Equivalence

9.1   Free fall

Since the dawn of human history, children and others with a playful or enquiring turn of mind have been dropping pairs of objects off branches and boulders with the aim of seeing which would hit the ground first. Aristotle thought that objects have a 'natural motion' that is faster for heavier objects, thus implying that the heavier object will win the child's game. This is true in a viscous medium such as air, but misses the essential point: the difference tends to vanish in the limit where the objects are heavy and have little air resistance. It was others such as John Philiponus (sixth century), Simon Stevin ($\sim$ 1586) and especially Galileo Galilei ($\sim$ 1610) who observed more carefully and realized that there is something universal about the motion of objects falling under gravity.

Galileo said later that he had first thought about this problem during a hailstorm, when he noticed that large and small hailstones hit the ground at the same time. If Aristotle had been right then these hailstones would have had to set out at substantially different times or from substantially different heights, which seems unlikely. Galileo studied the problem under controlled conditions by rolling balls down inclined planes and timing the rate at which their speed increased. In his writings he also described a thought experiment in which objects of different mass are dropped from a high tower such as the leaning tower of Pisa. Here is a lovely example of his reasoning:

Imagine that two objects, one light and one heavier than the other one, are connected to each other by a string. Drop this system of objects from the top of a tower. If we assume heavier objects do indeed fall faster than lighter ones (and conversely, lighter objects fall slower), the string will soon pull taut as the lighter object retards the fall of the heavier object. But the system considered as a whole is heavier than the heavy object alone, and therefore should fall faster. This contradiction leads one to conclude that the assumption is false.

Such reasoning makes the conclusion appear very convincing, but nevertheless to actually perform a controlled experiment is crucial. Let us examine another thought experiment—this time involving particles of differing charge-to-mass ratio moving in a uniform electric field. For this case we have the advantage of already knowing the answer: the particle with larger q/m will have the larger acceleration. What if we apply Galileo's argument here? We imagine a string attached between an object having $q/m = 1$ unit (say), and another having $q/m = 2$ units. The string will soon pull taut, and the composite object has an intermediate acceleration. There is no contradiction because the composite object has a net charge to mass ratio **not** equal to $1 + 2 = 3$ units, but to a value intermediate between 1 and 2 units. This makes it clear that Galileo's argument for masses falling under gravity turns on the fact that he already knew that masses add in a simple way when two objects come together to form a composite object.

In order to clear away all doubt on the matter, a mere thought-experiment is insufficient. The crucial experiment is the real experiment, and the outcome is now well known.

In Newtonian physics the fact that different objects have the same acceleration under gravity comes about by bringing together two ideas. The first is Newton's Second Law of Motion, which in modern notation reads

$$\mathbf{f} = \frac{\mathrm{d}}{\mathrm{d}t}(m_{\mathrm{i}}\mathbf{v}) \tag{9.1}$$

where m_{i} is inertial mass. The second is Newton's Law of Universal Gravitation, which reads, for the force on an object of passive gravitational mass m_{g},

$$\mathbf{f} = \frac{GMm_{\mathrm{g}}}{r^2}\hat{\mathbf{r}} \tag{9.2}$$

where the direction is such that the force is attractive. By combining these two equations, one finds that the acceleration under gravitational attraction to another body of active gravitational mass M is $(m_{\mathrm{g}}/m_{\mathrm{i}})GM/r^2$. Newton himself drew attention to the interesting fact that, in his theory, $m_{\mathrm{i}} = m_{\mathrm{g}}$, with the result that the acceleration is independent of m_{i}, so all bodies have the same acceleration in a given gravitational environment. Newton did not take this for granted, but tested it by experiment with a pendulum.

When we come to Special Relativity we encounter a problem: Newton's Law of Universal Gravitation is inconsistent with the Postulates of Relativity (it is not Lorentz-covariant), and there is no simple way to modify it to make it consistent. For example one cannot simply construct a gravitational version of Maxwell's equations because there is no suitable source quantity that is both conserved and Lorentz-invariant.[1] This means that either Newton's Law is wrong, or the Postulates of Relativity are wrong, or both, or the natural world is inconsistent. Following the hunch of every good scientist, we shall put the last possibility 'on the shelf' (guessing or hoping that it will not be needed) and explore the other possibilities. We shall find that Newton's Law has to be replaced by a much more complicated yet profoundly elegant and satisfying theory, and that the Main Postulates of Relativity can be retained, as long as we apply them to small regions of space and reinterpret what is meant by an

[1] However, General Relativity does lead to a close correspondence between electromagnetism and gravitation in the weak field limit.

inertial frame of reference. The Zeroth Postulate (concerning Euclidean geometry) will have to be abandoned, except as a limiting case. These conclusions are the subject of this chapter.

Since we shall be abandoning Newton's gravitational law, the question arises concerning whether we have any knowledge of gravity, at the outset, that we can still trust in conditions that have not been experimentally tested. Einstein recognised that the coming together of two ideas in Newton's theory to produce an outcome independent of m is a form of 'conspiracy' in the equations which we ought to question. If the fundamental principles of physics are mathematically elegant, then such conspiracies are a sign that we have been thinking about something the wrong way. In search of a new approach, he therefore thought through the physical implications of gravitational acceleration being the same for different bodies. We shall next do the same.

9.1.1   Free fall or free float?

A group of passengers in a lift suddenly feel the cabin judder as a loud sound of something snapping is heard. They feel in their stomachs a lurch and a change of motion, then notice that their feet are no longer pressed firmly against the floor. One passenger even begins to float up into the air.

'Oh no!' cry the passengers, 'we are falling!'

'Look Mummy!' excitedly cries a child, 'I am floating!'

Who is right? Are the passengers in trouble? Actually, they suffer no ill-effects whatsoever as long as the lift cabin falls freely. Their fear is justified only because they know that in ordinary buildings lift cabins are situated above hard ground, and when they hit the ground they will then experience large and damaging forces.[2] Is the child floating or falling? From the perspective of the cabin, the child's motion is smooth and simple, moving across the interior space of the cabin at uniform velocity, until she gently bumps into another passenger or a wall or ceiling or floor (the cabin all the while continuing its free descent). If she lets go of her notebook or bag of treasures then it will float across with her. In the freely-falling cabin, things move relative to other things just as they would move if there were no gravity and the cabin were at rest (apart from a detail we shall examine in a moment). But if all the motions are just as if there were no gravity, then what is the physical meaning of this thing called 'gravity'? If there are no effects of gravity, then physically speaking we might as well say there *is no gravity* in the freely-falling cabin.

For another example of the same idea, see Fig. 9.1. A house is built on a platform extended out from a cliff. An artist holding a cannister of spray-paint jumps across the room, spraying paint against the wall, and thus leaving a lovely parabolic arc on the wall. Now suppose the platform breaks just as the artist jumps. From the perspective of the cliff, the trajectory of the artist is unchanged: she follows exactly the same parabola as she did before. But now look at the line of paint on the

[2] Lifts in modern buildings have a large number of safety features that make their descent perfectly safe even under catastrophic failure such as a cable breaking. The story is for illustration only.

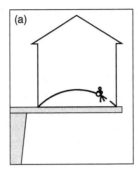
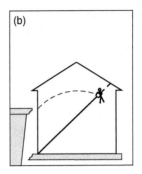

Fig. 9.1 Free fall or free float? An artist makes a big jump across the room of her house, spraying paint onto the wall from a paint cannister as she goes. In one case (a) the house is supported by a platform, and in the other (b) the platform has broken and the house falls. The artist describes exactly the same trajectory in the two cases. The line of paint on the wall reveals the trajectory relative to the house: it is a parabola in the first case, and a straight line in the second. We typically say that the artist is 'falling'—her downwards acceleration results in the parabolic curve. However, *exactly the same motion* could equally be regarded as 'floating'—her trajectory is perfectly straight when examined from another point of view. Einstein invites us to regard the second point of view—that of the freely-falling reference frame—as the best one in which to formulate the laws of physics. (Image concept copied from Taylor and Wheeler.)

wall: it is a straight line! Relative to the falling house, the motion of the artist's body is just as it would be if no forces acted on her.

These examples suggest a strange but wonderful idea: namely, that the effects of gravity are of a special kind, that can be made to disappear by a mere change of reference frame, as long as accelerating reference frames are allowed. This idea is one of the foundation stones of General Relativity. We shall find that it is true in sufficiently small regions of spacetime.

Tidal effects

The detail omitted from the above discussion was the fact that the acceleration due to gravity varies from one place to another, so the child's motion is not quite a straight line relative to the lift cabin. A useful way to consider this variation is to write the acceleration due to gravity in some region of space as a sum of two terms:

$$\mathbf{a}(x, y, z) = \mathbf{a}_0 + \mathbf{a}_\Delta(x, y, z) \tag{9.3}$$

where for simplicity we considered a static case. The first term $\mathbf{a}_0$ is independent of position and is the average of $\mathbf{a}$ over the region, and the second term accounts for the spatial dependence. Figure 9.2 shows a typical case: the field above a spherical object such as the Earth. If an

Fig. 9.2 Tidal effects of gravity. A group of rocks falls freely in the gravitational field of a spherical planet. The arrows show the acceleration due to gravity of each rock. This can be decomposed as shown, into the average acceleration of the group plus the departures from the average. The former causes the rocks to accelerate downwards without changing their relative locations, and the latter is called the 'tidal' contribution and causes relative acceleration within the group. In the example shown here the tidal effect is such as to squeeze the ring of rocks horizontally and stretch it vertically.

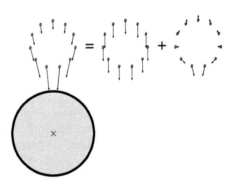

initially circular ring of non-interacting rocks is dropped in vacuum in such a field, then the $\mathbf{a}_0$ term causes all the rocks to accelerate together and therefore preserves the circular shape, while the $\mathbf{a}_\triangle$ term causes the top and bottom rocks to accelerate away from each other, and the side rocks to accelerate in, thus 'squashing' the circle into an egg-shape.

When as a child I first learned that the gravitational pull of the Moon is the cause of tides on the Earth, I struggled to understand why it was that high tide occurs *twice* and not once per day. It seemed natural to me that the pull of the Moon would make the water of the oceans pile up into a lump on the side facing the Moon, but I could never see why there would be a second lump on the side opposite from the Moon. Figure 9.2 shows the reason: the weaker gravitational force at positions further from the Moon is equally significant as the stronger gravitational force at positions closer to the Moon. The net effect is to provide the oceans with their average orbital motion and also to squeeze them into a shape with a lump of deeper water on both the nearer and the further side of the Earth, relative to the Moon. In general, the term $\mathbf{a}_\triangle(x, y, z)$ is called the *tidal* term, and the effects of the variation of gravity with respect to position are called *tidal effects*. For example, if a solid body such as a human being, rather than a circle of non-interacting rocks, were to fall in a gravitational field such as the one shown in Fig. 9.2, then the differing acceleration due to gravity at different parts of the body would tend to stretch it vertically and squeeze it horizontally. In Earth's field these effects are small, but near a very dense body such as a neutron star they can be large and painful!

We can estimate tidal effects in all but the most extreme cases by using Newtonian physics. If the gravitational acceleration is $\mathbf{g}(\mathbf{x})$ then the equation of motion of a freely-falling object is

$$\frac{\mathrm{d}^2\mathbf{x}}{\mathrm{d}t^2} = \mathbf{g}(\mathbf{x})$$

and that of a nearby object at $\mathbf{x} + \Delta\mathbf{x}$ is

$$\frac{\mathrm{d}^2}{\mathrm{d}t^2}(\mathbf{x} + \Delta\mathbf{x}) = \mathbf{g}(\mathbf{x} + \Delta\mathbf{x}).$$

Therefore the gap between the objects has the equation of motion

$$\frac{\mathrm{d}^2\Delta\mathbf{x}}{\mathrm{d}t^2} = \mathbf{g}(\mathbf{x} + \Delta\mathbf{x}) - \mathbf{g}(\mathbf{x})$$

$$\simeq \frac{\partial\mathbf{g}}{\partial x}\Delta x + \frac{\partial\mathbf{g}}{\partial y}\Delta y + \frac{\partial\mathbf{g}}{\partial z}\Delta z = (\Delta\mathbf{x} \cdot \nabla)\mathbf{g}, \qquad (9.4)$$

which can also be written

$$\frac{\mathrm{d}^2\Delta x_i}{\mathrm{d}t^2} = -\sum_{j=1}^{3} \frac{\partial^2\Phi}{\partial x_i\partial x_j}\Delta x_j \qquad (9.5)$$

where Φ is the gravitational potential, defined such that $\mathbf{g} = -\nabla\Phi$. Eqn (9.5) is called the equation of **geodesic deviation in Newtonian**

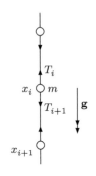

x_i ○ m

T_i

T_{i+1} $\mathbf{g}$

x_{i+1} ○

Fig. 9.3

physics. This terminology will be clarified in the next chapter. The essential idea is that the equation expresses how two initially close-together objects draw apart (or together) if both are in free fall.

As an example, consider a spherical body giving $\Phi = -GM/r$. Then, for particles situated close to the z axis at $\{0, 0, r\}$ the result is

$$\frac{\mathrm{d}^2}{\mathrm{d}t^2} \begin{pmatrix} \Delta x \\ \Delta y \\ \Delta z \end{pmatrix} = \frac{GM}{r^3} \begin{pmatrix} -1 & & \\ & -1 & \\ & & 2 \end{pmatrix} \begin{pmatrix} \Delta x \\ \Delta y \\ \Delta z \end{pmatrix}.$$

Thus there is attraction in the horizontal direction, tending to bring a set of test particles together, and expulsion in the vertical direction, tending to push the particles apart, as we have already seen in Fig. 9.2.

Now consider a single extended object such as a chain of massive rocks joined by light strings (Fig. 9.3). We suppose that the chain is oriented vertically and that the strings are to good approximation inextensible. In this case all the rocks in the chain have the same acceleration a, and if the chain is falling in the gravitational field of a large spherical object then the tidal gravitational forces are such as to put the strings into tension. The equation of motion of the ith rock is

$$mg(x_i) + T_{i+1} - T_i = ma.$$

Writing $g(x_i) \simeq g_0 + x_i \mathrm{d}g/\mathrm{d}x$, where g_0 is the gravitational acceleration at the centre of the chain, so $a = g_0$, we find

$$T_{i+1} = T_i - mx_i \frac{\mathrm{d}g}{\mathrm{d}x}.$$

This shows that as one moves down the chain from the top to the middle, each string has a tension exceeding that of the one before (since $\mathrm{d}g/\mathrm{d}x < 0$). If we now replace the chain by a single continuous rod, then one finds

$$\mathrm{d}T = -x\mathrm{d}m\frac{\mathrm{d}g}{\mathrm{d}x} = -x\rho A\mathrm{d}x\frac{\mathrm{d}g}{\mathrm{d}x} \quad \Rightarrow \quad \frac{\mathrm{d}T}{\mathrm{d}x} = -\rho A\frac{\mathrm{d}g}{\mathrm{d}x}x$$

where ρ is the density of the rod and A is its cross-sectional area. The tension falls to zero at each end of the rod. Integrating from one end to the centre of a rod of length L, one finds that the tension in the centre of the rod is

$$T_0 = \frac{1}{8}\rho AL^2\frac{\mathrm{d}g}{\mathrm{d}x} \tag{9.6}$$

where we have assumed that the gravitational gradient was uniform along the rod. The tidal forces will break the rod if this tension exceeds the tensile breaking strength of the rod.

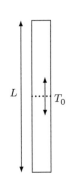

L T_0

Fig. 9.4

9.1.2   Weak Principle of Equivalence

So far we have seen that the fact that all bodies have the same acceleration in a given gravitational field lends a special character to motion under gravity. Einstein felt that it was so remarkable that it could not be a mere coincidence but must instead be somehow built in to the essential structure of a correct understanding of gravity. It is a valid insight into gravity that will survive the transition to a general theory. We state this formally as

The Weak Principle of Equivalence.
Version A: 'Universality of free fall'. All bodies experience the same acceleration when falling freely in the same gravitational field.
Version B: Gravitational mass and inertial mass are equal.

I prefer version A, but the alternative statement (version B) is included since the principle is often stated that way. The reason to prefer the 'universality of free fall' version is that it helps to avoid confusion between two possible meanings of the term 'gravitational mass'. One meaning is the *source* of gravity (sometimes called *active* mass)—that which gives rise to a gravitational field. This meaning makes sense in Newtonian physics but not in General Relativity, because in the latter the source is not a scalar quantity but a tensor quantity describing energy, momentum, internal pressure, and stress. The second meaning of 'gravitational mass' is a measure of the *response to* gravity (called *passive* mass), and this is the meaning intended in version B of the Weak Equivalence Principle. However, in General Relativity (and in any theory obeying the version A statement) this concept is superfluous. There is no need to introduce any such notion of 'gravitational mass' at all! In General Relativity the *response* to gravitation can (and should!) be written directly in terms of acceleration without regard to mass. This idea is at the heart of General Relativity, and its elucidation is one of the aims of this chapter.

9.1.3 The Eötvös–Pekár–Fekete experiment

A simple way to test the Weak Principle of Equivalence is to drop two different objects simultaneously down an evacuated tube. Depending on the precision of the timing and the care with which other effects such as magnetism are excluded, a moderate degree of precision can be obtained. A more accurate method, in the absence of modern technology, is to measure the periods of pendula of the same length having bobs of various different materials. In the modern era, much more precise methods have been used. An important step forward was taken at the beginning of the twentieth century by the Hungarian physicist Loránd Eötvös (1848–1919). He developed a clever experimental method that allowed him and his assistants Dezsö Pekár (1873–1953) and Jenö Fekete (1880–1943) to attain a precision around 5×10^{-9} in measurements of the ratio m_g/m_i. The basic idea is to investigate objects undergoing circular motion as they are carried around by the rotation of the Earth, the Earth's gravity providing the required centripetal force; see Fig. 9.5(a). The experimental ingenuity lay in discovering a method to measure directly and sensitively the *difference* between the gravitational acceleration of two such objects. The difference was obtained by means of a torsion balance, as follows.

Fig. 9.5 The Eötvös–Pekár–Fekete experiment. (a) A simple pendulum bob (shaded ball) experiences forces $\mathbf{F}_g$ and $\mathbf{T}$, and has acceleration a towards the rotation axis of the Earth. θ is the angle of latitude, and ϕ is the angle between the gravitational force and the direction of the pendulum when it is in equilibrium: i.e., not moving relative to the surface of the Earth. (b) Shows a pair of masses supported by a rigid cradle or bar, suspended from a single central wire, viewed from above. When at rest relative to the surface of the Earth, the whole apparatus has the acceleration shown in (a), and therefore a horizontal component of acceleration in the northerly direction, as shown. If the acceleration due to gravity of one bob differs from that of the other, then to obtain an equal total acceleration of both bobs the torsion pendulum must supply horizontal forces to make up the difference. It can do this by twisting away from its zero-torque position (shown dashed).

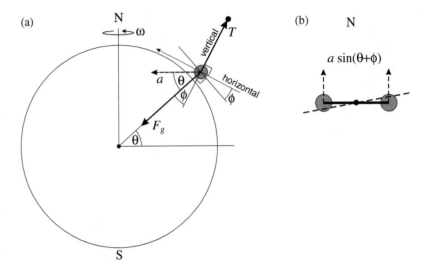

First, consider a single object suspended from a wire: a simple pendulum bob. The forces on it are the gravitational force of the Earth and the tension in the wire. If the object is at rest relative to a laboratory fixed to the surface of the Earth, then these forces must combine to provide the required acceleration $a = \omega^2 R \cos\theta$ towards the axis of rotation of Earth, where $\omega \simeq 2\pi/(24\,\text{hours})$ is the rotational frequency of the Earth, R is the radius of Earth (here assumed spherical), and θ is the angle of latitude. Let the equilibrium direction of the wire be called 'vertical' and the plane orthogonal to this be called 'horizontal', and let ϕ be the angle between the force $\mathbf{F}_g$ due to Earth's gravity and the vertical direction. Then the angle between the acceleration vector $\mathbf{a}$ and the vertical is $\theta + \phi$. Resolving the forces in the vertical and horizontal directions we obtain

$$F_g \cos\phi - T = m_i a \cos(\theta + \phi) \tag{9.7}$$

$$F_g \sin\phi + f = m_i a \sin(\theta + \phi) \tag{9.8}$$

where m_i is the inertial mass of the bob, and in the second equation we introduced a further non-gravitational horizontal force f whose value is zero for a simple pendulum. We now assume that the gravitational force F_g is given by a constant g that is the same for all types and size of bob, multiplied by a number m_g which may depend on both the type and amount of material in the bob. We can always quantify the amount of material by the inertial mass m_i, therefore we may write

$$F_g \equiv g m_i \Gamma \tag{9.9}$$

where g is the same for all bobs in a given gravitational environment, and Γ is a property of each bob. The equation serves to define Γ. The logic is that one sets $\Gamma = 1$ for one particular bob of known m_i, which permits g to be obtained, and then one aims to measure Γ for other

bobs. Better still (and this what the Eötvös experiment does), one aims
to measure the difference between Γ values of different bobs. In terms
of 'gravitational mass' we have

$$\Gamma \equiv \frac{m_g}{m_i} \tag{9.10}$$

so we expect Γ to be close to 1.

Solving eqn 9.8 for ϕ one obtains (for $f = 0$):

$$\phi = \tan^{-1} \left(\frac{a \sin \theta}{\Gamma g - a \cos \theta} \right) \simeq \frac{R\omega^2 \cos \theta \sin \theta}{\Gamma g} \tag{9.11}$$

where the second version assumed $g \gg a$ which is true for laboratories on
the surface of Earth $(g/R\omega^2 \simeq 290)$. It follows from this that one way to
measure Γ differences is to measure the angle (if any) between the wires
of different pendula. A better way is to place two bobs in a horizontal
cradle suspended from a single wire. In this arrangement (Figs 9.6 and
9.5(b)) the wire provides a single vertical force through its tension, and
a pair of opposed horizontal forces through torsion: that is, its response
to being twisted. In the situation where neither bob accelerates relative
to the laboratory, these forces, along with gravity, must provide both
bobs with the same acceleration. Therefore, from eqn (9.8) we have

$$g\Gamma_1 \sin \phi + \frac{f_1}{m_{i1}} = g\Gamma_2 \sin \phi + \frac{f_2}{m_{i2}}$$

where f_1 and f_2 are the torsional forces, related by $f_2 = -f_1 \equiv f$. The
total torque is $\mathcal{T} = fL_1 + fL_2$ where L_1 and L_2 are the distances of the
centres of the two bobs from the suspension wire, related by $m_{g1}L_1 =
m_{g2}L_2$ when the cradle is balanced. Therefore, after keeping only terms
of first order in $(\Gamma_1 - \Gamma_2)$, one finds that the torque is given by

$$\mathcal{T} = (\Gamma_1 - \Gamma_2)m_{i1}L_1 g \sin \phi. \tag{9.12}$$

A suitable wire will experience a twist through an angle in proportion to this torque. In practice one measures the oscillations about
equilibrium—for example, by fixing a mirror to the cradle and observing
it through a telescope. The angle the cradle would adopt in the absence of
torque is not accurately known, however. A crucial ingredient is to *rotate
the whole apparatus through 180° relative to the Earth's field* and then
repeat the observations. In this way the twist angle can be measured, or
an upper bound set, and hence an upper bound on $\mathcal{T}$ (the coefficient of
torsion of the wire having been measured separately).

The concept was subsequently refined by R. H. Dicke who used a fixed
apparatus, allowing the rotation of the Earth to rotate it relative to the
Sun's gravitational field, and with various other improvements reported
an accuracy of 10^{-11}. Modern experiments use a variety of methods,
including, for example, a test mass in free fall in a dedicated satellite.

These experiments test the Weak Equivalence Principle for ordinary
objects such as lumps of aluminium or gold. However, this allows a
variety of different types of mass-energy to be tested, since the mass

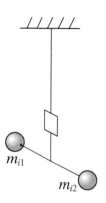

Fig. 9.6 Torsion pendulum: a pair of
masses suspended from a single wire.

of atoms is made up of contributions from the nuclear binding energies (approx 1% of the total), electrostatic energy ($\simeq 0.001$ to 0.01), and more exotic forms such as virtual positrons, as well as the masses of the constituent particles. So far all observations are consistent with the Weak Equivalence Principle, which has become one of the most well-attested ideas in physics. Such experiments remain important, however, since efforts to develop a quantum theory of gravity sometimes suggest that violations may be possible.

9.1.4   The Strong Equivalence Principle

In section 9.1.1 we considered some thought experiments involving free fall. We found that in a sufficiently uniform gravitational field, the local effects of gravity can be made to go away by adopting a suitably accelerating reference frame (one that fell with an acceleration equal to g, the local acceleration due to gravity). Next we shall consider a converse case: what if there is no gravity (e.g., we are far from all massive bodies), but our reference frame is accelerating?

Do not forget that a reference frame is a reference body—something that in principle could be present and would be made out of physical things such as rods and clocks, not abstractly defined notions of space and time. A fine example of a reference frame for present purposes is a rocket, complete with rocket motor and interior living space; see Fig. 9.7.

Since we have in mind a field-free region of space we already know all the laws of motion and dynamics we need to describe life on board such a rocket: they are those of Special Relativity, and associated dynamical theories such as electromagnetism. We suppose that the rocket motor is switched on, giving a constant acceleration, and that there are no other forces acting, so that objects released inside the rocket move uniformly relative to any nearby inertial frame of reference. An astronaut standing at a fixed place in the rocket finds that his feet are pressed firmly against the floor. This is because the floor has to push on him to give him the same acceleration as the rest of the rocket. An astronaut who releases an apple inside the rocket will find that the apple accelerates towards the bottom of the rocket (an inertial observer would say the bottom of the rocket accelerates towards the apple). We can imagine that the astronaut will become so accustomed to this state of affairs that he simply says 'everything behaves just as if there were a gravitational field pulling everything towards the bottom of the rocket.' Now we have another equivalence. If everything is *just as though* there were a gravitational field, then physically speaking does not this amount to saying there *is* a gravitational field? Or, arguing in the opposite direction, if the effects we normally ascribe to gravity are indistinguishable from effects of acceleration in the absence of gravity, then is not 'gravity' an unnecessary concept?

This second way of putting it is counter-intuitive but useful. Ultimately we shall not abandon the concept of gravity, because some of its properties, such as tidal effects and spacetime curvature, are absolute.

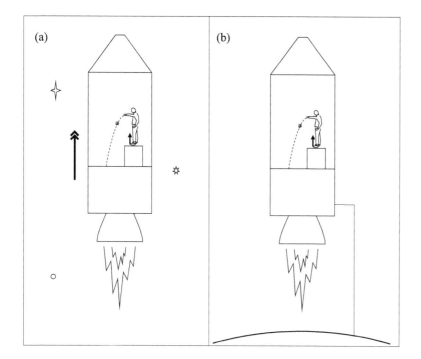

Fig. 9.7 (a) A rocket in outer space, far from other bodies. (b) A rocket hovering above the surface of a large planet. The rocket motor is on in both cases. In (a) the rocket is accelerating relative to the distant stars. An astronaut inside the rocket sees apples fall to the floor, and feels the floor pushing against his feet to support him. In (b) the rocket is not accelerating relative to the planet or the distant stars, but apples still fall and the floor pushes on the astronaut's feet. Most experiments inside the rocket cannot distinguish between case (b) and case (a). If the planet, initially absent, were suddenly to appear beneath the rocket, the astronaut simply would not and could not notice unless he looked out of a window or performed a careful survey of the variation of gravitational acceleration from place to place inside the rocket, or of the geometry of the space (e.g., by measuring the angles of triangles or the circumference of circles). Since the presence or absence of the planet is largely irrelevant to the experience of the astronaut, one may begin to wonder whether the planet exerts any gravitational influence at all! Of course it does, because it prevents the rocket from making progress towards its destination. Nonetheless it is very interesting to see from this example that much of what we ordinarily call 'the effects of gravity' are merely results of our refusal to adopt an inertial frame of reference.

However, as I sit in my chair writing this, feeling as if gravity is pulling me down towards the centre of the Earth, I could instead adopt the perspective that the pull I feel is really the result of my choice not to sit in an inertial frame of reference (i.e., a freely-falling one). I am accelerating upwards at 9.8 ms^{-1} relative to any local freely-falling frame. No wonder I feel the chair pushing on me: it is providing this acceleration. The squashing sensation I feel is not (or need not be called) the result of gravity pulling me down, it is (or may be considered) the result of my inertial mass resisting such upwards acceleration. The advantage of this perspective is that it makes it obvious and unavoidable that my passive 'gravitational mass' must equal my inertial mass.

In fact, there are differences in detail between the effects of real gravitational fields and the effects of acceleration: namely, the tidal effects. Suppose that our astronaut woke one morning to the sound of an alarm telling him that all outside sensors had failed and the shutters on the windows had closed. What can he tell about his situation? In particular, can he perform an experiment to tell between case A: the rocket is accelerating relative to the distant stars, and case B: the rocket is at a fixed distance from a massive object that is not accelerating relative to the distant stars? He feels his sense of weight, but this will happen in either case. He can check that the rocket motor is still functioning, but this also implies either case (either A: acceleration, or B: hovering above a planet). Then he notices that pendulum bobs suspended at opposite walls are not parallel but aligned towards a point several kilometres below the rocket. After first checking that the walls are not accelerating away from one another (he can do this, for example, by extending a piece of string across the cabin) he concludes that he is above a planet (or other massive object).³

The tidal effects thus allow a distinction to be made between gravitational effects and effects arising purely from a choice of reference frame. Another type of observation that the astronaut could perform is to measure the local curvature of spacetime (we shall learn how to do this in the next chapter). This property is a measurable quantity that is independent of the state of motion of the observer measuring it. It allows an observer to tell unambiguously whether or not he is in a gravitational field. Therefore it is not true to say that all effects of gravitational fields are completely equivalent to the inertial effects of accelerating reference frames in field-free space. However, the differences tend to vanish in the limit of small regions of spacetime, and this agreement between what otherwise might seem to be unrelated physical phenomena is of huge significance. It is the guiding principle in Einstein's formulation of the general theory of spacetime and gravity:

Strong Principle of Equivalence: In the limit of small extensions in space and small intervals of time, all physical phenomena have the same form in all freely-falling non-rotating cabins. In particular, the laws of physics are the same as those observed in a uniformly moving cabin in a field-free region of space.

The word 'cabin' here refers to a notional physical chamber or loose collection of bricks surrounding the region of interest; it can act as a reference body but is light enough to offer no significant gravitation of its own. The term '**local inertial frame**' or LIF is often used instead of 'cabin'. A LIF is the idealized limit of a small low-density reference body in free fall (a LIF can even fall through a solid object). The idea of the Strong Equivalence Principle (or EP for short) is to extend (or boldly extrapolate) the theme of the Weak Principle to all types of behaviour, not just acceleration of otherwise isolated bodies. For example, we are now making claims about the internal dynamics of atomic nuclei, and about chemical reaction rates, and about hydrodynamics inside stars, and so on. The EP says, in effect, we already know to first approximation what physics is like in the presence of gravity: in a LIF it is the familiar physics of Special Relativity and Minkowski spacetime.

For obvious reasons, the local inertial frames play the role in General Relativity which was played by the inertial frames of Special Relativity. In view of the relativity of time and distance with which we are already well acquainted, we shall have to be alert to the fact that these measures will need careful definition when gravity is present.

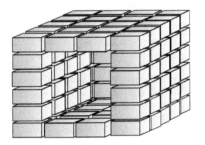

Fig. 9.8 A freely falling cabin.

9.1.5 Falling light and gravitational time dilation

The curvature of a light-ray in a gravitational field

A simple application of the Strong Equivalence Principle (EP) is to answer the question, whether or not the Weak Principle applies to electromagnetic waves. Does light experience gravitation? The answer offered by the Strong Principle is an unequivocal *yes*. For, imagine we shine a beam of light into a freely-falling horizontal tube. The Strong Principle says the light passes through the tube, travelling in a straight line in a reference frame freely-falling with the tube. It follows that, relative to a reference frame made of rods fixed to the massive object that creates the field (e.g., planet Earth), the light-ray is bent (see Fig. 9.9). Furthermore, the Equivalence Principle allows us to calculate the curvature of the light-ray *exactly*. This will be our first example of an exact result in General Relativity.

Definition of a static field. We treat a static field. This means that the gravitating object, called Earth for convenience, has no time dependence (including no rotation—so this is only approximately true for Earth). Since time is itself affected by gravity we need to define 'static' more fully; this may be done as follows. Imagine constructing a lattice of rods attached to Earth. We would like to consider a rigid lattice, but we need to define carefully what is meant by that. Imagine any well-defined sequence of physical operations that could be used to survey the lattice. For example, mount clocks throughout the lattice (they need not all have the same rate, but each clock should behave regularly) and carry out round-trip-time measurements for light-signals sent from one point to any nearby point and reflected back. Record the set of such measured round trip times at each lattice point. Then repeat the measurement after the clocks have ticked for a while. If the new set of values is the same as the old set (and this remains true whenever the measurements are repeated), then the conditions are said to be *stationary*. If the round-trip time for a signal sent around a closed loop (such as a polygon) is not only constant but also the same for both directions of travel around the loop, then the conditions are said to be *static*. (It is possible to have a stationary but not static situation when there is rotation.) If the procedure results in static conditions, then the lattice is said to be rigid, and it furnishes a natural choice of (non-inertial) reference frame—that in which the lattice is not moving.

We can now assign spatial coordinates to points on the lattice. We shall do that by comparison with the coordinates in a freely-falling cabin.

Let system F be a cabin in free fall, with coordinates x_f, y_f set up within the cabin, in the horizontal and vertical directions respectively, and time measured by a standard clock at the origin of F. We know from EP that such a clock functions normally. Imagine the cabin is launched upwards, and its internal clock is set to zero just as it reaches its maximum height, when the origin of the cabin is momentarily at rest relative to the rigid lattice fixed to Earth. Suppose that at that moment ($t_f = 0$) a light-flash sets off from the origin of F in the horizontal direction. Using the EP, we know its trajectory is given by

$$x_f = ct_f + O(t_f^3), \qquad y_f = 0 + O(t_f^3). \tag{9.13}$$

The $O(t_f^3)$ terms are needed because the EP concerns a limiting case where tidal effects are negligible. What we can claim to know from

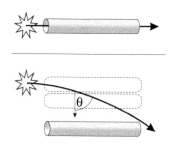

Fig. 9.9 Light passes along the axis of a non-moving tube in the absence of gravity (top picture); therefore (Equivalence Principle) light also passes along the axis of a falling tube in the presence of gravity (bottom picture). The curvature of the light-ray at each point can be calculated exactly from θ and the local acceleration due to gravity.

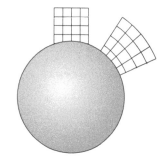

Fig. 9.10 A rigid lattice attached to a gravitating body. The lattice may be of rectangular or spherical or any other construction.

EP is that at the origin the ray is straight, and therefore $d^2x_f/dt_f^2 = d^2y_f/dt_f^2 = 0$ at $x_f = y_f = 0$.

Now consider what is observed relative to Earth. Let system S be the rigid lattice fixed to Earth, with coordinates x, y to be specified in the vicinity of the event P where the light-flash was emitted. *If* the lattice could be described by Euclidean geometry, then the transformation between coordinate systems (x_f, y_f, t_f) and (x, y) (we shall not need t) would be

$$x = x_f, \qquad y = y_f - \frac{1}{2}gt_f^2 \qquad (9.14)$$

where g is a number characteristic of the local field strength, called 'the acceleration due to gravity at P'. We shall now argue that the errors introduced by using this transformation are of third or higher order in x_f, y_f and t_f. For, by construction, the origin of the cabin was at rest relative to S at $t_f = 0$, so there are no terms linear in t_f; if there were a t_f^2 term in the x equation then we could make it go away by redefining what direction is called vertical; any adjustment to the quadratic term in the y equation is absorbed into the definition of g; finally, no matter what is the exact dependence of x, y on x_f, y_f at given t_f, the error introduced by using straight-line tangent approximations to the exact curves is of third order (see Fig. 9.11). The last step amounts to recognizing that eqn (9.14) can be used to *define* local distance (i.e., distance in the vicinity of P) in the lattice at rest in Earth's field.

Eliminating t_f, we find that the path of the light-ray in S is

$$y = -\frac{g}{2c^2}x^2 + O(x^3)$$

Therefore the curvature of the ray (see box below) is exactly

$$\kappa = \frac{d^2y/dx^2}{\left(1 + (dy/dx)^2\right)^{3/2}} = -\frac{g}{c^2}. \qquad (9.15)$$

For example, for $g = 9.8$ ms^{-2} the radius of curvature is about 1 light-year. If a spherical body were sufficiently massive to produce this amount of gravitational acceleration everywhere along a circular orbit of circumference approximately 2π light-years, then light emitted in the correct initial direction would orbit the body. Here the statement about κ is exact, but the statement about the circumference is rough because spacetime is non-Euclidean at a large scale, as we shall see. (In this example the body in question has to be very heavy, having the mass of a large galaxy, as a Newtonian estimate will tell you.)

The same calculation gives the curvature for an object travelling at any other velocity in the cabin (just replace c by the relevant velocity). It is also straightforward to generalize the calculation to an arbitrary initial direction of motion relative to the direction of **g**. One obtains

$$\kappa = -g\sin\theta/v^2. \qquad (9.16)$$

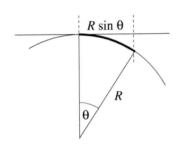

Fig. 9.11 Mapping from one coordinate system to another. The figure illustrates an assertion made in connection with eqn (9.14): namely, that the terms omitted from (9.14) are of third order not second order in x_f, y_f. The reason is that the approximation being made is of the same kind as is made when mapping distances on a curved surface onto a flat plane. In the illustration, the bold line shows a displacement on the surface of a sphere of radius R. The exact length of this displacement is the arc length $R\theta$. We would like to track movements on the curved surface by projecting them onto a flat plane. This can only be done for small regions, but it is convenient because on a flat plane Euclidean geometry applies. We always have the option of making the plane tangent to the curved surface near any given point of interest, as shown. The projected distance associated with the arc is then $R\sin\theta = R(\theta - \theta^3/6 + \cdots)$. The error term is of third order in θ. More generally, distances in a rigid lattice can be defined by making them agree locally with distances in a LIF, and the use of tangents to the exact curves always allows errors associated with non-Euclidean geometry to be reduced to third order.

Since by EP other objects travel in the cabin at less than c, we find that light-ray paths in a gravitational field are curved, but they have the smallest curvature of any path of a falling object in a given field for a given initial direction of motion.

Curvature of a line in plane Euclidean geometry. A general wiggly line specified by $y = y(x)$ may have a curvature that varies from place to place. A good way to specify the curvature at a point P on the line is to find the circle that touches the line at P, with matching first and second derivatives, and then *define* the radius of curvature of the line to be the radius of the touching circle. The equation of a general circle in the plane is $(x - x_0)^2 + (y - y_0)^2 = r^2$. Therefore the conditions are

$$(x - x_0)^2 + (y - y_0)^2 = r^2$$

$$2(x - x_0) + 2(y - y_0)\frac{\mathrm{d}y}{\mathrm{d}x} = 0$$

$$2 + 2\left(\frac{\mathrm{d}y}{\mathrm{d}x}\right)^2 + 2(y - y_0)\frac{\mathrm{d}^2 y}{\mathrm{d}x^2} = 0$$

where (x, y) are the coordinates of P. Solving for r in terms of derivatives of $y(x)$ one finds

$$r = \frac{\left(1 + (\mathrm{d}y/\mathrm{d}x)^2\right)^{3/2}}{\mathrm{d}^2 y/\mathrm{d}x^2}.$$

The sign is chosen by convention so that $r < 0$ signifies a bending downwards as x increases; $r > 0$ signifies bending upwards. The *curvature* κ is defined to be the inverse of this radius, $\kappa \equiv 1/r$.

Gravitational redshift and time dilation

Consider the same static field as in the previous calculation, and the same cabin falling from rest at $t_f = 0$. Now suppose a light-wave of frequency ν_0 sets off from an origin located on the ceiling of the cabin, and travels vertically downwards. By the EP the wave arrives at the floor with unchanged frequency ν_0, and if the cabin height is h then the wavefront arrives at the reception event at the floor at cabin time $t_f = h/c + O(h^3)$.

Now consider the observations made by observers fixed to the rigid lattice constructed on Earth. There are two observers to consider: the 'upper lattice observer' located near the emission event, and the 'lower lattice observer' located near the reception event. At the emission event the ceiling of the cabin is at rest relative to the lattice. Therefore the upper lattice observer observes the same frequency ν_0 as the one registered inside the cabin. To be precise, that is what we expect for small times. The lower lattice observer, on the other hand, finds that at

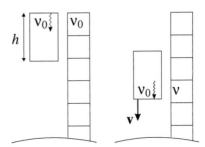

Fig. 9.12 Light emitted in a freely-falling cabin and measured by observers in the fixed lattice.

the reception event the cabin is moving relative to him in the downwards direction at a speed given, to first approximation, by $v = gt_{\mathrm{f}} = gh/c$. Therefore there is a Doppler shift between the frequency observed inside the cabin and the frequency observed by the lower lattice observer. The latter measures a frequency shift given to first order by

$$\nu - \nu_0 \simeq \frac{v}{c}\nu_0 = \frac{gh}{c^2}\nu_0. \tag{9.17}$$

Thus the lower observer finds that the frequency is higher than the one reported by the upper observer, for the same freely-falling light-wave. A similar argument applies to a light-wave propagating upwards: in the cabin LIF it agrees in frequency everywhere with the light-wave that we have just considered, so the lower lattice observer again finds its frequency to be higher than that at the upper lattice observer.

This effect is called the *gravitational redshift*. The word 'red' emphasizes the observation made by the upper observer, who observes the lower of the two frequencies. The emphasis is placed this way because gravitational fields from finite bodies reduce in strength as one moves 'upwards': i.e., away from the body. A lattice observer at infinity finds the local $g \to 0$, so he may legitimately regard his instruments as unaffected by gravity. A given type of source in free fall, such as a specific electronic transition in the hydrogen atom, always emits a fixed frequency in its LIF. When the emitted light arrives at the observer at infinity, he finds its frequency to be lower and lower as the emission point is situated closer and closer to a given gravitating object. In this sense the result can be regarded as a position-dependent redshift.

Near the surface of Earth g/c^2 has the value 10^{-16} per metre, so the effect is exquisitely small. It was first observed (to about 10% accuracy) in 1959 by R. Pound and G. A. Rebka, in an experiment involving gamma-rays propagating up and down a 22-m height difference. The method was to use a narrow resonance in a nuclear transition, and determine what relative velocity of the source and detector was needed to compensate the gravitational shift. The experiment took advantage of the Mössbauer effect to reduce the thermal broadening which otherwise would have swamped the shift. Gravitational redshift is now routinely observed in astronomy, and has been verified to very high accuracy by radar surveys of the solar system. The global positioning system has to take it into account at a level around one part in 10^{10}.

Eqn (9.17) can be written

$$\frac{\mathrm{d}\nu}{\mathrm{d}h} = \frac{g}{c^2}\nu \tag{9.18}$$

which is exact, because corrections to the calculation will be of higher order in h. If we now *define* a potential function Φ such that

$$\mathbf{g} = -\boldsymbol{\nabla}\Phi \tag{9.19}$$

(such a definition will be useful when $\mathbf{g}$ has zero curl), then eqn (9.18) is

$$\mathrm{d}\nu = -\frac{\nu}{c^2}\mathrm{d}\Phi \quad\Longrightarrow\quad \frac{\nu}{\nu_0} = e^{-(\Phi-\Phi_0)/c^2}. \tag{9.20}$$

This result should be read as a statement about the frequencies observed by observers at rest relative to the field at two different locations, in each case using a local standard clock to define time, when they both observe the same light-wave propagating freely (i.e., subject only to gravity) between them.

You will sometimes see this gravitational redshift described as if the light-wave itself changes frequency as it propagates in the field. However, this point of view is misleading since it misses the important fact that in each LIF traversed by the light, the frequency is completely unaffected. We shall now show that a valid interpretation is that the light is unchanged but the local clocks go at different rates. That is, the gravitational redshift should be regarded as a consequence of a change of reference frame from a LIF momentarily at rest at the emitter to a LIF momentarily at rest at the receiver. Since the emitter and receiver are at rest relative to one another, so are these LIFs—and yet there is a frequency shift. Its cause is a gravitational contribution to time dilation.

To see this, suppose that a given type of clock—say an atomic clock based on a caesium atom held in vacuum—is gently lowered from a high place A to a lower place B in a static gravitational field, all the while emitting a signal such as a microwave that oscillates in step with the atom. The clock is then held at B for a very long time, and finally gently raised to A again. An observer fixed at A receives the redshifted microwaves and counts the number of oscillations or 'ticks', as observed by herself, throughout the whole journey of the clock. She finds this number to be considerably smaller than the number of ticks of the same type of clock kept permanently by her at A. She must conclude that the lowered clock does not merely appear to run slower at B, but *actually does* run slower at B, because the sojourn at B could be made so long as to completely overwhelm the effect of the lowering and raising operations, and the number of ticks of the clock during its whole journey is absolute (compare this with the 'twin paradox'). It absolutely did tick fewer times than the clock at A, and all but a negligible part of the difference is due to its sojourn at B.

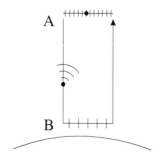

Fig. 9.13 Experiment to establish gravitational time dilation.

By this argument, eqn (9.20) is not only a statement about redshift; it is also a statement about *gravitation time dilation*. It shows the ratio between the rates of standard clocks located at rest in a static field at positions of differing gravitational potential. The ratio is measurable by a procedure such as the one outlined in the previous paragraph, and, for the removal of all doubt, that absolute measurement outcome is what is meant by statements such as eqn (9.20) concerning a comparison of the rates.

Gravitational time dilation is easy to state, but is nonetheless a subtle concept. Getting a good grasp of it is half the battle in understanding General Relativity. It can be linked to energy conservation, but this link can easily be misstated, so we shall investigate it only after first examining a very useful test case that can be treated using Special Relativity.

9.2   The uniformly accelerating reference frame

So far we have guessed (by postulating the Strong Equivalence Principle) that there is an intimate link between effects of gravity and effects of acceleration, and we have derived two consequences of that. Our next step will be to acquaint ourselves better with effects of acceleration in the absence of gravitational fields.[4] We can do this by using Special Relativity. There is a myth that seems to survive in the student physics community, to the effect that Special Relativity cannot handle accelerated reference frames, and you need General Relativity for that. This is quite wrong: Special Relativity can treat any sort of motion. It is limited (compared to General Relativity) merely by the Euclidean geometry postulate. The grain of truth in the myth is that the natural discourse of Special Relativity is inertial reference frames. The method to treat accelerated frames without adopting the mathematical tools of *general covariance* will be shown implicitly in the following.

9.2.1   Accelerated rigid motion

We saw in section 6.5 that the concept of rigidity has to be questioned in Special Relativity, for two reasons. First, the speed of sound is not infinite, so the effects of a force applied to one part of a body are not felt immediately throughout the body. Secondly, rigidity in the sense of fixed physical dimensions is not a Lorentz-invariant property, because for an accelerating body the size and shape may be constant in some inertial frames and nevertheless vary in others.

Having noted these facts we can nevertheless introduce a notion of rigid accelerated motion in the following way. We consider a composite object to be made of the particles composing it, and therefore to consist of a collection of worldlines. The particles may be undergoing accelerated motion, and we may *define* rigid motion to be motion such that:

[4] To be precise, in the absence of space-time curvature: we shall settle later whether or not purely inertial effects can be called 'gravitational'.

(1) At each instant there is an inertial frame in which all particles in the body are momentarily at rest.

(2) The distances between the particles, evaluated in the sequence of such instantaneous rest frames, are constant.

One can show that these conditions require the motion to be rectilinear.[5] Therefore we restrict attention to the case of rectilinear motion along the x direction, in which no particle overtakes or is overtaken by another (so the worldlines do not cross). Choose the particle at one end of the object, and suppose it has some arbitrary motion. Our task is to find, if possible, motions of all the other particles such that the above conditions are satisfied.

Let the end particle be called P, and let its position relative to some inertial reference frame S be given by $x = f(t)$ for some arbitrary function f. This defines the worldline of P. Its 4-velocity is $U = (\gamma c, \gamma v)$ where $v = df/dt$, and we suppressed the y and z components for convenience. Now pick some other particle Q, whose position relative to S is given by some function $h(t)$ to be discovered. Consider the pair of events $A = (t_A, f(t_A))$ and $B = (t_B, h(t_B))$—see figure 9.14. We choose B so that it is simultaneous with A in the instantaneous rest frame of P. The criterion for this is that the 4-displacement

$$\Delta X = (c(t_B - t_A), h(t_B) - f(t_A))$$

is orthogonal to U:

$$U \cdot X = 0 \quad \implies \quad c^2(t_B - t_A) = v(x_B - x_A) \quad (9.21)$$

where we introduced $x_A \equiv f(t_A)$ and $x_B \equiv h(t_B)$. Next we impose condition 2: that the distance between A and B, as measured in the instantaneous rest frame, does not depend on which event A was chosen. Since ΔX is along a line of simultaneity for the instantaneous rest frame, this distance is given by the Lorentz scalar length of ΔX. Therefore the condition is

$$(x_B - x_A)^2 - c^2(t_B - t_A)^2 = L_0^2 \quad (9.22)$$

where L_0 is the constant distance. Using eqn (9.21) to eliminate $t_B - t_A$ we find

$$x_B - x_A = \gamma L_0 \quad (9.23)$$

and therefore

$$t_B - t_A = \frac{v}{c^2} \gamma L_0. \quad (9.24)$$

These results allow us to find both x_B and t_B for given x_A, t_A, v and L_0, so they determine the worldline of Q if the worldline of P is given.

Next we need to check whether the worldline of Q is consistent with condition 1. This requires that the velocity of Q relative to S at event B is equal to the velocity of P relative to S at event A:

[5] A 'rigidly rotating' disc does not satisfy the definition because there is no inertial frame in which all parts of such a disc are at rest at any instant.

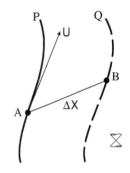

Fig. 9.14 Construction used to describe rigid motion.

$$\frac{\mathrm{d}x_B}{\mathrm{d}t_B} = \frac{\mathrm{d}x_A}{\mathrm{d}t_A} = v.$$

It requires some care to get this equation right. The idea is that the next event on the worldline of Q is at a time $\mathrm{d}t_B$ away and a distance $\mathrm{d}x_B$ away. From eqn (9.23) we have $\mathrm{d}x_B/\mathrm{d}t_A = \dot{\gamma}L_0 + v$, and from eqn (9.24) we have $\mathrm{d}t_B/\mathrm{d}t_A = (\dot{v}\gamma + v\dot{\gamma})L_0/c^2 + 1$, where the dot signifies $\mathrm{d}/\mathrm{d}t_A$ which is the same as $\mathrm{d}/\mathrm{d}t$ for the worldline of P. Therefore

$$\frac{\mathrm{d}x_B}{\mathrm{d}t_B} = \frac{\dot{\gamma}L_0 + v}{(\dot{v}\gamma + v\dot{\gamma})L_0/c^2 + 1}.$$

Setting this equal to v, one finds that two v terms cancel and L_0 factors out, so that the condition is

$$v\dot{v}\gamma = \dot{\gamma}(c^2 - v^2).$$

But from eqn (2.56) this is always true, so we conclude that condition 1 always holds for the motion imposed by condition 2.

This argument can be applied for a sequence of values of L_0, so that a complete rigid body can be specified with its particles at fixed proper separations as defined by the two conditions. Hence there is a sensible definition of rigid accelerated motion in Special Relativity. Fig. 9.15 shows an example.

There is one important limitation of this argument. For the argument to make sense we require $\mathrm{d}t_B/\mathrm{d}t_A > 0$, so that when time advances at one end of the body it also advances at the other end. This condition is

$$(\dot{v}\gamma + v\dot{\gamma})\frac{L_0}{c^2} + 1 \geq 0$$

which simplifies to

$$\gamma^3 \dot{v} L_0 \geq -c^2. \tag{9.25}$$

Now recall that $\gamma^3 \dot{v}$ is the proper acceleration in the case of straight-line motion (eqn (2.61)), so we have

$$a_0 L_0 \geq -c^2. \tag{9.26}$$

For positive a_0 this means that particle Q must not be too far to the left of particle P; for negative a_0 it means that particle Q must not be too far to the right of particle P. The physical origin of this limit is that Q must not be so far away that in order for the body to Lorentz-contract enough to maintain a fixed proper size, the particle Q has to move faster than light. The limiting value of L_0 is an example of an 'horizon'. This will be studied more fully in the next section.

9.2.2   Rigid constantly accelerating frame

It is now easy to construct an example of an interesting type of accelerating rigid body: one undergoing a constant proper acceleration. In this case the particle P undergoes the hyperbolic motion described in

Fig. 9.15 A body undergoing an example of rigid non-uniform motion. At every event the body has a rest frame agreed among all is constituent parts, and in the sequence of rest frames proper distances between parts of the body stay constant.

section 4.2.4, and the trajectory of any other particle Q in the body can be obtained from eqns (9.23) and (9.24).

To reduce clutter we shall adopt units such that $c = 1$ throughout this section.

Without loss of generality, we place the origin so that the worldline of P can be written $x^2 - t^2 = h_0^2$. Then we have the basic equations

$$\tau_0 = h_0\theta \tag{9.27}$$

$$a_0 = 1/h_0 \tag{9.28}$$

$$\mathsf{X} = (\sinh\theta,\ \cosh\theta)h_0 \tag{9.29}$$

$$\mathsf{U} = (\cosh\theta,\ \sinh\theta) \tag{9.30}$$

where τ_0 is the proper time along the worldline, a_0 is the proper acceleration, θ (rapidity) is a useful parameter, X is the displacement 4-vector with y and z components suppressed, and U is the 4-velocity.

To find the trajectory of any other particle Q, we could use eqns (9.23) and (9.24), but it is easier to argue directly from the two conditions for rigid motion, and use two notable properties of hyperbolic motion:

$$\mathsf{X} \cdot \mathsf{X} = h_0^2 = \text{const} \tag{9.31}$$

$$\mathsf{X} \cdot \mathsf{U} = 0 \tag{9.32}$$

The first property says that the spacetime interval from the origin is constant. The second property says that X is orthogonal to the worldline, and therefore *at each moment, the 4-vector from the origin to the particle is a line of simultaneity for the instantaneous rest frame*. As P moves along its trajectory, the lines of simultaneity form a set of straight lines through $(0,0)$; see Fig. 9.16. It follows that the 4-vector from the origin to the particle (i.e., X) is purely spatial in the instantaneous rest frame, so its length is the distance to the origin as observed by an inertial observer momentarily riding on the particle. But this length is constant! (It is equal to h_0.) So a sequence of such observers will find that for a particle undergoing hyperbolic motion, the origin is always a fixed distance away from the particle. The Lorentz contraction does just enough to bring this about.

Now suppose Q also undergoes hyperbolic motion, but at a different proper acceleration. The worldline of Q is then

$$\mathsf{X} = (\sinh\theta,\ \cosh\theta)h \tag{9.33}$$

for some constant $h \neq h_0$. This worldline satisfies the two conditions for rigid motion. A straight line through the origin is a line of simultaneity for both P and Q, so there is an agreed instantaneous rest frame (condition 1). Also, the distance in this frame from the origin to P is h_0 and from the origin to Q is h, so the distance from P to Q is $h - h_0$, which is constant (condition 2).

By choosing a sequence of values of h we can now construct a complete constantly accelerating rigid reference frame. It is called 'uniformly accelerating' to emphasize that each particle in the frame moves with

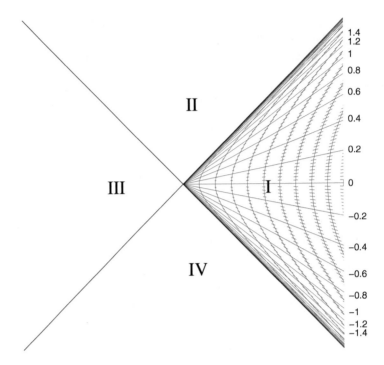

Fig. 9.16 The constantly accelerating reference frame. The figure shows a spacetime diagram. The hyperbolae are worldlines of particles separated by fixed proper distance, each having a constant proper acceleration. The straight lines are lines of constant θ, which are lines of simultaneity in the instantaneous rest frame. The tick marks are placed at equal increments of proper time along each worldline.

constant proper acceleration. However, the acceleration is not uniform with respect to distance: particles at different h have different proper acceleration.

This rigid frame is perfect to describe a uniformly accelerating rocket of the kind illustrated in Fig. 9.7a. The rocket is accelerating in the positive x direction. At early times its velocity is directed towards the origin, at late times its velocity is directed away from the origin, but its acceleration is always in the direction of positive x. Owing to singular behaviour at the origin, it is best to imagine that the rocket does not reach all the way to the lines $t = \pm x$, but its floor is situated at particle P with some finite value of h_0 which we may as well take as $h_0 = 1$. Observers in the rocket sit at $h \geq h_0$, but they have the right to reason about the region $h < h_0$, and to measure it by sending probes. The whole of the life of observers in the rocket is confined to the region I of the spacetime diagram shown in Fig. 9.16. Signals they emit can only reach regions I and II; signals originating from regions II and III never reach them. As far as they are concerned, region III might as well not exist! They can neither influence nor be influenced by it.

We shall now discuss the observations (i.e., the reasoning about space, time, and motion) made by observers fixed relative to the rocket. When in doubt we can settle disputes by appealing to the 'referee', who is an inertial observer in the sense of Special Relativity—one whose motion is not accelerated. The referee surveys spacetime by any suitable means (for example by using a set of standard clocks and rods, or by radar methods). He thus constructs a coordinate system (t, x) that can be used to assign coordinates to all events. This is the coordinate system we have been using already.

Each accelerated observer is fixed relative to the structure of the rocket, so naturally considers that he has a fixed position in the rocket frame. To quantify the notion of position in such a rigid accelerated frame, the most natural choice is to use the value of h (i.e., the distance from the origin in the instantaneous rest frame). That is, the 'height' of any given event in the rocket is *defined* as the value of h for the observer whose worldline passes through the event. This is a suitable definition, because its value is a constant for each observer fixed in the frame, and small changes in h agree with the 'standard rod' definition of distance. A 'standard rod' or 'ruler' is one that the referee would consider standard: that is, it has just the right Lorentz contraction to agree with a standard rod moving inertially at the same velocity. *In this way we can make a sensible definition of length for rigidly accelerating rulers.* Similarly, an accelerating clock is said to be 'standard' if during any small interval its ticking agrees with that of a standard clock moving inertially at the same velocity.

Similar arguments suffice to define what is meant by a 'standard rod' or 'standard clock' at rest in a gravitational field.

There are many ways in which the timing of events could be tracked in the accelerated frame. The two most natural ways are the 'local proper time' τ and the 'master time' θ. To set up the τ system, furnish every observer with a standard clock called that observer's 'proper clock', and pick one of the lines of simultaneity in Fig. 9.16. The most convenient such line is the x axis. Let all proper clocks be set to zero on this axis. Thereafter and before, each proper clock registers the amount of proper time along the observer's worldline. We can relate the inertial referee coordinates (t, x) of any event to the coordinates (τ, h) by using

$$\left. \begin{array}{l} t = h \sinh(\tau/h) \\ x = h \cosh(\tau/h) \end{array} \right\} \quad \Leftrightarrow \quad \begin{cases} \tau = (x^2 - t^2)^{1/2} \tanh^{-1}(t/x) \\ h = (x^2 - t^2)^{1/2} \end{cases} \tag{9.34}$$

This is an acceptable definition of position and time in the accelerating frame, but it is not the only possible one, nor the best. Another good method is to keep h for position, but instead of using τ let the 'time coordinate' be the parameter θ that we used to describe the worldlines. This has the advantage that it agrees with the notion of simultaneity that was used to define rigidity, and it agrees with τ for one of the clocks: the 'master' proper clock at $h = 1$. The transformation between (t, x) and the coordinates (θ, h) is

$$\left. \begin{array}{l} t = h \sinh(\theta) \\ x = h \cosh(\theta) \end{array} \right\} \quad \Leftrightarrow \quad \begin{cases} \theta = \tanh^{-1}(t/x) \\ h = (x^2 - t^2)^{1/2} \end{cases} \tag{9.35}$$

We can imagine that each observer owns not only a proper clock but also a 'master clock' that has been so built as to indicate θ. In fact, it is easy to construct these master clocks, because

$$\theta = \tau/h. \qquad (9.36)$$

Therefore a master clock is a standard clock with an extra 'multiplier cog' that multiplies the reading by the *constant* factor $1/h$.

Now let us compare the rocket to a gravitational field. If an observer in the rocket releases a test object such as an apple, he will observe it to accelerate downwards (i.e., towards smaller values of h). We already know the upwards proper acceleration, relative to the referee, of the part of the rocket at height h: it is $a = 1/h$ (eqn (9.28)). We expect that this is also the downwards acceleration of the referee, and therefore of anything moving inertially, such as a dropped apple, as observed by an observer fixed in the rocket. To be sure, however, we need to check that the system of measurement employed by the accelerating observers will not introduce some small correction. Consider, therefore, the situation near $t = 0$ when the whole rocket is momentarily at rest relative to the referee. At $t = 0$ the rods and clocks in the rocket have no contraction or dilation; at $t = dt$ the rocket speed at height h is $v = a dt = dt/h$ and therefore the Lorentz factor is $\gamma \simeq 1 + (dt)^2/2h^2$. Suppose an apple is released at $t = 0$. If there were no Lorentz contraction or time dilation, the downwards acceleration of the apple measured by a local rocket observer using his own rods and clocks would be a. The Lorentz factor introduces corrections which vanish in the limit $dt \to 0$, therefore its effects do not change the conclusion: the rocket observer finds the downwards acceleration of the apple relative to him, as indicated by his own instruments, to be $a = 1/h$. The same result would hold for an object released at any other time, since the rocket has constant proper acceleration and we can always Lorentz-transform to a frame in which it is momentarily at rest.

We have thus found that life in the rocket can be compared to life in a static gravitational field whose strength varies with height as

$$g = 1/h \qquad (9.37)$$

and which is independent of y and z. A gravitational field whose strength falls as $1/h$ would be produced by a long cylindrical planet (in the weak field, i.e. Newtonian, limit), so life on board the accelerating rocket is much like life in a tall tower resting on such a planet, as long as one only examines experiments in the xz plane. Using eqn (9.37) in eqn (9.19) we find that the gravitational potential function is

$$\Phi = \log h. \qquad (9.38)$$

Now let us examine light-signals sent vertically from the bottom towards the top of the rocket. Expressed in (t, x) coordinates, the worldline of a photon emitted by P at the event $\mathsf{X_0} = (\sinh\theta_0, \cosh\theta_0)$ and travelling in the vertical direction is $\mathsf{X} = \mathsf{X_0} + (t, t)$. This intersects the worldline of the rocket observer at height h at the event with master time θ given by

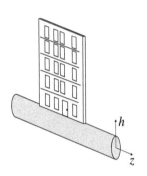

Fig. 9.17 A tall narrow tower on a cylindrical planet.

$$\sinh \theta_0 + t = h \sinh \theta,$$

$$\cosh \theta_0 + t = h \cosh \theta.$$

After subtracting in order to elliminate t, we find

$$h(\sinh \theta - \cosh \theta) = \sinh \theta_0 - \cosh \theta_0. \tag{9.39}$$

We are interested in the frequency of the light, so consider two wavefronts emitted from P at times $\theta_0, \theta_0 + d\theta_0$. They arrive at h at master times separated by $d\theta$, obtained by differentiating eqn (9.39):

$$\frac{d\theta}{d\theta_0} = \frac{\cosh \theta_0 - \sinh \theta_0}{h(\cosh \theta - \sinh \theta)} = 1.$$

Thus in 'master units' the received period agrees with the emitted period. Therefore in proper time units the periods are related by

$$\frac{1}{h} d\tau = \frac{1}{h_0} d\tau_0$$

and therefore the frequencies are related by

$$\frac{\nu}{\nu_0} = \frac{h_0}{h} = e^{-(\Phi - \Phi_0)} \tag{9.40}$$

where in the last step we used eqn (9.38). This is the gravitational redshift; see eqn (9.20). If two clocks of the same type (e.g., two hydrogen masers) are at different heights in the rocket, then waves emitted with the frequency of the lower clock will be observed, when they arrive at the upper clock, to have a frequency lower than that of the upper clock.

Fig. 9.18 shows the spacetime region I where life in the rocket takes place, plotted using each of our three coordinate systems. The worldlines of particles fixed in the rocket appear as hyperbolas in the (x, t) diagram but as vertical straight lines in the (h, τ) and (h, θ) diagrams; lines of constant θ which are sloping in the (x, t) diagram appear horizontal in the (h, θ) diagram. In the following we shall refer to these three diagrams as 'map 1', 'map 2', 'map 3'. They refer to the *same* region of spacetime, containing the *same* events, but mapped onto the page differently. The three maps present the (x, t), (h, τ), and (h, θ) 'point of view', respectively.

Fig. 9.19 reproduces the three maps, with some further worldlines added. With the aid of this figure we shall consider three experiments. For the moment we ignore the zigzag lines on the maps, and concentrate on the other worldlines.

First consider the redshift that we have just discussed. This is illustrated by the two photon worldlines setting off from $(x, t) = (2, 0)$ and $(2, 0.2)$. On map 1 these maintain a fixed separation in the t direction, so if they represent two successive wavefronts then the period is constant. In map 2 they separate as they go, their vertical separation doubling each time they double their height; this is the redshift. In map 3 we see again a fixed vertical separation (constant $d\theta$), but the small ticks remind us that the local proper clocks are ticking faster at higher heights, so

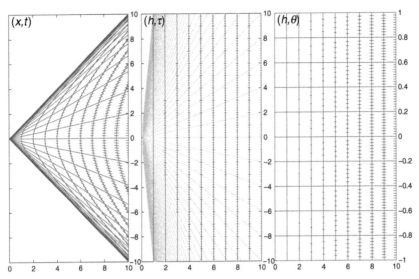

Fig. 9.18 Kinematics in the constantly accelerating frame. The diagram shows region I of spacetime as in Fig. 9.16, plotted using each of the three coordinate systems (x,t), (h,τ) and (h,θ). The vertical lines in the second and third maps are worldlines of particles at rest in the rocket; they appear as hyperbolas in the first map. The small tick-marks indicate equal increments of proper time along these worldlines. Lines of constant θ are also shown in all three maps; they converge at the origin in the first two maps, and are horizontal in the third.

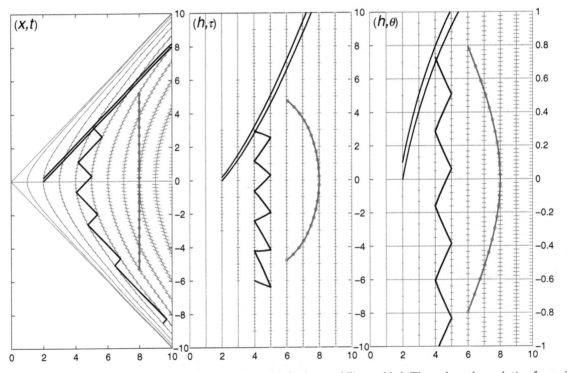

Fig. 9.19 The same three maps as shown in Fig. 9.18, but with further worldlines added. These show the evolution for various experiments discussed in the text.

the redshift is again indicated, along with its interpretation in terms of gravitational time dilation. This fact could also be 'read off' from map 1 by counting the number of small ticks between the events where the two photon lines reach any given hyperbola.

Next, suppose that a rocket observer throws a ball upwards. The ball rises up the rocket, and falls back down. This is inertial motion, given by the vertical straight worldline in map 1. In the other maps this worldline appears curved. The dots on the worldline indicate the passage of proper time for the thrown ball. The situation here may be compared to the twin paradox. In map 1 it is clear what is going on: the ball is the 'stay at home' twin, and the hyperbola at $h = 6$ is the 'travelling' twin. By counting ticks you can see that the travelling twin ages by $6 \times 0.2 \times 8 = 9.6$ units, and by measuring in the vertical direction, or by counting dots, you can see that the 'stay at home' twin (i.e., the thrown ball) ages by about 10.6 units ($4\sqrt{7}$ to be precise). The same conclusions about ageing can be obtained from the other maps. We shall discuss this further in a moment, after we have given a reason for preferring map 3 to map 2.

The third experiment illustrated on the maps consists of a 'photon clock' oriented in the vertical direction. By looking at map 1 you should be able to see that the zigzag line represents a photon moving to and fro between heights $h = 4$ and $h = 5$. On map 3 this line again appears zigzag in a sensible way, but on map 2 it looks a bit crazy. The zigzag is not regular, except at the turning points, and the photon starts off by travelling backwards in 'time' τ! Nothing unphysical is happening (as map 1 assures us), but unfortunately the (h, τ) coordinate system does not always do a good job (in the sense of an easily interpreted job) in describing situations involving movement from one height to another. The most convenient way to see why is to examine the invariant spacetime interval $ds^2 = -dt^2 + dx^2 + dy^2 + dz^2$ between any pair of neighbouring events (we reintroduced y and z to make it clear that this is the familiar interval of Special Relativity). From eqn (9.35) we have

$$dt = h\cosh(\theta)d\theta + \sinh(\theta)dh \qquad (9.41)$$

$$dx = h\sinh(\theta)d\theta + \cosh(\theta)dh \qquad (9.42)$$

and using $d\tau = hd\theta + \theta dh$ to replace the $hd\theta$ terms we obtain

$$dt = \cosh(\tau/h)d\tau + \left(\sinh(\tau/h) - \frac{\tau}{h}\cosh(\tau/h)\right)dh \qquad (9.43)$$

$$dx = \sinh(\tau/h)d\tau + \left(\cosh(\tau/h) - \frac{\tau}{h}\sinh(\tau/h)\right)dh \qquad (9.44)$$

Hence in the (θ, h, y, z) coordinate system the interval has the simple form:

$$ds^2 = -h^2 d\theta^2 + dh^2 + dy^2 + dz^2 \qquad (9.45)$$

whereas in the (τ, h, y, z) coordinate system the invariant interval has the more complicated form:

$$ds^2 = -d\tau^2 + 2\frac{\tau}{h}d\tau dh + \left(1 - \frac{\tau^2}{h^2}\right)dh^2 + dy^2 + dz^2. \qquad (9.46)$$

The expression for ds^2 is called the **line element** or the **metric equation**. The idea is that ds^2 is the 'way to measure' spacetime. The expression always has a form that allows it to be written $ds^2 = dX^T g dX$, where g is a 4×4 matrix (called the metric tensor) and $dX = (dx^0, dx^1, dx^2, dx^3)$ is the displacement 4-vector in the given coordinate system. The expressions (9.45) and (9.46) give, respectively,

$$g_{ab} = \begin{pmatrix} -h^2 & & & \\ & 1 & & \\ & & 1 & \\ & & & 1 \end{pmatrix}, \text{ and } g_{ab} = \begin{pmatrix} -1 & \tau/h & & \\ \tau/h & 1-\tau^2/h^2 & & \\ & & 1 & \\ & & & 1 \end{pmatrix}. \qquad (9.47)$$

(The subscripts a, b here are indices running over the range $0, 1, 2, 3$; they act as a reminder that g_{ab} is a matrix.) A photon worldline is null: i.e., it connects events separated by $ds^2 = 0$. For a photon travelling vertically upwards we obtain from eqn (9.45) the 'speed of light'

$$\frac{dh}{d\theta} = h$$

and from eqn (9.46) the 'speed of light'

$$\frac{dh}{d\tau} = \frac{1}{1+\tau/h}.$$

The latter is negative even for an upward-travelling photon when $\tau < -h$. The 'speed of light' indicated by these two results is not a speed in the sense intended in the Light Speed Postulate of Special Relativity (which is *not* broken here). Rather, it is a 'coordinate speed': that is, a ratio of small changes in quantities that are useful for mapping spacetime, but which are only indirectly related to the standard procedure for measuring distance and time interval. The standard procedure is to use a standard rod and standard clocks all at rest relative to one another in an inertial frame: the speed of light in vacuum (and more generally the maximum speed for signals) is a universal constant when measured in this standard way, and this remains true in General Relativity. In the present context it is easy to see what is 'wrong' with the (h, θ) system: the time is being measured by a master clock, not a proper clock. It is also easy to see how to correct the result: the master clock is running slow by a factor h so has given a speed measurement too high by a factor h. The problem with the (h, τ) system is more subtle. The clock is a proper clock, so indicates time correctly for events at given h—you can see this from the fact that $ds^2 = -d\tau^2$ when $dh = dy = dz = 0$ in eqn (9.46). However, in the (h, τ) system, the lines of constant τ are not orthogonal (in the spacetime sense) to the lines of constant h. Indeed, at $d\tau = 0$ we have $ds^2 = (1 - (\tau^2/h^2))dh^2$ (for $dy = dz = 0$) so the line of

'τ simultaneity' is not even spacelike when $|\tau| > h$. Therefore $dh/d\tau$ has no simple connection to any physical speed measured in the standard way.

The lesson here is that it is possible to make poor choices of coordinate system. Ultimately the physical predictions (ageing of twins, and so on) do not depend on what coordinate system is chosen, but the physical interpretation of the algebraic or graphical statements is more straightforward in some systems than in others. A crucial idea is:

The metric provides the 'key' that shows how to interpret the 'map' provided by the coordinate system.

The coordinate system (h, θ) that gives our 'map 3' is a good choice for the constantly accelerating frame; these coordinates are called **Rindler** coordinates.

Returning now to the twin paradox, we can give a beautiful interpretation of the situation in map 3. The worldline of the inertially moving ball is straight in map 1. This worldline has the special property that, of all worldlines between the given endpoints (the 'throw' and 'catch' events), it has the most proper time (see section 6.1). How does this property— the Principle of Most Proper Time—appear in map 3? The ball sets out with instructions to reach the given catch event having aged as much as possible. Since the high-up clocks are ticking faster, the ball tries to get up high in order to age more rapidly. However, it can only get there by moving relative to the rocket, and this motion introduces time dilation of the standard motion-related kind, quantified by the Lorentz factor γ. Therefore the ball has to make a compromise between the motional effect which makes it age more slowly and the gravitational one which gives it an opportunity to age more rapidly. That compromise is the curved trajectory shown in map 3: that trajectory is the worldline of most proper time. In the present case we already know this from map 1, but in view of the Equivalence Principle we can now make a beautiful (and correct) guess: *the Principle of Most Proper Time still applies when gravity is present.* That is, the motion of a freely-falling test object, in any gravitational field, is the one which maximizes the proper time along the worldline. *This is the equation of motion of test particles in General Relativity.*

In the next section we shall fill out this idea by showing explicitly how it leads to the Newtonian physics in the limit of weak fields. First, however, we shall examine one more experiment involving the rocket.

After living in the rocket for a while, the observers there begin to notice that something special is going on below them near $h = 0$. It is decided to send out a probe to investigate. The probe is released from the base of the rocket ($h = 1$) at time $t = 0$ (so also $\theta = 0$). The probe is in free-fall; its worldline is the straight line shown in Fig. 9.20a. It sends out a continuous stream of microwaves reporting on its experiences. These signals are the 45-degree photon worldlines shown in Fig. 9.20a. The

Fig. 9.20 A probe is released at $t = 0$ from the base of the rocket and falls towards $h = 0$, emitting a signal after every time interval $\Delta t = 0.1$. (a) In the (x, t) map this looks perfectly ordinary, but one can notice that owing to the motion of the rocket the signals do not all catch up with the rocket observers. (b) The (h, θ) map summarizes what is observed by the inhabitants of the rocket. They find that the signals arrive more and more infrequently, and they infer that the probe never reaches $h = 0$. That is, the observers would have to reach infinite age before the probe reached $h = 0$. Both maps agree on how many signals are emitted between the release and any given probe height, so the rocket observers agree that the probe can reach $h = 0$ in *its* lifetime— they would attribute this to the fact that the internal circuitry (clocks etc.) carried by the probe acquires a gravitational time dilation that tends to infinity.

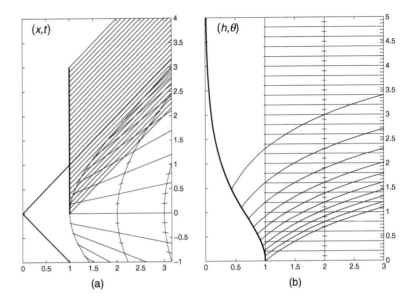

probe passes smoothly from $\{x, t\} = \{1, 0\}$ through $\{1, 1\}$ to $\{1, 2\}$, with nothing special happening to it. Notice that $\{x, t\} = \{1, 1\}$ is at $h = 0$, so this is the event where the probe reaches the place which observers on the rocket consider to be special.

Now look at the arrival of the signals at the rocket. The rocket observer who first receives the information is the one at $h = 1$. However, he finds that the signals arrive more and more infrequently. He can in principle deduce the location of each emission event, but he finds that as the probe approaches $h = 0$ the information arrives so slowly that he would have to wait forever to receive the signal emitted by the still happy and fully functioning probe as it passes through $h = 0$.

The line $h = 0$ in this scenario is an example of an *event horizon*, or *horizon* for short. In the context of gravity the above description describes a case where light takes longer and longer to emerge from a given region of space, and the gravitational time dilation tends to infinity. This is what happens when the gravitational field is strong enough, and the resulting region of space, from which light cannot escape, is called a **black hole**.

Our discussion of the uniformly accelerating reference frame has illustrated the following characteristic features of gravitation in General Relativity:

- Gravitational redshift and time dilation
- The use of more than one coordinate system
- The use of the spacetime metric ds^2
- Varying coordinate speed of light; fixed local relative speed of light
- Most Proper Time as the law of motion in free-fall
- Horizons

The discussion has been exact. The only major item missing, which the uniformly accelerating frame does not exhibit, is spacetime curvature.

9.3   Newtonian gravity from the Principle of Most Proper Time

By combining eqns (9.38) and (9.45) we can write the metric in (h, θ) coordinates as

$$ds^2 = -e^{2\Phi/c^2}c^2d\theta^2 + dh^2 + dy^2 + dz^2 \qquad (9.48)$$

where we have reinserted c (which was set equal to 1 in the previous section). For small Φ/c^2 this is

$$ds^2 \simeq -(1 + 2\Phi/c^2)c^2d\theta^2 + dh^2 + dy^2 + dz^2. \qquad (9.49)$$

We now assert the Equivalence Principle and claim that this same metric describes, to first approximation, the effect of any gravitational field whose potential function is Φ. This is an approximate statement not only because we made a linear approximation to $\exp(2\Phi/c^2)$, but also because most fields do not have the same spatial dependence as the one we found in the constantly accelerating rigid frame. The idea is to explore what would be the consequences if the claim were true.

In order to apply the metric to the case of a weak gravitational field in some region of space, we need to check the physical meaning of the quantities $d\theta$, dh, dy and dz. To this end, notice that for events at the same θ (i.e. $d\theta = 0$), ds^2 is always positive, hence a space-like interval, whereas for events separated only in the θ direction (i.e., $dh = dy = dz = 0$), ds^2 is always negative (since we are assuming $\Phi/c^2 \ll 1$), hence a time-like interval. It follows that we may interpret $d\theta$ as a measure of local time and (dh, dy, dz) as a measure of local displacement in space. Also, the scale factor for the space part is 1, and that for the time part is almost 1, so does not need adjusting by any further constant factor (the factor h which was needed in eqn (9.36) has been incorporated into Φ). With this in mind we shall now relabel the temporal coordinate as t and the height as x, so that the metric is

$$ds^2 \simeq -(1 + 2\Phi/c^2)c^2dt^2 + dx^2 + dy^2 + dz^2. \qquad (9.50)$$

This is merely a change of notation from that adopted in the previous section. In this section we shall only need one set of coordinates, and it is convenient to use the standard letters for symbols whose physical meaning is small increments of time and position.

Next we shall assert the Principle of Most Proper Time. That is, we claim that the motion of any test particle in the field is such that if the boundary conditions are specified by giving fixed start and finish events, then the worldline is the one having the most proper time, of all those worldlines that connect the events. The proper time increment is given by $cd\tau = (-ds^2)^{1/2}$, so we want the worldline with the maximal value of

$$\int_{(1)}^{(2)} c\,d\tau = \int_{(1)}^{(2)} \left(\left(1 + \frac{2\Phi}{c^2} \right) c^2 dt^2 - dx^2 - dy^2 - dz^2 \right)^{1/2}$$

$$= c \int_{t_1}^{t_2} \left(1 + \frac{2\Phi}{c^2} - \frac{v^2}{c^2} \right)^{1/2} dt$$

where $\mathbf{v} = d\mathbf{x}/dt$. Applying the binomial approximation, we find

$$\int_{(1)}^{(2)} c\,d\tau \simeq c \int_{t_1}^{t_2} 1 + \frac{\Phi}{c^2} - \frac{v^2}{2c^2}\,dt \tag{9.51}$$

$$= c(t_2 - t_1) - \frac{1}{mc} \int_{t_1}^{t_2} \frac{1}{2} m v^2 - m\Phi\,dt. \tag{9.52}$$

The condition that the proper time is maximal for given t_2, t_1 is therefore the condition that the last integral is minimal. But this is precisely the condition of **least action** of classical Lagrangian mechanics. Therefore our claim is justified: all the Newtonian gravitational effects are predicted correctly by using eqn (9.50), interpreting Φ as the Newtonian gravitational potential function, and making the assumption that particles move so as to maximize their proper time!

The two terms which in Newtonian mechanics are called kinetic energy and potential energy are now seen to come from two contributions to time dilation: $-v^2/2c^2$ comes from the Lorentz factor describing motional time dilation, and Φ/c^2 comes from gravity. All we are doing is adding up the ticking of a clock which is affected by both these contributions. These time dilation effects might appear to be small, but they lead to all the familiar phenomena of our everyday experience with gravity: a game of tennis exhibits gravitational time dilation just as surely as the most accurate test using atomic clocks in satellites.

To take another example, consider the orbit of the Earth around the Sun. In spacetime this looks like a helix. The radius of the helix is about 500 light-seconds (93 million miles), and the pitch (measured in the time direction) is one light-year, or 6×10^{12} miles. Therefore it is a very loosely wound helix. In six months the Earth moves between events situated at six months time separation and on opposite sides of the helix. If instead of following the helical worldline the Earth were to try some other nearby worldline—for example, by taking a more direct route and moving more slowly, or by zooming off towards Mars and then coming back—then the proper time would be smaller: in the first case the gravitational slow-down would win out, in the second the motional slow-down. The actual path makes the best compromise between these effects.

We have now made a profound shift in our physical understanding. The Newtonian view was that gravity acts by providing a force. The General Relativistic view is that gravity acts by slowing down the clocks. Since, in any given spacetime region, all clocks, of whatever construction, are slowed down by the same factor, we may say that time itself is slowed. Thus gravity acts by introducing changes into spacetime. Since space and time are interrelated, we must expect that space is also affected. The full theory has to handle the idea of a 'warped' spacetime.[6] We shall turn to that after first clarifying an easily misunderstood energy issue.

[6] Eqn (9.50) omits space curvature, and is only useful to treat short time intervals for slowly moving particles in a weak field.

9.4   Gravitational redshift and energy conservation

If we allow the use of $E = mc^2$ then it is not hard to show that even in Newtonian physics one must expect electromagnetic waves to lose energy as they move upwards in a gravitational field. If they did not, then one could create an infinite energy source, as follows (Fig. 9.21): (i) have an atom emit some waves of energy E in the upwards direction, (ii) absorb those waves using a similar atom at height h, (iii) swap the positions of the two atoms, (iv) repeat. If the waves did not lose energy then in step (iii) the top atom has a rest mass larger than that of the bottom atom by E/c^2, so in swapping their positions (e.g., using a rope running over a pulley to lower one while raising the other) we can obtain some work, an amount equal to $(E/c^2)gh$, at no cost. This is impossible, so the electromagnetic waves must have lost approximately this amount of energy in rising through a height h. If we now also employ the quantum mechanical relation $E = \hbar\omega$ then we shall find a gravitational redshift even in a Newtonian model of gravity. This is all very well, but it does not help very much in understanding the General Relativistic result, because of the great difference in physical interpretation.

In General Relativity we have from the EP that the light-waves do not change as they propagate in free fall. The frequency mismatch called redshift is caused by a mismatch in the natural frequency of the clocks at different heights. The interpretation is now:

(1) An atom of given type undergoing a given transition can serve as a standard of time, oscillating at a resonant frequency ν.

(2) Choose an observer at some arbitrarily chosen 'zero' height. Owing to gravitational time dilation, this observer finds that the resonant frequency of a similar atom situated at height z above him is

$$\nu = e^{\Delta\Phi/c^2}\nu_0 \simeq (1 + gz/c^2)\nu_0.$$

The sign is correct, since here g stands for the size of $\mathbf{g}$, and in contrast to eqn (9.20) we are referring to local clock rate, not observed frequency of received waves.

(3) Taking the Planck relation $E = h\nu$ to be unaffected by gravity (Strong Equivalence Principle), we have that the gap between the atomic energy levels is also affected by the multiplicative factor $\exp(\Delta\Phi/c^2)$. Extending the argument to all transitions and to particle creation/annihilation, we find that an atom of given type in a given internal state has a rest energy increase by the factor $\exp(\Delta\Phi/c^2)$ when it is moved from height zero to height z:

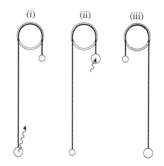

(i)   (ii)   (iii)

Fig. 9.21 An impossible 'energy pump' using atoms and light-waves. (i) A light-pulse passes from lower to upper atom. (ii) The upper atom absorbs the pulse and becomes heavier. (iii) Use the rope and pulley to swap the atoms, obtaining some work—then repeat *ad infinitum*.

$$E = e^{\Delta\Phi/c^2} E_0 \simeq (1 + gz/c^2)E_0. \qquad (9.53)$$

Note that the *rest* energy increases by the 'gravitational potential energy' mgz. This is an $O(\Phi/c^2)$ effect that can *not* be predicted by Newtonian physics, and it is completely different from the electromagnetic case. Electromagnetic forces are 'pure': they *preserve* the rest mass of the charged particles on which they act.

Although the above argument appealed to quantum mechanics and the Strong Equivalence Principle, this was merely for the sake of clarity and convenience. There is no need to use the quantum hypothesis in order to arrive at eqn (9.53). It is sufficient to argue that the energy of a pulse of light transforms, between one LIF and another at a given place, in the same way as its frequency. For, suppose an atom at rest at A emits a light-pulse of energy E_A in its rest LIF (this energy may or may not be the whole rest energy of the atom). The pulse arrives at some other height B with unchanged energy and frequency relative to the chosen LIF. If the pulse is then absorbed by (or reconstitutes) an atom of the same type at rest at the new height, the transformation of its energy from the LIF to a rest frame at the new height matches that of its frequency, so the acquired rest energy E_B of the atom at the new height is $E_B = \exp(\Delta\Phi/c^2)E_A$, which is eqn (9.53). (In chapter 14, eqn (14.30), we shall show this in more detail, but the present argument is already rigourous.)

The analysis of an 'energy pump' using atoms at two heights now goes as follows. We start with a collection of atoms on the 'floor' at height zero, all in an internal excited state, and a collection of similar atoms at height z, all in their internal ground state. Let the energy gap between two states of an atom on the floor be E_0. A photon of energy E_0 emitted by such an atom does not have enough energy to excite one of the upper atoms. Instead, therefore, let us suppose N_0 lower atoms emit N_0 photons. When these photons arrive at height z their combined energy $N_0 E_0$ is sufficient to excite $N = N_0 \exp(-\Delta\Phi/c^2)$ upper atoms. Now lower all these excited atoms down to height zero, while raising the same number N of de-excited atoms from the floor. We expect to get some work from this operation, since the atoms being lowered are objectively heavier than the ones being raised (there are the same number of them, but they are in a different internal state). Let the work obtained be W_0 as reckoned by an observer on the floor who receives the energy. We now have a situation almost the same as at the start, except that we received work W_0 and only N of the atoms on the floor are excited. To return to the starting situation, now excite $(N_0 - N)$ of the atoms on the floor, using up W_0 to provide the energy. To avoid an impossible infinite energy source, we must have

$$W_0 = (N_0 - N)E_0 \quad \Longrightarrow \quad \frac{W_0}{N} = \left(e^{\Delta\Phi/c^2} - 1\right)E_0 \simeq \frac{E_0}{c^2}gh$$

where the formula in terms of Φ is exact, and the approximate version helps to interpret it. We thus account for the phenomena exactly by finding that the rest energy of a given system (such as an atom) in a

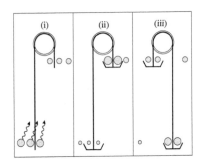

Fig. 9.22 A possible 'energy pump' using atoms and light-waves. The light emitted by the lower atoms is only sufficient to excite a smaller number of upper atoms. The gravitational work obtained is just sufficient to re-excite the remaining lower atom after stage (iii), so there is no net gain.

given state is a function of position in a gravitational field. In the limit $\Delta\Phi/c^2 \ll 1$ the amount of work obtained is the same as if the extra energy carried by each excited atom corresponded to a 'gravitational mass' of E_0/c^2. That can be a helpful way to remember the weak-field result, but one does not have to interpret it that way. In General Relativity there is never any need to introduce the concept of passive gravitational mass.

It is best to think of Φ as first and foremost a function to do with time dilation, which must then also be intimately related to gravitational acceleration, and from which it also follows that Φ can be regarded as a gravitational potential energy function per unit mass, as explored above. The interesting result is eqn (9.53). It says that an object of rest energy E_A at A will, after being slowly raised to B, be found to have rest energy

$$E_B = e^{(\Phi_B - \Phi_A)/c^2} E_A. \qquad (9.54)$$

The idea of a position-dependent rest energy is subtle, and can be confusing. When we say that an object has a mass of 1 kilogramme, what do we mean? We mean that its inertial mass would agree with that of a standard 1-kilogramme object kept in a vault in Sèvres, France, if both objects were brought to the same location without disturbing their inner constitution. Or perhaps we have in mind another definition which takes advantage of energy-frequency conversion via Planck's constant. In any case, eqn (9.54) E_B says that that same object will have a rest mass of half a kilogramme after it has been lowered down a sufficiently deep gravitational potential well, if we compare it with a standard object that was *not* lowered. However, if the standard object is lowered also, then their masses will agree again. So how shall we detect the reduction in rest mass? One way is through the contribution to gravity made by the lowered object, which we will explore in section 11.5.

The role of Φ in statements about energy is loosely comparable to the role of temperature in statements about heat in thermodynamics, in the following sense. A given amount of heat is more useful—can be used to accomplish more work—if it is delivered at high temperature. A given amount of material (say, a given number of hydrogen atoms) is more useful—can be used to accomplish more work—if it is delivered at high gravitational potential. To avoid ambiguity about rest energy, one can always invoke an invariant such as number of atoms—one may say 'this brick has the same rest energy as 10^{22} nearby hydrogen atoms'. Now everyone knows how much brick you have, without your needing to explain where it is relative to other things.

If we want to raise an object from A to B by standing at B and pulling on a rope, we shall have to provide the energy

$$W = E_B - E_A = \left(\frac{e^{\Phi_B/c^2}}{e^{\Phi_A/c^2}} - 1 \right) E_A = \left(1 - \frac{e^{\Phi_A/c^2}}{e^{\Phi_B/c^2}} \right) E_B. \qquad (9.55)$$

Conversely, this is also the energy we would extract at B by lowering the object. A gravitational field thus offers an efficient means to convert rest energy to energy in another form.

In the LIF that is momentarily at rest at B relative to the rigid lattice there, we can apply the Special Relativistic relation $\mathbf{f} \cdot \mathrm{d}\mathbf{r} = \mathrm{d}W$. Thus the force $\mathbf{f}_B$ that has to be exerted at B to move the string through a rest length $\mathrm{d}\mathbf{r}$ when the object is at A is

$$\mathbf{f}_B = -E_B \frac{\boldsymbol{\nabla}\Phi_A}{c^2} e^{(\Phi_A - \Phi_B)/c^2} = m_B g_A e^{(\Phi_A - \Phi_B)/c^2} = m_A g_A. \quad (9.56)$$

Another way to raise a body in a gravitational field is to provide the energy locally using a sequence of energy sources close to the body. The required force at each position $mg = -m\boldsymbol{\nabla}\Phi$, so the total energy that has to be supplied is $\int \boldsymbol{\nabla}\Phi \cdot \mathrm{d}\mathbf{r} = (\Phi_B - \Phi_A)m$ assuming the body does not change, so m is constant. This is different from eqn (9.55) (and equal to the classical result). However, a sum of energies located at different places should be interpreted with caution.

9.4.1   Equation of motion

The equation of free-fall motion in General Relativity is the equation of Most Proper Time. This can be treated in general using the Lagrangian methods described in chapter 14. However, for cases where the shape of the spatial trajectory of a freely-falling particle is known from symmetry considerations (e.g., straight down for a particle falling vertically in a uniform field), in order to get the worldline we only require to know the speed along the spatial trajectory, and this can be obtained from the energy, as follows.

We treat a static gravitational field, in which a free-fall trajectory can be written as a function of a single coordinate $r(\tau)$ (e.g., a radial coordinate in a spherically symmetric problem with the initial velocity in the radial direction). Let E_0 be the rest energy of the particle when it is far from other bodies. As it falls freely, its energy (relative to a distant observer) is constant. By contrast, the energy it would have if it were at rest relative to a rigid lattice fixed in the static gravitational field is given by eqn (9.53). Hence, at any event A on the worldline the falling particle has energy E_0, and a similar particle at rest in the rigid lattice would have energy

$$E_A = E_0 e^{\Phi_A - \Phi_0}$$

(in units where $c = 1$), where Φ_A is the gravitational potential at A, and Φ_0 is the gravitational potential at the top of the particle's trajectory. Now consider a LIF momentarily at rest relative to the lattice at A. Frequency and energy observations in this LIF match those of the lattice observer at A, and since we can apply Special Relativity in a LIF (strong equivalence principle), we must find

$$E_0 = \gamma E_A \equiv \frac{\mathrm{d}\tau_A}{\mathrm{d}\tau} E_A$$

where γ is the Lorentz factor, which is given by $\mathrm{d}\tau_A/\mathrm{d}\tau$, where $\mathrm{d}\tau_A$ is the proper time between successive events at the *same lattice position* r_A, and $\mathrm{d}\tau$ is the proper time between successive events on the particle's worldline. Hence we find

$$\frac{\mathrm{d}\tau_A}{\mathrm{d}\tau} = e^{\Phi_0 - \Phi_A}. \tag{9.57}$$

Let t be the time registered by a clock at rest in the lattice: i.e., one which undergoes gravitational time dilation compared to a clock at $\Phi = 0$. Then, from eqn (9.20) or (9.48),

$$\frac{\mathrm{d}\tau_A}{\mathrm{d}t} = e^{\Phi_A}.$$

Using this in eqn (9.57) we find

$$e^{2\Phi_A} \frac{\mathrm{d}t}{\mathrm{d}\tau} = e^{\Phi_0} \tag{9.58}$$

This equation is essentially a statement about conservation of energy. On the right-hand side is a constant, and on the left-hand side are quantities related to energy. This result is exact, and we will use it in chapter 11 to study free-fall motion near a black hole, for example (and see exercise 9.12). For illustration we now apply it to the approximate metric given in eqn (9.48), which we previously argued can be used to describe the Newtonian limit. Then we have (still using $c = 1$)

$$\mathrm{d}\tau^2 = e^{2\Phi}\mathrm{d}t^2 - \mathrm{d}r^2 \qquad \Rightarrow \qquad \frac{\mathrm{d}t}{\mathrm{d}\tau} = e^{-\Phi}\left(1 + \dot{r}^2\right)^{1/2}$$

where $\dot{r} = \mathrm{d}r/\mathrm{d}\tau$. Substituting this in eqn (9.58) and squaring gives

$$e^{2\Phi}(1 + \dot{r}^2) = e^{2\Phi_0} \tag{9.59}$$

$$\Rightarrow \qquad (1 + 2\Phi)(1 + \dot{r}^2) \simeq 1 + 2\Phi_0 \tag{9.60}$$

$$\Rightarrow \qquad m\Phi + \tfrac{1}{2}m\dot{r}^2 \simeq m\Phi_0 \tag{9.61}$$

where we used $\Phi/c^2 \ll 1$ and then multiplied by the particle mass m in the last line, in order to make the terms easy to recognize. On the left is the sum of potential energy and kinetic energy, and on the right is the total energy. By differentiating with respect to r one can also obtain $\mathrm{d}^2r/\mathrm{d}t^2 = -\mathrm{d}\Phi/\mathrm{d}r = g$, as expected.

Exercises

(9.1) Show that the tidal force on Earth due to the Moon is approximately twice that due to the Sun. (The mass of and distance to Moon and Sun are $(7.3 \times 10^{22}\,\text{kg}, 1.3\,\text{ls})$ and $(2 \times 10^{30}\,\text{kg}, 500\,\text{ls})$ respectively, where ls =light-second.)

(9.2) At what distance from a 1 solar-mass neutron star will a freely-falling steel cable of length 100 m be pulled apart by tidal forces, if the cable has density 8000 kg/m^3 and tensile strength 400 MPa? [*Ans.* 1.9×10^6 m.]

(9.3) The Earth was formed approximately 4.5 billion years ago. How much less has the core of the Earth aged since then, compared to rocks at the surface?

(9.4) The radius of the Sun is 7×10^8 m. Find the gravitational redshift of light of wavelength 500 nm emitted from the surface of the Sun. [*Ans.* $\Delta\lambda = 1.06$ pm]

(9.5) The equation of motion (9.31) (i.e., $-t^2 + x^2 = h_0^2$) is Lorentz-invariant. It follows that if one plots the worldlines of the particles in a constantly accelerating frame relative to some other inertial frame (moving in the x direction relative to the first), one will again obtain hyperbolas, and Fig. 9.16 will look the same. Show that the observer at h_1 finds the worldline at h_2 to be at a radar distance $h_1 \log(h_2/h_1)$ above him.

(9.6) A light-ray travelling in the y direction in Minkowski space has the worldline $x =$ const, $y = ct$, $z = 0$. Using a change of coordinates, show that in the space of the uniformly accelerating rocket, light-rays follow circular arcs. Show that eqn (9.16) gives an exactly correct prediction at any point on the arc.

(9.7) In what direction should a space-ranger in the uniformly accelerating rocket fire his laser gun in order to kill the enemy soldier standing across from him at the same height? Should he aim in the direction in which he sees the soldier?

(9.8) Consider the metric $ds^2 = -(2x - 1)dt^2 + (2x - 1)^{-1}dx^2 + dy^2 + dz^2$, where (x, y, z) are rectangular coordinates. Find the acceleration due to gravity and the coordinate speed of light. Identify the horizon. Introduce a change of coordinate to $X \equiv \sqrt{2x - 1}$. Do you recognize the resulting metric?

(9.9) Sketch the electric field lines due to a charged particle at rest in a uniform gravitational field.

(9.10) An object whose rest energy is E_{B1} when it is at location B is lowered slowly on a rope until it reaches location A, where its rest energy is $E_{A1} = \exp(-\Delta\Phi)E_{B1}$. The energy $E_{B2} \equiv E_{B1} - E_{A1}$ left at B by this process is then packed into a light bag and lowered to A, where its rest energy is $E_{A2} = \exp(-\Delta\Phi)E_{B2}$ and again some energy (E_{B3}) is left at B. This process is repeated until no energy is left at B. Confirm that all the energy arrives at A, i.e., $E_{A1} + E_{A2} + E_{A3} + \ldots = E_{B1}$.

(9.11) §*Climbing a ladder.* A man climbs a ladder in a gravitational field, and then jumps back down to the floor. Describe the energy changes. Do the energy level separations of the atoms of the man's body change? Is this a reference-frame-dependent question? Assuming the man emits no heat, does his rest mass change? Separate in your answer the the three stages of journey up, free fall, and landing on the floor.

(9.12) Use eqn (9.58) with the Rindler metric (9.45) to obtain the equation of motion

$$h^2\left(1 + \left(\frac{dh}{d\tau}\right)^2\right) = \text{const}$$

for an object in free fall in the constantly accelerating rocket. Hence, using $h^2 = x^2 - t^2$ (eqn (9.35)), deduce that in (x, t) coordinates the motion satisfies $(x - vt)^2 = \text{const}$.

(9.13) To obtain insight into least action (most proper time) methods, consider the conceptually simpler principle of *least optical path length* in optics (Fermat's principle). Optical path length is measured by the number of wavelengths along the path; material such as glass reduces the wavelength so increases the optical path length; light follows paths of stationary optical path length. By considering arrangements of prisms and mirrors, show that there can exist cases where there is more than one locally shortest optical path between given points.

(9.14) The previous question made an observation about optical paths in space. The corresponding fact about proper time along worldlines is that there can exist cases where there is more than one possible free-fall motion between given events. As Rindler has pointed out, the result is a 'twin paradox' where neither twin experiences proper acceleration. For example, consider twins born in a space station orbiting the Sun at 1 Astronomical Unit (not near the Earth). Let one twin be launched from the space station in a direction directly away from the Sun, with an initial velocity chosen such that it takes a year of free-fall motion for him to return to his initial position. Roughly estimate the gravitational and kinetic contributions to the time dilation for both twins, and hence establish which is younger when they meet again.

(9.15) A neutron is dropped from rest at a large distance from a compact star. It falls freely onto the star and collides with a neutron resting on the surface there. In the resulting collision, a neutral pion is produced by the process $n + n \rightarrow n + n + \pi^0$. If we suppose that both neutrons may be treated as free particles during the collision, what is the threshold value of the gravitational potential at the surface of the star for this process?

(9.16) A train of proper length $L_0 = 1$ light-second is initially at rest by a platform. The front of the train undergoes constant acceleration to $v = (24/25)c$ in 1 second. Its mean speed during this time, in the rest frame of the platform, is therefore $(12/25)c$, and the distance it travels is $12/25 = 0.48$ light-seconds. The Lorentz contraction of the train at its final speed is by a factor $\gamma = 25/7 \simeq 3.57$. Therefore the back of the train has to cover a further distance of $L_0(1 - 1/\gamma) = 18/25$ light-seconds. Hence its total distance travelled is $(12 + 18)/25 = 1.2$ light-seconds, so its mean speed is $1.2c$: i.e., faster than the speed of light. Or is it? Explain.

10 Warped spacetime

We now come to the final big idea which is needed to complete the theory of gravitation provided by General Relativity. This is the idea of *spacetime curvature*.

Curved or warped?

The basic idea is that if we liken spacetime in one spatial and one temporal dimension to a surface, then that surface is not flat. The notion of 'flatness' or otherwise can be extended to larger numbers of dimensions, and ultimately we will find that four-dimensional spacetime is warped or 'curved'.

The word 'curved' is used here in a technical sense. In everyday language when we say a surface is curved we usually have in mind an example such as the surface of a sphere. The surface is two-dimensional in the sense that two parameters (such as latitude and longitude) are enough to specify any point, but it is curved because it bends around into a third dimension. More generally, it could be curved into more complicated shapes. From a mathematical point of view, the question arises: is this idea of curvature essentially to do with the bending into the third dimension, or can it be understood purely in terms of measurements *along* the surface? The remarkable answer is that we do not need to appeal to the third dimension: it is sufficient to study only measurements along the surface. Such measurements are said to reveal *intrinsic* properties of the surface, and we will show in the following that they can reveal the curvature through the effects it has on intrinsic geometric quantities such as the sum of the angles of a triangle drawn on the surface, or the ratio of the circumference to the diameter of a circle on the surface. In these studies it is useful to say that the surface (even a curved surface) is two-dimensional, because two coordinates suffice to map it. That, in any case, is the standard terminology, and we shall adopt it. If the properties of geometric objects such as triangles and circles on the given surface do not match those of Euclidean geometry, we say the surface is 'curved', where the word now has a meaning closer to that conveyed in everyday language by 'warped'. That is, we can imagine the surface curving away into a third dimension if we like, but we do not have to use that image: we could instead just say that the 'fabric' of the surface is warped, with precise algebraic consequences, and not appeal to a third dimension.

In a similar way, we shall say that spacetime is four-dimensional and is curved or warped. You can, if you like, infer that there is a fifth dimension into which this four-dimensional object can curve, but such a fifth dimension would only serve as an aid to the imagination: it is not a dimension into which any worldline can ever depart out of four-dimensional spacetime. Also, not all types of warping that make good algebraic sense can be modelled accurately by an appeal to a fifth dimension. The same is true at lower numbers of dimensions: we shall meet, for example, a two-dimensional surface which can be mapped by a pair of periodic coordinates such as latitude and longitude, and which has everywhere the same positive curvature. Such a surface is called a spherical surface, and can be regarded as the surface of a sphere in an ordinary (unwarped) three-dimensional space. However, we can also define algebraically a surface which is mapped by the same pair of coordinates but which has everywhere the opposite sign of curvature. There is no way to picture this as the surface of any imaginable shape (complete and without singularities) in unwarped three-dimensional space. We say the former surface can be 'embedded' in flat three-dimensional space, but the latter cannot. Having made this cautionary note with respect to 'embedding', we shall nonetheless use our knowledge of surfaces that can be embedded to learn about the general problem.

Our aim is to understand what it means to say we live in a warped four-dimensional spacetime, and especially to discover what the measurable physical consequences would be. To this end, we shall first examine the spatial part of the problem. That is, we do not for the moment care about the time taken for any process we shall consider; we concentrate purely on distance and angle measurements. We begin with two-dimensional surfaces, since these are easier to imagine, then generalize to three spatial dimensions, and then incorporate time at the end.

10.1 Two-dimensional spatial surfaces

Suppose there is a species of intelligent bugs who live on a surface. The surface might be flat as in Fig. 10.1 or curved as in Fig. 10.2. We can see immediately the shape of these surfaces, but we suppose the bugs have no direct experience of the third dimension: they cannot see, but have to find their way around by feeling. They can, however, make accurate surveys of their world by making careful distance and angle measurements *within* the surface where they live.

We shall also consider a surface that looks flat to us, Fig. 10.3, but which has a special property which makes it feel to the bugs just like the spherical surface of Fig. 10.2. This is a device to show that curvature of a space can be modelled in more than one way. The special property is that this surface carries at every point an 'expansion field' f which causes any object placed there to immediately expand to f times its normal length. The factor f is a function of position on the plate, but since this field

Fig. 10.1 A bug living on a flat surface.

Fig. 10.2 A bug living on a spherical surface.

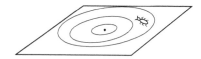

Fig. 10.3 A bug living on a 'hot-plate': a flat surface possessing an expansion field that causes all objects placed on it to expand by a locally determined factor (e.g., thermal expansion caused by the temperature of the plate, which is here coldest at its centre).

affects everything equally, including the bodies of the bugs themselves, everything they possess, and their constituent particles, the bugs are not directly aware of it. If a bug carries a standard rod called a ruler, and finds that his own body has a length of 3 standard rods when he is standing somewhere on the plate, then he will find his body remains of length 3 standard rods wherever he goes, since he and the rod both expand or contract together. The bugs *define* distance in their world by using standard rods of given construction. The plate with the expansion field is sometimes called a 'hot-plate' model, since we can imagine that the expansion field is like a temperature which causes things to expand by thermal expansion. Let (ρ, ϕ) be plane polar coordinates on the plate, as measured by an onlooker such as ourselves whose measuring sticks are not subject to the expansion. We shall treat a plate where the expansion factor varies with position as

$$f = 1 + k^2 \rho^2 \tag{10.1}$$

where k is a constant. This means that the infinite plate is ordinary in the middle and at increasing 'temperature' as one moves out from the middle.

Now suppose the bugs begin to do some geometry. We suppose that the legs of bugs move in a very regular way, taking small steps of the same size, so that by counting steps the bugs have a good way of measuring distance along any given path. A bug Euclid proposes that a straight line can be defined as one connecting two points with the smallest length. By running too and fro between two given points the bug on the flat plane soon learns which is the straight line between them. The bug will also find that if he carefully ensures that at every step he moves a left and right leg simultaneously, so that both sides of his body move through the same distance, then he follows the very path that he previously found to be straight by the criterion of least length. This is no surprise, because the bug's legs are here probing one pair of nearby paths: these paths must have the same length if they closely bracket a path of least length. It also shows that the definition of a straight line in terms of least distance agrees with an alternative definition in terms of 'parallel transport': the second definition is 'if you walk straight ahead without turning then your path will be straight'. The second definition is similar to the way by which tracked vehicles such as tanks and bulldozers move: they steer by making one track travel further than the other. If both tracks move the same amount and there is no slipping, then a bulldozer moves in a straight line.

How do things go for the bug on the sphere? He also finds a path of least length: it is an arc of a great circle (Fig. 10.4). He also finds that if he walks straight ahead (like a bulldozer) then his path agrees with the least-length one between any given pair of points on it.

The bug on the 'hot-plate' can also find 'straight lines' by either of the criteria (least distance or no-turning), and the lines he thus identifies appear curved to us: Fig. 10.5. This is because to minimize the number of steps taken between a given pair of points, the bug should wander

Fig. 10.4 A geodesic between two points on a sphere is an arc of the unique great circle on which they both lie.

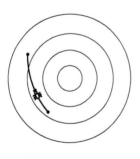

Fig. 10.5 Geodesic on the hot-plate.

How to draw a geodesic.

Here is how to draw a geodesic on an arbitrary surface. First triangu-
late the surface. That is, approximate it by a collection of small flat
triangular faces. Now start drawing the geodesic using a straight line
on the starting face. When you approach an edge, fasten a small strip
of tape over the edge and let the straight line pass onto this tape up
to the edge. Next, remove the tape and flatten it. Now you can see
how the geodesic line should continue: make it a straight line when
the tape is flattened. Finally, fold the tape back over the edge you
are treating, lining it up with the geodesic traced so far, and continue
the line onto the next face. Keep going with this procedure until the
geodesic is as long as you require. For an exact treatment, do the
same but with the surface initially triangulated into infinitely many
infinitessimal triangles.

somewhat outwards from the centre of the hot-plate, to regions where
the expansion factor is larger, so that fewer steps (or fewer metre sticks)
are needed to make progress towards the goal. Also, when travelling
'straight ahead' the outermost leg of the bug moves further (as far as
we are concerned) than the innermost one, again causing a trajectory
whose shape agrees with the 'least distance' definition.

These lines, which are most generally defined as paths of no turning,
and which are also paths of least length, are called **geodesics**.

Next the bugs define a circle as the locus of points at given distance
from some centre point. The bugs can then measure the ratio of circumference to radius of their circles. The circumference is not a geodesic,
but the bug can easily measure it by walking around it. On the flat
plane the bug finds the ratio is close to 6.2831853—a number we shall
agree to call 2π. On the sphere, the bug finds a different answer. For
a given circumference C he measures the radius to be somewhat *larger*
than $C/(2\pi)$, because on his journey out from the circle's centre he has
to follow the surface of the sphere that defines his world.

To make a quantitative study of this we introduce the **metric**. This
is a mathematical summary of what the bugs find by their distance
measurements over small distances at any point. The bug on the flat
plane, for example, might propose to map the plane using plane polar
coordinates (r, ϕ). At any point labelled by (r, ϕ), by walking to nearby
points $(r + \mathrm{d}r, \phi)$ and $(r, \phi + \mathrm{d}\phi)$ the bug can discover how far away
they are. For example, a short walk along the r direction at constant ϕ
has length $\mathrm{d}s = \mathrm{d}r$, and a short walk along the ϕ direction at constant
r has length $\mathrm{d}s = r\mathrm{d}\phi$. Since these line segments meet at right angles,
the square of the total distance from (r, ϕ) to $(r + \mathrm{d}t, \phi + \mathrm{d}\phi)$ can be
obtained from Pythagoras' theorem:

$$\mathrm{d}s^2 = \mathrm{d}r^2 + r^2\mathrm{d}\phi^2. \tag{10.2}$$

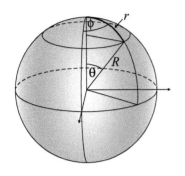

Fig. 10.6 (r, ϕ) coordinates on the surface of a sphere.

This equation is called the 'line element', or 'metric equation' or 'metric' for short. It is precisely the same concept we have already met in eqns (9.46) and (9.47). The metric depends on the coordinate system employed, and on the intrinsic properties of the surface in question.

Now consider the bug on the sphere. We might naturally map a spherical surface by using the angles (θ, ϕ) of spherical polar coordinates, and the bug could do the same, but let us choose a coordinate system more natural to a bug walking about on the surface. Let the bug pick an arbitrary point on the surface to be the origin, and introduce polar coordinates (r, ϕ), where r is the distance *along the surface* (i.e., the arc length) from the origin to the point in question (following a 'straight line', i.e., an arc of a great circle), and ϕ is the azimuthal angle. By definition, then, a short walk along the direction of increasing r at constant ϕ has length $\mathrm{d}s = \mathrm{d}r$. A short walk along the direction of increasing ϕ at constant r has length $\mathrm{d}s = R \sin \theta d\phi$, where R is the radius of the sphere (known to us, but unknown, in the first instance, to the bug) and $\theta = r/R$ is the spherical polar angle. (We introduced θ merely for convenience of computation; it has now served its purpose and will not be needed again.) The coordinate lines are again everywhere orthogonal, so we may use Pythagoras' theorem to deduce

$$\mathrm{d}s^2 = \mathrm{d}r^2 + R^2 \sin^2(r/R)\mathrm{d}\phi^2. \tag{10.3}$$

This is the metric of the spherical surface. It is a summary of *precisely what the bug would find* by his measurements of distances in the surface.

Now consider the bugs' measurements of the radius and circumference of a circle. For convenience of calculation we place the origin of coordinates at the centre of the circle. To find the distance of an arbitrary point (r, ϕ) from the origin, calculate $\int \mathrm{d}s$ along a suitable geodesic, which in this case is a line of constant ϕ. One finds using either metric (10.2) or (10.3) that a set of points at a given distance from the origin is a set at fixed r, and the distance from the origin (i.e., the radius of the circle, as observed by the bug) is

$$\int \mathrm{d}s = \int_0^r \mathrm{d}r = r \tag{10.4}$$

in both cases. Having used radial geodesics to find the edge of the circle, perhaps marking it by dropping seeds there, the bugs can now walk around the circumference. This circumference is *not* a geodesic (this is obvious for the case of the flat plane), but the length along it is still given by the metric. The circumference of the circle is given by $\int \mathrm{d}s$ calculated for the path around the circle. This is a path at fixed r, so one finds

$$C = \int \mathrm{d}s = \int_0^{2\pi} r\mathrm{d}\phi = 2\pi r \qquad \text{using (10.2)}$$

$$C = \int \mathrm{d}s = \int_0^{2\pi} R \sin(r/R)\mathrm{d}\phi = 2\pi R \sin(r/R) \qquad \text{using (10.3).}$$

The second result is approximately

$$C \simeq 2\pi r \left(1 - \frac{1}{6}\frac{r^2}{R^2}\right). \tag{10.5}$$

Therefore, for the bug on the sphere, the circumference tends to $2\pi r$ for small circles, but is 'too small' for larger circles. The bug could quantify this by defining a 'radius excess':

$$r_{\text{excess}} \equiv r - \frac{C}{2\pi}. \tag{10.6}$$

This is the amount by which the measured radius exceeds the one which would have been expected on the basis of Euclidean geometry and the measured circumference. In this example one finds $r_{\text{excess}} \simeq r^3/(6R^2)$; the main point is that distance measurements purely *within* the surface can reveal and quantify the departure of the surface from Euclidean geometry. A commonly used measure is called *Gaussian curvature*, and for a sphere of radius R is defined $K = 1/R^2$. The bug can obtain K for his world by evaluating

$$K = \lim_{r\to 0} \frac{6r_{\text{excess}}}{r^3}. \tag{10.7}$$

The bug on the hot-plate also finds the ratio of circumference to radius of a circle to be smaller than 2π. It is easy to see this by choosing a circle centred at the centre of the plate, then the bug-steps used to measure the radius are on average shorter (to us) than the bug-steps used to measure the circumference, so the radius found by the bug using the 'number of steps' measure is larger than $C/2\pi$. For a quantitative statement, let us map the plate using plane polar coordinates (ρ, ϕ), where ρ is the distance from the origin according to our privileged measurements that are *not* subject to the expansion field. The bug's measuring rulers expand by the factor f, so if two points are observed by us to be separated by a distance ds_0, then the bug finds the 'ruler distance' between them to be $ds = ds_0/f$. The metric observed by the hot-plate bug is therefore the one for plane polar coordinates (10.2) divided by the square of the expansion factor:

$$ds^2 = \frac{1}{(1 + k^2\rho^2)^2}\left(d\rho^2 + \rho^2 d\phi^2\right). \tag{10.8}$$

The ruler distance around the circumference of a circle of given ρ is obviously $C = 2\pi\rho/f$, and the radius is

$$r = \int_0^\rho \frac{1}{f}\,d\rho = \frac{1}{k}\tan^{-1}(k\rho).$$

Therefore $\rho = (1/k)\tan kr$ and

$$C = \frac{2\pi\rho}{1 + k^2\rho^2} = 2\pi\frac{1}{k}\frac{\tan kr}{\sec^2 kr} = 2\pi\frac{1}{2k}\sin 2kr.$$

This is precisely the same as the result for a spherical surface with $R = 1/(2k)$. Therefore, by this measure, the bug cannot tell whether he is on

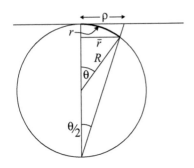

Fig. 10.7 Relationship between various coordinates that can be used to map the surface of a sphere. $r = R\theta$, $\bar{r} = R\sin\theta$. Let $\zeta = \tan(\theta/2)$, then $\sin\theta = 2\zeta/(1+\zeta^2)$ and $\rho = 2R\zeta$.

the spherical surface or the hot-plate. More generally, one can prove that the two cases are exactly equivalent, for all geometrical measurements by the bugs, because by a change of coordinates one can bring the metric for the hot-plate to the same form as the metric for the sphere (Fig. 10.7; exercise 10.2). The expansion factor was here chosen in such a way as to make this happen; in this example the 'hot-plate' is a stereographic projection of a sphere. You can see that other choices of expansion factor could represent other types of surface, not all of them easily pictured.

The idea of radius excess can be refined in a useful way by considering instead of a full circle a small arc of a circle. That is, suppose the bug marks off two geodesics issuing from the same point with a small angle between them, travels the same distance r along each, and then measures the arc distance η between the end points. This quantity measures the *geodesic deviation*: i.e., the amount by which two geodesics spread out. On a flat surface one would expect them to spread linearly with distance r (Fig. 10.8). For a given choice of a pair of geodesics issuing from a point, η is some fixed small fraction ϵ of the circumference of a circle. Therefore, for the case of a bug on a sphere, we may use eqn (10.5) to find

$$\eta \simeq \epsilon 2\pi r \left(1 - \frac{1}{6}Kr^2\right). \tag{10.9}$$

Differentiate this twice with respect to r and one finds, to leading order in r,

$$\frac{\mathrm{d}^2\eta}{\mathrm{d}r^2} = -K\eta. \tag{10.10}$$

If we now apply this equation to any shape of surface, then it amounts to a way of *defining* curvature K. That is, the bug's job now is to find the second rate of change of η for a pair of neighbouring geodesics (and it turns out they do not need to intersect at the point of interest), and

Fig. 10.8 Curvature examples: a cylindrical, spherical, and saddle-shaped surface. When we try to 'squash' a section of these surfaces flat, we find in the non-flat cases that there is either not enough or too much circumference. Geodesic lines diverge linearly in the flat case, or sub-linearly, super-linearly, respectively, in the other cases. The curvature K is hence said to be zero, positive, negative, respectively.

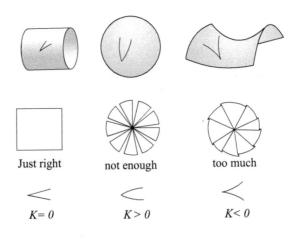

compare it with η. We have arranged by definition that this measure of curvature agrees with the previous definition in terms of radius excess.

Another simple geometrical experiment is for the bugs to create triangles and measure the sum of their interior angles. A triangle can be defined as a shape composed of three geodesic line segments joining three given points. To define angle the bugs can use the ratio of arc length to radius for circular arcs, in the limit where the radius of the circular arc tends to zero. In that limit they will all agree that there are 2π radians in a complete circle. This is because a small enough region of any curved surface is to good approximation flat: the bugs always find that the radius excess of a circle tends to zero as $r, C \to 0$.

The bug on the flat plane finds that the interior angles of any triangle sum to π radians, or 180°. The bug on the sphere finds that the angles sum to more than 180°. For example, one can find triangles composed of three right angles: start at the 'equator' and walk in a 'straight line' (i.e., geodesic) to the 'north pole', turn through a right angle and walk back to the equator, then turn through a right angle again and walk back to the starting point (see Fig. 6.17). The resulting triangle has a total of 270° interior angle. For smaller triangles the sum is smaller. It turns out that the excess angle (i.e., the extra 90° for the special case) is proportional to the area of the triangle (see box below). Angle excess thus provides another measure of the departure of the underlying space from flat geometry.

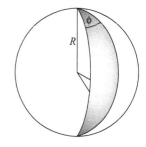

Fig. 10.9 A lune. Its surface area is $2R^2\phi$.

Angle excess. Draw two semicircular arcs (lines of 'longitude') on a sphere of radius R as in Fig. 10.9. The lozenge-shaped region on the sphere's surface is called a *lune*. The area of such a lune increases in proportion to ϕ, since the arc length around every latitude is proportional to ϕ. Therefore, the area is $2R^2\phi$. Now consider the sphere shown in Fig. 10.10. The angles α, β, γ are associated with three lunes, of areas

$$\sigma_\alpha = 2\alpha R^2 \qquad \text{(lune } ABA'CA)$$

$$\sigma_\beta = 2\beta R^2 \qquad \text{(lune } BAB'CB)$$

$$\sigma_\gamma = 2\gamma R^2 \qquad \text{(lune } CAC'BC)$$

These sum to

$$\sigma_\alpha + \sigma_\beta + \sigma_\gamma = 2(\alpha + \beta + \gamma)R^2.$$

By inspecting the diagram you can see that the sum of these lunes covers the areas

$$(ABC + BCA') + (ABC + ACB') + (CB'A' + A'B'C')$$

$$= (ABC + BCA' + ACB' + CB'A') + (ABC + A'B'C')$$

which is the whole front hemisphere, plus the triangles ABC and $A'B'C'$. The areas of the latter are equal by symmetry, so we have

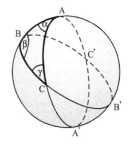

Fig. 10.10 A triangle on a sphere.

$$\sigma_\alpha + \sigma_\beta + \sigma_\gamma = 2\pi R^2 + 2\sigma_{ABC}$$

Hence

$$\alpha + \beta + \gamma = \pi + \frac{1}{R^2}\sigma_{ABC}$$

so the angle excess of the triangle ABC is

$$\alpha + \beta + \gamma - \pi = K\sigma_{ABC} \qquad (10.11)$$

where $K = 1/R^2$ is the curvature. This proves that the angle excess is proportional to the area of the triangle, and can be used to extract the curvature of the surface. This result is exact on a sphere, and also applies for a small enough triangle on any surface, since a small region can be approximated by a spherical surface.

You can show that a similar excess angle appears in the case of the bug on the hot-plate (the easiest case to consider is a triangle with one vertex at the centre of the plate).

For the spherical surface and the hot-plate model we have a radius and angular *excess* compared with Euclidean geometry. When the surface exhibits such an excess it is said to have *positive curvature*. It is easy to see that one can also have a radius and angular *deficit*: this would be the case for a saddle-shaped surface (Fig. 10.8) and for a hot-plate model with the opposite sign in the expansion factor (e.g., replace k^2 by $-k^2$). Such surfaces are said to have *negative curvature*. In general, the size and the sign of the Gaussian curvature can vary from place to place (think of the surface of a pear, for example).

A general formula for the Gaussian curvature of a surface described by the metric

$$ds^2 = h^2 dr^2 + f^2 d\phi^2,$$

where (r, ϕ) are coordinates in the surface and $h(r, \phi)$, $f(r, \phi)$ are arbitrary smooth functions, is

$$K = \frac{1}{h^3 f^3}\left(f^2\left[\frac{\partial h}{\partial r}\frac{\partial f}{\partial r} - h\frac{\partial^2 f}{\partial r^2}\right] + h^2\left[\frac{\partial h}{\partial \phi}\frac{\partial f}{\partial \phi} - f\frac{\partial^2 h}{\partial \phi^2}\right]\right). \qquad (10.12)$$

Although we will not prove this general result here, you are invited to obtain the result for some simpler cases (without assuming eqn (10.12)) in the exercises. Note that the curvature can be obtained directly from the metric, and no other information is needed.

The curvature revealed by these geometric measurements is called *intrinsic curvature*. Some surfaces which appear 'curved' in the ordinary sense of the word can nevertheless have zero intrinsic curvature. Examples are the surface of a cylinder or a cone. Because it can be 'unrolled' to a flat surface without distortion, the intrinsic geometric measurements in a cylindrical surface reveal strictly zero radius excess and zero angular excess. The kind of curvature it possesses could still be revealed by global measurements, and is called *extrinsic curvature*. The spacetime curvature associated with gravity is not merely of the latter kind, it is intrinsic.

So far we have described intrinsic curvature; we have not yet described the connection to gravity. The basic idea is that the worldlines of particles that experience no force other than gravity are geodesics in spacetime. Since the shape of a geodesic is a property of the underlying space, not the particle, this immediately explains why all small enough systems at a given place have the same acceleration due to gravity, irrespective of their mass and internal constitution. Geodesics can have interesting shapes when the underlying space is curved; gravity is the name we give to this phenomenon—it is essentially a consequence of, or a name for, the curvature of spacetime.

10.1.1   Conformal flatness

The metric tensor g is defined such that the line element can be written $ds^2 = dX^T g X$. For example, eqns (10.2) and (10.8) give, respectively,

$$g_{ab}^{\text{flat}} = \begin{pmatrix} 1 & 0 \\ 0 & r^2 \end{pmatrix}, \qquad g_{ab}^{\text{hotplate}} = \begin{pmatrix} (1+k^2\rho^2)^{-2} & 0 \\ 0 & \rho^2(1+k^2\rho^2)^{-2} \end{pmatrix}.$$

We have retained the coordinate labels (r, ϕ) and (ρ, ϕ) in displaying these; but of course, if we take the view of a bug on just one of these surfaces, he does not care about the names of coordinate labels on the other. In particular, the bug on the hot-plate could pick the letter r for the coordinate we called ρ, without changing any deductions. In other words, the second metric could equally be written

$$g_{ab}^{\text{hotplate}} = \begin{pmatrix} \frac{1}{(1+k^2r^2)^2} & 0 \\ 0 & \frac{r^2}{(1+k^2r^2)^2} \end{pmatrix} = \frac{1}{1+k^2r^2} g_{ab}^{\text{flat}}. \qquad (10.13)$$

When one metric tensor can be obtained from another simply by multiplying by a function, as here, then the metrics are said to be *conformally equivalent* (or we say they are related by a *conformal transformation*). More generally, the relationship between conformally equivalent metrics has the form $g_{ab}^{(1)} = \Omega g_{ab}^{(2)}$, where Ω may be a function of all the coordinates. A metric that is conformally equivalent to the metric of flat space is said to be *conformally flat*. Thus eqn (10.13) is a demonstration that the 2-sphere (the two-dimensional spherical surface) is conformally flat. Note that this does *not* mean it is flat. It *does* mean there exists a mapping from the curved surface to a flat surface that preserves angles. You should be able to see that the two statements

(1) 'surface Σ can be described using a flat space with a scalar expansion factor (i.e., one which does not depend on orientation of the ruler)',

(2) 'surface Σ is conformally flat',

are strictly equivalent, since both translate mathematically to the same statement about the metric.

A remarkable property of two-dimensional surfaces is that *all two-dimensional surfaces are conformally flat*. To prove this in general would take us further than we want (or need) to go in the present book. However, we shall present the proof for surfaces of revolution, which suffices to convey the idea.

A surface of revolution is formed when one takes a line in a Euclidean plane, $y = y(x)$, and rotates it about the x axis to form a surface. We will assume that the line meets the x axis at exactly two points, and at right angles, so that we have a single smooth surface. Such a surface has cylindrical symmetry, so there always exists a coordinate choice, such as cylindrical polar coordinates x, ϕ, in which the metric has no dependence on the azimuthal coordinate ϕ and no cross terms (no $\mathrm{d}x\mathrm{d}\phi$ terms). Hence the line element is

$$\mathrm{d}s^2 = a^2\mathrm{d}x^2 + b^2\mathrm{d}\phi^2$$

where $a(x)$ and $b(x)$ are smooth functions of x alone. Now consider a function $r(x)$. Since this is a function of a single variable, we have $\mathrm{d}r = r'\mathrm{d}x$ (where $r' = \mathrm{d}r/\mathrm{d}x$), so

$$\mathrm{d}s^2 = \frac{a^2}{r'^2}\mathrm{d}r^2 + b^2\mathrm{d}\phi^2. \qquad (10.14)$$

Clearly this metric is conformally flat if

$$a(x) = b(x)\frac{\mathrm{d}r}{\mathrm{d}x}.$$

This is a first-order differential equation for $r(x)$ which in principle has a solution. Hence by changing coordinates from x to r we can bring the metric to a conformally flat form.[1] Another possible choice is $ra = br'$, which gives $\mathrm{d}s^2 = (b/r)^2(\mathrm{d}r^2 + r^2\mathrm{d}\phi^2)$.

The conformal flatness property enables one to deduce a further property: as far as intrinsic geometry is concerned, all surfaces are isotropic at each point. This means that near any given point, triangles of the same area all have the same angle excess, no matter how they are oriented in the surface, and geodesic deviation is independent of the direction of the pair of geodesics used to measure it. This shows that the intrinsic curvature of these surfaces can be quantified by a *single* parameter K. Consider, for example, an ellipsoid of revolution, in the form of a long thin needle. From our three-dimensional perspective we may think it is 'more curved' in one direction than another. However a bug on the surface has *no way of telling that!* (except possibly by global measurements). It is true that the line curvature of lines along the surface of the needle, when measured using its embedding in three-dimensional space, is large in one direction and small in another. However, this does not lead to anisotropy of the intrinsic geometric properties: near any given point the surface is to first approximation flat, and to next approximation 'spherical' (i.e., curved but locally isotropic). One can prove this most easily by writing the metric in the conformally flat form, leading to the scalar 'expansion field' picture.

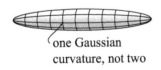

one Gaussian
curvature, not two

Fig. 10.11

To see it intuitively, take a small section of whatever surface is under investigation, and 'squash it flat' as in the middle three diagrams of Fig. 10.8. Depending on the surface, one will then find either a sequence of gaps where the surface has to be cut in order to flatten it (because of too little circumference), or a sequence of folds (too much circumference). One can always arrange that these gaps or folds lie equally spaced around the circumference. It follows that in two dimensions the geodesic deviation is isotropic.

10.2   Three spatial dimensions

Curvature of a three-dimensional space is less easy to imagine in direct pictorial terms, because there is no easy way to draw a four-dimensional picture. However, it can be treated algebraically, and the 'expansion field' model can also be helpful. There could be an 'expansion field' at every point in three-dimensional space which affected all length measurements. The consequence would be departures from Euclidean geometry, a consequence which we may choose to say is owing to 'space curvature'—but if you do not like the word you do not have to use it. In three dimensions a scalar expansion factor is not sufficient to capture all the possibilities: the factor may now depend on the orientation in space of the ruler, because three-dimensional surfaces are not guaranteed to be conformally flat.

One can detect and quantify curvature of a three-dimensional space just as we did in two dimensions: construct circles and triangles and measure their sizes and angles, or use geodesic deviation. The sense and size of any disagreement with Euclidean geometry gives a measure of the curvature of the three-dimensional space. However, since a plane can be oriented in more than one way in three dimensions, there is now the possibility that the value of the Gaussian curvature K obtained from radius or angular excess will have a range of values, depending on the orientation in space of the plane used to measure it. In three dimensions there are three independent directions for planes, suggesting that we should seek three Gaussian curvatures. However, this does not exhaust the possibilities: we can now enquire into parallel transport of a vector, around a trajectory in a given plane, for a vector not necessarily in that plane. It turns out that the complete information about K can be specified in terms of $n^2(n^2 - 1)/12$ 'curvature components', where n is the number of dimensions of the space. This evaluates to $(1, 6, 20)$ components for a space of $(2, 3, 4)$ dimensions respectively. The fourth power of n arises because the most general question one can ask about curvature involves four vectors, each having n components: 'what is the net change $\delta \mathbf{w}$ in a vector $\mathbf{w}$ when it is parallel-transported around a small parallelogram with sides given by vectors $\mathbf{u}$, $\mathbf{v}$?' (per unit size of the vector and the parallelogram).

The idea of parallel transport was introduced informally above and in section 6.7.2. Now we shall present a precise definition. In principle,

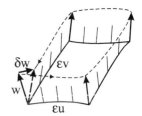

Fig. 10.12 Transporting a vector around a small closed loop.

any path can be considered, but we begin with the case of a geodesic path.

Definition 10.1   *A vector* **w** *carried along a geodesic G undergoes a parallel transport if and only if the angle θ between* **w** *and G does not change as the vector is carried along.*

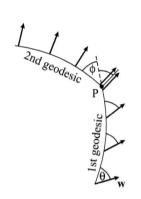

Fig. 10.13 Parallel transport of a vector **w** along a pair of geodesics.

Next consider a path made of two geodesic segments meeting at a point P (Fig. 10.13). **w** undergoes parallel transport in this case if and only if θ is constant along each segment, and changes by φ at the join (so that **w** does not rotate as the path changes direction abruptly at P). An arbitrary path can be treated by dividing it into many geodesic segments and finding the total 'steer' $\phi(s) = \sum_i \phi_i$ of the path, but for our purposes it is sufficient to treat paths with a finite number of geodesic segments. You can now connect parallel transport to angle excess (exercise 10.10).

A useful measure of average curvature at a point (i.e., averaged over orientation in space) can be obtained by comparing the surface area of a sphere to the Euclidean expected value $4\pi r^2$. One can then define a radius excess by

$$r_{\text{excess}} \equiv r - \sqrt{A/4\pi}. \tag{10.15}$$

The value of this excess radius gives a measure of the orientation-averaged curvature of the three-dimensional space at the chosen point.

For an example of a curved three-dimensional space, consider ordinary three-dimensional space in which rulers are subject to an expansion field as given by eqn (10.1), where ρ is now the radial coordinate in spherical polar coordinates. The metric is

$$ds^2 = \frac{1}{(1 + K\rho^2/4)^2}\left(d\rho^2 + \rho^2(d\theta^2 + \sin^2\theta\, d\phi^2)\right). \tag{10.16}$$

In such a space one would find that any plane through the origin has exactly the same properties as the two-dimensional 'hot-plate' of Fig. 10.3, so is a surface of constant curvature. The whole space is called a 'hypersphere' or '3-sphere'. Warning: this is hard to imagine, as we are not talking about an ordinary three-dimensional sphere, but a three-dimensional universe which is everywhere warped. In this case the radius excess obtained from eqn (10.15) is the same as the one obtained from measurements of a circle, eqn (10.6), and therefore the curvature is given by eqn (10.7). Imagine taking a long walk in a straight line in such a universe. You would find, just like a bug on a spherical surface, that after a while you arrive back at where you started! The universe would have no boundary, and yet it would have a finite volume (just as a spherical surface has a finite area), equal to $V_{\text{universe}} = 2\pi^2 K^{-3/2}$ (exercise 10.6). Such a universe is *finite but unbounded*. It is possible that our own universe may be like this at a large scale: i.e., a hypersphere, albeit an expanding one.

10.3   Time and space together

Spacetime is a four-dimensional type of 'space' that can also be curved. It differs from 4-space (i.e. a purely spatial region in four dimensions) because time is not the same as space. The measure of 'distance in spacetime' is the *invariant interval,* given by by $ds^2 = -c^2 dt^2 + dx^2 + dy^2 + dz^2$ for neighbouring events in some small region. The minus sign in the signature makes the difference between time and space. Geodesics can be defined as lines of either maximum or minimum 'length', and now there are three kinds: time-like, space-like, and null. These labels make sense because it can be shown that the sign of ds^2 is everywhere the same along a geodesic. A time-like geodesic is a line of most proper time. A space-like geodesic is a line of stationary interval that is everywhere space-like. The interval along a space-like geodesic may be minimal with respect to some variations, such as purely spatial ones, and maximal with respect to others. A null geodesic is a line of no turning (in a spacetime sense, i.e. constant velocity) along which the interval is everywhere zero.

To measure curvature in spacetime one can perform either purely spatial measurements—sizes and angles of geometric figures, for example—or one can combine spatial and temporal measurements. For example, to construct a triangle in spacetime one could use two time-like geodesics and one space-like geodesic, or two null geodesics and one time-like geodesic. For the first type of triangle one might use a pair of freely falling clocks released at the same event with different initial velocities, and make a ruler measurement of the distance between them as a function of proper time. This reveals the geodesic deviation. For the second type one could use a radar echo, recording the time between emission of the outgoing pulse and reception of the reflected pulse using a clock in free fall. In fact, precisely this sort of test has been carried out by radar reflection experiments between Earth and other planets of the solar system, and General Relativity confirmed experimentally to high accuracy. (There is no need in practice to use a clock in free fall; one may use a clock at rest on Earth and calculate the expected result for such a clock, whose worldline is not geodesic.)

Geodesic plane and coordinate 'plane'

There is a pitfall in reasoning about curvature that we do well to avoid, and that arises only in more than two dimensions. It has two aspects.

First, suppose we start with a flat 3-space, such as Euclidean space in three dimensions. It does not follow that two-dimensional subspaces inside this space are necessarily flat. Of course not! Take a tub of ice cream, and scoop out of it a ball using an ice cream scoop: the tub contained a flat 3-space filled with ice cream, but the surface scooped out is curved. Perhaps nobody would make this particular mistake, but a related mistake in General Relativity is to assume that the curvature of *space* is given by the metric alone. It is not—it also depends on how spacetime is 'sliced up' or 'foliated' into 'time' and 'space', which can be done in more than one sensible way when conditions are not static. One can have a curved space inside a flat spacetime, or a flat space inside a curved spacetime. Both possibilities arise in cosmological models of our universe, and another example of the former is a rotating reference frame in Minkowski spacetime.

The second aspect of the pitfall is to assume that the Gaussian curvature for a given direction in a three- or higher-dimensional space can be found simply by dropping terms in the line element: i.e., keeping one or more coordinates constant. This does not work. The set of events at one value of a coordinate might be called a 'coordinate plane', and it is certainly two-dimensional if just two coordinates remain, but its curvature has no simple relation to the curvature components of the higher space. Instead of a coordinate plane, one must use a *geodesic plane*: this is the set of events that can be reached from a given event P by all geodesics from P that set off in a direction $\lambda U + \mu V$ where U and V are fixed vectors and λ, μ can vary. Such a plane is guaranteed to contain at least two geodesics of the higher space, and these can be used to find one component of its curvature at P by using geodesic deviation.

10.4   Gravity and curved spacetime

The gravitational time dilation factor that we deduced from the Equivalence Principle is an example of a factor affecting spacetime measurements in the temporal direction. Its influence can be compared to the expansion factor in the hot-plate model, in the sense that by influencing time measurement it influences the shape of geodesics. For most functional forms of the variation of time dilation with position, it implies a non-zero curvature of spacetime. The study of the constantly accelerating reference frame showed the exception to this rule: the spacetime treated in that example was Minkowskian and therefore had zero curvature. This means that gravitational time dilation is not directly connected to curvature, but the form of its variation as a function of position is.

It should by now be abundantly obvious that a gravitational time dilation factor must imply the possibility of a gravitational space expansion or contraction factor, since temporal and spatial separations can be mixed up by a change of reference frame. We do not have to think of gravity directly in terms of curvature: we can think of it as an 'expansion field' which acts in an underlying flat spacetime (like the flat hot-plate) and influences all types of clock and ruler. Of course, such an 'underlying flat spacetime' is not directly accessible to us, but it is a useful aid to the imagination. With this point of view you have the right to insist that spacetime is utterly flat, but has in it a field (the gravitational field) whose effect is to cause time dilation and length contraction of all processes and all types of object by the same factor. For some types of calculation, such as 'gravitational lensing' (see the next chapter), this way of thinking about gravity is clearest. However, you must allow other people the right to point out that since the result is indistinguishable from a curved spacetime, we may equally say that there is no expansion field but spacetime is curved.

The essential idea is that **matter influences the metric of space-time**, and this influence is called gravity. We have learned how to use the metric to measure distance and its generalization: spacetime interval. However, we should notice that the very possibility of achieving this is a remarkable thing. Suppose we pick a point and then locate a set of points at constant distance from the first (as indicated by the metric). We can then further use the metric to measure the radius and the circumference of the circle so defined. In a flat space we should get the answer 2π for the ratio between those numbers. But wait: in view of the fact that the coordinate system, and therefore the metric, is almost completely arbitrary, how does it come about that the ratio is always the same? It must be that the value of the metric at the points on the circumference is not completely independent of its value along the radius. *There must be a differential equation that brings about this consistency.* This differential equation (a second-order differential equation for the metric tensor g_{ab}) is the Einstein field equation!

We can now present the essential elements of the complete theory of gravitation offered by General Relativity. There is a single grand idea, which may be summarized in J. A. Wheeler's phrase, 'matter tells space how to curve, space tells matter how to move'.[2] The most important and central equation is the **Einstein field equation**. This relates a measure of average spacetime curvature called the Einstein tensor to a measure of the local energy density, momentum density, pressure, and stress of matter and non-gravitational fields, called the stress-energy tensor. These tensors have sixteen components, of which ten are independent. Even in free space (i.e., away from matter) the equation is rather complicated. The difference $(g_{ab} - \eta_{ab})$, where η_{ab} is the Minkowski metric, serves as a sixteen-component 'potential', and the Einstein field equation is a non-linear second-order differential equation relating this 'potential' to energy-density; it is a (very specific and non-trivial) generalization of Poisson's equation.

For the sake of completeness the equation is presented in the box below, using a notation developed in chapter 12, but we shall not need its full details here. We shall provide what may be called the essence of the field equation, by quoting a quantitative statement of the predicted connection between mass-energy and spacetime curvature. We treat a region of space where conditions are static and isotropic. In this case the Einstein field equation predicts that the local Gaussian curvature of space is isotropic, and is given by

$$K = \frac{8\pi G}{3c^2}\rho_0 \tag{10.17}$$

where ρ_0 is the proper mass density (i.e., the mass density observed in a local inertial rest frame, in which energy flux and momentum flux are zero).

[2] One could describe electromagnetism in a similar way: *charge tells field how to diverge; field tells charge how to accelerate.*

Einstein field equation

$$\partial_b \Gamma^\lambda_{\lambda a} - \partial_\lambda \Gamma^\lambda_{ab} + \Gamma^\lambda_{a\mu}\Gamma^\mu_{b\lambda} - \Gamma^\lambda_{\mu\lambda}\Gamma^\mu_{ab} = -\frac{8\pi G}{c^4}\left(T_{ab} - \tfrac{1}{2}g_{ab}T^\lambda_\lambda\right)$$

where

$$\Gamma^a_{bc} = \frac{1}{2}g^{a\lambda}(\partial_b g_{c\lambda} + \partial_c g_{\lambda b} - \partial_\lambda g_{bc}),$$

g_{ab} is the metric, T_{ab} is the stress-energy tensor. The equation may also usefully be written in the form

$$R^\lambda_{a\lambda b} - \tfrac{1}{2}R^{\lambda\mu}_{\ \ \lambda\mu}g_{ab} = -\frac{8\pi G}{c^4}T_{ab}.$$

where R^a_{bcd} is the Riemann curvature tensor, whose components describe Gaussian curvature and related parallel-transport results, and which is related to the metric by

$$R^a_{bcd} \equiv \partial_d \Gamma^a_{bc} - \partial_c \Gamma^a_{bd} + \Gamma^a_{d\lambda}\Gamma^\lambda_{bc} - \Gamma^a_{c\lambda}\Gamma^\lambda_{db}.$$

Note that the reader is not expected nor required to appreciate the mathematical details displayed in this box in order to handle the next chapter. They are provided merely for general interest and to back up the statements made in the text.

Using eqn (10.17) in eqn (10.7) we find that for a spherical region small enough that ρ_0 is uniform throughout the region,

$$r - \sqrt{A/4\pi} = \frac{GM}{3c^2}, \tag{10.18}$$

where r is the radius of the spherical region (as measured by standard rulers), A is its surface area, and M is the total mass enclosed. The left-hand side of this equation is the radius excess defined in eqn (10.15). The equation states that the excess radius of a sphere of uniform density is proportional to the mass of the sphere, the proportionality constant being $G/3c^2$, where G is a universal constant called the gravitational constant (the same one that appears in Newton's Law of Gravitation). The value of G has been measured as $G \simeq 6.674 \times 10^{-11}\,\mathrm{m^3\,kg^{-1}\,s^{-2}}$, so $G/3c^2 \simeq 2.475 \times 10^{-28}$ m per kg.

The derivation of the beautifully simple result (10.17) from Einstein's full equation is presented in volume 2.

In the vacuum outside a spherically symmetric body of mass M, the field equation predicts that the orientation-averaged curvature is zero, but the curvature of any given geodesic plane is not. In this case the Gaussian curvature of any vertical plane is given by eqn (11.20).

Eqn (10.18) permits an example quantitative calculation, as follows. Let us treat a sphere having the same mass and radius as the Earth, but having a uniform density (so that the equation can be applied). The formula says that the excess radius is about 1.5 mm. That is, the Earth has approximately 1.5 mm of extra radius in addition to what one would expect from its surface area. Or, putting it the other way around, a sphere of uniform density having the same mass and radius r as the Earth would be found to be 'missing' some surface area. On the basis of Euclidean geometry one would expect the surface area to be $4\pi r^2$, but owing to spacetime warping it would be found to be smaller by about $0.24\,\text{km}^2$. This is approximately equal to the area of thirty-three football fields (soccer pitches)—a non-negligible amount of missing area!

The second half of Wheeler's statement, 'space tells matter how to move', is the remaining central idea. In the case of electromagnetism one has a differential equation for the tensor field (Maxwell's equations), but this does not in itself state how test-particles move. The equation of motion of test particles—the Lorentz force equation—has to be introduced as an additional axiom, or else obtained by assuming energy-momentum conservation.[3] In the case of General Relativity the motion of test-particles is not an additional axiom, but follows from the field equation. This is a rather subtle and technical point (it is related to energy conservation) but the result is simply stated: freely falling test-particles (i.e., particles subject to no force other than gravity) follow worldlines satisfying the Principle of Most Proper Time. In geometric language the worldline of a massive spinless test-particle, in the absence of effects other than gravity, is a time-like geodesic, and the worldline of a massless spinless test particle is a null geodesic. You can adopt whichever language you prefer: the idea of proper time, or the notion of a geodesic.

When the equation of motion (most proper time) is applied to a simple case—a small region with isotropic pressure and density—the result again has a simple expression. If a set of test-particles are initially at rest relative to one another and are situated on a small spherical shell in such a region, then it may be shown that they move initially in such a way that the volume V of the shell varies as

$$\frac{\mathrm{d}^2 V}{\mathrm{d}t^2} = -4\pi G(M + 3pV/c^2) \tag{10.19}$$

where M is the mass inside the shell (the test-particles themselves being of negligible mass) and p is the pressure. This exact result should be compared with the Newtonian geodesic deviation formula (9.5). We have to imagine that the test-particles can move relative to the other material: e.g., it is a dilute gas or electromagnetic radiation. Several consequences of this formula are easy to see. First, gravity is attractive: the shell shrinks for $M > 0$. Secondly: in vacuum ($M = p = 0$) the tidal forces are such that the initial motion may change the shape but not the volume of the shell (the volume may change over longer time-scales). Thirdly, the presence of pressure *increases* the effective active gravitational mass. This may seem surprising at first (because pressure pushes things apart), but the formula refers only to the gravitational effects (and in any case, a particle suspended in a uniform fluid experiences no net force from the pressure around it). Pressure in a fluid similarly enhances the

3 See section 16.5.

Table 10.1 The main experimental tests of General Relativity. 'Early data' refers to the date of an early experimental investigation where suggestive or better evidence for the effect was obtained. 'arc-s/cent' means arc-seconds per Julian century. The final column gives the precision of the most accurate measurements to date. Modern methods, such as radar surveys or very-long-baseline interferometry, typically test several effects simultaneously. The precision of tests of the universality of free fall (weak equivalence principle) depends on the nature of the material and the distance scale investigated. Torsion balance experiments employing ordinary matter have attained precision of order 10^{-13}.

Effect	eqn	Gravitating object	size of effect	early data	current precision
Universality of free fall	(9.12)	Numerous	0	1890	
Mercury perihelion	(11.41)	Sun	42.98 arc-s/cent	1859	9×10^{-4}
Deflection of light by Sun	(11.51)	Sun	1.75 arc-s	1919	3×10^{-4}
Gravitational red shift	(9.18)	Earth	2.4×10^{-15}	1959	
Shapiro radar echo delay	(11.46)	Solar system	$220 \, \mu s$	1968	2×10^{-5}
de Sitter precession	(11.29)	Sun	1.9 arc-s/cent	1987	1×10^{-2}
Spin-up of PSR 1913+16		binary pulsar	$72 \, \mu s$/year	1974	5×10^{-3}
Periapsis shift of ,,	(11.41)	,,	$4.2°$/year		
Gravitational lensing	(11.51)	numerous			
Cosmological models					

inertia of the fluid—a special relativistic effect that will be explained in chapter 16.

Pressure typically has a value similar to the volume density of *kinetic* energy, so for most systems $pV \ll Mc^2$. The exception is thermal radiation, where $p = u/3$ for energy density u, hence $M + 3pV/c^2 = 2M$. Therefore a ball of thermal radiation gravitates twice as strongly as would the same amount of energy in the form of rest mass of non-moving particles.

Either of eqns (10.18) or (10.19) allow, in principle, the whole of the Einstein field equations to be deduced by insisting that they be generally covariant: i.e., one can write them in a tensor notation which allows that coordinates may refer to position and duration relative to any reference body, no matter how moving, including an arbitrary non-rigid accelerating body such as a jellyfish. The field equations predict that the direction-averaged curvature in free space (i.e., away from material bodies, electromagnetic fields, etc.) is zero. This does not necessarily mean that the twenty curvature components are all zero, but it means that if the curvature is positive in some directions then it must be negative in others. The curvature components (or at least some of them) are non-zero near material bodies, and fall to zero as one moves away from material bodies (and other forms of localized energy or pressure).

Einstein's theory has passed the tests both of mathematical elegance and experimental verification. Table 10.1 lists the main types of experiment which have been able to explore the departures from Newtonian predictions, and which also serve to compare General Relativity with other candidate theories. General Relativity is believed to be correct over a very wide range of distance and energy scales: from elementary particles to the cosmos as a whole. At very small distance scales it is likely that its range of validity runs out, and some sort of quantum field theory, or a merger of quantum field theory and spacetime geometry, is needed. This is a famous open problem in physics.

Exercises

(10.1) A tank (i.e., a tracked vehicle) has gearbox trouble: the two tracks are locked in synchrony, such that both advance together by the same amount. It is approaching a circular hill (not in a radial direction). Will the tank's path steer it towards or away from the top of the hill? Does it depend on the shape of the hill? What about a circular depression?

(10.2) Prove (by making a suitable change of coordinates) that the metrics (10.3) and (10.8) describe the same surface.

(10.3) Use the metric (10.3) to show that the area of a circle of radius r on the surface of a sphere of radius R is

$$2\pi R^2(1 - \cos(r/R)).$$

If we call $r = 0$ the 'north pole', then at what value of r does one reach the 'south pole'? How is this indicated by the metric? (Hint: seek a value for r such that points at different ϕ are very close together.) Hence use the above result to confirm the well-known formula $4\pi R^2$ for the surface area of a sphere.

(10.4) Repeat the previous exercise, but using the metric (10.8).

(10.5) Explain why, for a diagonal metric (in any number N of dimensions) the volume element is the product of coordinate differentials with the square root of the metric determinant: $\sqrt{|\det g_{ab}|}dx_1 dx_2 \ldots dx_N$. (In fact this result is also valid for non-diagonal g_{ab}, but you are not asked to prove that.) Verify that in 3D flat space this reduces to the familiar expressions $dx dy dz$ and $r^2 \sin\theta dr d\theta d\phi$ in Cartesian and spherical coordinates.

(10.6) A '3-sphere' is a three-dimensional space having everywhere the same positive curvature K. If we map such a space using spherical polar coordinates, then the metric can be obtained by replacing $d\phi^2$ in either eqn (10.3) or (10.8) by $d\Omega^2 = d\theta^2 + \sin^2\theta d\phi^2$ (where Ω is solid angle). Use this to find the volume of a general spherical region in such a space, and hence to show that the volume of the whole space is $2\pi^2 R^3$ where $R = K^{-1/2}$.

(10.7) Show that the metric of the 2-sphere can also be written $ds^2 = (1 - K\bar{r}^2)^{-1}d\bar{r}^2 + \bar{r}^2 d\phi^2$.

(10.8) §(i) Show that for a space with metric $ds^2 = dr^2 + f^2(r)d\phi^2$ the Gaussian curvature is $K = -f''/f$. (Hint: first explain why a line at constant ϕ must be geodesic, and then use geodesic deviation.)
(ii) Show that if $ds^2 = h^2(x)dx^2 + f^2(x)dy^2$ then $K = (h'f' - hf'')/(h^3f)$. (Hint: change coordinates; see eqn (10.14).)
(iii) Find the Gaussian curvature of the surface of revolution formed by rotating the line $y = y(x)$ about the x axis. (Hint: arc length along the line is $ds^2 = dx^2 + dy^2 = (1 + y'^2)dx^2$.)

(10.9) Find the Gaussian curvature of the ellipsoid of revolution formed by rotating the curve $x^2 + ky^2 = R^2$ about the x axis, where k and R are constants. Hence show that $K = 1/R^2$ at $x = 0$ and $K = k^2/R^2$ at $x = \pm R$.

(10.10) An N-sided polygon on a curved spatial 2-surface is formed by connecting N points with geodesics. A vector is parallel-transported around such a polygon. Show, using simple geometric arguments at the corners of the polygon, that the net rotation of the vector, after a circuit around the polygon, is by an amount equal to the angle excess of the polygon. Indicate the sense of rotation.

(10.11) Estimate the radius excess of the Sun.

11 Physics from the metric

11.1 Example exact solutions

We shall finish this introduction to General Relativity by examining some exact solutions to the Einstein field equation and extracting some of the associated physics. The solutions take the form of metrics, and most of the chapter is devoted to the last one:

(1) Flat spacetime.
   Using rectangular coordinates:

$$ds^2 = -c^2dt^2 + dx^2 + dy^2 + dz^2. \tag{11.1}$$

   Using spherical polar coordinates:

$$ds^2 = -c^2dt^2 + dr^2 + r^2\left(d\theta^2 + \sin^2\theta d\phi^2\right) \tag{11.2}$$

(2) Rotating reference frame in flat spacetime, in cylindrical coordinates:

$$ds^2 = -c^2dt^2 + dr^2 + dz^2 + r^2(d\phi + \omega dt)^2. \tag{11.3}$$

(3) Static field in one spatial dimension, using rectangular coordinates:

$$ds^2 = -\alpha^2 c^2 dt^2 + dx^2 + dy^2 + dz^2 \tag{11.4}$$

   where $\alpha(x)$ is some function of x.

(4) Schwarzschild solution: spacetime outside a spherically symmetric body.

   Using Schwarzschild coordinates:

$$ds^2 = -\left(1 - \frac{r_s}{r}\right)c^2dt^2 + \frac{dr^2}{1 - r_s/r} + r^2\left(d\theta^2 + \sin^2\theta d\phi^2\right) \tag{11.5}$$

   where r_s is a constant related to the mass of the body, called the Schwarzschild radius.

   Using an alternative radial coordinate $\bar{r}$ defined through

$$r = \left(1 + \frac{r_s}{4\bar{r}}\right)^2 \bar{r} \tag{11.6}$$

   gives the isotropic form (exercise 11.1)

$$ds^2 = -\left(\frac{1 - r_s/4\bar{r}}{1 + r_s/4\bar{r}}\right)^2 c^2 dt^2$$

$$+ \left(1 + \frac{r_s}{4\bar{r}}\right)^4 \left(d\bar{r}^2 + \bar{r}^2\left[d\theta^2 + \sin^2\theta d\phi^2\right]\right). \tag{11.7}$$

The first example (flat spacetime) is included for completeness: it is the simplest possible solution to the Einstein field equation, and plays a special role because spacetime is always locally flat. That is, any small enough region of a curved spacetime must have a metric that is locally indistinguishable from eqn (11.1). This means there always exists a choice of coordinates which can be used to map the small region such that the exact metric has the Minkowski form (11.1) to lowest order in the displacements dt, dx, dy, dz in the region. Such a choice is a local inertial frame (LIF). The polar form is also shown so that the reader will learn to recognize it, and to emphasize that it is possible for two metric equations to have a substantially different functional form and nonetheless describe exactly the same spacetime.

The next example (rotating reference frame in flat spacetime) is included chiefly in order to warn the reader of its subtleties. Particles on a rigidly rotating disc have helical worldlines relative to an ordinary inertial coordinate system. The rotating reference frame introduces a coordinate system which attempts to 'announce' that these worldlines are 'purely temporal', by assigning fixed $\{r, \phi, z\}$ to each such particle. However, this results in a situation where intervals in the ϕ direction are not orthogonal to intervals in the t direction. This requires careful interpretation; it means that t does not straightforwardly represent 'time' (recall the similar situation in the (h, τ) coordinate system on Fig. 9.19); events at the same t but different ϕ are not simultaneous to an observer fixed on the disc. A lattice constructed on the disc is said to be *stationary* but *not* static, owing to the fact that the time for a light-signal to traverse a closed 'spatial' loop can depend on the direction of travel around the loop. This is related to the presence of a cross term dtdϕ in the metric.

The next example (static field in one dimension) can be used to treat the constantly accelerating reference frame, and also, depending on the functional form of α, other one-dimensional cases. An example is discussed below. The final case (Schwarzschild solution) is an important basic example in gravitation physics, comparable to the case of the Coulomb field in electrostatics.

11.1.1   The acceleration due to gravity

The acceleration due to gravity, g, is of course not an absolute quantity (except that it is zero in any LIF), but in the case of a static spacetime mapped by a static metric there is a natural choice of non-inertial reference frame relative to which it is useful to know g. This is any frame defined by a rigid lattice (recall the definition of a rigid lattice in section 9.1.5: it does not change with time when surveyed by light-signals). When a metric has the general form

$$ds^2 = -\alpha^2 c^2 dt^2 + d\boldsymbol{\sigma}^2$$

where $\alpha(x, y, z)$ and the spatial part of the metric d$\boldsymbol{\sigma}^2$ (not necessarily flat) are independent of t, then it is clear that if one takes the slice through spacetime defined by $dt = 0$, then the space thus defined has properties that do not depend on when the slice is taken. The coordinates x, y, z can therefore be understood to be labelling positions on a rigid spatial structure, which we call a *lattice*. It is also clear that t labels a time coordinate, because when $dx = dy = dz = 0$ one finds that $ds^2 < 0$: i.e., a change in t on its own makes a time-like interval. For such changes we find

$$\sqrt{-ds^2} = \alpha c \, dt \qquad \Rightarrow \; d\tau = \alpha \, dt.$$

It follows that α is the gravitational time dilation factor. In other words, α is the function we previously (eqn (9.20)) called e^{Φ/c^2}. Therefore, the metric can be written

$$ds^2 = -e^{2\Phi/c^2} c^2 dt^2 + d\boldsymbol{\sigma}^2 \tag{11.8}$$

where $\Phi(x, y, z)$ is *by definition* the time dilation factor appearing in eqns (9.20) and (9.19).

Using eqns (9.20) and (9.19) we can now extract from any given α both the gravitational redshift and the local acceleration due to gravity:

$$\nu(x) = \nu(x_0)\alpha(x_0)/\alpha(x), \tag{11.9}$$

$$\mathbf{g} = -\frac{c^2}{\alpha}\nabla\alpha, \tag{11.10}$$

where in the second equation the gradient is with respect to ruler distance (because we originally derived the equation by arguing from the equivalence principle using a LIF). Ruler distance is related to the coordinates by the spatial part of the metric. If the spatial part is Euclidean as in eqn (11.4), then this is straightforward: for example,

$$g = -\frac{c^2}{\alpha}\frac{d\alpha}{dx}$$

if α depends only on x.

Notice what happens to these equations if the spacetime is unchanged but it is mapped in a different way by introducing a change of coordinates. A general linear change of the time coordinate, such as $t' = at + b$ for some constants a, b, results in $dt = (1/a)dt'$ so changes the α function merely by an overall factor $1/a$. *This has no effect on the redshift predicted by eqn (11.9), nor on the gravitational acceleration predicted by eqn (11.10).* For metrics of type (11.4) an arbitrary change in the x coordinate—i.e., $x = f(h)$, where h is the new coordinate and f is some function of h alone—makes the metric take the form

$$ds^2 = -\alpha(x)^2 c^2 dt^2 + (f'(h))^2 dh^2 + dy^2 + dz^2$$

where in the first term $\alpha(x) = \alpha(f(h))$ and in the second term the prime indicates the first derivative of the function. The redshift prediction is again unaffected and *so is the g prediction!* For, writing dl for ruler distance, given by ds at $dt = 0$, we have

$$g = -\frac{c^2}{\alpha}\frac{d\alpha}{dl} = -\frac{c^2}{\alpha}\frac{d\alpha}{dh}\frac{dh}{dl} = -\frac{c^2}{\alpha}\frac{\alpha' f'}{f'} = -\frac{c^2\alpha'}{\alpha},$$

i.e., g is equal to the same function as before the coordinate change. More general coordinate changes can result in a change of ν and g. One then finds that either eqns (11.9) and (11.10) are still valid, or else time-dependence or a cross-term (e.g., $dxdt$) is introduced into the metric equation. The metric is then said to be *non-static*, and the definition of concepts such as 'at a given place' has to be reconsidered before gravitational time dilation can be meaningfully defined.

In summary: in the static case the gravitational time dilation and acceleration are unaffected by mere rescaling of the temporal or spatial coordinates, because they are defined in terms of proper time and ruler distance. However, they can be affected by some changes that preserve staticity, such as a change of reference frame to one which is uniformly accelerating relative to the first (e.g., the constantly accelerating frame in Minkowski spacetime).

As an example simple case, take $\alpha = x$ (i.e. $\Phi = c^2 \ln x$), then we find $\nu \propto 1/x$ and $g = -c^2/x$. This describes the constantly accelerating rigid reference frame in flat spacetime that was discussed in section 9.2.2. For $\alpha = x$ the metric (11.4) is equivalent to the metric (9.45). The only difference is one of labelling: the parameters we called θ and h before are now being called t and x. We have already shown that there exists another choice of coordinates which makes the metric revert exactly and everywhere to the Minkowski form (11.1), so in this case there is no spacetime curvature. Two observations follow: first, this is certainly a possible solution to the Einstein field equation in free space; and second, we can, if we prefer, attribute all the effects of $\Phi(x)$ to Special Relativistic time dilation and space contraction, plus the effects of inertial forces that are present because the (t, x, y, z) reference frame is accelerating relative to an inertial frame. It is here a matter of taste as to whether or not the effects of $\nabla\Phi$ are called gravitational.

Other choices of α do give spacetime curvature. An example is where Φ is a linear function of position: $\Phi = kx$ for some constant k: i.e.,

$$\alpha = e^{kx/c^2}. \tag{11.11}$$

Using eqn (11.10) we then find that the acceleration due to gravity is everywhere the same:

$$\mathbf{g} = -\nabla\Phi = -k\hat{\mathbf{x}}.$$

Hence this describes what may reasonably be called a uniform gravitational field. In this case there is spacetime curvature for the tx 'plane' (exercise 11.12). Since curvature cannot be transformed away by a change of coordinates, there is now no special choice of reference frame in which gravitational effects vanish everywhere. Indeed, since the curvature is essentially that of a plane not a higher-dimensional entity, it cannot here attain an average of zero even when averaged over orientations. It follows that the metric does not describe a free-space solution to the field equation. It is an artificial case that serves merely to illustrate what would be observed in a uniform gravitational field if one could be produced, for example, inside a suitably arranged solid object. Positive k corresponds to the case where the gravitational acceleration is towards negative x, so x is a measure of 'height'. Let $g = |\mathbf{g}|$ be the magnitude of $\mathbf{g}$, then $\Phi = gx$. We can immediately extract predictions such as the gravitational redshift of light-waves propagating in such a uniform field: eqn (11.9) gives

$$\nu(x) = \nu(0)e^{-gx/c^2}.$$

Note that there is no singular behaviour here: all finite values of x give finite ν. This shows that there is no horizon for this spacetime, in contrast to the horizon at $h = 0$ that we discussed in section 9.2.2. Consult the exercises for further properties.

11.2   Schwarzschild metric: basic properties

The Schwarzschild metric is important because it is an exact solution of the free-space field equation for a physically important case: namely, spacetime outside a spherically symmetric body. This will be proved in volume 2. It should be compared to the case of the Coulomb field in electrostatics, and indeed has some similarities. The main difference is that we do not have a superposition principle in the case of General Relativity because the field equation is non-linear. The field due to two nearby stars is not equal to the sum of the fields due to each star on its own.

Schwarzschild radius

The spatial region $r = r_s$ is a spherical surface. Outside this surface (the only region we shall consider for the moment) the metric is static. Therefore the gravitational time-dilation factor (gravitational potential) is

$$\Phi = \frac{c^2}{2}\ln(1 - r_s/r) \tag{11.12}$$

in Schwarzschild coordinates: i.e., using the metric (11.5). The acceleration due to gravity is

$$g = |\mathbf{\nabla}\Phi| = \frac{d\Phi}{dl} = \frac{d\Phi}{dr}\frac{dr}{dl} = \frac{c^2 r_s}{2r^2(1 - r_s/r)^{1/2}} \tag{11.13}$$

where dl refers to the proper distance. This is obtained from the metric by considering a displacement dr with $dt = 0$: $dl = ds = (1 - r_s/r)^{-1/2}dr$ so

$$\frac{dl}{dr} = \frac{1}{(1 - r_s/r)^{1/2}}. \tag{11.14}$$

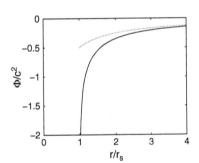

Fig. 11.1 Φ/c^2 verses r/r_s (full curve). The dashed curve shows $-r_s/2r$ for comparison.

Note that g tends to infinity at the so-called 'Schwarzschild radius' $r = r_s$, and that this is because $d\Phi/dr$ tends to infinity faster than does dl/dr. This is an example of an **horizon**, to be compared with the one we explored in section 9.2.2. We shall postpone discussion of it until section 11.5.

The value of the Schwarzschild radius for a given gravitating body can be obtained by looking at the weak-field limit at large r. Here we know that the field must become Newtonian, so, using (11.13) at $r \gg r_s$,

Schwarzschild radius

$$\frac{c^2 r_s}{2r^2} = \frac{GM}{r^2} \qquad \Rightarrow \quad r_s = \frac{2GM}{c^2} \qquad (11.15)$$

where M is the mass of the body. To remember this formula, the following mnemonic may be useful: consider the escape velocity for a particle in a Newtonian gravitational field. This is obtained by setting the kinetic energy equal to the binding energy: $(1/2)mv^2 = GMm/r$. Apply this formula with $v = c$ and one finds that r_s is the radius at which the escape velocity equals c in a purely Newtonian model: this is purely coincidence but easy to remember.

The values of r_s for some example objects are given in the table:

Earth	8.87 mm
Sun	2.95 km
Galaxy	0.03 light-year

where the last case is approximate (it treats a spherically symmetric body of mass $10^{11} M_\odot$). Note that all these objects are of a size much exceeding their Schwarzschild radius, so for them r_s is simply a length scale associated with gravity. *The Schwarzschild metric only applies outside the body, where there is no matter.* Inside a material body some other metric applies—one which is smooth and without extreme behaviour.

Birkhoff's theorem

The Schwarzschild spacetime is unique. That is, it is the only solution to the field equation having both spherical symmetry and complete time-independence. In 1923 Birkhoff discovered an important extension to its validity: it turns out that *even without the assumption of staticity, the Schwarzschild solution is the unique solution with spherical symmetry.* The proof uses the idea that even if one introduces time-dependence into the functions in front of dt^2 and dr^2 in the metric, it can be transformed away by a change of coordinates. It follows that even if the central spherical body (a star or planet) were collapsing, exploding, or pulsating in some spherically-symmetric way, the external field would show no change whatsoever. The same is true in electromagnetism for a pulsating charged sphere (this is easily proved by applying Gauss's theorem and symmetry arguments to the electric and magnetic contributions).

It follows that the spacetime inside a spherically symmetric cavity must be flat (it must be Schwarzschild's solution with $r_s = 0$, because the absence of any matter at $r = 0$ means that nothing special can be happening there). The surrounding material can even be moving radially. Also, if one has radially moving spherically symmetric material outside some spherical surface Σ, with a spherical star at the centre and vacuum in between, then between the star's surface and the surface Σ the spacetime must be Schwarzschildian. In particular, if the rest of spacetime is isotropic and radially expanding, then the planetary orbits about the star know nothing of that global expansion. Solar systems, and even galaxies, do not in general become larger as the universe expands.

11.3   Geometry of Schwarzschild solution

Two models can help to get a feeling for the geometry of spacetime implied by the Schwarzschild metric. For $r > r_s$ the t coordinate is clearly temporal (its coefficient in the metric is negative-definite) and the other coordinates are spatial (their coefficients are positive-definite) so a time-slice can be taken as $dt = 0$. This reveals a three-dimensional space whose metric is

$$ds^2 = (1 - r_s/r)^{-1}dr^2 + r^2 \left(d\theta^2 + \sin^2\theta d\phi^2\right).$$

In view of the spherical symmetry, the geometry of the whole space is summarized by the geometry of any plane through the origin. We then have a two-dimensional space to think about. Taking the plane at $\theta = \pi/2$ for convenience, the metric is

$$ds^2 = (1 - r_s/r)^{-1}dr^2 + r^2 d\phi^2. \tag{11.16}$$

Adopting the 'expansion field' viewpoint, we can now pretend that the space is really flat but that rulers in it behave strangely, shrinking by the factor $(1 - r_s/r)^{1/2}$ in the radial direction (but not in the ϕ direction). Alternatively, adopt the $\bar{r}$ coordinate defined in eqn (11.6) and then the rulers shrink by a factor $(1 + r_s/4\bar{r})^{-2}$ independent of the direction in which they are laid down (consult the metric (11.7)). This is shown in Fig. 11.2. We say 'shrink' rather than 'expand' because it makes sense to compare the rulers to their behaviour at large r where the geometry tends to Euclidean.

The other useful model is that in which the surface is 'embedded' in a three-dimensional Euclidean space. In other words, we let r, ϕ serve as coordinates in a plane, and we introduce a height z in such a way that distances on the surface $z(r, \phi)$ exactly match ruler distances along the corresponding tracks in the Schwarzschild space. To obtain z, first recall that the distance along a line $y = y(x)$ in two dimensions is $dl = (dx^2 + dy^2)^{1/2} = (1 + dy/dx)^{1/2}dx$. Therefore, on a surface of revolution around the z axis, distances are given by the metric

$$ds^2 = \left(1 + \frac{dz}{dr}\right) dr^2 + r^2 d\phi^2. \tag{11.17}$$

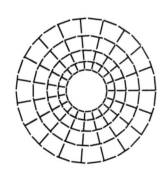

Fig. 11.2 'Expansion field' picture of the Schwarzschild geometry. The figure shows a geodesic plane through the origin with a set of rods lying in the plane, all having the same proper length. The 'isotropic' coordinate system $(\bar{r}, \theta, \phi)$ has been adopted so that the length of the rods is independent of their orientation.

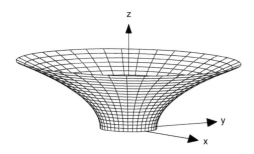

Fig. 11.3 Flamm's paraboloid, eqn (11.18).

Comparing this to eqn (11.16) we obtain a differential equation for $z(r)$ whose solution is (exercise 11.2)

$$z^2 = 4r_s(r - r_s). \tag{11.18}$$

This surface, called *Flamm's paraboloid*, is shown in Fig. 11.3. From the fact that any small part of it is saddle-shaped, we deduce that the Gaussian curvature is negative, in agreement with eqn (11.20) below.

Flamm's paraboloid (also called an *embedding diagram*) is an exact (and useful) representation of the *spatial* geometry. Sometimes people use the idea of a ball rolling on this surface to illustrate the way curvature leads to bending of orbits; but this is a mistake, because worldlines are governed by curvature of time and space together. Indeed, even if this surface were exactly flat then particles would still orbit the central body, as eqn (9.50) and following shows.

To find the Gaussian curvature of our two-dimensional space, let us adopt the method of geodesic deviation, eqn (10.10). To apply the method we need to identify a pair of neighbouring geodesics in the space. Finding a general geodesic is a non-trivial task, but fortunately we have a situation where one set is easy to spot: the radial lines. It is clear by either definition (stationary length or non-turning (= parallel transport)) that these are geodesic. Pick the line at $\phi = 0$, then a neighbouring geodesic line is at $\phi = $ const for some small ϕ, and in the limit $\phi \to 0$ the ruler distance between neighbouring points on this pair of geodesics is given by $\eta = r\phi$. In eqn (10.10) r refers to ruler distance along a geodesic, not the Schwarzschild coordinate r; therefore, we need a different symbol for the former. We shall call the ruler distance l, so that eqn (10.10) reads

$$K = -\frac{1}{\eta}\frac{d^2\eta}{dl^2}. \tag{11.19}$$

Using $\eta = r\phi$ we have

$$\frac{d^2\eta}{dl^2} = \phi\frac{d}{dl}\left(\frac{dr}{dl}\right) = \phi\left(\frac{d}{dr}\frac{dr}{dl}\right)\frac{dr}{dl}$$

and the metric gives $dr/dl = (1 - r_s/r)^{1/2}$. After carrying out the differentiation and substituting in eqn (11.19), one finds

$$K = -\frac{r_s}{2r^3}. \tag{11.20}$$

Radial distance and 'Tardis effect'

Using the metric (11.5) it is easy to see that the distance around a circumference at $\theta = \pi/2$ is $C = 2\pi r$. This means that the radial coordinate r in the Schwarzschild metric can be conveniently interpreted as $r \equiv C/2\pi$, where C is the circumference, as measured by standard rulers, of the circle defined by the locus of points at given r. However, r is not the ruler distance from the origin, nor are changes in r directly equal to radial distances. The radial distance from the origin is not a well-defined concept (see later on black holes), but the radial distance from the horizon ($r = r_s$) is. It is given by the integral of $\mathrm{d}l/\mathrm{d}r$ if we choose to measure it using standard rulers at rest relative to the rigid lattice (the radar distance is different). Introducing $x \equiv r/r_s$ for convenience, this is

$$\int_1^x \frac{1}{\sqrt{1 - 1/x}}\, \mathrm{d}x = \sqrt{x(x-1)} + \log\left(\sqrt{x} + \sqrt{x-1}\right). \tag{11.21}$$

This function is plotted in Fig. 11.4. For example, the surface $r = 2r_s$ is at a distance $\simeq 2.2956\, r_s$ from the horizon. Take a look at Flamm's paraboloid to obtain a visual impression of this.

In the popular science-fiction television series *Dr Who*, the hero owns a time-travel machine called the 'Tardis'. This Tardis has a wonderful property that always captures the imagination of viewers: it is larger on the inside than on the outside. From the outside its dimensions are those of an old-fashioned police box (or telephone box): a square prism with a diameter of approximately 1 metre, and a height of approximately 2 metres. Upon opening the door one steps into a large, roughly spherical room of radius 6 metres.

General Relativity says that something approximating to this is possible. Suppose we have a spherical box containing within it a massive body whose surface lies near but just outside its own Schwarzschild radius, which is 1 metre. For example, suppose the surface lies at $r = 1.0217\,\mathrm{m}$ in Schwarzschild coordinates. Let the wall of the box be a spherical surface at $r = 2\,\mathrm{m}$ in Schwarzschild coordinates. A visitor approaches the box and finds its surface area to be $4\pi r^2 \simeq 50\,\mathrm{m}^2$. She guesses, therefore, that it has a radius of 2 m. Upon opening the door she steps inside and notices first that the wall has the same area on its interior as on its exterior surface: no surprises yet. Then she walks towards the centre of the spherical room that she finds herself in, until she has traversed a distance of 2 m. From her previous observation of the exterior surface she might expect that this would bring her to the centre of the room, but she finds instead that she has only arrived at the surface of a sphere contained within the room (perhaps it is the control panel of this Tardis). The central sphere is quite substantial, having a surface area of $\simeq 13\,\mathrm{m}^2$—plenty of room for all the control systems. The distance to the centre of the sphere is a further 1.19 m (calculated by taking into

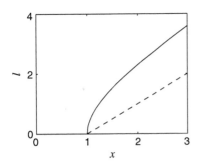

Fig. 11.4 Radial distance from the horizon, eqn (11.21) (with the line $x - 1$ shown dashed, for comparison).

account its radius excess which, from eqn (10.18), is $r_s/6 \simeq 17\,\text{cm}$). We have ignored the tremendous forces our visitor meanwhile experiences (exercise 11.4). The visitor concludes that the volume inside the Tardis is larger than it appeared (on Euclidean expectations) from the outside.

11.3.1   Radial motion

A test particle (i.e., one whose own gravitational effect is negligible) moving in the Schwarzschild spacetime has a variety of possible orbits, roughly comparable to the scattered, orbiting, or absorbed orbits of the special relativistic problem (section 4.2.6). They are derived by solving the appropriate equation of motion (which is equivalent to finding the shape of geodesics in the spacetime). The general problem requires some further mathematical apparatus that will be presented in chapter 14, but we already have the tools needed to treat straight-line radial motion and to find the circular orbits. That there exist such solutions is obvious by symmetry.

A particle initially moving in the radial direction will continue to move in a straight line at constant θ, ϕ. The radial position as a function of proper time can be obtained from the metric and eqn (9.58). From the metric at $d\theta = d\phi = 0$ we have

$$c^2 d\tau^2 = e^{2\Phi/c^2} c^2 dt^2 - e^{-2\Phi/c^2} dr^2$$

$$\Rightarrow \qquad e^{4\Phi/c^2} \left(\frac{dt}{d\tau} \right)^2 = e^{2\Phi/c^2} + \frac{1}{c^2} \left(\frac{dr}{d\tau} \right)^2 .$$

Using eqn (9.58) we find that in free fall the left-hand side of this equation is constant; therefore, so is the right-hand side. Hence the motion satisfies

$$\left(1 - \frac{r_s}{r} \right) + \frac{1}{c^2} \left(\frac{dr}{d\tau} \right)^2 = \text{const} \equiv E. \qquad (11.22)$$

Substituting $r_s c^2 = 2GM$ (eqn (11.15)) and multiplying by the mass m of the particle, we find

$$-\frac{GMm}{r} + \tfrac{1}{2} m \dot{r}^2 = \frac{E-1}{2} mc^2 \qquad (11.23)$$

which is reminiscent of the Newtonian result, except that the dot signifies $d/d\tau$ and r is the Schwarzschild coordinate. This equation makes it easy to find the velocity of the falling particle at any given r, for given initial conditions, and it can be integrated to find $r(\tau)$ (exercise 11.6(i)). Differentiating the equation w.r.t., r gives

$$\frac{d^2 r}{d\tau^2} = -\frac{GM}{r^2} \qquad (11.24)$$

which is also easy to remember. To find the trajectory relative to the Schwarzschild lattice, we can either find $t(\tau)$ by solving eqn (9.58) after substituting the known $r(\tau)$ into the expression for Φ, or use $dr/dt = \dot{r}/\dot{t}$

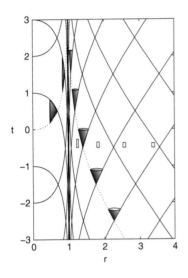

Fig. 11.5 Schwarzschild spacetime in the rt geodesic plane (in units with $r_s = c = 1$). The lines are null geodesics (photon worldlines); the dashed line shows an example ingoing worldline to enable the reader to follow it across the horizon at $r = 1$. Note the light-cone behaviour. At the horizon the cones become narrow, and the time-like direction changes abruptly. The boxes in the $r > 1$ region are shown at equal increases of ruler distance from the horizon, and all have the same ruler width and proper duration.

[1] It helps to picture the local lattice as having a rectangular construction here.

to get a first-order differential equation for the trajectory $r(t)$. The result is given in exercise 11.6(ii).

For photons we cannot use proper time along the path, but these are easily treated by setting $ds^2 = 0$ in the metric equation. This picks out a null cone. On its own this constraint is not enough to determine the null geodesics, but by symmetry we know there exist geodesics having $d\theta = d\phi = 0$, and these further constraints suffice to find them. Hence a radial photon worldline is described by

$$c\,dt = \pm \frac{dr}{1 - r_s/r}$$

which integrates to

$$\pm ct = r + r_s \log|r - r_s| + \text{const.} \qquad (11.25)$$

Examples are plotted on Fig. 11.5.

11.3.2   Circular orbits

To treat a circular orbit, we combine eqns (11.13) and (9.15) for gravity and particle track curvature κ, taking care to interpret the latter correctly. κ here refers not to the Gaussian curvature of the spacetime, but to the rate of bending, relative to a local rigid lattice,[1] of the path in space traversed by a particle in free fall. For a particle moving at speed v relative to the local lattice, we have, at any point where its motion is perpendicular to $\mathbf{g}$ (i.e., anywhere on the orbit under consideration),

$$\kappa = \frac{g}{v^2} = \frac{c^2 r_s}{2r^2(1 - r_s/r)^{1/2} v^2}. \qquad (11.26)$$

The relation between κ and r is now to be found by geometric arguments from the metric. What we need to do is to find the flat space that matches the spatial part of the Schwarzschild metric locally at some point in the orbit (a so-called *tangent surface*), and find how a line at constant r looks in this Euclidean space. A neat trick suffices to accomplish this. Drop a cone onto Flamm's paraboloid (Fig. 11.6a) so that the cone touches the surface at r and is tangent to it there. Then any small part of this cone, near the coordinate position r, is the flat space we seek. However, the cone has the nice feature that it is a flat surface in its entirety (excluding the vertex): it can be 'unrolled' to a flat plane without distortion (Fig. 11.6b). When thus unrolled the path on the cone is clearly part of a circle of radius

$$a = \frac{r}{\cos \beta}$$

where the angle β (not necessarily small) is given by

$$\tan \beta = \frac{dz}{dr} = (r/r_s - 1)^{-1/2}.$$

Using the trigonometric identity $\tan^2 \beta + 1 = 1/\cos^2 \beta$ we find

$$a = r(1 - r_s/r)^{-1/2}.$$

The line curvature measure κ is, by definition, the inverse of this radius (since we are now examining a flat surface), so we have $ag = v^2$. This is an equation for r which is easily solved, giving

$$r = r_s \left(1 + \frac{c^2}{2v^2}\right). \tag{11.27}$$

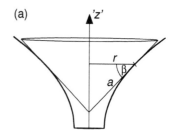

Note that Fig. 11.6b makes it obvious that although the worldline is a geodesic of the spacetime, the path in *space* is *not* a geodesic of the space. It is not true to say that 'the Earth travels in a straight (i.e., geodesic) line' as is sometimes asserted in popular treatments of General Relativity; rather, its worldline is straight (i.e., geodesic).

Eqn (11.27) shows that for circular orbits, as v increases the radius decreases (as one would expect from familiarity with the Newtonian problem). At small v the result matches the Newtonian prediction, and at $v = c$ we have $r = (3/2)r_s$ (the method of calculation has been valid for zero rest-mass as well as for massive particles). Thus we find at $(3/2)r_s$ a spherical surface in which photons can orbit the central body: a so-called 'light-sphere'. Sufficiently dense neutron stars can possess such a light-sphere, and so do all black holes. For radii smaller than $(3/2)r_s$ there are no circular orbits.

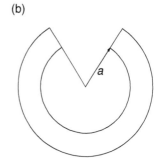

Fig. 11.6 Geometric construction used to find the radius of a circular orbit. The axis 'z' refers to the embedding coordinate introduced in Flamm's paraboloid (see Fig. 11.3).

It is useful to connect the particle speed v in eqn (11.27) to a description of the orbit using Schwarzschild coordinates. Care is needed. At any moment the speed v in eqn (11.27) is the speed of the particle relative to a certain LIF: namely, the LIF which is momentarily at rest relative to the lattice as the particle passes by. Let the time coordinate in this (Minkowskian) LIF be t', then

$$v = r\frac{d\phi}{dt'}$$

since $rd\phi$ is the distance along the orbit, in the LIF under consideration. Let $\gamma = (1 - v^2/c^2)^{1/2}$ be the Lorentz factor, then

$$\frac{dt'}{d\tau} = \gamma$$

where τ is proper time along the worldline. To be certain that this familiar result applies here, just consider adjacent events on the worldline: they will be separated by dt', dx', 0, 0 in the LIF whose x'-axis has been aligned along $\mathbf{v}$ so that $v = dx'/dt'$; the result follows. Thus we find

$$\gamma v = r\frac{d\phi}{d\tau}. \tag{11.28}$$

de Sitter precession

Suppose a particle following a circular orbit possesses intrinsic angular momentum: for example, consider a gyroscope orbiting the Sun. Ignoring

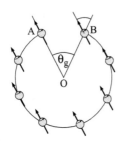

Fig. 11.7 Behaviour of an orbiting gyroscope, neglecting Thomas precession. The figure shows the gyroscope moving in the flat space of the unrolled cone. The effect of space curvature is that lines OA and OB are in fact identical: they are at the same location in the (r, θ, ϕ) Schwarzschild lattice. When the gyroscope arrives at OB it has also arrived at OA, but now it makes an angle θ_g with the radial line.

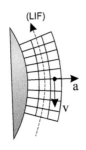

Fig. 11.8 Acceleration and velocity of the Schwarzschild lattice relative to a LIF that follows a circular orbit in the anticlockwise sense. In this case the Thomas precession of the *lattice* relative to the LIF is in the clockwise sense.

Thomas precession for the moment, such a gyroscope will maintain a fixed orientation as it orbits in the flat space shown in Fig. 11.6b. Therefore, the angle between the gyroscope's angular momentum vector and a radial vector increases as the gyroscope orbits the central mass. However, it does not increase by 2π when a single orbit is completed, because of the opening angle of the flattened cone. When the gyroscope completes the travel distance $2\pi r$ around the circular arc of radius a, it is 'surprised' to find itself back where it started (Fig. 11.7), and its angular momentum has therefore precessed in the prograde sense by

$$\theta_g = \frac{2\pi(a-r)}{a} = 2\pi\left(1 - \frac{1}{\sqrt{1 + 2v^2/c^2}}\right) \simeq \frac{2\pi v^2}{c^2}$$

where 'g' denotes 'geometric' and the approximate version is for $v \ll c$.

The gyroscope is in free fall, so does not experience proper acceleration and therefore does not Thomas precess. However, the fixed Schwarzschild lattice is experiencing a constant outward radial acceleration relative to an orbiting LIF, and therefore relative to such a LIF it Thomas precesses in the retrograde sense by $\theta_{\text{Thomas}} = 2\pi(\gamma - 1) \simeq \pi v^2/c^2$ per orbit. Therefore, relative to the lattice the gyroscope's angular momentum vector precesses in the prograde sense in total by

$$\theta_{\text{deS}} = \theta_g + \theta_{\text{Thomas}} = 3\pi v^2/c^2 \tag{11.29}$$

per orbit. This effect is called **de Sitter precession**. It has been measured for the Earth–Moon 'gyroscope' in orbit around the Sun to about 1% accuracy.

11.3.3 General orbits and the perihelion of Mercury*

To treat a general orbit in Schwarzschild spacetime we need the equations of motion, which are the equations of time-like geodesics. The mathematical apparatus to obtain them is presented in chapter 14. Here we shall discuss their solution.

For a general motion we have four unknowns, $t(\tau)$, $r(\tau)$, $\theta(\tau)$, $\phi(\tau)$, so we need four equations. We will show in section 14.4 that a suitable set is

$$c^2(1 - r_s/r)\frac{\mathrm{d}t}{\mathrm{d}\tau} = \text{const} \equiv E, \tag{11.30}$$

$$r^2\frac{\mathrm{d}\phi}{\mathrm{d}\tau} = \text{const} \equiv L, \tag{11.31}$$

$$\theta = \pi/2, \tag{11.32}$$

$$\frac{1}{2}\dot{r}^2 + \left(1 - \frac{r_s}{r}\right)\frac{L^2}{2r^2} - \frac{GM}{r} = \frac{E^2 - c^4}{2c^2}. \tag{11.33}$$

The first equation is the same as eqn (9.58), but now it is more general because it is not restricted to radial motion only. It reveals a constant of the motion E that tends to energy per unit mass in the Newtonian limit, so we call it energy. The second constant of the motion looks like angular momentum per unit mass, so that is what we call it. The equation $\theta = \pi/2$ states that the whole motion stays in the 'equatorial' plane in Schwarzschild coordinates. This is obvious by symmetry: we can always orient axes so that the initial velocity is in the plane $\theta = \pi/2$, and then the gravitational acceleration never steers the velocity out of that plane. The last equation is obtained immediately from the Schwarzschild line element (= metric equation). It is a statement of $(ds/d\tau)^2 = -c^2$, as you should verify. The equation looks much like an equation for energy. By analogy with the classical problem one may roughly interpret the terms on the LHS as radial kinetic energy, rotational energy, and potential energy. Writing the RHS as $\frac{1}{2}(E/c^2 + 1)(E - c^2)$ one sees that in the Newtonian limit it is the difference between total energy and rest energy.

Eqn (11.33) may be treated just as we treated the special relativistic motion in a Coulomb field; see section 4.2.6. We have $\ddot{r} = -dV_{\text{eff}}/dr$, where the effective potential is

$$V_{\text{eff}} = \left(1 - \frac{r_s}{r}\right)\frac{L^2}{2r^2} - \frac{c^2 r_s}{2r}. \tag{11.34}$$

See Fig. 11.9. First we consider the general form of the motion. There is a 'centrifugal barrier' associated with angular momentum, and an attractive $1/r$ potential. The $-1/r^3$ term dominates at small r, so $V_{\text{eff}} \to -\infty$ at the origin; therefore, at any given angular momentum there exists a threshold energy above which an incident particle is not scattered but surmounts the centrifugal barrier and is 'sucked in' (i.e., spirals in) to the origin. In units where $r_s = c = 1$ the effective potential is

$$V_{\text{eff}} = \left(1 - \frac{1}{r}\right)\frac{L^2}{2r^2} - \frac{1}{2r}, \tag{11.35}$$

so it is clear that it is essentially a function of two parameters: L and r.

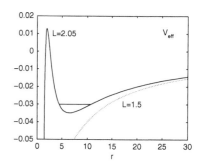

Fig. 11.9 Effective potential for Schwarzschild orbits, for two values of L (eqn (11.35)). The ends of the horizontal line indicate the turning points of an example quasi-elliptical orbit.

Circular orbits and accretion disc

The stationary points $dV_{\text{eff}}/dr = 0$ give the possible radii of circular orbits. They are located at

$$L^2 = \frac{c^2 r_s r^2}{2r - 3r_s} \quad \Rightarrow \quad r = \frac{L^2 \pm \sqrt{L^4 - 3c^2 r_s^2 L^2}}{c^2 r_s}. \tag{11.36}$$

From the definition (11.31) and (11.28) we can deduce

$$L = \gamma r v.$$

Using this you can check that eqns (11.36) and (11.27) agree with one another.

When $L < \sqrt{3} r_s c$ there is no barrier so no orbital motion: all incident particles either fall into the black hole or hit the central body if there is one. When there are stationary points (i.e., for $L > \sqrt{3} r_s c$) they give the radii of possible circular orbits. The inner stationary point is a maximum giving an unstable circular orbit, and the outer is a minimum giving a stable circular orbit. From the fact that $L = \sqrt{3}r_s c$ gives $r = 3r_s$, and higher L is required to produce a minimum, one can deduce that stable circular orbits must have $r > 3r_s$. The circular orbits at $r < 3r_s$, including the light-sphere at $r = (3/2)r_s$, are unstable.

For stable orbits r increases monotonically with L; for unstable ones it decreases monotonically with L. Therefore, the tightest orbit is obtained at $L \to \infty$, giving $r \to (3/2)r_s$: i.e., the light-sphere.

The tightest *stable* orbit, at $r = 3r_s$, has $V_{\text{eff}} = -c^2/18$, and therefore $E = 2\sqrt{2}/3c^2 \simeq 0.943c^2$. This means that tremendous energies can be released when particles orbit black holes. An infalling particle typically has orbital angular momentum, and therefore approaches on some quasi-hyperbolic or elliptical trajectory. It may fall straight into the black hole, in which case the black hole acquires the energy and none is released to the rest of the world. However, it may encounter other material already in orbit—a so-called *accretion disc*. Collisions will tend to bring its trajectory into agreement with that of the other material. If it eventually arrives at the tightest circular orbit it must have given up almost 6% of its rest energy. This should be compared to fusion reactions inside a star like the Sun, which liberate only $\sim 1\%$ of the rest energy. The energy is released to thermal kinetic energy in the accretion disc, and thence to emitted thermal radiation. In consequence, a black hole can be sitting inside a very brightly emitting disc.

It can also be shown that for circular orbits one has

Fig. 11.10 Accretion disc.

$$r^3 \left(\frac{\mathrm{d}\phi}{\mathrm{d}t}\right)^2 = GM. \tag{11.37}$$

This is Kepler's Third Law!

Advance of the perihelion

Around the minimum in the effective potential (when there is one) is the region where quasi-elliptical orbits exist, with the bound particle moving between two turning points at $V_{\text{eff}} = E/m - c^2$ (so $\dot{r} = 0$). The orbit looks similar to the rosette shown in Fig. 4.6.

To determine this rosette shape we adopt the same approach as in section 4.2.6. We change variable from r to $u \equiv 1/r$. Because we have in L the same type of constant of motion as in the Special Relativistic calculation, we can employ eqn (4.62) (with $m = 1$) to obtain

$$\frac{\mathrm{d}^2 u}{\mathrm{d}\phi^2} = -u + \frac{3}{2}r_s u^2 + \frac{c^2 r_s}{2L^2}. \tag{11.38}$$

In the classical case we would have simple harmonic motion of u as a function of ϕ. In the present case we still have oscillatory motion, but now in yet another effective potential well, given by

$$V(u) = \frac{1}{2}u^2 - \frac{r_s}{2}u^3 - \frac{c^2 r_s}{2L^2}u. \tag{11.39}$$

The exact solution can be expressed in terms of elliptic integrals. We will use an approximation to treat orbits of small eccentricity. In this case the motion is almost circular, so it remains close to the minimum of the potential $V(u)$. But since any potential well is approximately quadratic close to its minimum, the motion $u(\phi)$ will be simple harmonic. All we need do is find the minimum of $V(u)$—it is at

$$u = \frac{1}{3r_s}\left(1 - \sqrt{1 - 3(cr_s/L)^2}\right) \tag{11.40}$$

—and then evaluate d^2V/du^2 at this minimum. The result is

$$\frac{d^2V}{du^2} = 1 - 3r_s u = \sqrt{1 - 3(cr_s/L)^2}.$$

The simple harmonic motion therefore has angular frequency

$$\tilde{\omega} = (1 - 3(cr_s/L)^2)^{1/4} \simeq 1 - \frac{3}{4}\frac{c^2 r_s^2}{L^2}.$$

This is smaller than 1, so when the azimuthal angle advances by 2π the radial oscillation is not quite complete. After setting out from a minimum value of r, for example (called perihelion in the case of an orbit around the Sun), the radius next reaches a minimum when $\tilde{\omega}\phi = 2\pi$; therefore $\phi = 2\pi + \delta$, where

$$\delta = 3\pi\frac{c^2 r_s^2}{2L^2} = 3\pi\frac{r_s}{r}, \tag{11.41}$$

where in the last step r is the mean radius of the orbit, obtained from eqn (11.40); in terms of the standard orbit parameters it is $r = (1 - e^2)s$, where e is the eccentricity and s is the semi-major axis.[2]

This δ is the famous **advance of the perihelion** first obtained by Einstein. It is six times larger than the Special Relativistic result (4.66). This was the first great success of General Relativity (beyond the purely abstract success of its innate beauty), because despite its small magnitude this precession had already been noticed, in the case of Mercury's orbit around the Sun, by means of centuries of careful astronomical

[2] See exercise 11.16 for a treatment of orbits of any ellipticity.

Table 11.1 Data for Mercury's orbit.

Aphelion	r_{ap}		69,816,900 km
Perihelion	r_{per}		46,001,200 km
Semi-major axis	s	$\frac{1}{2}(r_{ap} + r_{per})$	57,909,100 km
Eccentricity	e	$\frac{r_{ap} - r_{per}}{r_{ap} + r_{per}}$	0.205 630
Specific angular momentum	L	$\sqrt{(1 - e^2)GMs}$	2.713×10^{15} m^2/s
Orbital period			87.9691 days
Solar mass	M		1.9891×10^{30} kg
Advance of perihelion			570.87 arc-s/century
Newtonian			527.9 arc-s/century
GR		$3\pi\frac{c^2 r_s^2}{2L^2}$	42.98 arc-s/century

observation. It was known to be unaccounted for in Newtonian theory. Mercury's perihelion is observed to advance by 570.9 arc-s/century. Most of this is expected from Newtonian theory: each other planet such as Venus presents a mass distribution which, averaged over long times, can be regarded as a ring of material at the orbital radius of the planet. The gravitational pull of this ring modifies the field near the Sun. The effect of Venus accounts for 277" of Mercury's perihelion advance, Jupiter 153", etc. The total Newtonian prediction is 528 arc-s/century, leaving a discrepancy with the observed motion of 42".98 ± 0.04, which is precisely the GR prediction; see table 11.1.

11.3.4   Photon orbits*

For massless particles we use a parameter λ to measure arc length along the worldline (since proper time is zero), and one finds

$$(1 - r_s/r)\frac{\mathrm{d}t}{\mathrm{d}\lambda} = \text{const} \equiv \tilde{E} \tag{11.42}$$

$$r^2\frac{\mathrm{d}\phi}{\mathrm{d}\lambda} = \text{const} \equiv \tilde{L} \tag{11.43}$$

and

$$\dot{r}^2 = \tilde{E}^2 - V_{\text{eff}}(r); \qquad V_{\text{eff}} \equiv (1 - r_s/r)\frac{\tilde{L}^2}{r^2}. \tag{11.44}$$

The equations for the conserved quantities look the same, but now the dot signifies differentiation with respect to λ, so the physical meaning is different; to keep this in mind we use a tilde. The equation of motion then follows directly from the line element, with $(\mathrm{d}s/\mathrm{d}\lambda)^2 = 0$. Clearly $\tilde{L}$ has no effect other than to change the height of the effective potential, but by rescaling the λ parameter we can change it back again, so we may as well set $\tilde{L} = 1$. This is except for the radial paths ($\tilde{L}=0$), which are trivially straight lines. V_{eff} is shown in Fig. 11.11. It is zero at $r = r_s$ and has a single maximum at $r = 3r_s/2$. A ray with sufficient $\tilde{E}$ to get past this maximum is never turned around. This means that any incoming ray that hits the light-sphere at $(3/2)r_s$ is subsequently swallowed by the black hole (or hits the surface of the central body if there is one). Similarly, an outgoing ray that makes it as far as the light-sphere has guaranteed its own freedom. For $\tilde{E}$ values close to $V_{\text{eff}}^{(\text{max})} = 4/27 r_s^2$ there is spiraling motion; at other values there is scattering or absorption. There are no stable orbits. Solution of the equation of motion permits an exact treatment of aberration and lensing, and in some cases this is algebraically simpler than the effective refractive index method to be discussed below.

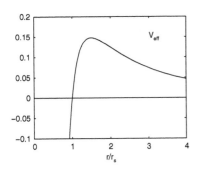

Fig. 11.11 Effective potential for photon orbits, eqn (11.44).

11.3.5   Shapiro time delay

Consider a pulse of light travelling past a star such as the Sun, at a distance large compared to r_s. The pulse will travel in an almost straight line, and will experience on its journey the effects of gravitational time dilation and space contraction. The slight bending of the path is treated in section 11.4; here we shall examine the time of travel.

It is convenient to use the form of the metric given in eqn (11.7). By comparing this with the flat space metric (11.2) we see that the spatial part corresponds to a flat metric multiplied by a function of $\bar{r}$ alone. It follows that the spatial behaviour is isotropic in this coordinate system. When written in rectangular coordinates, the spatial part is

$$dl^2 = (1 + r_s/4\bar{r})^4(dx^2 + dy^2 + dz^2) \qquad (11.45)$$

with $\bar{r} = (x^2 + y^2 + z^2)^{1/2}$.

In first approximation the path of the light-signal can be taken as the straight line $y = b$, $z = 0$ (see Fig. 11.12), with x running from $-X_1$ to X_2. b is the distance of closest approach; $\bar{r} = (x^2 + b^2)^{1/2}$. For a light-ray we set $ds = 0$ in eqn (11.7), so with also $dy = dz = 0$ we have

$$c\frac{dt}{dx} = \frac{(1 + r_s/4\bar{r})^3}{(1 - r_s/4\bar{r})} \simeq 1 + \frac{r_s}{\bar{r}} = 1 + \frac{r_s}{(x^2 + b^2)^{1/2}}.$$

Integrating this between 0 and X gives the coordinate time elapsed between the point of closest approach and the point $x = X$:

$$c\Delta t = X + r_s \ln \frac{X + \sqrt{X^2 + b^2}}{b} \simeq X + r_s \ln(2X/b),$$

where the second step used $X \gg b$. If the signal travels from X_1 on one side to X_2 on the other then the total coordinate time elapsed is

$$c\Delta t_{12} = (X_1 + X_2) + r_s \ln \frac{4X_1 X_2}{b^2}. \qquad (11.46)$$

The logarithmic term is the *Shapiro time delay*. To measure it one may use a radar reflection experiment between Earth and a planet such as Mercury or Venus. The method is to compare the echo time under ordinary conditions when the path does not approach to the Sun, with the time at close conjunction where it shows a pronounced increase. Shapiro first proposed and carried out such experiments in the 1960s. For example, in the case of Mercury an effect of approximately $220\,\mu s$ is expected. In principle this is easily detectable, but the experiments are complicated by the fact that planets are not smooth mirrors, and the Sun has an outer 'atmosphere' of free electrons that extends into the region where the effect is largest, which presents a refractive index that has to be taken into account. More recent surveys use a space probe to reflect the signal, and a computer program to adjust for the masses of the planets, oblateness of the Sun, etc. Excellent agreement (to $\sim 10^{-3}$) with General Relativity is found. Note that this tests both the temporal and

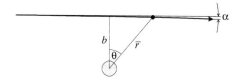

Fig. 11.12 A particle or light-pulse passing the Sun on an almost unde-flected trajectory. The small angle α is the subject of section 11.4.

spatial parts of the metric, since one can gather data under a variety of conditions and perform a best fit. (These measurements even furnished an improved estimate of the mass of the asteroid Ceres. The old value proved to be in error by 15%.)

11.4   Gravitational lensing

Next we consider the propagation of light in the vicinity of a spherically symmetric massive body.

Using the isotropic metric (11.7), the coordinate speed of light (d/dt of coordinate location at $ds = 0$) is given by

$$v = c\frac{1 - r_s/4\bar{r}}{(1 + r_s/4\bar{r})^3}$$

independent of the direction of travel of the light. Therefore we can define an effective refractive index:

$$n \equiv \frac{c}{v} = \frac{(1 + r_s/4\bar{r})^3}{1 - r_s/4\bar{r}}. \tag{11.47}$$

This allows a very nice method of treating the propagation of light around spherically symmetric bodies. Introduce a flat spacetime. We will call it the 'shadow' spacetime and use it as a mathematical device. When particles move around the real spacetime their mathematical shadows move around the shadow spacetime, such that an event at $(t, \bar{r}, \theta, \phi)$ in the coordinate system adopted for eqn (11.7) has its shadow event at $(t, \bar{r}, \theta, \phi)$ in the shadow spacetime. We have learned that light moves in the real spacetime with coordinate speed $v = c/n$, as given above. It follows that the shadow light moves around the shadow spacetime with exactly this same speed. Therefore we can calculate the shadow light-ray paths by using Euclidean geometry and a refractive index that varies with position. This is an exact method (see proof in section 14.4) which will predict ray paths that can immediately be mapped back into the real curved spacetime. The net result is that the curved spacetime behaves, as far as light-propagation is concerned, exactly as if it were flat but the vacuum possessed a refractive index. When we look up into the sky, therefore, it is as though we are looking through a sheet of bobbly glass. We can expect refraction and focusing.

Light propagating in a region of non-uniform refractive index is called 'gradient index' or 'graded index' optics. You can buy graded index lenses, for example, which are flat but have an index which is a function of distance from the axis. Another well-known example is that of a mirage in the desert or near a hot road. Fig. 11.13 shows how a flat wavefront is expected to behave in a region of linear index gradient. Using Huygens' construction, we find that the bottom of the wavefront propagates further than the top in any small time interval dt, with the result that the wavefront turns through an angle

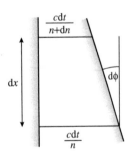

Fig. 11.13 Huygens' construction used to find the deflection of a plane wavefront in a region of non-uniform refractive index. The x direction is taken perpendicular to the ray, and dn refers to $(\partial n/\partial x)dx$. The gradient of n is not necessarily in the x direction.

$$d\phi = \frac{c\,dt}{n}\frac{dn}{n}\frac{1}{dx},$$

where x is the direction perpendicular to the ray, and no assumption has been made about the direction of ∇n. The ray turns towards the direction of increased refractive index (e.g., the light above a hot road turns towards the colder, denser air). dn in the calculation is given by $dn = (\partial n/\partial x)dx$; the variation of n along the ray affects the component of velocity in that direction, but contributes to the deflection only at higher order.

Let $dv_\perp$ be the change in the perpendicular component of the velocity, then clearly we have

$$\frac{dv_\perp}{v} = d\phi \qquad \Rightarrow \qquad \frac{dv_\perp}{dt} = \frac{c^2}{n^3}\nabla_\perp n. \qquad (11.48)$$

More generally, Fermat's Principle of Least Time can be used (exercise 14.1 of chapter 14) to show that the ray path $\mathbf{r}(s)$, where s is distance along the ray, satisfies the differential equation

$$\frac{d}{ds}\left(n\frac{d\mathbf{r}}{ds}\right) = \nabla n. \qquad (11.49)$$

However, the result (11.48) is all we shall need in the following.

We shall calculate the deflection of a light-ray passing near to a star, but in the weak field region, such that $\bar{r} \gg r_s$. To this end, first we present the classical prediction for a fast-moving particle in a Newtonian inverse-square-law field. When the speed is high the trajectory is almost straight, so can be modelled to first approximation as a straight line, as in Fig. 11.12. As it passes along this line, the particle of mass m, speed v receives an impulse given by

$$\Delta p = \int \frac{GMm}{\bar{r}^2}\cos(\theta)dt = \int_{-\pi/2}^{\pi/2} \frac{GMm}{b^2}\cos\theta\frac{b}{v}d\theta = \frac{2GMm}{vb}.$$

This impulse is in the direction perpendicular to $\mathbf{p}$; the net impulse in the parallel direction is zero by symmetry. The effect of the impulse is to steer the momentum (and therefore the velocity) through an angle

$$\alpha = \frac{\Delta p}{p} = \frac{2GM}{v^2 b}. \qquad \text{(classical)}$$

Now we carry out the relativistic calculation.

For $\bar{r} \gg r_s$ we have

$$(1 + r_s/4\bar{r})^2 \simeq \frac{1 + r_s/4\bar{r}}{1 - r_s/4\bar{r}}$$

so the metric takes the form

$$ds^2 \simeq -e^{2\Phi/c^2}c^2 dt^2 + e^{-2\Phi/c^2}(dx^2 + dy^2 + dz^2) \qquad (11.50)$$

and therefore

$$n \simeq e^{-2\Phi/c^2} \simeq 1 - \frac{2\Phi}{c^2}.$$

Substituting this in eqn (11.48) gives

$$\frac{dv_\perp}{dt} = -\frac{2}{n^3}\boldsymbol{\nabla}_\perp\Phi \simeq -2\boldsymbol{\nabla}\Phi.$$

The right-hand side is exactly twice the result for the classical case. Therefore, the calculation of the deflection goes precisely as before, except that the result is multiplied by 2. The deflection angle for light passing near a star is therefore

$$\alpha = \frac{4GM}{c^2 b} = \frac{2r_s}{b}. \tag{11.51}$$

In this formula, $b \gg r_s$ is the impact parameter or the distance of closest approach (they are the same in the approximation which has been assumed). For example, for the Sun at grazing incidence the prediction is $\alpha = 1.75$ arc seconds. The factor of 2, compared with the classical calculation, can be regarded as owing to the curvature of space—the factor $\exp(-2\Phi/c^2)$ in front the spatial term in the metric. We showed in section 9.3 that Newtonian gravitational theory is obtained by neglecting precisely that term. Slow-moving particles in a weak field do not care about the spatial curvature, because their worldlines stay very close to the time axis of their initial rest frame. Not so for light, which cannot help but explore the spatial and temporal curvature together.

In 1919 Arthur Eddington famously led an expedition to measure the deflection angle during a solar eclipse. He understood the need to show not merely a deflection, but that the deflection angle was a factor of 2 larger than might be expected from classical physics. His team published reasonable evidence that the angle was as predicted by Einstein's theory. The case was widely accepted at the time, though his complete dataset was not sufficiently unambiguous to give a clear test. However, since then much more accurate observations have been possible. In the 1990s the Hipparcos satellite measured the positions of $\sim 10^5$ stars at various times of the year, with milliarcsecond precision. The effects of light-deflection by the Sun were apparent all over the sky, and amply supported General Relativity.

The situation of 'gravitational lensing' is broadly as shown in Fig. 11.14. From the fact that the deflection angle (11.51) decreases as a function of impact parameter b, one may deduce that a bundle of rays issuing from a point source and all passing the same side of the lensing

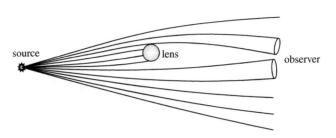

Fig. 11.14 Gravitational lensing

centre is caused to diverge more. In this sense we have a diverging lens. Nonetheless, rays arriving in a collimated beam and passing at a given distance either side of the star are brought together or 'focused'. The focal length of an ordinary lens is defined as the distance from the lens at which a collimated incident beam is brought to a focus. We can apply this idea to a gravitational lens as shown in Fig. 11.15. One finds the focal length:

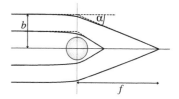

$$f \equiv \frac{b}{\alpha} = \frac{c^2 b^2}{4GM}.$$

Fig. 11.15 Defining the 'focal length' $f(b)$.

This is a strong function of b: rays incident at small radius are brought to a focus closer to the star than rays incident at large radius. In optics one would say there is a large amount of 'spherical aberration'. In fact there is so much aberration that no self-respecting optics manufacturer would offer such a lens on the market. In the astronomical situation, however, there is a convenient consequence. One does not have the ability to choose the distance from Earth of either the source or the lensing object, but owing to the dependence of f on b there is always a value of b for which the rays from the source are focused at Earth. Using the 'thin lens formula' $1/d_1 + 1/d_2 = 1/f$, where d_1, d_2 are the source-lens and lens-image distances respectively, and solving for b, one finds the

Einstein radius

$$b_E = \sqrt{\frac{4GM}{c^2} \frac{d_1 d_2}{(d_1 + d_2)}}. \qquad (11.52)$$

This is the apparent radius, at the lensing star, of the ring of light which is observed by the receiver, when the source, lens, and receiver are aligned. Numerous examples of these rings have now been found by astronomers. The more general scenario when the alignment is imperfect gives rise typically to a pair of images smeared into arcs, as Fig. 11.16 explains. Fig. 11.17 shows a spectacular display of this phenomenon in a famous image obtained by the Hubble Space Telescope.

Since astronomical distances vastly exceed typical Schwarzschild radii, the Einstein radius is in practice much greater than the Schwarzschild

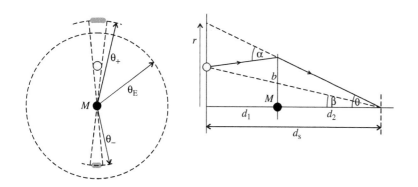

Fig. 11.16 Geometry of gravitational lensing in the general case, for a spherically symmetric lens of mass M. The apparent position r of the source is given by $r = \theta d_s$ and by $r = \beta d_s + \alpha d_1$, where $d_s \equiv d_1 + d_2$ is the distance from the observer to the source. Equating these expressions yields a formula for θ in terms of β and α. Using $b = d_2 \theta$ in (11.51) then gives $\beta = \theta - \frac{4GMd_1}{d_2 d_s c^2 \theta}$. This relationship is known as the **lens equation**. Solving for θ we have $\theta = \frac{1}{2}(\beta \pm \sqrt{\beta^2 + 4\theta_E^2})$, where $\theta_E \equiv b_E/d_2$ is the angular radius of the Einstein ring.

Fig. 11.17 Results of gravitational lensing in an image from the Hubble Space Telescope. (W. Couch, R. Ellis, NASA STScI-1995-14.)

radius. Its significance is that it indicates roughly the 'radius of the lens': i.e., the size of the region around the lensing body within which it deflects rays sufficiently to cause significant modification of the appearance of the source.

When the lensing mass is not spherically symmetric—for example, because it is a galaxy—more images can appear. Lensing has become an important tool in astronomy. It can be used, for example, to constrain the mass of the lensing object, and even to detect the presence of lensing objects that are too faint to be seen directly.

11.5   Black holes

The Schwarzschild metric has yielded great riches. It has a further treasure still to offer. When the gravitating body does not extend beyond its own Schwarzschild radius, a new type of object is born: a *black hole*. The physical phenomena outside the Schwarzschild radius are just as they were before, so an isolated black hole at a distance is just like any other simple body (be it of small mass and called a particle, or of stellar mass, or a 'supermassive' black hole). However, if it approaches closely, a black hole wreaks havoc wherever it goes.

That there can exist material bodies whose mass is compacted into a region smaller than their Schwarzschild radius is suggested by an order-of-magnitude estimate, as follows. Ignoring curvature effects, the Schwarzschild radius of a body of uniform density ρ and radius r is

$$r_s = \frac{2GM}{c^2} \simeq \frac{8\pi G}{3c^2}\rho r^3$$

so

$$\frac{r}{r_s} \simeq \frac{3c^2}{8\pi G\rho r^2}.$$

Therefore, at any given density we can attain $r < r_s$ just by making r big enough. For example, the density of a neutron star is of order $4 \times 10^{17}\,\text{kg m}^{-3}$ (similar to the density of an atomic nucleus). At this density $r \simeq r_s$ if $r \simeq 20\,\text{km}$, implying a mass of about 7 solar masses. Therefore, if such an object can form (and there is plenty of evidence that it can) then $r < r_s$ is possible. Once the matter radius shrinks below r_s a new process sets in that rapidly collapses the material down to $r = 0$, as we shall see.

Another interesting hypothetical case discussed by Rindler is that of a roughly spherical galaxy: i.e., a cloud of about 10^{11} stars, each assumed to have mass and density similar to that of the Sun. Then $r_s \simeq 0.03$ light-year $\simeq 4 \times 10^5 r_\odot$. Therefore, for a galaxy which collapsed to the point where its radius equals r_s, the volume of the galaxy is $64 \times 10^{15}/10^{11} \simeq 10^6$ times larger than the combined volume of the stars it contains. In other words the stars are still far apart when the whole system shrinks within its Schwarzschild radius. Most galaxies are saved from this fate by their angular momentum and non-spherical shape, but the argument serves to suggest that the possibility of r falling below r_s is not unphysical.

History of black holes. In 1783 the geologist John Michell argued from Newtonian gravity theory that a sufficiently massive object would have an escape velocity larger than c and therefore would prevent emitted light from escaping. Consequently, it would appear black to observers sufficiently far away. Laplace later promoted this idea of a 'dark star'. In 1915 Karl Schwarzschild and, independently, Johannes Droste obtained the solution to Einstein's vacuum field equations that subsequently became known as the Schwarzschild solution. However, the nature of the surface at $r = r_s$ was not under-stood until considerably later. In 1924 A. Eddington showed that the singular behaviour at $r = r_s$ disappears after a change of coordinates. In the 1930s S. Chandrasekhar and others calculated the mass above which known types of matter must suffer gravitational collapse. The objects that resulted were called 'frozen' stars, owing to the infinite gravitational time dilation as $r \to r_s$. It was not until a 1950 paper of J. L. Synge that the scientific community began to understand the nature of spacetime at $r = r_s$ more fully. It was subsequently further clarified by D. Finkelstein and M. Kruskal. In 1963 R. Kerr found the exact metric for a rotating black hole, and the discovery of pulsars in 1967, subsequently shown to be rapidly rotating neutron stars, helped to convince people that extreme types of astrophysical object were not only possible but existed. The term 'black hole' is usually attributed to J. Wheeler, in that he used it in a public lecture in 1967, but it was in circulation from at least 1964. In 1971 astronomical observations of the X-ray source Cygnus X-1 showed (L. Braes, G. Miley, R. M. Hjellming, C. Wade) that it was associated with a large star that by itself would be incapable of emitting the observed quantities of X-rays, and also (L. Webster, P. Murdin, C. T. Bolton) that it had a non-visible companion whose mass was too high to be a neutron star. This companion became the first strong candidate black hole, with an accretion disk accounting for the X-ray emission. Since then, many others have been identified.

11.5.1   Horizon

In a (non-inertial) reference frame at rest relative to the rigid lattice outside r_s, it is readily seen from the Schwarzschild metric that the following quantities (among others) tend to infinity as one approaches the Schwarzschild radius:

time dilation, gravitational redshift, dl/dr, acceleration due to gravity.

Also, the coordinate speed of light falls to zero, so light propagating in the radial direction does not emerge (nor does it fall in, in a finite amount of coordinate time). The first thing to settle is whether or not

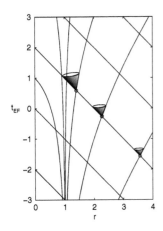

Fig. 11.18 Radial null geodesics (photon worldlines) in Schwarzschild spacetime in Eddington–Finkelstein coordinates. Here there is no discontinuity in the light-cone structure at the horizon.

[3] A minor point: the coordinate transformation from Schwarzschild to Eddington–Finkelstein coordinates is itself singular at $r = r_s$, but this does not matter. The important point is that the resulting metric describes the same spacetime, so is still a solution of the Einstein field equation for free space. How we arrived at it is immaterial.

these properties imply that spacetime is irregular at r_s, or whether this is an example of a *horizon* whose extreme properties can be made to go away by a change of reference frame, like the horizon illustrated in Fig. 9.20. That it is an example of a horizon can be proved by adopting the **Eddington–Finkelstein coordinates**, in which ct is replaced by

$$c\bar{t} \equiv ct + r_s \log |r/r_s - 1|. \tag{11.53}$$

The idea behind this coordinate change is that it makes the incoming radial null lines straight: see eqn (11.25). The coordinate change gives $c\mathrm{d}t = c\mathrm{d}\bar{t} - (1 - r_s/r)^{-1}\mathrm{d}r$, yielding the metric

$$\mathrm{d}s^2 = -\left(1 - \frac{r_s}{r}\right)c^2\mathrm{d}\bar{t}^2 + \frac{2cr_s}{r}\mathrm{d}\bar{t}\mathrm{d}r + \left(1 + \frac{r_s}{r}\right)\mathrm{d}r^2$$
$$+ r^2(\mathrm{d}\theta^2 + \sin^2\theta\mathrm{d}\phi^2). \tag{11.54}$$

This is regular for all $r > 0$, which shows that at $r = r_s$ there is ordinary smooth spacetime. The singularity at r_s in the Schwarzschild metric is known as a *coordinate singularity:* it can be transformed away. Having satisfied ourselves of this, we can now continue to use Schwarzschild metric both at $r > r_s$ and $r < r_s$, but not at $r = r_s$. Worldlines passing the horizon are best treated in another system such as Eddington–Finkelstein coordinates or a local inertial frame.[3]

At $r \to 0$ there remains a singularity which is a genuine singularity involving infinite curvature, which cannot be transformed away. The result is that we can apply our reasoning down to arbitrarily small r, but we may suspect that the theory breaks down eventually at the centre of a black hole, and that the spacetime there does something we have not yet determined.

The horizon is a point of no return, but nothing special happens to a body such as an astronaut as it passes through the horizon (in free fall). The tidal effects merely grow continuously, just as they do as one approaches other massive bodies. This can be illustrated by an analogy with flowing water. If there is laminar flow through a large pipe of decreasing diameter, then the flow velocity increases as a function of distance along the pipe, and one can imagine that at some point z_s the flow velocity exceeds the speed of sound in water. Then sound waves emitted from $z > z_s$ will never propagate to $z < z_s$. However, fish swimming in the water will notice nothing remarkable at $z = z_s$.

Within the horizon something important happens to the Schwarzschild metric: the coefficient in front of $\mathrm{d}t^2$ becomes positive, and that in front of $\mathrm{d}r^2$ becomes negative. Therefore, intervals in the t direction are spacelike and intervals in the r direction are time-like (the horizon itself is null). In short, despite the letter, t now represents a spatial quantity and r represents time. Particle worldlines remain time-like (after all nothing special is happening at the horizon, as Eddington–Finkelstein coordinates have taught us), but the central singularity is still at $r = 0$. The conclusion is that motion forward in time *is* motion towards smaller r. An object entering the horizon is carried down to $r = 0$ just as surely as you and I are carried into next week. This is clarified by a spacetime

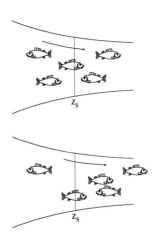

Fig. 11.19 The fish do not notice z_s.

diagram such as Fig. 11.5 or Fig. 11.18. Inside the horizon (or perhaps we should say *after* the horizon) the light-cones, and therefore all time-like intervals, and therefore all particle worldlines, tip over towards $r = 0$. It follows that once a star or other body manages to get completely inside its own Schwarzschild radius, it must collapse all the way to $r = 0$. No opposing force can be strong enough to prevent it.

Now let us consider an astronaut explorer who goes to visit a black hole and falls in. According to her own proper time, the explorer can soon arrive in the vicinity of the horizon. Any light emitted at r_s in the outward radial direction as she falls in stays at the horizon, according to outer observers, but travels at c relative to the astronaut. Therefore, in the astronaut's rest frame the horizon moves outwards at c. Her proper time increments in a regular way as she crosses the horizon. For example, if she falls straight down then eqn (11.61) for $r(\tau)$ applies, and there is no special behaviour at $r = r_s$; the equation applies all the way to $r = 0$. It does not take long for her to reach $r = 0$. To discover this we do not even need the equation of motion, because remarkably the total length of *any* worldline inside the horizon is at most $\pi r_s/2c$ (whereas in ordinary spacetime, worldlines can extend forever). For, integrating along an arbitrary worldline we have the total proper time

$$c d\tau = \int \left[\frac{dr^2}{r_s/r - 1} - (r_s/r - 1)c^2 dt^2 - r^2(d\theta^2 + \sin^2\theta d\phi^2) \right]^{1/2} \quad (11.55)$$

Any variation in t, θ, or ϕ only decreases this integral ('proof by twin paradox'), so it is maximal at $t, \theta, \phi = $ const, and at that maximum is readily integrated (e.g., use the substitution $r = r_s \sin^2 u$):

$$d\tau = \frac{1}{c} \int_0^{r_s} (r_s/r - 1)^{-1/2} dr = \frac{\pi r_s}{2c}.$$

This time is just 0.3 ms for a 10-solar-mass black hole.

Nevertheless, the time taken for the astronaut to reach the horizon in the first place is infinite according to outside observers. To be precise, according to any reasonable definition of simultaneity (such as the radar definition), the attempt to say which tick of some other clock, located outside the horizon, is simultaneous with each event on the astronaut's worldline is doomed to become meaningless as the succession of astronaut events passes $r = r_s$. In a static metric such as Schwarzschild's, the outer clock has to tick forever until it reaches the event at its location that is considered simultaneous with the astronaut's arrival at $r = r_s$. However, it is not necessary (nor useful) to agonize about the precise meaning of the Schwarzschild time coordinate near $r = r_s$; much better is to consider the observable information. Fig. (9.20) tells you essentially all you need to know: what outside observers see is that signals sent out by the explorer become less and less frequent, and redder and dimmer (all the Special Relativistic effects are there). Also, since the falling object only emits a finite number of wavefronts (or other signals) before it passes the horizon, there is only a finite number that outside observers can receive, and it takes a finite, and usually not very long, time to receive them (the chance of one being emitted exactly on the horizon being vanishingly small).

The spacetime inside the horizon is very non-Euclidean, non-Newtonian, and counter-intuitive, because it is non-static: the factors of $1 - r_s/r$ in the Schwarzschild metric now represent time-dependence, and furthermore it cannot be transformed away by a change of coordinates. One can see this by the fact that all time-like worldlines go to $r = 0$, and there they finish. There is no further spacetime at $r < 0$, and they cannot return to larger values of r because that would mean to go backwards in time. Because free-fall motion has the longest proper time, the set of freely-falling frames is the best replacement for the concept of a static rigid lattice; and they exhibit odd features, such as diverging ruler distance from each such frame to the next as they all zoom down to the singularity.

So far we have implied that a voyage inside an horizon is quite comfortable, if rather limited in duration. In fact this is far from the case, owing to the tidal forces. Near the horizon, tidal forces may already be very large; as one approaches the singularity they become enormous, stretching any object vertically and compressing it horizontally—a situation informally referred to as 'spaghettification'.

11.5.2   Energy near an horizon

We discussed in section 9.4 what happens when objects are lowered into gravitational potential wells by using a rope. Applying those ideas to the field outside a black hole gives an interesting result.

Although the acceleration due to gravity tends to infinity at the horizon, another useful measure of the gravitational force does not. This is the *surface gravity*, defined as the force required at large r to dangle a particle of unit rest mass near the horizon using a massless string. Let f be the force applied to the top of such a string, when suspending a unit mass particle at r. Eqn (9.56) gives this as

$$f = g(r)\frac{e^{\Phi(r)/c^2}}{e^{\Phi(\infty)/c^2}} = \frac{c^2 r_s}{2r^2} \tag{11.56}$$

using eqns (11.12) and (11.13). Therefore, the surface gravity is

$$\kappa \equiv f(r_s) = \frac{c^2}{2r_s} = \frac{c^4}{4GM} \tag{11.57}$$

using eqn (11.15).

Now consider the work required to raise an object from r to infinity. Let m_0 be the rest mass of the object when it is far from the black hole. Using eqn (9.55) the work required is

$$W = \left(1 - \sqrt{1 - r_s/r}\,\right) m_0 c^2. \qquad (11.58)$$

This is finite even when raising from the horizon, for which case $W = m_0 c^2$. The work is not infinite, because this amounts to the whole rest energy, therefore lowering the particle back to the horizon must reduce its rest mass to zero. By slowly lowering the particle in this way, we extract at a location far from the horizon all its rest energy—a 'perfect power station'. The particle can then be released into the black hole, *whose mass does not then increase.*

To see the latter point, use the following argument from Rindler, based on Birkhoff's theorem. For convenience, consider a spherical shell of matter rather than a point particle. As the shell (mass $m_0 - \Delta m$) is lowered from B to A we accumulate energy at B equal to the mass reduction Δm (conservation of energy). Therefore the total mass within any radius larger than r_B is unchanged, and therefore the spacetime outside r_B must be unchanged. Spacetime within r_A is also unchanged. Between r_A and r_B there is the complicating effect of the tension forces in the (now extending in spherically symmetric fashion) strings. When r_A reaches the horizon, the shell is released and the forces vanish. Spacetime outside r_B is still unchanged. Spacetime inside r_B must be unchanged also. If it were not, there would be a difference at r_B which could be made arbitrarily large by repeating the process (we use the acquired energy to reconstitute the shell, and repeat *ad lib.*) *Therefore, the energy received at B has been extracted from the lowered body, not from the field.*

It also follows that such a process of slow lowering of a particle into a black hole does not change the mass of the black hole. Particles entering by free fall, on the other hand, do add to the mass of the black hole (Birkhoff's theorem again). One implication is that whereas the mass of a black hole is a well-defined concept, the 'number of particles inside' cannot be quantified and may have no physical meaning.

More generally, the lesson is that tremendous energies are made available by the gravitational effects near neutron stars and black holes, as we have already mentioned in section 11.3.3. The material orbiting a black hole can be extremely energetic, and can emit copious amounts of X-rays.

11.6   What next?

We are almost ready to finish the introduction to General Relativity that has been our theme for three chapters. The treatment has been restricted to static metrics, but within that restriction it has been exact. The intention has been to provide the general physicist with an accurate grounding in the subject, and to smooth the way for those who would like to take it further. The next step for the latter group would be to learn some ideas and techniques of tensor analysis and differential geometry. These are needed to appreciate how the Einstein field equation is constructed, and they greatly simplify tasks such as obtaining the Schwarzschild solution.

We will conclude by sketching the chief further phenomena that arise in non-static problems.

Moving bodies (and momentum density in general) generate a further contribution to spacetime curvature that is sometimes loosely referred to as a 'dragging' of spacetime. In the weak field limit a better, and quantitatively accurate, picture is to speak of a 'gravimagnetic' contribution to the acceleration due to gravity relative to a suitably chosen lattice. One may write

$$\mathbf{a} \simeq -\boldsymbol{\nabla}\Phi + \mathbf{v} \wedge \mathcal{B}$$

where $\mathcal{B} = \boldsymbol{\nabla} \wedge \mathbf{W}$ is the 'gravimagnetic field' and $\mathbf{W}$ is a 'gravitational vector potential' given by

$$\mathbf{W} \equiv 4 \int \frac{G\rho\mathbf{u}dV}{c^2 r} \tag{11.59}$$

where ρ and $\mathbf{u}$ are the mass density and velocity of the source. This results in phenomena such as precession of a gyroscope fixed near a rotating sphere (the Lens–Thirring effect).

The Einstein field equation allows wave-like solutions in vacuum, which propagate at the speed of light; these are called gravitational waves. If a cloud of dust is floating above a steel plate, then a passing gravitational wave will make the dust particles oscillate relative to the plate, by exerting oscillating tidal forces. With a suitable choice of coordinates, in the weak field limit the Einstein field equation itself takes the form of a wave equation, again closely analogous to electromagnetism.

The Schwarzschild black hole is the exception rather than the rule for black holes generally. Matter collapsing to form a black hole almost always possesses angular momentum. The resulting black hole is spinning, and the metric of the surrounding spacetime reflects this: it is not a static metric anywhere. There is still a spherical event horizon, but also a region where escape is possible but standing still is not. That is, near the event horizon but outside it is a second surface in the shape of an oblate spheroid, where the dt term in the metric changes sign. Inside this surface is a region called the *ergosphere* where, if it is to be time-like, a worldline must have dϕ/d$t > 0$: i.e., particles and light must move around the black hole in the same sense as its rotation. Visitors wishing to traverse the ergosphere can do so, but not in a purely radial direction. This solution of the Einstein field equations is called the Kerr metric, and the associated black hole is called a Kerr black hole.

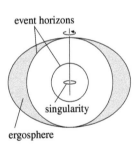

event horizons

singularity

ergosphere

Fig. 11.20 A Kerr black hole.

11.6.1   Black-hole thermodynamics

The Kerr black hole is associated with a striking physical phenomenon called the **Penrose process**.

The outermost event horizon of a Kerr black hole is at a radius given by

$$r_{\text{horizon}} = \tfrac{1}{2}\left(r_s + \sqrt{r_s^2 - 4a^2}\right)$$

in a suitable coordinate system, where $r_s = 2GM/c^2$ is the Schwarzschild radius and $a = J/Mc$ is a distance scale associated with the angular momentum J of the black hole. Thus the rotation of the hole makes the horizon *smaller* than would be the case if it did not rotate. The Penrose process is a process whereby energy can be *extracted from* the black hole! The essential idea is that a system of two parts, say two rocks, falls into the ergosphere and there splits, such that one piece is thrown against the sense of rotation of the hole, and subsequently falls past the event horizon, while the other piece escapes. Penrose showed that the escaping piece can emerge with a mass energy greater than the sum of the initial mass energies of both. Meanwhile, the hole has its angular momentum J reduced, and also its mass M. This is possible because the rotational energy of the hole is located outside the horizon.

It is found that in the Penrose process the mass of the hole falls but the area of the horizon does not: the reduction in a always compensates the reduction in r_s. This is an example of a more general idea, proved by Hawking and Penrose: namely, that under the action of classical (i.e., not quantum) physical processes, whereas the mass of a black hole may or may not decrease, the area can never decrease. This **area theorem** is reminiscent of the Second Law of Thermodynamics, and led to the idea that it might be appropriate to associate an entropy with the area of a black-hole horizon. This entropy is an important and not yet fully understood idea. There are strong reasons to assert that it is given by

$$S = \frac{1}{4} A \frac{c^3}{G\hbar} k_{\text{B}}$$

where A is the horizon area, equal to $4\pi r_s^2$ for a Schwarzschild black hole. In a further remarkable development, Hawking showed that quantum processes near a horizon led to the emission of radiation with a thermal spectrum, travelling up from the horizon. This *Hawking radiation* has a temperature given by

$$T = \frac{\hbar\kappa/c}{2\pi k_{\text{B}}} \tag{11.60}$$

where κ is the surface gravity; eqn (11.57). This gives 61 nK for a solar-mass black hole. The Hawking radiation is accompanied by a reduction in the mass and horizon area of the black hole, but not of the total entropy of the universe.

Exercises

(11.1) Prove that the change of coordinate from r to $\bar{r}$ given by eqn (11.6) causes the Schwarzschild metric (11.5) to take the isotropic form shown in eqn (11.7).

Prove that the acceleration due to gravity obtained from this isotropic metric is the same as given by eqn (11.13).

(11.2) From eqn (11.17) obtain the differential equation $\frac{dz}{dr} = (-1 + 1/(1 - r_s/r))^{1/2}$ and show that the solution is eqn (11.18).

(11.3) Use eqn (11.20) to find the radius excess of the London Eye (a vertical circle of radius 60 m resting on Earth's surface). Hence show that the area of the London Eye is smaller than $C^2/4\pi$, by 230 nm^2, where C is its circumference.

(11.4) Obtain the acceleration due to gravity experienced by the visitor to the spherical Tardis. If her locally measured rest mass is 50 kg, (i.e., it is fifty times larger than a group of 5×10^{25} nearby carbon-12 atoms), what are the sizes of the forces she experiences? (The Schwarzschild radius given in the example is unrealistic, as no known material could withstand the gravitational forces; however, similar reasoning would apply near a neutron star at the distance scale of tens of kilometres rather than metres and $r \simeq 2r_s$.)

(11.5) A 1-kg brick is dropped from far away onto a 1-solar-mass neutron star of radius 6 km. How much energy is released when the brick hits the surface of the star and comes to rest? If this energy is all radiated away, how much is received by distant detectors?

(11.6) **Radial trajectory.** (i) Show that $E = 1$ in eqn (11.22) represents the case of a particle falling from rest at $r \to \infty$. In units where $c = r_s = 1$, show that for such a particle the proper time as a function of r is

$$\tau = \pm \frac{2}{3} \left(r_0^{3/2} - r^{3/2} \right) \qquad (11.61)$$

where r_0 is the position at $\tau = 0$ and the plus sign gives the infalling case. (ii) Show that in terms of Schwarzschild time, the trajectory satisfies

$$\frac{dr}{dt} = \frac{1-r}{r^{3/2}}$$

and hence

$$t = \left[-\frac{2}{3}\sqrt{r}(3+r) + \log\left(\frac{\sqrt{r}+1}{\sqrt{r}-1} \right) \right]_{r_0}^{r} \qquad (11.62)$$

(The substitution $u = 1/r$ helps to do the integral).

(11.7) A circular orbit in Schwarzschild spacetime does *not* follow a geodesic of Flamm's paraboloid (Fig. 11.3). Explain.

(11.8) Check that eqns (11.48) and (11.49) agree.

(11.9) Show that the relationship between the acceleration due to gravity at a light-sphere and the circumference of the sphere is $C = 2\pi c^2/\sqrt{3}g$.

(11.10) (i) Show that the surface at ruler distance b outside the horizon of a Schwarzschild black hole is at radial coordinate $r = r_s + b^2/4r_s$ for $b \ll r_s$.
(ii) An object of rest mass m_0 is slowly lowered on a rope from infinity towards a Schwarzschild black hole, until it reaches a ruler distance b from the horizon, at which point it is released and then the rope is retrieved. Show that if $b \ll r_s$ then the mass of the black hole grows by $\Delta M = m_0 b/2r_s$.

(11.11) Find the focal length of the Sun for rays at grazing incidence.

(11.12) Show that for a metric of the form (11.4), the Gaussian curvature in the tx 'plane' is $K = \alpha''/\alpha$.

(11.13) Obtain the effective refractive index in the uniform gravitational field. Using eqn (11.48), show that the downwards acceleration of a light-ray is equal to g at the moment when the ray is horizontal, in agreement with the Equivalence Principle.

(11.14) Suppose an astronaut in the Rindler frame (constantly accelerating frame in flat spacetime) only has access to the line $y = z = 0$. Explain precisely what observations he might use in order to discover that his spacetime is flat.

(11.15) In a science fiction story by Ursala le Guin it is proposed that astronauts could travel into the far future in comfort by 'parking' their spaceship in a fast orbit around a dense star, thus taking advantage of time dilation. Could the method work?

(11.16) **Advance of the perihelion.** If we change variable to $w \equiv (2L^2/c^2 r_s)u$, show that eqn (11.38) becomes

$$\ddot{w} + w = 1 + \epsilon w^2$$

where the dot signifies $d/d\phi$ and $\epsilon = 3r_s^2 c^2/4L^2$. For $\epsilon \ll 1$ this equation can be solved by perturbation theory. Try the form $w = w_0 + \epsilon w_1$, where $w_0 = 1 + e\cos\phi$. Show that, to first order in ϵ,

$$\ddot{w}_1 + w_1 = 1 + e^2/2 + 2e\cos\phi + (e^2/2)\cos 2\phi$$

We already have enough degrees of freedom in w_0 to satisfy boundary conditions, so we only need a particular solution (not the general solution) for w_1. Show that $w_1 = A + B\phi\sin\phi + C\cos 2\phi$ is a solution with $B = e$. The $\phi\sin\phi$ term dominates, since it grows with ϕ, so we have $w \simeq 1 + e(\cos\phi + e\phi\sin\phi)$. Show that this is

$$w = 1 + e\cos(\phi(1 - \epsilon)) + O(\epsilon^2)$$

and hence obtain eqn (11.41), where now the calculation is valid for any ellipticity.

(11.17) Given that neutron stars have a radius of order twice their Schwarzschild radius, could the star considered in exercise 9.15 of chapter 9 be a neutron star?

(11.18) Using a change of coordinates to u, r, θ, ϕ with $u \equiv ct + r - r_s \log(r/r_s - 1)$, show that the Schwarzschild line element becomes

$$ds^2 = -(1 - r_s/r)du^2 + 2dudr$$
$$+ r^2(d\theta^2 + \sin^2\theta d\phi^2).$$

Confirm that in these Eddington–Finkelstein coordinates a radial null line has either $du/dr = 0$ or $du/dr = 2r/(r - r_s)$, and indicate which is the future direction on these lines.

(11.19) A rigid circular scaffold is constructed around a large black hole, some distance outside the horizon, so that it is prevented by its compressive reaction forces from falling towards the hole. An astronaut attaches one end of a rope to the scaffold, then jumps into hole. She allows the rope to play out behind her on a spool so that it does not restrict her free fall. What happens to the rope?

(11.20) Confirm that in the absence of pressure, both eqns (9.16) and (10.19) agree with the predictions of Newtonian gravity. In view of the fact that these exact General Relativistic equations agree with Newtonian physics, how does it come about that the General Relativistic predictions for trajectories in gravitational fields do not match Newtonian predictions?

Part III

Further Special Relativity

Tensors and index notation

12

In this chapter we shall introduce some methods of tensor algebra, which are needed to take the subject further. The study of tensors and their manipulation is a rich field of mathematics in its own right, and this can be daunting for a physics student meeting the ideas for the first time. For Special Relativity, however, we do not need to invoke all the methods. We will take a 'gentle' approach that is intended to bridge the gap between the 4-vectors we have met so far, and the complicated multi-dimensional objects whose treatment requires a whole new notation. We shall concentrate our attention mostly on scalars, 4-vectors, and the type of tensor quantity that was introduced in section 7.5—the second rank tensor. Occasionally, higher-rank quantities appear in equations, but only as a stepping stone to a simpler result.

Tensor analysis is used extensively in General Relativity. Most of the results of this chapter can be used in General Relativity. We will point out the main occasions where Special Relativity is assumed.

An important theme is the introduction of a new notation, called index notation. This notation is needed for some of the more advanced results. However, we will not completely abandon matrix notation, but use whichever notation is more convenient for any given calculation. We shall also display many results in both notations. This will help to clarify the meaning of some of the tensor equations that are hard to read when one first sees them written down.

12.1   Index notation in a nutshell

The essential elements of index notation are as follows (the list is followed by explanatory comments):

(1) *Displaying an element.* We display a 4-vector or tensor by writing down a representative element. For example,

$$\mathsf{A}^a$$

signifies the element a of the 4-vector A. Since a could take any of the values $0, 1, 2, 3$, by writing down a representative element, we implicitly show the whole 4-vector. Similarly, a second rank tensor is displayed thus:

$$\mathbb{T}^{ab}$$

where the indices a, b both take on values $0, 1, 2, 3$, so there are sixteen elements in full. A scalar quantity needs no indices.

(2) *Lorentz transformation.* The Lorentz transformation matrix is written $\Lambda^{a'}_{a}$.

(3) *Summation rule.* It often happens that a sum is involved, for example, when one takes an inner product, or when a matrix multiplies a vector:

$$A^{a'} = \sum_{\lambda=0}^{3} \Lambda^{a'}_{\lambda} A^{\lambda}.$$

We introduce the summation rule: *when an index is repeated in any given term or product of terms, the sum over all values of that index is taken.* That is, instead of writing the summation $\sum_{\lambda}$ explicitly, it is assumed to be there. For example, the above expression is written as

$$A^{a'} = \Lambda^{a'}_{\lambda} A^{\lambda}.$$

(4) *Metric tensor.* The metric tensor g_{ab} is written with lowered indices and is defined to be such a tensor that, for any pair of 4-vectors A^a, B^b, the combination

$$A^{\mu} g_{\mu\nu} B^{\nu} \tag{12.1}$$

is Lorentz-invariant. (Note that there are two repeated indices here, so two sums, giving a scalar result.)

(5) *Index lowering.* If A^a is a 4-vector, then we define a new quantity A_a by

$$A_a \equiv g_{a\lambda} A^{\lambda}. \tag{12.2}$$

Thus the placement up or down of the index is significant, and this operation is called index lowering.

(6) *Index raising.* Let δ^a_b be the Kronecker delta, whose value is 1 for $a = b$ and 0 otherwise. (When written out as a matrix, δ^a_b is the 4×4 identity matrix.) Then the matrix g^{ab} is defined by

$$g^{a\lambda} g_{\lambda b} \equiv \delta^a_b. \tag{12.3}$$

Interpreted as a matrix equation, this says that the matrix g^{ab} is the inverse of g_{ab}. By using the definition, one can easily show (see below) that

$$A^a = g^{a\lambda} A_{\lambda}. \tag{12.4}$$

This operation is called index raising.

(7) *Differential operator.* The differential operator is defined

$$\partial_a \equiv \frac{\partial}{\partial x^a} \tag{12.5}$$

where (x^0, x^1, x^2, x^3) is the coordinate system of the reference frame under consideration. For example, for rectangular coordinates we would have

$$\partial_a = \left(\frac{1}{c} \frac{\partial}{\partial t}, \frac{\partial}{\partial x}, \frac{\partial}{\partial y}, \frac{\partial}{\partial z} \right).$$

By raising the index one can also define

$$\partial^a = g^{a\lambda} \partial_\lambda.$$

In rectangular coordinates, and assuming the Minkowski metric, this is

$$\partial^a = \left(\frac{-1}{c} \frac{\partial}{\partial t}, \frac{\partial}{\partial x}, \frac{\partial}{\partial y}, \frac{\partial}{\partial z} \right).$$

Comments

(1) Since the number of indices can always be used to specify what type of quantity one is dealing with, it is no longer necessary to adopt a different font. However, we will mostly retain the use of special fonts, as in A and $\mathbb{T}$, so that we can move between index and matrix notation when it suits us.

(2) The Lorentz boost matrix is symmetric, so we do not need to separate the indices horizontally in order to know which refers to the column and which the row: it does not matter. More generally, however (e.g., rotations, and General Relativity), the transformation matrix is not always symmetric.

(3) The summation rule takes care of matrix multiplication, but it is more general. In a matrix multiplication such as $\mathbb{MN}$ there is, built in to the definition, the rule that we must sum over the last index of the left matrix ($\mathbb{M}$) and the first index of the right matrix ($\mathbb{N}$). In index notation this rule does not have to be obeyed: one can sum over any index. For example, the equation

$$\mathsf{A}^{ab} = \mathbb{M}^{a\mu} g_{\mu\nu} \mathbb{N}^{\nu b}$$

can be easily 'translated' to the matrix equation

$$\mathsf{A} = \mathbb{M} g \mathbb{N}$$

but, be careful, the equation

$$\mathsf{A}^{ab} = \mathbb{M}^{\mu a} g_{\mu\nu} \mathbb{N}^{b\nu}$$

corresponds to

$$\mathsf{A} = \mathbb{M}^T g \mathbb{N}^T. \tag{12.6}$$

Why? Because $\mathbb{M}^{\mu a} = (\mathbb{M}^T)^{a\mu}$, so we have

$$\mathsf{A}^{ab} = (\mathbb{M}^T)^{a\mu} g_{\mu\nu} (\mathbb{N}^T)^{\nu b}.$$

Now the repeated indices are adjacent in both products (last index of left-hand term, first index of right-hand term), which means they are in the places assumed for matrix multiplication, and we can translate to the matrix equation (12.6) as claimed.

The result can also be expressed as

$$\mathbb{A}^T = \mathbb{N} g^T \mathbb{M}. \tag{12.7}$$

We could obtain this form by rearranging the terms appearing in the double sum:

$$\mathbb{M}^{\mu a} g_{\mu\nu} \mathbb{N}^{b\nu} = \mathbb{N}^{b\nu} g_{\mu\nu} \mathbb{M}^{\mu a}.$$

This is correct, since multiplication of scalars is commutative, so in any sum each term can be rearranged; for example, $su + vw = us + wv$. Then, using $g_{\mu\nu} = g^T_{\nu\mu}$ we have

$$\mathbb{A}^{ab} = \mathbb{N}^{b\nu} g^T_{\nu\mu} \mathbb{M}^{\mu a}.$$

The right-hand side has the repeated indices adjacent, so is easy to translate into matrix notation; but now we have to take care to notice that the indices a and b are in reverse order on the right-hand side compared to the left (a refers to a row of $\mathbb{A}^{ab}$, but to a column of $\mathbb{M}^{\mu a}$, etc.), leading to eqn (12.7).

(4) This definition of the metric tensor is valid in General Relativity. By using the definition of index lowering, the invariant quantity can be written

$$\mathsf{A}^\lambda \mathsf{B}_\lambda \tag{12.8}$$

and you can see that this is precisely the inner product that we have written $\mathsf{A} \cdot \mathsf{B}$ up until now. The metric tensor is always symmetric.

(5) Symbols with all upper indices are called *contravariant*, those with all lower indices are called *covariant*, and those with some indices up and some down are called *mixed*.

(6) To prove eqn (12.4), let $\bar{\mathsf{A}} \equiv g\mathsf{A}$. Then, by pre-multiplying by the inverse of g,

$$\mathsf{A} = g^{-1}\bar{\mathsf{A}}$$

QED. Alternatively, for a little practice with index notation, use

$$\mathsf{A}_b = g_{b\mu} \mathsf{A}^\mu$$
$$\Rightarrow \quad g^{ab} \mathsf{A}_b = g^{ab} g_{b\mu} \mathsf{A}^\mu = \delta^a_\mu \mathsf{A}^\mu = \mathsf{A}^a$$

where we used eqn (12.3) (the defining equation for g^{ab}), and in the last step, notice that a sum involving δ^a_μ has the effect of changing the name of the index that remains (you should convince yourself that this is so by thinking about all values of the indices and doing the sum).

(7) The 4-gradient operator that we have used until now should be displayed with an upper index, so that $\square$ becomes $\square^a \equiv \partial^a$. In a rectangular coordinate system in Special Relativity, ∂_a and ∂^a differ only by a sign in the first term; more generally, however, the relationship between them is more complicated, and one should start from definition (12.5).

A final remark on index letters. In principle one can use any symbol as an index, including, for example, all the letters of the Roman and Greek alphabets. However, it is useful to adopt the convention that the Roman letters i, j, k range only over values $1, 2, 3$, so that these are used for 3-vector analysis, while other letters are used to indicate the full range $0, 1, 2, 3$. Also, it is helpful to make repeated indices easy to notice. This can be done by reserving early letters such as α, β for repeated indices, or by using Roman letters for non-repeated indices and Greek letters for repeated ones. I shall mostly adopt the latter practice. Note that when an index is repeated, and so being summed over, it is a 'dummy' variable whose name can be changed with impunity. For example:

$$\mathsf{A}^\lambda \mathsf{B}_\lambda = \mathsf{A}^\mu \mathsf{B}_\mu = \mathsf{A}^a \mathsf{B}_a$$

etc. (The last example did not adopt the convention of using Greek letters for repeated indices, to show that this is perfectly allowable.)

12.2 Tensor analysis

Tensors in general are mathematical objects expressed by a set of components that change in a given way under a change of coordinate system. In particular, the 4-vector or 'first-rank tensor' transforms as $\mathsf{A} \to \Lambda \mathsf{A}$, and higher-rank objects transform in the same way as outer products of 4-vectors. (Not all tensors can be written as an outer product—those that can are called 'pure'—but they can always be written as a sum of outer products, so the outer product is sufficient to tell us how they behave.)

Prime notation. It is useful to indicate two different coordinate systems (associated with two different reference frames) by the use of a prime, as in $\{t, x, y, z\}$ and $\{t', x', y', z'\}$. So far, when referring to a 4-vector quantity in either frame we have used A and A'. In index notation we have two choices: the prime can be attached to the main letter ('kernel') as in A'^a or to the index as in $A^{a'}$. The latter choice is arguably more logical, since when transforming from one coordinate system to another the 4-vector does not itself change, but its components change because the basis vectors change. Therefore we will use $A^{a'}$ in the following.

We already know how a 4-vector changes from one frame to another:

$$\mathsf{A}^{a'} = \Lambda^{a'}_{\ \lambda} \mathsf{A}^\lambda.$$

By considering the outer product, we also found how second-rank tensors transform, eqn (7.38), which in index notation is

$$\mathbb{M}^{a'b'} = \Lambda^{a'}_{\ \mu} \mathbb{M}^{\mu\nu} (\Lambda^T)^{\ b'}_{\nu} = \Lambda^{a'}_{\ \mu} \Lambda^{b'}_{\ \nu} \mathbb{M}^{\mu\nu}$$

where the first version is a direct translation from our earlier matrix result, and the second version is the way it is usually written. One could also obtain the latter directly in index notation by considering the outer product:

$$\mathsf{A}^{a'}\mathsf{B}^{b'} = (\Lambda^{a'}_{\ \mu}\mathsf{A}^{\mu})(\Lambda^{b'}_{\ \nu}\mathsf{B}^{\nu}) = \Lambda^{a'}_{\ \mu}\Lambda^{b'}_{\ \nu}\,\mathsf{A}^{\mu}\mathsf{B}^{\nu}.$$

12.2.1   Rules for tensor algebra

We can now extend the ideas to tensors in general. We *define* a tensor of any rank to be an entity which transforms in the right way: namely, in the same way as an outer product, which means that for a contravariant (upper index) tensor of any rank, one factor of $\Lambda^{a'}_{\ b}$ should be used to convert each index. We *allow* operations that take tensors to tensors (not necessarily of the same rank). The legal operations are: *sum, outer product, contraction,* and *index permutation*. By using the metric we obtain two further operations: index lowering and raising.

The *valence* of a tensor refers to the number of upper and lower indices. The *rank* is equal to the total number of indices.

The *sum* of two tensors of the same valence is defined as

$$C^{ab\cdots}_{cd\cdots} = A^{ab\cdots}_{cd\cdots} + B^{ab\cdots}_{cd\cdots},$$

i.e., just add corresponding elements. It is easy to prove that this is a tensor if A and B are being evaluated at the same event. Note, however, that when summing tensors at different points in the coordinate space (i.e., different events in spacetime) the sum is a tensor when the transformation is linear,[1] as, for example, the Lorentz transformation, but not always if it is non-linear, as in General Relativity.

The *outer product* of two tensors is obtained by forming the product of their representative components, as in

$$A^a B^b = C^{ab} \quad\text{or}\quad A^a_b B^c_{de} = C^{ac}_{bde}.$$

Contraction consists in replacing one upper and one lower index by a dummy index, and summing over it. For example, the scalar product of a pair of 4-vectors is obtained by first forming their outer product, then lowering an index, and then contracting, so as to obtain:

$$A^\lambda B_\lambda, \quad\text{cf.}\quad \mathsf{A}^T g \mathsf{B} = \mathsf{A}\cdot\mathsf{B}.$$

More generally one could have combinations such as

$$C^{ab} = A^{a\lambda}B_\lambda^{\ b} \quad\text{cf.}\quad \mathbb{C} = \mathbb{A}g\mathbb{B}$$

and

$$B^{ab}_c = A^{a\lambda b}_{\lambda c}.$$

Contraction reduces each valence by 1, and therefore the rank of the tensor by 2. Contracting all the way down to a scalar results in an invariant, so this is an important operation. Examples include

$$F^\lambda_\lambda \quad \text{and} \quad F^{\lambda\mu}G_{\lambda\mu}. \tag{12.9}$$

To calculate the first, starting from F^{ab}, multiply by the metric and then take the trace. To calculate the second, use the metric if necessary to calculate $G_{\lambda\mu}$, and then multiply corresponding elements and sum.

Index permutation consists in reordering *either* the upper *or* the lower indices (of all terms in a sum). This is a generalized form of the transpose operation (for a second-rank tensor, it is equivalent to a transpose of its matrix representation).

Index lowering/raising can be used to show the result of multiplying an equation by the metric or its inverse.

12.2.2   Contravariant and covariant

The rules we have supplied for index notation are all that one requires to carry out legal manipulations. However, it is helpful to have some idea of what the upper/lower index signifies. A good way to explore this is to examine the question, whether the metric is itself a tensor. We have been calling it a tensor, so you will not be surprised to learn that it is one—but can we prove it?

The proof is easy in matrix notation. The metric is *defined* to be that quantity such that the combination

$$\mathsf{A}^T g \mathsf{B}$$

is invariant. This means that

$$\mathsf{A}^T g \mathsf{B} = \mathsf{A}'^T g' \mathsf{B}'$$

where $\mathsf{A}' = \Lambda\mathsf{A}$, $\mathsf{B}' = \Lambda\mathsf{B}$, and g' is to be discovered. We have

$$\mathsf{A}^T g \mathsf{B} = (\Lambda\mathsf{A})^T g' (\Lambda\mathsf{B}) = \mathsf{A}^T (\Lambda^T g' \Lambda)\mathsf{B}$$

and the result is valid for all 4-vectors A, B. It follows that

$$g = \Lambda^T g' \Lambda \quad \Rightarrow \quad g' = (\Lambda^{-1})^T g \Lambda^{-1}. \tag{12.10}$$

Compare this with the rule for a contravariant tensor:

$$\mathbb{T}' = \Lambda \mathbb{T} \Lambda^T. \tag{12.11}$$

We have found that g does *not* transform in the same way as a contravariant tensor, but it *does* transform in a way that is easily obtained from the Lorentz transformation: to transform g, treat it like a tensor, but instead of Λ use $(\Lambda^{-1})^T$.

In the case of the Lorentz transformation, the definition (6.64) results in $(\Lambda^{-1})^T g \Lambda^{-1} = g$, so $g' = g$—which we have in fact assumed until now in this book. For more general transformations (General Relativity) one would find $g' \neq g$.

A tensor which transforms in the standard way is called *contravariant*. An entity which transforms in the other way, like g, is also called a tensor and is said to be *covariant*.

Contravariant vector or contravariant components?

Some types of argument require care to distinguish between a vector and the set of components which may be used to describe it. Be warned: this can be confusing. My advice is: do not worry about it but just learn the rules of index notation. However, for General Relativity, greater clarity is required. Here is a brief discussion.

A vector is not described by a set of components alone, but by a set of components and another set of vectors: namely, the basis vectors. The notation A^a refers to each of the components when the basis is the standard one: i.e., unit vectors along the directions of the axes of some chosen reference frame. The notation A_a refers to a different set of components for the *same* vector; this is possible because a given vector can be expressed in terms of more than one basis. When using index notation, one does not need to know what this second basis is; it suffices to know that $A^\lambda A_\lambda$ is a Lorentz scalar. However, if you want to take an interest in the basis vectors, consult Fig. 12.1.

When we change reference frame, any given 4-vector such as a 4-momentum does not change, but the basis vectors do change, and usually we would prefer to know the components in terms of the new basis vectors. A matrix equation such as $A' = \Lambda A$ should be regarded as shorthand for the index notation version, $A^{a'} = \Lambda^{a'}_\lambda A^\lambda$. This makes it clear that we are here talking about each component of the vector, not the vector itself. The idea of 'a contravariant 4-vector' or 'a covariant 4-vector' is meaningless, according to this stricter use of terminology. Rather, the set of components A^a is contravariant, and the set A_a is covariant.

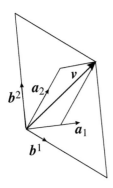

Fig. 12.1 Contravariant and covariant components. A given vector **v** can be expressed in more than one way: either using the set of basis vectors $\{\mathbf{a}_i\}$, or the set of basis vectors $\{\mathbf{b}^i\}$. Thus $\mathbf{v} = v^1 \mathbf{a}_1 + v^2 \mathbf{a}_2 = v_1 \mathbf{b}^1 + v_2 \mathbf{b}^2$ where $\{v^i\}$ and $\{v_i\}$ are the respective sets of components. Neither set of basis vectors need itself be orthonormal. However, the ith basis vector in the second set ($\mathbf{b}^i$) is chosen to be orthogonal to all of the first set except the ith member ($\mathbf{a}_i$), and it is given a length such that its inner product with $\mathbf{a}_i$ is 1. Thus $\mathbf{a}_i \cdot \mathbf{b}^j = \delta_{i,j}$. This permits the inner product between any pair of vectors **u**, **v** to be written $\sum_\lambda u^\lambda v_\lambda$ as you can confirm by expanding $(u^1 \mathbf{a}_1 + u^2 \mathbf{a}_2) \cdot (v_1 \mathbf{b}^1 + v_2 \mathbf{b}^2)$. If $\mathbf{a}_i$ are along the coordinate axes then the components v^i are said to be contravariant. The other set of components, v_i, are said to be covariant.

Once we have discovered one covariant tensor, it is easy to form others. Consider, for example, $(g A)$, where A is contravariant. Under a change of reference frame,

$$g A \to g' A' = (\Lambda^{-1})^T g \Lambda^{-1} \Lambda A = (\Lambda^{-1})^T (g A)$$

so $(g A)$ is a covariant 4-vector. This fact is indicated in index notation by index lowering: $(g_{a\lambda} A^\lambda) = A_a$. The upper/lower index signifies the contravariant/covariant nature of the entity.

All the 4-vectors and second-rank tensors we have been using in matrix notation have been contravariant, with the single exception of the metric. The terminology seems strange at first: why are the 'ordinary' ones called contravariant, and the 'contrary' ones called covariant? The reason is partly historical accident, but is connected to behaviour of the metric: the word 'covariant' has the connotation 'transforming in the same way as the metric tensor'.

There is a simple geometric interpretation of the contravariant and covariant sets of components of a vector, illustrated in Fig. 12.1. This makes it clear that one must regard the contravariant and covariant forms as two versions of the *same* object, not two different objects.

The metric tensor can now be understood to play two roles. Its primary role is to show how to calculate an invariant 'distance' in spacetime, and its secondary role is to allow easy conversion between contravariant and covariant forms of tensors.

12.2.3   Useful methods and ideas

The combination $\mathsf{A} \cdot \mathbb{F} \cdot \mathsf{B}$ (i.e. $\mathsf{A}_\mu \mathbb{F}^{\mu\nu} \mathsf{B}_\nu$) is a scalar and therefore is unaffected by a transpose, hence

$$\mathsf{A} \cdot \mathbb{F} \cdot \mathsf{B} = \mathsf{B} \cdot \mathbb{F}^T \cdot \mathsf{A}. \qquad (12.12)$$

If the tensor is antisymmetric, then this gives $\mathsf{A} \cdot \mathbb{F} \cdot \mathsf{B} = -\mathsf{B} \cdot \mathbb{F} \cdot \mathsf{A}$, and therefore

$$\mathsf{A} \cdot \mathbb{F} \cdot \mathsf{A} = 0 \qquad \text{for antisymmetric } \mathbb{F}. \qquad (12.13)$$

We have already used this in eqn (7.44) in order to argue that the Faraday tensor should be antisymmetric in order to produce a pure force, and we shall use it again in chapter 15. More generally, you can prove that if $\mathbb{F}$ is antisymmetric and $\mathbb{S}$ is symmetric, then

$$\mathbb{S}_{\mu\nu} \mathbb{F}^{\mu\nu} = 0 \qquad \text{for sym, antisym.} \qquad (12.14)$$

Some tips for tensor manipulations are shown in the box below. By using $g^{a\lambda} g_{\lambda b} = \delta^a_b$ one can prove the useful 'see-saw' rule, as follows. We have $\mathsf{A}_{a\cdots} = g_{a\lambda} \mathsf{A}^\lambda_{\cdots}$ and $\mathsf{B}^{a\cdots} = g^{a\mu} \mathsf{B}_\mu^{\cdots}$ for any A, B, where the dots signify other indices (which may more generally be up or down and in any order). Therefore

$$\mathsf{A}_{a\cdots} \mathsf{B}^{a\cdots} = \mathsf{A}^\lambda_{\cdots} \mathsf{B}_\mu^{\cdots} g_{\lambda a} g^{a\mu} = \mathsf{A}^\lambda_{\cdots} \mathsf{B}_\mu^{\cdots} \delta^\mu_\lambda = \mathsf{A}^\lambda_{\cdots} \mathsf{B}_\lambda^{\cdots}$$

where the first step used that g is symmetric.

The order of the indices of a given tensor in index notation does matter and must be respected. For example, $\mathbb{A}^{ab}$ is not necessarily equal to $\mathbb{A}^{ba}$, and $\mathbb{A}^a_{\ b}$ is not necessarily equal to $\mathbb{A}_b^{\ a}$. This point is sometimes treated rather loosely in the literature. The only exception is when a tensor is symmetric. The Kronecker delta can safely be written δ^a_b without bothering to indicate which index is first, which second, because[2] $\delta^a_b = \delta_b^{\ a}$.

In order to 'read' a tensor M^{ab} as a matrix, it is helpful to think of the indices a, b as a two-digit number, and to 'read' the matrix in the way one reads text in most western languages: i.e., across the top row from left to right, then down to the next row, etc. As the 'two-digit number' a, b increments, the second digit b changes fastest, and this corresponds to moving along a row in the matrix.

Lowering a first index of a tensor corresponds to pre-multiplying the matrix by g, thus changing the sign of the first *row*. Lowering a second index corresponds to post-multiplying by g, thus changing the sign of the first *column*. Lowering both indices changes the sign only of the time-space part (i.e., the $0, 0$ element and the lower right block are unaffected).

[2] Note that here we are *not* comparing δ^a_b with δ^b_a, which one would obtain from the former by lowering one index and raising another, a quite different operation, involving the metric twice.

Tips for manipulating tensor equations

(1) Name your indices sensibly; make repeated indices easy to spot.

(2) Look for scalars: e.g., $\mathbb{F}_{\lambda\mu}\mathbb{A}^a_b\mathbb{F}^{\lambda\mu}$ is $s\mathbb{A}^a_b$ where $s = \mathbb{F}_{\lambda\mu}\mathbb{F}^{\lambda\mu}$.

(3) You can always change the names of dummy (summed over) indices; if there are two or more, you can swap names.

(4) The 'see-saw rule'

$$A_\lambda B^\lambda = A^\lambda B_\lambda \quad \text{(works for any rank)}$$

(5) ∂_a behaves like $\partial/\partial x$

(6) In the absence of ∂_a, everything commutes.

(7) $\partial_a x^b = \delta^b_a; \quad \partial_a x_b = g_{ab}.$

(8) $\partial_a k^\lambda x_\lambda = k_a$ for constant k_a.

Non-tensors. The transformation matrix $\Lambda^{a'}_b$ is not itself a tensor: it cannot be written down in any one reference frame, but rather acts as the 'bridge' between reference frames. There can be other matrix-like quantities that are not tensors (because they do not transform in the right way) but are nonetheless useful.

Consider now the equation used to define g^{ab}, eqn (12.3). The combination on the left-hand side could also be written $g_a{}^b$, since it can be read as g^{ab} with the first index lowered, so we have

$$g_a{}^b = \delta^b_a.$$

and therefore $g^\lambda_\lambda = \delta^\lambda_\lambda = 4$.

In Special Relativity using rectangular coordinates, one finds that g_{ab} and g^{ab} are represented by the same matrix. More generally, this need not be the case, but eqn (12.3) is always true. We met some examples in chapter 9. The first metric in eqn (9.47) reads $g_{ab} = \text{diag}(-h^2, 1, 1, 1)$. Taking the inverse, we find $g^{ab} = \text{diag}(-1/h^2, 1, 1, 1)$.

It is interesting to ask whether the Kronecker delta δ^a_b is a tensor. If it is, then the placement of the indices implies that it is of mixed rank, so it ought to transform as Λ for one index, $(\Lambda^{-1})^T$ for the other. Using matrix notation, this means the transformed version, $\delta^{a'}_{b'}$ is the matrix given by

$$\Lambda I \Lambda^{-1}$$

where I is the identity matrix. This matrix product evaluates to I: i.e., we get back the Kronecker delta once again. This shows that δ^a_b respects the rules: it is a tensor of the type indicated by the placement of its indices. This can also be proved from the quotient rule.

Quotient rule. The *quotient rule* states that if an expression of the form

$$B^{a\lambda}C_\lambda \tag{12.15}$$

yields a 4-vector whenever C is a 4-vector, then B must be a tensor (of the type indicated by the placement of its indices), and similar statements apply at all ranks. This idea can be expressed 'if when something multiplies a vector it always yields a vector, then the thing is a tensor'. This is familiar in 3-vector analysis, where, for example, one has expressions such as

$$\mathbf{j} = C\mathbf{E}$$

where $\mathbf{j}$ is the current density produced by an electric field $\mathbf{E}$ in a crystalline material whose conductivity is given by the 3×3 tensor C. The i, j element of C tells how much current density in the ith direction is produced by an electric field in the jth direction. Further examples are polarizability, magnetic susceptibility, and moment of inertia—whenever the conditions are not isotropic these are tensor rather than scalar quantities.

We shall now prove the quotient rule. In matrix notation, we have

$$A = BgC$$

where A and C are 4-vectors, and we want to prove that B transforms as a tensor. Under a change of frame, $A \to \Lambda A$, $B \to B'$, $g \to (\Lambda^{-1})^T g \Lambda^{-1}$ and $C \to \Lambda C$, so

$$\Lambda(BgC) = B'(\Lambda^{-1})^T g \Lambda^{-1} \Lambda C$$

$$\Rightarrow \qquad BgC = \Lambda^{-1} B'(\Lambda^{-1})^T g C$$

and this is true for all C. It follows that $B = \Lambda^{-1} B'(\Lambda^{-1})^T$, and therefore

$$B' = \Lambda B \Lambda^T.$$

This is the defining property of a tensor, therefore B is a (contravariant) second-rank tensor. The proof for tensors of higher rank proceeds similarly.

At first sight the index notation is not very appealing: it appears to offer rather cluttered expressions, and one needs to look hard to keep track of the indices. However, it comes into its own when differentiation is needed, and it makes it easy to construct the all-important invariants, by contracting tensors of even rank. It also makes some basic calculations easier.

Example   Find the transformation of electric and magnetic fields under a change of inertial reference frame.

Solution
We use the Faraday tensor (7.46), and calculate

$$\mathbb{F}^{a'b'} = \Lambda^{a'}_{\ \mu} \Lambda^{b'}_{\ \nu} \mathbb{F}^{\mu\nu}. \tag{12.16}$$

From the antisymmetry of $\mathbb{F}^{ab}$ and the symmetry of Λ we know that $\mathbb{F}^{a'b'}$ is antisymmetric, so the diagonal elements are zero. Next consider the element $\mathbb{F}^{0'1'}$: i.e., $a' = 0$, $b' = 1$. In the double sum over μ and ν there are sixteen terms, but for frames in standard configuration only two are non-zero, because $\Lambda^{0'}_{\ \mu}$ is only non-zero for $\mu = 0, 1$, $\Lambda^{1'}_{\ \nu}$ is only non-zero for $\nu = 0, 1$, and $\mathbb{F}^{\mu\nu}$ is only non-zero for $\mu \neq \nu$. Hence only the combinations $(\mu, \nu) = (0, 1)$ and $(1, 0)$ survive, and we have

$$\mathbb{F}^{0'1'} = \Lambda^{0'}_{\ 0}\Lambda^{1'}_{\ 1}\mathbb{F}^{01} + \Lambda^{0'}_{\ 1}\Lambda^{1'}_{\ 0}\mathbb{F}^{10} = (\gamma^2 - \gamma^2\beta^2)\mathbb{F}^{01} = \mathbb{F}^{01}.$$

Therefore $\mathbf{E}'_\| = \mathbf{E}_\|$. Proceeding similarly, one finds

$$\mathbb{F}^{0'2'} = \Lambda^{0'}_{\ 0}\Lambda^{2'}_{\ 2}\mathbb{F}^{02} + \Lambda^{0'}_{\ 1}\Lambda^{2'}_{\ 2}\mathbb{F}^{12} \quad \Rightarrow \quad E'_y/c = \gamma(E_y/c - \beta B_z)$$

and $\quad \mathbb{F}^{0'3'} = \Lambda^{0'}_{\ 0}\Lambda^{3'}_{\ 3}\mathbb{F}^{03} + \Lambda^{0'}_{\ 1}\Lambda^{3'}_{\ 3}\mathbb{F}^{13} \quad \Rightarrow \quad E'_z/c = \gamma(E_z/c + \beta B_y),$

therefore $\mathbf{E}'_\perp = \gamma(\mathbf{E}_\perp + \mathbf{v} \wedge \mathbf{B})$. It is also easy to see that $\mathbb{F}^{2'3'} = \mathbb{F}^{23}$, therefore $\mathbf{B}'_\| = \mathbf{B}_\|$, and the equation for $\mathbf{B}_\perp$ follows from $\mathbb{F}^{1'2'}$ and $\mathbb{F}^{1'3'}$ (exercise for the reader).

The above method of calculation is simpler than evaluating the matrix product $\Lambda\mathbb{F}\Lambda^T$.

A summary of some of the merits or otherwise of the two notations is given in the following table. Note that both notations allow you to write nonsense such as $A^a \stackrel{?}{=} B^{a\lambda}C^\lambda$ or $\mathsf{A} \stackrel{?}{=} \mathbb{B}\mathsf{C}$ (in both cases the symbol on the left seems to be a 4-vector, but the combination on the right is not). To avoid nonsense it is up to you to obey the rules!

index notation	vectors and matrices
number of indices tells you the rank	font or underline tells you the rank
lots of fiddly indices	less clutter
use further labels with caution	labels are ok, e.g. $\mathsf{P}_{tot} = \sum_i \mathsf{P}_i$
upper, lower index to take care of g	use · or remember g
all ranks	only rank 0 to 2
handles everything	restricted
identify invariants easily	invariants less obvious
longer derivations easier	good for the simplest derivations

12.2.4   Parity inversion and the vector product

We still have not exhibited a 4-vector quantity similar to the well-known vector product $\mathbf{a} \wedge \mathbf{b}$ for 3-vectors. The reason is connected with the fact that $\mathbf{a} \wedge \mathbf{b}$ is not quite a 'perfectly proper' vector. It is (quite rightly) called a vector because it behaves the right way under rotations, but it gets up to no good when you try reflections or inversions through the origin (parity transformation). Consider a rotating object and its angular momentum $\mathbf{L} = \sum \mathbf{r} \wedge \mathbf{p}$ for example. The angular momentum vector is defined *by convention* to point along the axis of rotation, with a direction such that the rotation is right-handed. Now imagine placing an ordinary arrow-shaped rod next to a rotating wheel, with the arrow pointing in the direction of the angular momentum **L** of the wheel. Place a mirror next to them. First suppose that the axis of rotation of the wheel is vertical and so is the mirror surface (Fig. 12.2). Now look in the mirror: the arrow, seen in reflection, is still pointing in the same direction, but what has happened to the wheel? Its reflection is rotating in the opposite sense, so its **L** vector has reversed direction! The angular momentum vector and the arrow have done opposite things: one reversed direction, the other did not.

Now lay the mirror flat, in a horizontal plane (Fig. 12.3). This time the arrow changes direction but the rotation does not.

We have in $\mathbf{r} \wedge \mathbf{p}$ a quantity that behaves like a vector under rotations, but has exactly the 'wrong' behaviour under reflections. Such a quantity is called a *pseudovector*. Alternatively, the ordinary vectors are called *polar* vectors, and ones like angular momentum are called *axial* vectors. A polar vector is one that changes sign under **spatial inversion** (also called parity transformation); an axial vector is one that does not. Spatial inversion is a reversal of all three coordinate axes. Under such an inversion an ordinary vector changes sign—what you would expect—but an axial vector does not (Fig. 12.4).

Axial vectors might seem to be an invention that should have been avoided, but once you are aware of them you will find them throughout physics. We already mentioned one important example, the angular momentum, and another is the magnetic field vector **B**. The electric field, on the other hand, is a 'straightforward' polar vector. The vector product of two polar vectors (e.g. $\mathbf{r} \wedge \mathbf{p}$) gives an axial vector. The scalar product of a polar vector with an axial vector produces a scalar that changes sign under parity inversions; it is called a *pseudoscalar*.

We mentioned this business of polar and axial vectors in order to introduce the fact that the vector product has to be reconsidered before we can generalize it to more than three dimensions.

If we examine the vector product

$$\mathbf{r} \wedge \mathbf{p} = (r_y p_z - r_z p_y)\mathbf{i} + (r_z p_x - r_x p_z)\mathbf{j} + (r_x p_y - r_y p_x)\mathbf{k}, \quad (12.17)$$

we find all combinations of elements of **r** with elements of **p**, except for those along the same direction. This suggests that we should consider the outer product $\mathbf{r}\mathbf{p}^T$. After trying that, and also trying $\mathbf{p}\mathbf{r}^T$, one discovers that the interesting combination is

$$\mathbb{L} \equiv \mathbf{r}\mathbf{p}^T - \mathbf{p}\mathbf{r}^T = \begin{pmatrix} \cdot & r_x p_y - r_y p_x & \cdot \\ \cdot & \cdot & r_y p_z - r_z p_y \\ r_z p_x - r_x p_z & \cdot & \cdot \end{pmatrix} \quad (12.18)$$

where we have written out three of the elements, and the dots signify the other elements, which can be obtained by using the fact that the whole matrix is antisymmetric. $\mathbb{L}$ is, by construction, a (three-dimensional) second-rank tensor. Notice that the x component of the vector **L** appears in the 'y, z' slot (second row, third column) of the tensor $\mathbb{L}$, and L_y appears in the 'z, x' slot, and L_z appears in the 'x, y' slot:

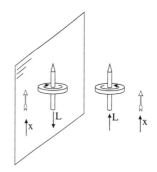

Fig. 12.2

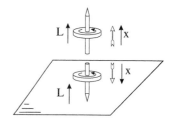

Fig. 12.3

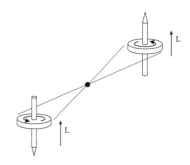

Fig. 12.4 Inversion through the origin: spatial displacements and linear velocities reverse direction, but angular momentum does not.

$$\mathbb{L} = \begin{pmatrix} \cdot & L_z & \cdot \\ \cdot & \cdot & L_x \\ L_y & \cdot & \cdot \end{pmatrix}.$$

The 4-vector generalization of the vector product is, then, a second-rank tensor defined by

$$\mathsf{A} \wedge \mathsf{B} \equiv \mathsf{A}\mathsf{B}^T - \mathsf{B}\mathsf{A}^T. \tag{12.19}$$

For example, an important antisymmetric tensor is the angular momentum tensor defined by

$$L^{ab} \equiv \mathsf{X}^a \mathsf{P}^b - \mathsf{X}^b \mathsf{P}^a \tag{12.20}$$

for a particle whose position and momentum are given by X and P.

The tensor $\mathbb{L}$ behaves the same way under parity changes as other second-rank tensors: it does not change sign. Picking out some elements of this second-rank tensor and calling them a vector is a 'trick' that only works in three dimensions. It works because those elements do transform as a vector under rotations, and this is partly because an antisymmetric second-rank tensor in three dimensions has just three non-zero independent elements. In four dimensions we can construct the tensor; it is antisymmetric and so now has six independent non-zero elements, but that is two too many to have any hope of making a 4-vector out of them! Instead we can find two 3-vectors, one polar and one axial: recall eqn (7.42). For example, for angular momentum the definition (12.20) gives the top row $L^{0i} = ct\mathbf{p} - (E/c)\mathbf{x}$ which is clearly polar, and the space-space part is equal to the 3-angular momentum tensor, which we already showed is related to an axial vector. *It follows that a polar and an axial vector can be similarly extracted from any antisymmetric tensor, since they all transform in the same way.*

12.2.5   Differentiation

We have already defined the 4-gradient $\Box\phi$, the 4-divergence $\Box \cdot \mathsf{F}$, and the d'Alembertian $\Box^2 \equiv \Box \cdot \Box$. The quantity

$$\partial_\lambda \mathbb{A}^{\lambda b} \qquad \text{or} \quad \Box \cdot \mathbb{A}$$

is a 'divergence of a tensor'; it yields a 4-vector.

Two more derivatives naturally suggest themselves:

$$\Box \mathsf{A}^T \qquad \text{and} \qquad \Box \wedge \mathsf{A} \equiv \Box \mathsf{A}^T - (\Box \mathsf{A}^T)^T,$$

$$(\partial^a \mathsf{A}^b \qquad \text{and} \qquad \partial^a \mathsf{A}^b - \partial^b \mathsf{A}^a) \tag{12.21}$$

where we displayed them in both notations (the index notation is clearer for these outer products). The first of these was exhibited in eqn (7.47). It should be read as a sort of 'gradient' of a vector field, but now the gradient has to say how every component of the vector changes in every direction. For example, you should confirm that $\partial_a \mathsf{X}^b = \delta_a^b$ and $\partial_a \mathsf{X}_b = g_{ab}$.

The second quantity in eqn (12.21), a sort of '4-curl', gives an antisymmetric tensor, therefore a set of 6 independent non-zero elements. You can read the result as $\Box \wedge A = $ '(thing) $-$ (transpose of thing)' which makes it clear that the outcome is antisymmetric.

The 'gradient of a vector' idea is quite useful in 3-vector analysis too. Compare, for example, the horrible

$$\nabla(\mathbf{u} \cdot \mathbf{v}) = (\mathbf{u} \cdot \nabla)\mathbf{v} + (\mathbf{v} \cdot \nabla)\mathbf{u} + \mathbf{u} \wedge (\nabla \wedge \mathbf{v}) + \mathbf{v} \wedge (\nabla \wedge \mathbf{u}) \quad (12.22)$$

with the much more elegant

$$\nabla(\mathbf{u} \cdot \mathbf{v}) = (\nabla \mathbf{u}^T)\mathbf{v} + (\nabla \mathbf{v}^T)\mathbf{u}. \quad (12.23)$$

(You can prove the latter without much difficulty by converting to components in a rectangular coordinate system.)

Further information on derivatives is contained in appendix C.

12.3 Antisymmetric tensors and the dual

Most tensors one encounters in physics are either symmetric or antisymmetric. A symmetric tensor has ten independent elements (six for the upper triangle, which also gives the lower triangle, plus four more on the diagonal). An antisymmetric tensor in four dimensions has six independent elements (if these form the upper triangle then the lower triangle is the negative of this and the diagonal is zero).

Since tensors of given rank all transform the same way, we can obtain the transformation of 3-angular momentum (for example) by substituting $\mathbf{E} \to c\mathbf{w}$, $\mathbf{B} \to \mathbf{L}$ into the equations for the transformation of the electromagnetic field, where $\mathbf{w} = ct\mathbf{p} - (E/c)\mathbf{x} = \gamma mc(\mathbf{v}t - \mathbf{x})$.

Some relationships between tensors can be found by using the *Levi-Civita symbol* or 'permutation symbol' ϵ_{abcd}. This is defined as

$$\epsilon_{abcd} = \begin{cases} +1 & \text{if } abcd \text{ is an even permutation of } 0123 \\ -1 & \text{if } abcd \text{ is an odd permutation of } 0123 \\ 0 & \text{otherwise} \end{cases} \quad (12.24)$$

This is a four-dimensional object (there are versions for any number of dimensions), but only $4! = 24$ of its elements are non-zero—half of them $+1$ and half -1; see table 12.1. It is defined to be invariant (it is what it is: it does not matter what reference frame you are working in), but it is easy to prove that it always converts tensors to pseudotensors, so it is itself a pseudotensor, by the quotient rule. Note that $\epsilon_{abcd} = -\epsilon^{abcd}$, which is a source of ambiguity in the literature: some authors chose ϵ^{0123} to be 1.

Consider the combination $\epsilon_{ab\lambda\mu}F^{\lambda\mu}$. Choose, for example, $a = 2, b = 3$. You can see that the $(2, 3)$ element of the result is made from F^{01} and F^{10}, the latter subtracted from the former. If $\mathbb{F}$ is symmetric then this is zero, if it is antisymmetric then this is $2\mathbb{F}^{01}$.

Table 12.1 Evaluating the Levi–Civita symbol.

$$\epsilon_{01cd} = \begin{pmatrix} 0 & & & \\ & 0 & & \\ & & 0 & 1 \\ & & -1 & 0 \end{pmatrix}, \quad \epsilon_{02cd} = \begin{pmatrix} 0 & & & \\ & 0 & & -1 \\ & & 0 & \\ & 1 & & 0 \end{pmatrix}, \quad \text{etc.}$$

Eqn (12.18) can be written

$$\mathbb{L}_{ij} = \epsilon_{ijk} L^k \tag{12.25}$$

where $\mathbf{L} = \mathbf{r} \wedge \mathbf{p}$ is the 3-angular momentum (axial), and ϵ_{ijk} is the permutation symbol in three dimensions.

For antisymmetric $\mathbb{F}$, the tensor (actually a pseudotensor)

$$\tilde{\mathbb{F}}_{cd} \equiv \frac{1}{2} \epsilon_{cd\mu\nu} \mathbb{F}^{\mu\nu} \tag{12.26}$$

is called the *dual* of $\mathbb{F}$. It does not take long to check the six terms to find that the matrix for $\tilde{\mathbb{F}}_{cd}$ looks like the one for $\mathbb{F}^{cd}$ but with $\mathbf{a}$ and $\mathbf{b}$ swapped, where $\mathbf{a}, \mathbf{b}$ are the polar and axial vectors forming $\mathbb{F}$ by the recipe of eqn (7.42). It follows that $\tilde{\mathbb{F}}^{cd}$ can be obtained from $\mathbb{F}^{cd}$ by the substitutions $\mathbf{a} \to -\mathbf{b}$, $\mathbf{b} \to \mathbf{a}$. For an example, see eqns (7.46) and (13.10).

Exercises

(12.1) If A^{ab} and B^{abc} are tensors, then which of the following expressions yield a tensorial result: $A^{a\lambda} B^{bc}_\lambda$; $A^a_\lambda B^{b\lambda c}$; $A^{ab} + A^{ba}$; $A^{ab} + A^{cd}$; $A^{ab} + A_{cd}$; $A^{ab} A_{cd}$; $B^{\lambda b}_\lambda$? Give the valence of any that are tensorial.

(12.2) Describe the relationship between the tensor A^{ab} and $A^a_{\ b}$, $A_b^{\ a}$, $A_a^{\ b}$, $A^b_{\ a}$.

(12.3) Prove eqn (12.14) (by using that $\mathbb{F}^{\mu\nu} = -\mathbb{F}^{\nu\mu}$ if $\mathbb{F}$ is antisymmetric, and $\mathbb{S}^{\mu\nu} = \mathbb{S}^{\nu\mu}$ if $\mathbb{S}$ is symmetric).

(12.4) Consider the effect of parity transformation (spatial inversion) on Maxwell's equations and the Lorentz force equation. Show that all phenomena of classical electromagnetism are invariant under spatial inversion. (It follows that if a given electromagnetic phenomenon is possible, then so is its mirror image.)

(12.5) Evaluate $\Box(\mathsf{K} \cdot \mathsf{X})$ and $\partial^a(\mathsf{K}^\lambda \mathsf{X}_\lambda)$ where K is a constant 4-vector—and decide which notation you prefer!

(12.6) In a given inertial frame, whose 4-velocity is U, a tensor $\mathbb{T}^{ab}$ has just one non-vanishing component: $\mathbb{T}^{00} = c^2$. Find a way of writing such a tensor in terms of a pair of 4-vectors. Hence find the components of this tensor in an arbitrary frame moving at velocity $\mathbf{v}$ relative to the first.

(12.7) Use eqn (12.25) to prove that ϵ_{ijk} is a pseudo-tensor.

(12.8) Confirm the statements made after eqn (12.26).

(12.9) Prove that any second-rank tensor can be written as a sum of a symmetric and an antisymmetric tensor: $M^{ab} = S^{ab} + A^{ab}$, and give expressions for S^{ab} and A^{ab} in terms of M^{ab}. State how the result extends to higher ranks.

(12.10) Show that $\epsilon^{a\lambda\mu\nu} \epsilon_{b\lambda\mu\nu} = -6\delta^a_b$.

Rediscovering electromagnetism

13

We are now ready to 'reinvent' electromagnetism. The approach taken in chapter 7 was to introduce the electric and magnetic fields in terms of the forces exerted on charged particles, and to reason from Lorentz transformations, from easily analysed basic phenomena, and from the Maxwell equations. We mentioned that the electric and magnetic fields should be regarded as two parts of a single entity, and at the end of the chapter we briefly introduced that entity: the antisymmetric second-rank tensor called the Faraday tensor $\mathbb{F}$.

In chapters 7 and 8 (section 8.2.3) we also examined the claim that the whole theory of electromagnetism can be derived from Coulomb's Law and Lorentz covariance. This claim seemed attractive at first, but on further consideration it turned out to be far too sweeping. It is based on several tacit assumptions, some of which are quite subtle. It is an important skill in physics to be able to identify what non-trivial assumptions have in fact been invoked in any given argument.

In this chapter we shall obtain the Lorentz force equation and the Maxwell equations from an explicit set of assumptions, restricting ourselves as far as possible to the simplest possible assumptions that are consistent with Lorentz covariance, and that give rise to some sort of field theory (i.e., a theory of point-like entities called particles interacting via extended entities called fields). In this way we will show that electromagnetic theory can be considered to be one of the most simple possible field theories. The mathematical language of tensors guides us very quickly to the right formulation.

13.1 Fundamental equations

Suppose we want to construct a field theory with two basic physical elements. These will be fields (whose nature is to be discovered) and material particles. By a field we mean simply something that exerts a force on a particle, and we shall assume that the particles in turn give rise to the field through a property we shall call their charge. For simplicity, we take it that the charge is a scalar invariant. Do not forget that we are in the process of inventing a theory, so we can hypothesize anything we like; we are constrained only by the language of tensors (to maintain covariance) and the policy of simplicity.

We shall therefore further assume that the force is pure: i.e., rest-mass preserving: that is a great simplification if we can achieve it.

Now let us consider whether the 4-force exerted by the field at a given event might depend on anything else in addition to the charge of the particle. Suppose, for example, that it is independent of the particle's 4-velocity U and 4-acceleration $d\mathsf{U}/d\tau$, etc. A field theory can be built from such an assumption, but it is not the one we are looking for because it cannot give rise to a pure force. For a pure force we require $\mathsf{F} \cdot \mathsf{U} = 0$, but if the force is independent of U, then $\mathsf{F} \cdot \mathsf{U}$ can only vanish for all U if F is itself zero. We conclude that we shall need some dependence of F on U.

The next simplest assumption would seem to be that the 4-force is proportional to the charge and to the 4-velocity of the particle, but is independent of its 4-acceleration:

$$\mathsf{F} \overset{?}{=} q\phi\mathsf{U} \tag{13.1}$$

for some scalar field ϕ. This is no good, however: still not pure: $\mathsf{F} \cdot \mathsf{U} = q\phi\mathsf{U} \cdot \mathsf{U} \neq 0$. Next we try

electromagnetic force equation

$$\mathsf{F}^a = q\mathbb{F}^{a\mu}\mathsf{U}_\mu \qquad\qquad [\, \mathsf{F} = q\mathbb{F} \cdot \mathsf{U} \qquad (13.2)$$

where $\mathbb{F}$ is an object that describes the field. It is a second-rank tensor. This is the simplest thing (other than a scalar) that can take a 4-vector as 'input' and give back a 4-vector force. So is the force pure now? We have already given the answer in the discussion following eqn (7.44): the force is pure for all U if $\mathbb{F}$ is antisymmetric (and you can easily prove this condition is necessary as well as sufficient, by finding a U for which $\mathsf{F} \cdot \mathsf{U} \neq 0$ if $\mathbb{F}$ is not antisymmetric). The conclusion is *requiring $\mathbb{F}$ to be antisymmetric is both necessary and sufficient to guarantee a pure force.*

We now know our tensor is antisymmetric. That is good, because this is the least complicated type of second-rank tensor. It can be regarded as being composed of two 3-vectors, so our tensor field can be interpreted as a linked pair 3-vector fields. We have already shown in eqn (7.45) that the spatial part of eqn (13.2) gives the Lorentz force equation, and hence $\mathbb{F}$ is as given in eqn (7.46). We immediately know how the fields transform under a change of reference frame; see eqn (12.16), which gives our old friend eqn (7.13). Note that, as before, we have obtained the field transformation by using the force equation without needing to evoke the field equations.

So far we have established how our field $\mathbb{F}$ affects particles, and we have learned that we can, if we so chose, interpret it as a linked pair of 3-vector fields. It remains to propose how the particles might generate the field. We shall assume that some sort of differential equation is needed, so we take an interest in $\partial_\lambda \mathbb{F}^{\lambda b}$, which is a sort of divergence of the tensor field. This reduces the rank of the object from 2 to 1; it is arguably the simplest differential operator we could use. It is certainly one of the simplest anyway, so let us try it.

We have already proposed that the effect of the field on the particles is proportional to their charges and their velocities. Some sort of general notion of a 'third law' (action and reaction), which we know will be needed to respect momentum conservation, leads us to guess that the particles should in return affect the field also through their charges and their velocities, so we guess

First field equation

$$\partial_\lambda \mathbb{F}^{\lambda b} = -\mu_0 \rho_0 U^b \qquad\qquad [\Box \cdot \mathbb{F} = -\mu_0 \rho_0 U \quad (13.3)$$

where μ_0 is a proportionality constant, and ρ_0 is the proper charge per unit volume: i.e., for any given event it is the charge density in the reference frame in which the local charge is at rest.

So far we only assumed the charge was Lorentz invariant. Our field equation (13.3) gives us something more: it can be valid only if the charge is conserved. This is the well-known connection between the completion of Maxwell's equations and the conservation of charge. To prove it, we investigate the 4-divergence of the 4-vector on the right-hand side of eqn (13.3):

$$\partial_\mu(\rho_0 U^\mu) = \partial_\mu \partial_\lambda \mathbb{F}^{\lambda\mu}$$
$$= \partial_\lambda \partial_\mu \mathbb{F}^{\mu\lambda} \qquad\qquad \text{swap } \lambda, \mu$$
$$= -\partial_\lambda \partial_\mu \mathbb{F}^{\lambda\mu} \qquad\qquad \text{antisymmetric } \mathbb{F}$$
$$= -\partial_\mu \partial_\lambda \mathbb{F}^{\lambda\mu} \qquad \text{commute partial differentiation}$$
$$\implies \partial_\lambda(\rho_0 U^\lambda) = 0.$$

In the first step we simply swapped the indices: this is valid because they are dummy indices (being summed over): we can call them what we like. You can imagine that λ was first changed to σ, then μ to λ, then σ to μ. In the second step we invoked the antisymmetry of $\mathbb{F}$. In the third we invoked the symmetry of second partial derivatives: $\partial_\lambda \partial_\mu f = \partial_\mu \partial_\lambda f$ for any well-behaved scalar f, and thus for all the elements of $\mathbb{F}$. The whole argument is essentially the same as the one leading to eqn (7.45), but now we have $\Box \cdot (\Box \cdot \mathbb{F})$ instead of $U \cdot \mathbb{F} \cdot U$.

Defining the 4-vector $J \equiv \rho_0 U$, we can write the conclusion $\partial_\lambda J^\lambda = 0$. This is the continuity equation (previously we wrote it $\Box \cdot J = 0$), so we have deduced that the quantity whose flow is described by J—i.e., the charge—is conserved!

This conservation law greatly cheers us. In fact, one might argue that for a simple theory one should insist on such a conservation, and this is further evidence that eqn (13.3) is a unique choice: it is the only one that is remotely simple and that is consistent with charge conservation.

We have already seen in eqn (7.49) that eqn (13.3) is the Maxwell equations M1 and M4 in tensor notation.

Eqn (13.3) is our first field equation. It does not yet fully describe the field, because it is only a 4-vector equation: i.e., it contains four equations, while we need six altogether. The problem is that the divergence of a field does not in itself fully characterize the field. The natural next step is to consider the 'curl' of the field—some sort of derivative that would generate a third-rank tensor. There are many possibilities. However, we can keep the problem under control by noticing that we have not yet taken advantage of another feature that can arise in field theories: the concept of a potential. Therefore we shall assume next that $\mathbb{F}$ can be derived from a potential. We can soon convince ourselves that a scalar potential will not suffice, so we try a vector potential A, and propose

$$\mathbb{F}^{ab} = \partial^a \mathsf{A}^b - \partial^b \mathsf{A}^a. \tag{13.4}$$

It follows that

> **Second field equation**
>
> $$\partial^a \mathbb{F}^{bc} + \partial^c \mathbb{F}^{ab} + \partial^b \mathbb{F}^{ca} = 0, \tag{13.5}$$

as you can verify. Now, it can be shown that eqn (13.5) is not only a necessary but also a sufficient condition that $\mathbb{F}$ can be obtained from a 4-potential as in eqn (13.4). Therefore we can describe either of eqn (13.4) or eqn (13.5) as our second field equation. The form of eqn (13.5) represents sixty-four equations, and yet only four of them are independent, so the index notation is introducing a lot of unwanted redundancy. However, there is something attractive about having a set of equations only in terms of $\mathbb{F}$ and the charges, and the tensor technique is still playing its crucial role of guaranteeing Lorentz covariance.

With the benefit of the assumption that $\mathbb{F}$ is completely determined by a 4-vector potential (so there are only four unknowns), now our first field eqn (13.3) becomes sufficient to determine the field. Substituting from eqn (13.4) it becomes

$$\partial_\lambda \partial^\lambda A^b - \partial^b \partial_\lambda A^\lambda = -\mu_0 J^b. \tag{13.6}$$

This is $\square^2 \mathsf{A} - \square(\square \cdot \mathsf{A}) = -\mu_0 \mathsf{J}$, which we previously wrote in component form in eqns (7.26) and (7.27).

We now have a complete theory, consistent unto itself. It remains to extract predictions and compare with experiment, and of course we know well that we shall be richly rewarded with experimental confirmation. The foundational equations are summarized in the box below. We added the equation of motion $\mathsf{F} = d\mathsf{P}/d\tau$ in order to provide a complete story: the field equations say how the fields move, the equation of motion says how the particles move.[1] All of classical physics except gravitation is included in this box!

[1] See section 16.5 for a discussion of whether or not the Lorentz force equation is axiomatic.

Electromagnetic field theory

Force equation

$$F^a = q\mathbb{F}^{a\lambda}U_\lambda$$

(pure force $\Leftrightarrow \mathbb{F}$ is antisymmetric).
Field equations

$$\partial_\lambda \mathbb{F}^{\lambda b} = -\mu_0 J^b \tag{13.7}$$

$$\partial^c \mathbb{F}^{ab} + \partial^a \mathbb{F}^{bc} + \partial^b \mathbb{F}^{ca} = 0. \tag{13.8}$$

The first $\Rightarrow \partial_\lambda J^\lambda = 0$, charge is conserved.
The second $\Leftrightarrow \mathbb{F}^{ab} = \partial^a A^b - \partial^b A^a$, the field can be derived from a potential.
Equation of motion of a test particle

$$m\frac{dU^a}{d\tau} = q\mathbb{F}^{a\lambda}U_\lambda. \tag{13.9}$$

Variations

Variations which give rise to other sensible and reasonably simple theories are mainly of two types. We can give up the requirement of a pure force and try a simpler potential, such as a scalar potential (i.e., a Lorentz scalar, not part of a 4-vector). An example of this is the Yukawa scalar meson theory, which was a forerunner of some aspects of the Standard Model of particle physics. Or we can introduce further sources to gain more symmetry between the electric and magnetic parts, at the expense however of losing the 4-potential.

Consider the 'dual' field tensor defined by

$$\tilde{\mathbb{F}}_{ab} = \frac{1}{2}\epsilon_{ab\mu\nu}\mathbb{F}^{\mu\nu} = \begin{pmatrix} 0 & B_x & B_y & B_x \\ -B_x & 0 & E_z/c & -E_y/c \\ -B_y & -E_z/c & 0 & E_x/c \\ -B_z & +E_y/c & -E_x/c & 0 \end{pmatrix}. \tag{13.10}$$

The second field equation (13.5) can be written

$$\partial_\lambda \tilde{\mathbb{F}}^{\lambda b} = 0, \tag{13.11}$$

i.e., the dual field is 'source free.' This suggests that one natural modification of the Maxwell theory is to introduce a magnetic current density J^b_{mag}, which would represent a density and flux of magnetic monopoles. One replaces the second field equation by $\partial_\lambda \tilde{\mathbb{F}}^{\lambda b} \propto J^b_{mag}$. The magnetic 'charge' of a monopole would be invariant and conserved (like electric charge). This might seem like a modest modification, and one which Nature might have adopted. However, it has profound consequences, because it results in a loss of symmetry under space inversion (parity transformation) and under time reversal. Extensive searches for magnetic monopoles have so far yielded null results, which suggests that they do not exist and Maxwell's theory is the correct one. Nevertheless, such searches will continue, in part because of an ingenious quantum mechanical argument due to Dirac, which suggests that the existence of magnetic monopoles would allow one to infer that electric charge must be quantized. The argument examines the motion of a particle moving in the field of a magnetic monopole, and the quantization condition emerges as a consistency requirement.

The wonderful succinctness of eqs. (13.7) and (13.8) does not mean that the equations are simple: they remain precisely the full Maxwell equations, with all their complexity and richness. However, we have shown that we cannot expect to find anything much simpler than this. Furthermore, the introduction of $\mathbb{F}$ gives us a sense that we are getting to grips with what the electromagnetic field really *is*. It is a 'tensor thing' that exists throughout spacetime. At each event in spacetime there is this four-dimensional 'thing' that looks like two 3-vectors when you pick any given reference frame. (It is four-dimensional in the same sense that a moment of inertia 3-tensor is three-dimensional.) It exerts forces and, as we shall explore in chapter 16, it carries energy and momentum. It may be right to say that it is part and parcel of the structure of spacetime itself, or else that spacetime is 'made of' things like this: this is the type of question that attempts to unify quantum field theory and General Relativity are trying to resolve.

13.2   Invariants of the electromagnetic field

Tensor analysis yields up some fruit straight away. We know we can obtain at least one scalar invariant from any tensor, and from an antisymmetric second-rank tensor we can get two. The first is easy:

$$D \equiv \frac{1}{2}\mathbb{F}_{\mu\nu}\mathbb{F}^{\mu\nu} = B^2 - E^2/c^2 \qquad (13.12)$$

(obtain this by summing the squares of the elements, with a minus sign for the time-space part coming from the presence of $g\mathbb{F}g$). The second is found by using the 'dual' field tensor given in eqn (13.10). Using this we can form the invariant

$$\alpha = \frac{1}{4}\tilde{\mathbb{F}}_{\mu\nu}\mathbb{F}^{\mu\nu} = \mathbf{B}\cdot\mathbf{E}/c. \qquad (13.13)$$

Our two invariants D (for 'difference') and α (for 'angle' or 'alignment') allow some general observations about the fields. For example, if the fields are orthogonal at some event in one reference frame, then they are orthogonal at that event in all reference frames ($\alpha = 0$). If the magnitudes are 'equal' (i.e., $cB = E$) in one frame, then they are in all frames ($D = 0$). If the magnetic field vanishes in one frame then the field cannot be purely electric in another (since $D < 0$), and *vice versa* (when $D > 0$). If the angle between $\mathbf{E}$ and $\mathbf{B}$ is acute in one frame

$(\alpha > 0)$ it cannot be obtuse in another. There can be a frame in which one of the fields vanishes only if $\alpha = 0$: i.e., the fields are orthogonal in other frames.

Supposing $\alpha = 0$, then are we guaranteed to be able to find a frame in which the magnetic field vanishes? Clearly only if $D < 0$, but suppose that it is. Then we can use the field transformation equations (7.13) to find a frame in which $\mathbf{B}' = 0$, as follows.

Let the fields in some frame S be $\mathbf{E}$, $\mathbf{B}$. To get $\mathbf{B}'_{\parallel} = 0$ we need $\mathbf{B}_{\parallel} = 0$ so we are clearly going to have to pick a frame S' with velocity $\mathbf{v}$ perpendicular to $\mathbf{B}$. Then we have $\mathbf{B}_{\perp} = \mathbf{B}$, so

$$\mathbf{B}' = \gamma(\mathbf{B} - \mathbf{v} \wedge \mathbf{E}/c^2).$$

For this to be zero, we need $\mathbf{B}$ and $\mathbf{E}$ to be perpendicular—but they are (if $\alpha = 0$), so there exists a solution: the component of $\mathbf{v}$ perpendicular to $\mathbf{E}$ must be $c^2 B/E$, and the component v_E along $\mathbf{E}$ does not matter. It remains to check that this solution can have $v < c$: we require

$$v_E^2 + c^4 B^2/E^2 < c^2 \quad \Rightarrow \quad v_E^2 E^2 < c^2(E^2 - c^2 B^2)$$

but $D < 0$ by hypothesis, so the right-hand side is positive and there exists a solution for a range of values of v_E. Note that all the frames determined by this analysis have velocity in the plane perpendicular to $\mathbf{B}$ and with the same component of velocity perpendicular to $\mathbf{E}$. This means they have a common direction of motion relative to one another; see Fig. 13.1. The simplest case is where $\mathbf{v}$ is perpendicular to $\mathbf{E}$ (as well as to $\mathbf{B}$), then

$$\mathbf{v} = c^2 \frac{\mathbf{E} \wedge \mathbf{B}}{E^2}, \qquad \mathbf{E}' = \mathbf{E}/\gamma, \;\; \mathbf{B}' = 0. \tag{13.14}$$

A similar argument allows one to find a set of frames in which $\mathbf{E}'$ vanishes if $\alpha = 0$ and $D > 0$. The simplest case is

$$\mathbf{v} = \frac{\mathbf{E} \wedge \mathbf{B}}{B^2}, \qquad \mathbf{B}' = \mathbf{B}/\gamma, \;\; \mathbf{E}' = 0. \tag{13.15}$$

It can also be shown that when $\alpha \neq 0$ there is a continuum of frames in which $\mathbf{E}$ is parallel to $\mathbf{B}$. One such frame moves in the direction $\mathbf{E} \wedge \mathbf{B}$ with speed βc given by the smaller root of the quadratic $\beta^2 - b\beta + 1 = 0$ where $b = (E^2 + c^2 B^2)/|\mathbf{E} \wedge \mathbf{B}c|$.

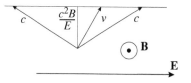

Fig. 13.1 If $\mathbf{E}$ and $\mathbf{B}$ are perpendicular in one frame, with $cB < E$, then there is a set of frames in which $B = 0$. Their velocities $\mathbf{v}$ relative to the first frame are in the plane perpendicular to $\mathbf{B}$, and have the same component in the direction perpendicular to $\mathbf{E}$, therefore they have a common direction of motion relative to one another. Similar statements apply to frames in which $\mathbf{E}$ vanishes when $\mathbf{E} \cdot \mathbf{B} = 0$ and $E < cB$.

13.2.1   Motion of particles in an electromagnetic field

The equation of motion of a particle moving in an electromagnetic field can be written either

$$m\frac{\mathrm{d}}{\mathrm{d}t}(\gamma\mathbf{v}) = q(\mathbf{E} + \mathbf{v} \wedge \mathbf{B}) \tag{13.16}$$

or

$$m\frac{d\mathsf{U}^a}{d\tau} = q\mathbb{F}^a_{\;\lambda}\mathsf{U}^\lambda. \tag{13.17}$$

The first equation offers a way to find the motion as a function of reference frame time, the second as a function of proper time along the worldline.

Static, uniform field

In chapter 4 we treated the motion of a charged particle in a static uniform purely electric field and in a static uniform purely magnetic field, using the three-vector approach. The discussion in the preceding section shows one way to treat the case of a general static uniform field. If the electric and magnetic fields are orthogonal, then first identify a frame in which one or the other is zero, solve the motion in that frame, then transform back. One may as well pick a frame that is moving orthogonally to both the fields, then its speed has to be E/B to null the $\mathbf{E}$ field, or $c^2 B/E$ to null the $\mathbf{B}$ field.

For $\mathbf{E} \cdot \mathbf{B} = 0$, $cB > E$, the motion is helical in the special frame, and therefore a combination of the drift velocity $\mathbf{E} \wedge \mathbf{B}/B^2$ and the (Lorentz transformed) helix in the original frame. Since the drift is perpendicular to $\mathbf{B}$, it is in the plane of the circular motion.

For $\mathbf{E} \cdot \mathbf{B} = 0$, $cB < E$, the motion is hyperbolic (with increasing momentum along the direction $\mathbf{E}'$) in the special frame, and therefore a combination of the drift velocity $c^2 \mathbf{E} \wedge \mathbf{B}/E^2$ and the (Lorentz-transformed) hyperbolic motion the original frame.

If the fields are not orthogonal, then a simplification is obtained by adopting a frame in which they are parallel. However, this does not save much labour compared with the 'brute force' approach of setting out and solving the equations of motion in the original frame in rectangular coordinates.

The motion as a function of proper time is best obtained by solving eqn (13.17). To this end, it is convenient to write the equation using matrix notation:

$$m\frac{d\mathsf{U}}{d\tau} = q(\mathbb{F}g)\mathsf{U}. \tag{13.18}$$

This can be solved by noticing that for a uniform constant field, $\mathbb{F}$, is independent of space and time, and therefore the equation is precisely the same as the one obtained in a classical normal modes problem, and can be solved by the same methods. One proposes a solution of the form $\mathsf{U} = \mathsf{U}_0 \exp(\Gamma\tau)$, where U_0 is a constant 4-vector, and then the equation becomes an eigenvalue equation, with eigenvalues $\lambda = m\Gamma/q$.

We are looking for eigenvalues and eigenvectors of the matrix $(\mathbb{F}g)$. This matrix is is not symmetric so the right-eigenvectors are not the same as the left-eigenvectors. We only need the right-eigenvectors here. Without loss of generality we can take the z-axis along $\mathbf{B}$ and $\mathbf{E}$ in the xz plane. One finds that the eigenvalues are

$$\lambda^2 = -\frac{D}{2} \pm \sqrt{D^2/4 + \alpha^2} \tag{13.19}$$

where D and α are the invariants defined in eqns (13.12) and (13.13). Using these values one can find the four corresponding eigenvectors $\mathsf{U}_{0(i)}$, and the general solution is

$$\mathsf{U}(\tau) = \sum_{i=1}^{4} a_i \mathsf{U}_{0(i)} \, e^{q\lambda_i \tau / m} \qquad (13.20)$$

where a_i are constant coefficients that are given by the initial conditions. Note that it is allowable to include eigenvectors $\mathsf{U}_{0(i)}$ that would be unphysical on their own (for example, having $v > c$), as long as the solution 4-velocity $\mathsf{U}(\tau)$ is physical at all times.

When $\alpha = 0$ (orthogonal fields) there is a zero eigenvalue. For $E < cB$ this is the well-known case $\mathsf{U} =$ constant when $\mathbf{f} = 0$; for $E > cB$ it corresponds to a solution with $v > c$, which is unphysical on its own, but may be needed as part of an allowed solution.

Arbitrary field

To treat more general problems, one may use Lagrangian methods to find constants of the motion (chapter 14), or simply write down and try to integrate eqns (13.16) or (13.17). It is often useful to treat the former using eqn (4.13), which give

$$m_0 \gamma^3 \frac{d^2 \mathbf{r}_\|}{dt^2} = q\mathbf{E}_\|, \qquad m_0 \gamma \frac{d^2 \mathbf{r}_\perp}{dt^2} = q(\mathbf{E}_\perp + (\mathbf{v} \wedge \mathbf{B})_\perp), \quad (13.21)$$

not forgetting that the parallel and perpendicular directions change when $\mathbf{v}$ rotates. If the moving body radiates significantly, these equations are still valid but the electromagnetic field will not be known at the outset, and it is necessary to reconsider whether or not a point-like model for the body is appropriate. For a non-point-like body the charge on one part of the body experiences a field which is in part produced by the charge on another part of the body, and when the body accelerates the sum of such interactions over the body does not in general cancel (see exercise 8.5 of chapter 8). This *self-force* or *radiation reaction* will be discussed in volume 2.

Exercises

(13.1) **Change of metric.** The form of some of the basic equations of electromagnetism depends on which Minkowski metric $((+ - - -)$ or $(- + + +))$ is assumed. The effect of a change of metric on equations using index notation is summarized by the rule 'introduce one sign change for each index in the "sensitive" place', where a 'sensitive' place is down for an ordinary vector, and up for the gradient operator. Thus A_a and ∂^a change sign, A^a and ∂_a do not. However, thought is required because the presence of a gradient operator may be hidden, as may the presence of $\mathsf{U}_\lambda \mathsf{U}^\lambda$ (which becomes either $-c^2$ or c^2). Also some equations serve as definitions which are independent of the metric. For example, we take the force equation (13.2) to be the definition of the field tensor $\mathbb{F}$ no matter which metric is adopted.

Confirm that the main changes required if $(+ - - -)$ is adopted instead of $(- + + +)$ are as follows:

(i) the relation of $\mathbb{F}$ to the electric and magnetic fields changes sign;

(ii) $\partial_\lambda F^{\lambda b} = \mu_0 J^b$. Thus, for given $(\mathbf{E}$ and $\mathbf{B})$ fields, $\mathbb{F}$ and g_{ab} change sign.

(13.2) Derive eqn (13.11): e.g., by using Maxwell's equations.

(13.3) Show that $\det(\mathbb{F}) = (\mathbf{E} \cdot \mathbf{B})^2/c^2$. (Hint: save yourself a lot of trouble by choosing a coordinate axis direction in a helpful way.) Hence show that $\det(\mathbb{F})$ is Lorentz-invariant.

(13.4) §Introduce the complex 3-vector $\mathbf{K} \equiv \mathbf{E} + ic\mathbf{B}$. Show from the known transformation properties of $\mathbf{E}$ and $\mathbf{B}$ that this transforms as

$$\mathbf{K}' = \begin{pmatrix} 1 & 0 & 0 \\ 0 & \cosh\rho & i\sinh\rho \\ 0 & -i\sinh\rho & \cosh\rho \end{pmatrix} \mathbf{K}. \quad (13.22)$$

for frames in standard configuration. This is a rotation through a complex angle. Deduce that $E^2 - c^2 B^2$ and $\mathbf{E} \cdot \mathbf{B}$ are invariants, and that no other invariants can be made from $\mathbf{E}$ and $\mathbf{B}$.

(13.5) If $(\mathbf{E}, \mathbf{B})$ and $(\mathbf{E}', \mathbf{B}')$ are two different electromagnetic fields, prove that $\mathbf{E} \cdot \mathbf{E}' - c^2 \mathbf{B} \cdot \mathbf{B}'$ and $\mathbf{E} \cdot \mathbf{B}' + \mathbf{B} \cdot \mathbf{E}'$ are invariants.

(13.6) §In a certain frame there is a uniform electric field E in the y direction and a uniform magnetic field $B = 5E/3c$ in the z direction. A particle of charge to mass ratio q/m is released from rest on the x axis. Show that the particle returns to the x axis after a time $37\pi mc/16qE$. (Hint: consider the situation in a frame where one field vanishes.)

(13.7) Prove the statement about frames with parallel fields made after eqn (13.15).

(13.8) Derive eqn (13.19) for a particle moving in a constant uniform electromagnetic field. Consider the case $\alpha = 0$. What does this tells us about the fields? Interpret the solution corresponding to a zero eigenvalue. Find $U(\tau)$ for a particle initially at rest in a uniform purely electric field, and for a particle moving in a plane perpendicular to a uniform purely magnetic field. (Hint: first show that the right eigenvectors of the matrix

$$\begin{pmatrix} 0 & 1 \\ -1 & 0 \end{pmatrix} \text{ may be written } \begin{pmatrix} 1 \\ i \end{pmatrix}, \begin{pmatrix} i \\ 1 \end{pmatrix}.)$$

(13.9) Physically interpret the tensor $M^{ab} \equiv X^a J^b - X^b J^a$ where J is the electric 4-current density in some region of space. Hence deduce that *polarization* and *magnetization* (density of electric and magnetic dipole moment) form the components of an antisymmetric second rank tensor, and transform in the same way as electromagnetic fields with the replacement $\mathbf{E} \to \mathbf{P}, \mathbf{B} \to \mathbf{M}$.

Lagrangian mechanics

14

It is assumed that the reader has met the Principle of Least Action in classical mechanics, and the related concepts of the Lagrangian, the Hamiltonian, and the Euler–Lagrange equations. In this chapter we shall examine their Special Relativisitic generalisation. We begin with a summary of the classical results, both as a reminder, and to introduce notation.

14.1 Classical Lagrangian mechanics

Students usually first meet classical mechanics in the setting of Newton's laws, and the formula

$$\mathbf{f} = \frac{\mathrm{d}\mathbf{p}}{\mathrm{d}t},$$

which we shall write in the form

$$-\boldsymbol{\nabla}V = m\frac{\mathrm{d}\dot{\mathbf{x}}}{\mathrm{d}t}.$$

The basic idea of Lagrangian mechanics is to replace this vector treatment by a treatment based on a scalar quantity called the Lagrangian, which allows vector equations to be extracted by taking derivatives (just as $\boldsymbol{\nabla}V$ is a vector extracted from the potential energy V). This approach proves to be more flexible and it simplifies many problems in mechanics.

At any given instant of time, the state of a physical system is described by a set of n variables q_i called *coordinates*, and their time-derivatives $\dot{q}_i$ called *velocities*. For example, these could be the positions and velocities of a set of particles making up the system, though later we shall allow a more general notion of a coordinate.

Define a function $\mathcal{L}$ called the *Lagrangian*, given by

$$\mathcal{L} = T - V \tag{14.1}$$

where T and V are the kinetic energy and potential energies of the system. The Lagrangian is therefore a function of the positions and velocities, and it can be a function of time. This is indicated by the notation $\mathcal{L} = \mathcal{L}(\{q_i\}, \{\dot{q}_i\}, t)$, which we shall abbreviate to $\mathcal{L} = \mathcal{L}(q, \dot{q}, t)$.

For particle motion with no external time-dependent fields, the Lagrangian has no explicit dependance on time. The phrase 'no explicit dependence on time' means it has no dependence on time over and above that which is already implied by the fact that q and $\dot{q}$ may depend on time. For example, a single particle undergoing simple harmonic motion has the Lagrangian $\mathcal{L} = (1/2)m(\dot{x}^2 - \omega^2 x^2)$. An example motion of the particle is $x = x_0 \sin(\omega t)$, and for this motion the Lagrangian can also be written $(m\omega^2 x_0^2/2)\cos 2\omega t$—a function of time. However, the latter form hides the dependence on x and $\dot{x}$, which is what we are chiefly interested in, and furthermore in general the Lagrangian cannot be deduced from the motion, but the motion can be deduced from the Lagrangian when the latter is written as a function of coordinates and velocities. For this reason the variables $\{q_i, \dot{q}_i\}$ are said to be the 'natural' or 'proper' variables of $\mathcal{L}$. (A similar issue arises in the treatment of functions of state in thermodynamics.)

The time integral of the Lagrangian along a path $q(t)$ is called the *action S*:

$$S[q(t)] = \int_{q_1,t_1}^{q_2,t_2} \mathcal{L}(q, \dot{q}, t)\mathrm{d}t \tag{14.2}$$

The *Principle of Least Action* states that the path followed by the system is the one that gives an extreme value (maximum or minimum) of S with respect to small changes in the path. (The title 'Least' action comes from the fact that in practice a minimum is more usual than a maximum.) The path is to be taken between given starting and finishing 'positions' q_1, q_2 at times t_1, t_2.

To find the extremum of S we need to ask for a zero derivative with respect to changes in all the variables describing the path. The calculus of variations may be used to show that the result is that S reaches an extremum for the path satisfying

Euler–Lagrange equations

$$\frac{\mathrm{d}}{\mathrm{d}t}\left(\frac{\partial \mathcal{L}}{\partial \dot{q}_i}\right) = \frac{\partial \mathcal{L}}{\partial q_i}. \tag{14.3}$$

The physical interpretation of this set of equations is found by discovering its implications. The end result of such a study may be summarized:

$$\frac{\mathrm{d}}{\mathrm{d}t} \qquad \left(\frac{\partial \mathcal{L}}{\partial \dot{q}}\right) \qquad = \qquad \left(\frac{\partial \mathcal{L}}{\partial q}\right)$$
(rate of change of)   ('momentum') = ('force')

The 'force' here is called a *generalized force*, and the 'momentum' is called *canonical momentum*, defined by

$$\tilde{p}_i \equiv \left(\frac{\partial \mathcal{L}}{\partial \dot{q}_i}\right). \tag{14.4}$$

In the simplest cases such as motion of a free particle, or a particle subject to conservative forces, the canonical momentum may be equal to a familiar momentum such as linear momentum or angular momentum, but this does not have to happen. A counter-example occurs for the motion of a particle in a magnetic field, as we shall see.

The *Hamiltonian* of a system is defined as

$$\mathcal{H}(q, \tilde{p}, t) \equiv \left(\sum_i^n \tilde{p}_i \dot{q}_i \right) - \mathcal{L}(q, \dot{q}, t) \qquad (14.5)$$

where the $\dot{q}_i$ are to be written as functions of the q_i and $\tilde{p}_i$, so that the result is a function of coordinates and canonical momenta (the natural variables of the Hamiltonian). For conservative forces one finds that the sum in eqn (14.5) evaluates to twice the kinetic energy, and then $\mathcal{H} = T + V$, which is clearly the total energy of the system.

The Euler–Lagrange equations imply

Hamilton's canonical equations

$$\frac{\mathrm{d}q_i}{\mathrm{d}t} = \frac{\partial \mathcal{H}}{\partial \tilde{p}_i}, \qquad \frac{\mathrm{d}\tilde{p}_i}{\mathrm{d}t} = -\frac{\partial \mathcal{H}}{\partial q_i}. \qquad (14.6)$$

Thus the Hamiltonian with the canonical equations offer an alternative to the Lagrangian with the Euler–Lagrange equations. In practice, both are useful.

14.2   Relativistic motion

In generalizing Lagrangian mechanics to Special Relativity, we shall proceed in two steps. First we ask the question: are the Euler–Lagrange equations (and their counterparts, the canonical equations) still valid? The answer is *yes, as long we use the right Lagrangian*. However, such a formulation is only partially useful. It can correctly generate 3-vector equations such as $-\boldsymbol{\nabla}V = \gamma m\mathbf{v}$, but it does not immediately give the 4-force. Therefore the second step will be to reconsider the action and Lagrangian from a more thoroughly 'four-dimensional' (spacetime) point of view.

14.2.1   From classical Euler–Lagrange

First we consider the argument based on the classical formula for the action, eqn (14.2). We restrict attention to a single particle, and write the Lagrangian

$$\mathcal{L} = \mathcal{L}_{\text{free}} + \mathcal{L}_{\text{int}} \qquad (14.7)$$

where $\mathcal{L}_{\text{free}}$ is the Lagrangian for a free particle, and $\mathcal{L}_{\text{int}}$ is the part describing interaction with something else such as an electromagnetic field.

For a single particle the complete path of the system (i.e., the specification of $q_i(t)$ for all the coordinates) is simply the worldline of the particle. In this case it is straightforward to write the action integral as an integral with respect to proper time τ along the worldline:

$$S[q(t)] = \int_{q_1, t_1}^{q_2, t_2} \mathcal{L}(q, \dot{q}, t)\mathrm{d}t = \int_{(1)}^{(2)} \mathcal{L}\gamma \mathrm{d}\tau \qquad (14.8)$$

where we have used the by now familiar $dt/d\tau = \gamma$. We already know an important property of free motion: it maximizes the proper time. This suggests that the Lagrangian for free motion should be such that $\gamma \mathcal{L}_{\text{free}}$ is a constant. With this hint, we propose

$$\mathcal{L}_{\text{free}} = -mc^2/\gamma = -mc^2(1 - v^2/c^2)^{1/2}. \qquad (14.9)$$

You can check that this gives the canonical momenta $\partial \mathcal{L}_{\text{free}}/\partial v_i = \gamma m v_i$: i.e., the three components of the relativistic 3-momentum.

Next let us treat the case of electromagnetic interactions. We propose (or guess) the interaction term $\mathcal{L}_{\text{int}}$ and then prove that it gives the right equation of motion of the particle. Consider, then,

$$\mathcal{L}_{\text{int}} = q\mathsf{U} \cdot \mathsf{A}/\gamma = q(-\phi + \mathbf{v} \cdot \mathbf{A}). \qquad (14.10)$$

After adding this to $\mathcal{L}_{\text{free}}$ one obtains the three canonical momenta

$$\frac{\partial \mathcal{L}}{\partial v_i} = \gamma m v_i + q A_i \qquad (14.11)$$

which can be expressed as

$$\tilde{\mathbf{p}} = \gamma m \mathbf{v} + q\mathbf{A}. \qquad (14.12)$$

This equation commonly causes confusion. It does *not* mean that the momentum of the particle has changed. The *momentum* (i.e., that which is conserved in collisions and influenced by forces) is still $\gamma m \mathbf{v}$. The *canonical momentum* (i.e., that which has a rate of change given by the gradient of $\mathcal{L}$) is $\gamma m \mathbf{v} + q\mathbf{A}$.

Now write the Euler–Lagrange equations:

$$\frac{\mathrm{d}}{\mathrm{d}t}(\gamma m \mathbf{v} + q\mathbf{A}) = q\left(-\boldsymbol{\nabla}\phi + \boldsymbol{\nabla}(\mathbf{v} \cdot \mathbf{A})\right) \qquad (14.13)$$

The $d\mathbf{A}/dt$ term on the left has two parts, because a change in $\mathbf{A}$ along the worldline is made of the time change of the field, plus a part owing to the fact that the moving particle visits a different place:

$$\frac{\mathrm{d}\mathbf{A}}{\mathrm{d}t} = \frac{\partial \mathbf{A}}{\partial t} + (\mathbf{v} \cdot \boldsymbol{\nabla})\mathbf{A} \qquad (14.14)$$

(see eqn (14.25)). Substituting this in eqn (14.13) gives

$$\frac{\mathrm{d}}{\mathrm{d}t}(\gamma m \mathbf{v}) = -q\left(\boldsymbol{\nabla}\phi + \frac{\partial \mathbf{A}}{\partial t}\right) + q\left(\boldsymbol{\nabla}(\mathbf{v} \cdot \mathbf{A}) - (\mathbf{v} \cdot \boldsymbol{\nabla})\mathbf{A}\right)$$

$$= q(\mathbf{E} + \mathbf{v} \wedge \mathbf{B}) \qquad (14.15)$$

where we have used the vector identity

$$\mathbf{v} \wedge (\boldsymbol{\nabla} \wedge \mathbf{A}) = \boldsymbol{\nabla}(\mathbf{v} \cdot \mathbf{A}) - (\mathbf{v} \cdot \boldsymbol{\nabla})\mathbf{A}.$$

Eqn (14.15) is the correct equation for relativistic motion in an electromagnetic field, so we have confirmed that our choice of Lagrangian was correct and also that the Euler–Lagrange equations are valid as they are: they do not need to be modified, and they take the same form in all inertial frames of reference. They are covariant, but not manifestly covariant. The only drawback of the present approach is that one must pick a frame of reference before starting the calculation of the motion in any given case. In practice the mathematics is often easier if one does that anyway, so it is not much of a drawback. Nevertheless, we should like to see, if we can, a frame-independent formulation: i.e., a manifestly covariant formulation. That is the subject of the next section.

The Hamiltonian is obtained from eqn (14.5). We have

$$\mathcal{H} = (\gamma m\mathbf{v} + q\mathbf{A}) \cdot \mathbf{v} + \frac{mc^2}{\gamma} + q(\phi - \mathbf{v} \cdot \mathbf{A}) = \gamma mc^2 + q\phi.$$

This is what one might expect: a sum of motional energy and potential energy. However, we should express the result in terms of the canonical momenta. To this end, use $\gamma mc^2 = (m^2c^4 + p^2c^2)^{1/2}$, where $\mathbf{p} = \gamma m\mathbf{v} = \tilde{\mathbf{p}} - q\mathbf{A}$. We find

$$\mathcal{H} = \left((\tilde{\mathbf{p}} - q\mathbf{A})^2 c^2 + m^2c^4\right)^{1/2} + q\phi. \tag{14.16}$$

14.2.2   Manifestly covariant

The 'problem' with the Lagrangian presented in eqns (14.9) and (14.10) is that it is not a Lorentz scalar. However, it gives a hint to what Lorentz scalar Lagrangian we could try:

$$\mathcal{L}(\mathsf{X}, \mathsf{U}) = -mc(-\mathsf{U} \cdot \mathsf{U})^{1/2} + q\mathsf{U} \cdot \mathsf{A}. \tag{14.17}$$

We use this in the action integral

$$S[\mathsf{X}(\tau)] = \int_{(\mathsf{X}_1)}^{(\mathsf{X}_2)} \mathcal{L}(\mathsf{X}, \mathsf{U}, \tau)\mathrm{d}\tau \tag{14.18}$$

which is also a Lorentz scalar.

The inclusion of eqn $\mathsf{U} \cdot \mathsf{U}$ in eqn (14.17) raises a subtle point that merits a comment. We know that the velocity is a 'unit vector' with $\mathsf{U} \cdot \mathsf{U} = -c^2$, so why not write $\mathcal{L} = -mc^2 + q\mathsf{U} \cdot \mathsf{A}$? The problem with this version is that when substituted into the relativistic Euler–Lagrange equations it does not result in the correct equation of motion. We have lost the information about the kinetic energy of the particle. One can get around this problem in more than one way, but the most convenient is to insist on the form $mc(-\mathsf{U} \cdot \mathsf{U})^{1/2}$ and keep in mind that the Lagrangian is not to be regarded as a property of the particle, but as a function whose 'job' is to tell us how the action changes if there are changes in the path. We shall comment further on this at the end.

One way to handle the minimisation of the action (14.18) is to change variables back to t in the integral, and then look for a minimum with respect to variations in the path. It is immediately clear that we shall regain the same Euler–Lagrange equations as before, and the same equations of motion. Nonetheless, we shall pursue the manifestly covariant formulation a little further, to see if we can learn anything new.

Use of a parameter to minimize the action. There is an important difference between eqns (14.18) and (14.2), although they appear at first glance to be similar. The difference is that in eqn (14.2) we know from the outset the values of t_1, t_2 (as well as q_1, q_2) at the beginning and end of any path. This is important, as the variational calculation requires that they are fixed: i.e., the same for all paths. In eqn (14.18), as it stands, the end points are defined by two events, but the value of the integration variable τ at those events will will be different from one path to another. Therefore the calculus of variations cannot be applied to the integral as it stands. This situation is handled by introducing a parameter λ that increases monotonically along the path, and whose start and end values can be fixed at some λ_1 and λ_2. The action integral then reads

$$\int \mathcal{L}(\mathsf{X}, \dot{\mathsf{X}}, \tau)\mathrm{d}\tau = \int_{\lambda_1}^{\lambda_2} \mathcal{L}\frac{\mathrm{d}\tau}{\mathrm{d}\lambda}\mathrm{d}\lambda.$$

This version has fixed limits, and now the Lagrangian is

$$\tilde{\mathcal{L}} = \mathcal{L}\frac{\mathrm{d}\tau}{\mathrm{d}\lambda}$$

where $\tilde{\mathcal{L}}$ should be written and treated as a function of X and $\mathrm{d}\mathsf{X}/\mathrm{d}\lambda$. In our case we have $\mathrm{d}\tau^2 = \mathrm{d}t^2 - (\mathrm{d}x^2 + \mathrm{d}y^2 + \mathrm{d}z^2)/c^2$ so

$$\frac{\mathrm{d}\tau}{\mathrm{d}\lambda} = \frac{1}{c}\left(-g_{\mu\nu}\frac{\mathrm{d}\mathsf{X}^\mu}{\mathrm{d}\lambda}\frac{\mathrm{d}\mathsf{X}^\nu}{\mathrm{d}\lambda}\right)^{1/2}. \tag{14.19}$$

The minimization procedure can now go through, and we have the Euler-Lagrange equations

$$\frac{\mathrm{d}}{\mathrm{d}\lambda}\frac{\partial\tilde{\mathcal{L}}}{\partial\dot{\mathsf{X}}^a} = \frac{\partial\tilde{\mathcal{L}}}{\partial\mathsf{X}^a}, \tag{14.20}$$

where the dot signifies $\mathrm{d}/\mathrm{d}\lambda$. Owing to the presence of $\mathrm{d}\tau/\mathrm{d}\lambda$ the new Lagrangian looks rather cumbersome, but fortunately, by a good choice of the parameter λ, we can now simplify the equations. One possible choice is to define λ as the value of τ *along the solution worldline*. For that worldline, and for that worldline only (but it is the only one we are interested in from now on), we must then find $\mathrm{d}\tau/\mathrm{d}\lambda = 1$ and $\tilde{\mathcal{L}} = \mathcal{L}$ and $\dot{\mathsf{X}}^a = \mathsf{U}^a$. Then the Euler–Lagrange equations become the very equations (14.21) that we would have written had we been ignorant of this issue!

There is one limitation to this 'trick', however. If the original Lagrangian has no dependence on one of the variables *and* its velocity, then the set (14.21) will include an equation reading $0 = 0$, which is true but not helpful. Then we must return to eqn (14.20) and make some other choice of λ. For example, setting λ equal to one of the variables X^a is often a good choice.

By minimizing the action with respect to variations of the worldline (see box), one finds the manifestly covariant Euler–Lagrange equations

$$\frac{\mathrm{d}}{\mathrm{d}\tau}\frac{\partial \mathcal{L}}{\partial \mathsf{U}^a} = \frac{\partial \mathcal{L}}{\partial \mathsf{X}^a}. \tag{14.21}$$

Now we extract the relativistic canonical momentum

$$\frac{\partial \mathcal{L}}{\partial \mathsf{U}^a} = \frac{mc}{(-\mathsf{U}\cdot\mathsf{U})^{1/2}}\mathsf{U}_a + q\mathsf{A}_a \tag{14.22}$$

where we have used

$$\frac{\partial}{\partial \mathsf{U}^a}(\mathsf{U}\cdot\mathsf{U}) = \frac{\partial}{\partial \mathsf{U}^a}(\mathsf{U}^\lambda g_{\lambda\mu}\mathsf{U}^\mu) = \delta^\lambda_a g_{\lambda\mu}\mathsf{U}^\mu + \mathsf{U}^\lambda g_{\lambda\mu}\delta^\mu_a$$

$$= g_{a\mu}\mathsf{U}^\mu + \mathsf{U}^\lambda g_{\lambda a} = 2\mathsf{U}_a$$

and have assumed A is independent of U. (A is the potential *experienced by* the particle, not the one produced by the particle.)

Now we can safely replace $\mathsf{U}\cdot\mathsf{U}$ by $-c^2$ because we no longer need partial derivatives of this quantity with respect to components of U, so we find

$$\tilde{\mathsf{P}}_a \equiv \frac{\partial \mathcal{L}}{\partial \mathsf{U}^a} = m\mathsf{U}_a + q\mathsf{A}_a. \tag{14.23}$$

(compare with eqn (14.12)).

Evaluation of $d\mathsf{A}/d\tau$

For any function that depends on position and time, we may write

$$\mathrm{d}f = \left(\frac{\partial f}{\partial t}\right)_{x,y,z}\mathrm{d}t + \left(\frac{\partial f}{\partial x}\right)_{t,y,z}\mathrm{d}x + \left(\frac{\partial f}{\partial y}\right)_{t,x,z}\mathrm{d}y$$

$$+ \left(\frac{\partial f}{\partial z}\right)_{t,x,y}\mathrm{d}z$$

$$\Rightarrow \quad \frac{\mathrm{d}f}{\mathrm{d}\tau} = \left(\frac{\partial f}{\partial t}\right)\frac{\mathrm{d}t}{\mathrm{d}\tau} + \left(\frac{\partial f}{\partial x}\right)\frac{\mathrm{d}x}{\mathrm{d}\tau} + \left(\frac{\partial f}{\partial y}\right)\frac{\mathrm{d}y}{\mathrm{d}\tau} + \left(\frac{\partial f}{\partial z}\right)\frac{\mathrm{d}z}{\mathrm{d}\tau}$$

$$= (\partial_\lambda f)\frac{\mathrm{d}x^\lambda}{\mathrm{d}\tau}.$$

Since this result applies to all f, we may write

$$\frac{\mathrm{d}}{\mathrm{d}\tau} = \frac{\mathrm{d}x^\lambda}{\mathrm{d}\tau}\partial_\lambda \tag{14.24}$$

and this may be applied to all the components of any tensor. For example,

$$\frac{\mathrm{d}\mathsf{A}}{\mathrm{d}\tau} = \frac{\mathrm{d}x^\lambda}{\mathrm{d}\tau}\partial_\lambda\mathsf{A} = \mathsf{U}^\lambda\partial_\lambda\mathsf{A}. \tag{14.25}$$

Notice that our manifestly covariant Lagrangian (14.17) differs by a factor γ from the one we used in the previous section, yet we obtain the same canonical momentum: $\tilde{\mathsf{p}}$ in eqn (14.12) is the spatial part of $\tilde{\mathsf{P}}$. The reason is that the relation between Lagrangian and action is different: in the first case we had an integral with respect to reference frame time t, now we have an integral with respect to proper time τ. This resulted in a different set of Euler–Lagrange equations: (14.21) instead of (14.3). To confirm the agreement between the two approaches, one can manipulate eqn (14.21), replacing $\mathrm{d}/\mathrm{d}\tau$ on the left-hand side by $(\mathrm{d}t/\mathrm{d}\tau)\mathrm{d}/\mathrm{d}t = \gamma\mathrm{d}/\mathrm{d}t$, and writing $\mathsf{U}^a = \gamma\mathrm{d}\mathsf{X}^a/\mathrm{d}t$, then one regains eqn (14.3) as long as one makes the replacement $\mathcal{L} \to \mathcal{L}/\gamma$.

The right-hand side of the Euler–Lagrange equation (14.21) is $q\partial_a(\mathsf{U}^\lambda\mathsf{A}_\lambda) = q\mathsf{U}^\lambda\partial_a\mathsf{A}_\lambda$, so the equation reads

$$\frac{\mathrm{d}}{\mathrm{d}\tau}\left(m\mathsf{U}_a + q\mathsf{A}_a\right) = q\mathsf{U}^\lambda\partial_a\mathsf{A}_\lambda.$$

This is like eqn (14.13). Now use

$$\frac{\mathrm{d}\mathsf{A}_a}{\mathrm{d}\tau} = U^\lambda\partial_\lambda\mathsf{A}_a$$

(see eqns (14.25) and (14.14)), giving

$$m\frac{\mathrm{d}\mathsf{U}_a}{\mathrm{d}\tau} = q\left((\partial_a\mathsf{A}_\lambda) - (\partial_\lambda\mathsf{A}_a)\right)\mathsf{U}^\lambda \tag{14.26}$$

$$\text{or} \qquad \frac{\mathrm{d}\mathsf{P}}{\mathrm{d}\tau} = q(\square \wedge \mathsf{A}) \cdot \mathsf{U} \tag{14.27}$$

We have found that the 4-force associated with the potential A is $q(\square \wedge \mathsf{A}) \cdot \mathsf{U}$. You can verify that this gives once again the correct equation for motion in an electromagnetic field, or else just recognize from eqn (13.4) that we have the equation of motion under the Lorentz force, eqn (13.9).

Further comment on $\mathsf{U} \cdot \mathsf{U}$

Since the combination $\mathsf{U} \cdot \mathsf{U} = -c^2$, one may wish to adopt the Lagrangian $-mc^2$ for free motion. This can be done, but then the information that $\mathsf{U} \cdot \mathsf{U} = -c^2$ has to be incorporated into the action minimisation procedure. One has a *constrained* minimization.

Keeping $\mathsf{U} \cdot \mathsf{U}$ in the Lagrangian leads to an easier solution, but one may be uneasy about the meaning of terms such as $\partial\mathcal{L}/\partial\mathsf{U}^a$, because this quantity refers to a change in the Lagrangian when one component of U is changed *while keeping other components of* U *fixed*. One might argue that it is not possible to change one component of a 4-velocity while keeping all the other components fixed. If one component changes on its own, the size of the 4-velocity will change. To maintain the size fixed, another component must change to compensate.

This objection muddles two different things: namely, path variations considered in the calculus of variations, and the evolution actually followed by the system. Consider a more familiar and simpler example:

classical motion in a circle. When a particle moves in a fixed uniform magnetic field, the speed remains constant. Therefore, throughout the motion, changes in v_x are accompanied by changes in v_y, with the result that $(v_x^2 + v_y^2)$ is independent of time (for a B field in the z direction). However, this does not mean that it is illegal to consider $\partial \mathcal{L} / \partial v_x$ or $\partial \mathcal{L} / \partial v_y$. By considering the effect of such 'excursions' while minimizing the action, one arrives at the very equation (Euler–Lagrange) which ensures that the v_x and v_y changes are coupled in the right way. Similarly, in the relativistic case, one may postpone applying the constraint on the size of U, because after the whole procedure yields a prediction for the worldline, one finds that the worldline satisfies the constraint anyway!

14.3   Conservation

Lagrangian mechanics reveals an important connection between symmetry and conservation. The Euler–Lagrange equations show that if $\mathcal{L}$ is independent of a coordinate, then the corresponding canonical momentum is constant in time: i.e., conserved. It is very reasonable to postulate that the Lagrangian describing an isolated body ought to be independent of the position of that body relative to other bodies. For, if the body is isolated—i.e., not interacting with anything else—then who cares where it is? For a Lagrangian having this symmetry (independence of translation in space), the corresponding canonical momentum is the total linear momentum. It follows that we can replace the conservation-of-momentum postulate introduced in section 1.2.1 by a (very reasonable) symmetry postulate, namely that $\mathcal{L}$ must be translation-invariant for isolated systems. Conservation of energy then follows, as we showed in chapter 5, or it can be obtained from time-independence of the Lagrangian.

14.4   Equation of motion in General Relativity*

Since the equation of motion of a test particle in GR is given by the Principle of Most Proper Time (equivalently, a time-like geodesic), it can be conveniently treated using Euler–Lagrange equations. We wish to find a path having a maximum value of

$$\tau = \int_{(1)}^{(2)} \mathrm{d}\tau = \int_{\lambda_1}^{\lambda_2} \frac{\mathrm{d}\tau}{\mathrm{d}\lambda} \mathrm{d}\lambda$$

where λ is a parameter increasing monotonically along the worldline, and $\mathrm{d}\tau / \mathrm{d}\lambda$ is given by eqn (14.19) in which g_{ab} is now the GR metric tensor. The Lagrangian is clearly

$$\mathcal{L} = (1/c) \left(-g_{\mu\nu} \dot{x}^{\mu} \dot{x}^{\nu} \right)^{1/2}$$

where x^a is the position 4-vector of the particle. Adopting the 'trick' explained in the box, *after* completing the variational calculation we choose λ equal to the value of τ *along the solution worldline*. Then the Lagrangian can also be written $\mathcal{L} = (1/c)(-U \cdot U)^{1/2}$; see eqn (14.17). For this choice we must find $\mathcal{L} = 1$ at points on the path, since the path length is $\int d\tau = \int \mathcal{L} d\lambda$. The property "$\mathcal{L} = 1$" is a statement about the value of the function $\mathcal{L}(x^a, \dot{x}^a)$ at a certain locus of events, it says nothing about derivatives with respect to the coordinates and velocities, which need not be zero. However, it does imply that, along the solution path, $d\mathcal{L}/d\tau = 0$.

With this in mind let us consider a different variational problem: namely, one in which the Lagrangian is

$$\tilde{\mathcal{L}} \equiv -g_{\mu\nu} \dot{x}^{\mu} \dot{x}^{\nu} = c^2 \mathcal{L}^2. \tag{14.28}$$

We then have the Euler–Lagrange equations

$$\frac{d}{d\tau} \left(\frac{\partial \tilde{\mathcal{L}}}{\partial \dot{x}^a} \right) = \frac{\partial \tilde{\mathcal{L}}}{\partial x^a} \tag{14.29}$$

$$\Leftrightarrow \quad \frac{d}{d\tau} \left(2\mathcal{L} \frac{\partial \mathcal{L}}{\partial \dot{x}^a} \right) = 2\mathcal{L} \frac{\partial \mathcal{L}}{\partial x^a}$$

$$\Leftrightarrow \quad \frac{d}{d\tau} \left(\frac{\partial \mathcal{L}}{\partial \dot{x}^a} \right) = \frac{\partial \mathcal{L}}{\partial x^a},$$

where the last step used $d\mathcal{L}/d\tau = 0$. We thus find that the new Lagrangian $\tilde{\mathcal{L}}$ yields Euler–Lagrange equations that are satisfied if and only if the Euler–Lagrange equations for $\mathcal{L}$ are satisfied. Since $g_{\mu\nu} \dot{x}^{\mu} \dot{x}^{\nu}$ (which can also be written $U \cdot U$) is considerably simpler to work with than its square root, we much prefer $\tilde{\mathcal{L}}$ to $\mathcal{L}$, so we adopt it to treat geodesics in GR (see exercises 14.3 and 14.4).

For example, if the metric is of the static form

$$ds^2 = -e^{2\Phi/c^2} c^2 dt^2 + d\sigma^2$$

where the spatial part $d\sigma^2$ is time-independent but not necessarily flat, then we have

$$\tilde{\mathcal{L}} = e^{2\Phi/c^2} c^2 \dot{t}^2 - \dot{\sigma}^2$$

Since this is time-independent, we have that $\partial \tilde{\mathcal{L}}/\partial \dot{t}$ is a constant of the motion: i.e.,

$$e^{2\Phi/c^2} \dot{t} = \text{const.} \tag{14.30}$$

This is eqn (9.57). We have thus provided the further details that were promised in section 9.4, with regard to the arguments about time dilation and energy.

The Schwarzschild line element, eqn (11.5), yields the Lagrangian

$$\tilde{\mathcal{L}} = \left(1 - \frac{r_s}{r}\right)c^2\dot{t}^2 - \left(\frac{1}{1 - r_s/r}\dot{r}^2 + r^2\dot{\phi}^2 + r^2\sin^2\theta\dot{\theta}^2\right) = c^2 \quad (14.31)$$

Since there is no dependence on r or ϕ, the Euler–Lagrange equations for those variables yield the constants of the motion shown in eqns (11.30) and (11.31). Since the orbit stays in one plane (see section 11.3.3) we may set $\theta = \pi/2$ and $\dot{\theta} = 0$. Finally, $\tilde{\mathcal{L}} = c^2$ gives eqn (11.30).

The Euler–Lagrange equations remain valid also for non-static metrics, and can be used to find space-like as well as time-like geodesics. They also apply to null geodesics if one uses a parameter (not proper time) to measure arc-length along the line. In this case $\mathcal{L}$ is still constant (equal to zero) along the solution worldline.

Refractive index method for photons

Consider a spacetime which can be described by a static isotropic metric:

$$ds^2 = -\alpha^2 c^2 dt^2 + \alpha^2 n^2 \left(dx^2 + dy^2 + dz^2\right) \quad (14.32)$$

where $\alpha(x, y, z)$ and $n(x, y, z)$ are functions of position. Then the Lagrangian is (dropping the tilde):

$$\mathcal{L} = \alpha^2 \left(c^2\dot{t}^2 - n^2(\dot{x}^2 + \dot{y}^2 + \dot{z}^2)\right) \quad (14.33)$$

From the time-independence we have

$$\frac{\partial \mathcal{L}}{\partial \dot{t}} = 2\alpha^2 \dot{t} = \text{const} \quad (14.34)$$

and the canonical momentum associated with $\dot{x}$ is

$$\frac{\partial \mathcal{L}}{\partial \dot{x}} = -2\alpha^2 n^2 \dot{x}. \quad (14.35)$$

Its rate of change is governed by

$$\frac{\partial \mathcal{L}}{\partial x} = 2\alpha \frac{\partial \alpha}{\partial x} \frac{\mathcal{L}}{\alpha^2} - \alpha^2 2n \frac{\partial n}{\partial x} v^2 = -2\alpha^2 n \frac{\partial n}{\partial x} v^2 \quad (14.36)$$

where we introduced $v^2 \equiv \dot{x}^2 + \dot{y}^2 + \dot{z}^2$ and in the second step we used $\mathcal{L} = 0$, so we are treating null worldlines. Therefore the Euler–Lagrange equation for $\dot{x}$ reads

$$\frac{\mathrm{d}}{\mathrm{d}\lambda}\left(n^2\alpha^2\dot{x}\right) = n\alpha^2 v^2 \frac{\partial n}{\partial x}. \quad (14.37)$$

After using our constant of the motion, eqn (14.34), to replace α^2, and using $\dot{t} = nv$ (which follows from $\mathcal{L} = 0$), this reads

$$\frac{\mathrm{d}}{\mathrm{d}\lambda}\left(n\frac{\dot{x}}{v}\right) = v\frac{\partial n}{\partial x}. \quad (14.38)$$

Now let the parameter λ be equal to the coordinate distance along the path. Then $v = 1$ and (after similar reasoning for y, z) we have eqn (11.49). This shows that, for a static isotropic metric, the refractive index method reproduces the null geodesics exactly.

Exercises

(14.1) **Graded index optics.** In optics, Fermat's Principle of Least Time says that light-rays follow paths for which the travel time $t = \int n \, ds/c$ is stationary with respect to small changes in the path, where $n(x, y, z)$ is the refractive index and s is distance along the path. By introducing a suitable parameter θ, show that this corresponds to a least action calculation with Lagrangian

$$\mathcal{L} = n(x, y, z)(\dot{x}^2 + \dot{y}^2 + \dot{z}^2)^{1/2}$$

where the dot signifies $d/d\theta$. Hence derive eqn (11.49).

(14.2) (i) Show that the manifestly covariant Lagrangian (14.17) leads to the Hamiltonian

$$\mathcal{H} = (\tilde{\mathsf{P}} - q\mathsf{A})^2/m + c\sqrt{-(\tilde{\mathsf{P}} - q\mathsf{A})^2}.$$

(ii) Show that the associated Hamilton's equations are

$$\frac{\mathrm{d}\mathsf{X}^a}{\mathrm{d}\tau} = \frac{\tilde{\mathsf{P}}^a - q\mathsf{A}^a}{m}, \qquad \frac{\mathrm{d}\tilde{\mathsf{P}}^a}{\mathrm{d}\tau} = \frac{q}{m}(\tilde{\mathsf{P}}_\lambda - q\mathsf{A}_\lambda)\partial^a \mathsf{A}^\lambda,$$

and that these are equivalent to the Euler–Lagrange equation (14.26). Note, however, that because this Hamiltonian always evalu-

ates to zero (prove it!), problems can arise when using it, because some commonly used methods assume non-zero partial derivatives.

(14.3) Show that in an arbitrary variational problem, if a Lagrangian $\mathcal{L}$ is constant along the solution path: i.e., $\mathrm{d}\mathcal{L}/\mathrm{d}s = 0$ where s is a parameter along the path, then replacing $\mathcal{L}$ by $\tilde{\mathcal{L}} = f(\mathcal{L})$ in the Euler–Lagrange equations, for an arbitrary function f, leads to the same predicted path.

(14.4) Show that the following Lorentz-invariant Lagrangian leads to the same equations of motion as were obtained from eqn (14.17):

$$\mathcal{L} = \tfrac{1}{2}m\mathsf{U} \cdot \mathsf{U} + q\mathsf{U} \cdot \mathsf{A}.$$

Show that the Hamiltonian is now $\mathcal{H} = \tfrac{1}{2}(\tilde{\mathsf{P}} - q\mathsf{A})^2/m$.

(14.5) Use the Euler–Lagrange method to find the equation of motion in three dimensions in the Rindler spacetime (constantly accelerating reference frame).

(14.6) Is it possible to write down a manifestly covariant Lagrangian that is invariant under displacements in space but not time? What does this tell us about conservation laws?

Angular momentum*

For a particle whose position and momentum are given by X and P, the angular momentum tensor is defined by

$$L^{ab} \equiv \mathsf{X}^a \mathsf{P}^b - \mathsf{X}^b \mathsf{P}^a = \left(\begin{array}{c|c} 0 & -\mathbf{w}/c \\ \hline \mathbf{w}/c & L^{ij} \end{array} \right) \qquad (15.1)$$

where $\mathbf{w} = \mathbf{x}E - \mathbf{p}c^2 t$ The two 3-vectors associated with this are $\mathbf{w}$ (polar) and the 3-angular momentum $\mathbf{L} = \mathbf{x} \wedge \mathbf{p}$ (axial).

15.1   Conservation of angular momentum

The conservation of angular momentum can be investigated by defining the total angular momentum of a system of particles by

$$L_{\text{tot}}^{ab} = \sum_{\text{particles(i)}} L_{(i)}^{ab} \qquad (15.2)$$

where the sum is over the angular momenta of the different particles. Just as was the case in the discussion of conservation of momentum, we need to check whether this sum over tensors evaluated at different events is itself a tensor. First we check that any single freely moving particle has constant L^{ab}. Using that P is constant for a free particle,

$$\frac{dL^{ab}}{d\tau} = \frac{d\mathsf{X}^a}{d\tau}\mathsf{P}^b - \frac{d\mathsf{X}^b}{d\tau}\mathsf{P}^a = \mathsf{U}^a m_0 \mathsf{U}^b - \mathsf{U}^b m_0 \mathsf{U}^a = 0, \qquad (15.3)$$

so L^{ab} is constant along the worldline.

Next we examine the effect of a collision involving several particles. In this case the angular momenta are all being evaluated at the same event so the X's factor out of the sum (for those particles participating in the collision):

$$L_{\text{collide}}^{ab} = \mathsf{X}^a \sum \mathsf{P}^b - \mathsf{X}^b \sum \mathsf{P}^a.$$

But the 4-momenta are conserved, so the sums are not changed by the collision, so neither is L_{tot}^{ab}. It follows that the total angular momentum we have defined for a composite system is indeed conserved under internal interactions, and hence that L_{tot}^{ab} is a valid tensor (see section 5.2).

We deduce from the spatial part that the 3-angular momentum is conserved in any given reference frame. Also, from the time–space part we have that

$$\mathbf{p}_{\text{tot}} c^2 t - \sum_i \mathbf{x}_i E_i \tag{15.4}$$

is constant in time. The sum

$$\mathbf{x}_{\text{C}} \equiv \frac{\sum_i \mathbf{x}_i E_i}{E_{\text{tot}}} \tag{15.5}$$

can be interpreted as the position of the 'centre of energy' or *centroid* (loosely speaking, this is the 'centre of mass'). So, upon differentiating eqn (15.4) with respect to reference frame time t, we have

$$\frac{\mathrm{d}\mathbf{x}_{\text{C}}}{\mathrm{d}t} = \frac{\mathbf{p}_{\text{tot}} c^2}{E_{\text{tot}}} = \mathbf{v}_{\text{CM}}.$$

Therefore, the centroid moves uniformly, at a speed equal to that of the centre of momentum frame. This is an interesting and non-trivial result. When a composite body moves freely with a tumbling motion, most or all of its particles undergo non-inertial motion. It is not surprising that there is a uniformly moving point which remains always somewhere near the middle of the body, but it is surprising, or at least interesting, that such a point can be found, at all times, by means of the simple sum presented in eqn (15.5). This guarantees that, in the absence of external forces, the particles of the body will not all veer to one side simultaneously, or all get ahead or lag behind, even though the body may change shape in complicated ways. This also shows that our policy of treating composite objects as single entities with well-defined properties (velocity, momentum, energy etc.) continues to make good sense.

15.2 Spin

Next we investigate a form of angular momentum called intrinsic angular momentum or *spin*. For a composite system (i.e., a set of interacting particles, such as a 'rigid' body) the intrinsic angular momentum is defined to be the angular momentum of the system in the centre of momentum frame. We shall show that this definition makes sense. This connects naturally to the familiar idea in classical mechanics of a body rotating about its centre of mass, while the latter may also undergo translational motion. We shall be interested to know whether this angular momentum can be related to a 4-vector.

The issue also arises of whether point-like particles can possess intrinsic angular momentum. When experimental evidence of angular momentum of particles began to emerge in the early twentieth century, this caused some controversy, because for a very small extended object to possess significant angular momentum associated with rotation about its centre of mass, it must rotate very fast. Consider, for example, the electron, which possesses an intrinsic angular momentum of order $\hbar \simeq 10^{-34}$ Js. An attempt to model the electron as a particle of finite size might propose a radius of order 10^{-15} m, but then to produce the observed angular momentum the outer part of this notional electron would have to be moving at about 400 times the speed of light. More generally, consider a ring of radius r and rest mass m_0 rotating with angular frequency ω, giving it angular momentum $L = \gamma r^2 m_0 \omega$. The speed of a point on the ring is $v = \omega r$. If this speed cannot exceed c then it appears the angular momentum cannot exceed $L_{\max} = rmc$ where $m = \gamma m_0$ is the mass that the ring presents to 'the rest of the world', i.e. it is E_{tot}/c^2 of the particles in the ring, evaluated in the centre of momentum frame. This is finite as long as $v < c$. The conclusion is that if $r \to 0$ with finite m, then $L_{\max} \to 0$. Therefore infinitesimally small particles have infinitesimally small angular momentum associated with any rotation they may undergo about their centre of mass-energy.

Fig. 15.1

However, the Poincaré group (i.e., the group of translations and Lorentz transformations (including rotations)) allows one to investigate in general what sort of quantities can be rotated, translated and Lorentz-boosted, and it turns out that an angular-momentum-like property that can be associated with point particles naturally arises in the mathematical description of the Poincaré group. This does not necessarily imply that particles in Nature will be found to possess such a property, but it shows that if they did then a sensible mathematical treatment is available. This treatment introduces the notion of a *spinor* that we shall present in volume 2.

It is found that most elementary particles do possess such a property, called *spin*. The property is not associated with a rotation of the particle, it is an intrinsic property like mass or charge, but instead of a scalar it has an axial vector-like character.

Historically, spin was discovered at the same time as quantum mechanics, and this has led to some confusion over whether spin is an essentially quantum mechanical (or 'non classical') property. Ultimately, all physical properties such as momentum, position, mass, etc. are quantum mechanical, but their behaviour in the classical limit matches the behaviour of the corresponding quantities in classical physics. The same can be said of spin. That is to say, there is a classical theory of spin as well as a quantum-mechanical one. Given that the concept of spin arises naturally in mathematical analysis of the Poincaré group, one may say (but one does not have to say) that it is a relativistic concept. The connection with quantum mechanics is the following. Whereas one can have a classical relativistic mechanics either with spin or without spin, it appears to be impossible to construct a quantum-relativistic mechanics without spin. So in quantum mechanics, spin is not an optional extra: it is an essential property of elementary particles.

15.2.1   Introducing spin

Recall the definition (15.2) of the angular momentum tensor for a collection of particles. In the sum each term is given by eqn (15.1) where $\mathsf{X} = (ct, \mathbf{r})$ is the 4-displacement of the particle from the origin 0, so we have defined the total angular momentum about the origin. Here the origin (of the chosen inertial frame) is serving as the 'pivot' for the definition of angular momentum. The angular momentum about any other pivot R is given by

$$L_{tot}^{ab}(R) = \sum(X^a - R^a)P^b - (X^b - R^b)P^a$$

$$= L_{tot}^{ab}(0) - (R^a P_{tot}^b - R^b P_{tot}^a). \qquad (15.6)$$

In the centre of momentum frame the 3-momentum part of P_{tot} is zero, so we find the space-space part of $L_{tot}^{ab}(R)$ equals that of $L_{tot}^{ab}(0)$. In other words, *the 3-angular momentum in the CM frame is independent of the pivot*. This means that we can regard the angular momentum in the CM frame as the 'intrinsic' or 'spin' angular momentum of the system.

So far we have discussed only the angular momentum associated with *motion* of point particles, but we have shown that a composite system possesses a 3-angular momentum in its centre of momentum frame that is independent of the pivot.

Now let us drop the subscript 'tot' on our labels, and take it for granted that the angular momentum under discussion describes a system of one or more particles. Also, we define J^{ab} to be the total angular momentum about the origin. Then in the absence of intrinsic spin of the particles, $J^{ab} = L^{ab}(0)$ and we can rewrite eqn (15.6) as

$$J^{ab} = L^{ab}(R) + (R^a P^b - R^b P^a).$$

In the case of a point particle at R we would recognize the second term on the right-hand side as the angular momentum 4-tensor about the origin. More generally, we can choose the case $R = X_C$, the displacement from the origin to the centroid of the system, and define

$$S^{ab} \equiv L^{ab}(X_C), \qquad (15.7)$$

$$L_C^{ab} \equiv X_C^a P^b - X_C^b P^a, \qquad (15.8)$$

so that we have

$$J^{ab} = S^{ab} + L_C^{ab}. \qquad (15.9)$$

This makes perfect sense: the total angular momentum about the origin is the sum of a part associated with rotation about the centroid and a part associated with movement of the centroid. The former is called 'spin angular momentum', the latter 'orbital angular momentum'.

The tensor S^{ab} is antisymmetric, and therefore in principle it has two 3-vectors associated with it. However, the definition ensures that

$$S^{0b} = L^{0b}(X_C) = 0 \qquad (15.10)$$

(obtain this by using (15.6) to calculate $L^{ab}(X_C)$, with $X_C = (ct, \mathbf{x}_C)$). Therefore the polar vector is zero and we are left with just the axial vector. The definition of S^{ab} makes the axial vector $\mathbf{S}$ equal to the 3-angular momentum about the centroid (and see box below).

Further remark. Eqn (15.10) states that the first row and column of the spin tensor S^{ab} is zero, but how can this property survive a change of reference frame? Upon applying a Lorentz transform we shall find $S^{0'b'} \neq 0$, unless some further physics intervenes. It does: the centroid must also be transformed. The property (15.10) should be read $L^{0b}(\mathsf{X}_C) = 0$; it holds for the angular momentum about the centroid, in any frame. Equation eqn (15.10) is not in itself tensorial, but it leads to the Fokker-Synge equation:

$$\mathsf{P}_\lambda L^{\lambda b}(\mathsf{X}_{PC}) = 0 \tag{15.11}$$

where PC is the *proper centroid* (the centroid in the CM frame). This equation is tensorial and correct in the CM frame, therefore in all frames. When introducing intrinsic spin angular momentum of particles, we *assume* it too satisfies eqn (15.11).

We can incorporate intrinsic spin of *particles* into this notation by claiming that it is a property that makes a further contribution to S^{ab} without contributing to L^{ab}. We merely remark that there is nothing mathematically wrong with such a concept[1]. We shall assume that J^{ab} is conserved in collisions. For those collisions that conserve L_C^{ab} it follows that S^{ab} is also conserved; in general, however, there can be a transfer of angular momentum between orbital and spin forms.

[1] Sometimes it is useful to reserve the letter S for the intrinsic spin contribution alone, then eqn (15.9) would be written $J^{ab} = \tilde{L}^{ab} + L_C^{ab}$ with $\tilde{L}^{ab} = L^{ab}(\mathsf{X}_C) + S^{ab}$.

15.2.2   Pauli–Lubanski vector

We have already commented that 4-angular momentum is, in its essential character, a second-rank tensor not a 4-vector. However, this does not stop us from enquiring into 4-vectors that can be related to angular momentum. Consider, for example, the total angular momentum in the CM frame, $\mathbf{s}_0$, for some system that has a CM frame (i.e., any system having non-zero rest mass). Let us *define* a four-vector S^a to be that 4-vector whose components in the CM frame are

$$\mathsf{S}^a = (0, \mathbf{s}_0) \tag{15.12}$$

where the subscripted zero emphasizes that $\mathbf{s}_0$ is the angular momentum in the rest frame. This '4-spin' provides a useful way to discuss spin. The Lorentz-invariant scalar quantity associated with S is

$$\mathsf{S} \cdot \mathsf{S} = s_0^2. \tag{15.13}$$

i.e., *the Lorentz-invariant size of the 4-spin is equal to the size of the 3-spin in the rest frame.* We also find

$$\mathsf{S} \cdot \mathsf{U} = 0 \tag{15.14}$$

(since this is the result in the CM frame and the equation is covariant). That is, *the 4-spin is orthogonal to the 4-velocity of the CM frame.* This means the 4-spin of a particle is parallel to the plane of simultaneity of the particle. One can study the evolution of the 4-spin by proposing an equation of motion, and interpret the results by examining the situation in the CM frame. However, this definition does not offer any immediate physical interpretation of the components of S^a in other frames.

We can get some more information by exploring further ways of constructing 4-vectors. Let us try combining the angular momentum tensor with other quantities such as position and momentum. For example $J^{a\lambda}X_\lambda$ is a 4-vector and so is $J^{a\lambda}P_\lambda$. The first of these depends on position and time, so it is not associated with any property (such as intrinsic spin) that is independent of position and time for an isolated system. The second looks more promising, but it does not reduce to an angular-momentum-like property in the low-velocity limit. In that limit we have $P \to (E/c, 0)$ so

$$J^{a\lambda}P_\lambda \to (0, \mathbf{x}E^2/c^2).$$

This product of energy-squared and position vector has no particular physical significance.

A more thorough search yields up a result. A useful 4-vector associated with angular momentum can be obtained by combining the dual of J^{ab} with the 4-momentum of the particle (or total 4-momentum for a composite system):

$$W_a \equiv -\tilde{J}_{a\lambda}P^\lambda = -\tfrac{1}{2}\epsilon_{a\lambda\mu\nu}J^{\mu\nu}P^\lambda = \tfrac{1}{2}\epsilon_{\lambda a\mu\nu}P^\lambda J^{\mu\nu}. \qquad (15.15)$$

You can confirm that

$$\epsilon_{a\lambda\mu\nu}L_C^{\mu\nu}P^\lambda = 0$$

and therefore only the spin part of J contributes to W: i.e.,

$$W_a = \tfrac{1}{2}\epsilon_{\lambda a\mu\nu}P^\lambda S^{\mu\nu} \qquad (15.16)$$

This is called the *Pauli–Lubanski spin 4-vector*. We have already deduced one important property: it has nothing to do with orbital angular momentum. You can also verify that it is independent of the pivot, and in any frame the spatial part is equal to $(E/c)\mathbf{s}$, where $\mathbf{s}$ is the axial vector associated with the spin tensor S^{ab} (i.e., the angular momentum about the centroid). To obtain this, recall that the time-space part of S^{ab} is zero, so in the calculation of W_x, for example, only two terms contribute:

$$W_x = \tfrac{1}{2}\left(\epsilon_{0123}S^{23}P^0 + \epsilon_{0132}S^{32}P^0\right)$$

$$= \frac{1}{2}\left(s_x E/c - (-s_x)E/c\right) = (E/c)s_x \qquad (15.17)$$

and similarly for the other components. These are the spatial components of the covariant form W_a. It follows that the contravariant form is $W^a = (W^0, (E/c)\mathbf{s})$. You can verify that the zeroth component is $W^0 = \mathbf{s} \cdot \mathbf{p}$, so the summary is

$$W = (\mathbf{s} \cdot \mathbf{p}, \ (E/c)\mathbf{s}) \qquad (15.18)$$

where **s** is the 3-spin: i.e., the angular momentum about the centroid. In the CM frame this reduces to $W = (0, mc\mathbf{s}_0)$. Therefore $W = mcS$ for systems or particles possessing rest mass. The Pauli–Lubanski spin 4-vector is somewhat more general than S, since it remains well-defined for particles having no rest mass and no rest frame, such as photons. We can use eqn (15.18) to interpret the components of S in an arbitrary frame:

$$S = W/mc = (\gamma \mathbf{s} \cdot \mathbf{v}/c, \ \gamma \mathbf{s}). \tag{15.19}$$

We shall now revert to the language of particles and speak of 'rest frame', 'position', 'velocity', 'energy' and 'momentum'; for a composite system it is understood that these refer to CM frame, centroid, velocity of the CM frame, total energy, and total momentum respectively.

Under the action of a Lorentz boost, the Pauli–Lubanski vector transforms like any other 4-vector. To investigate this we shall perform a Lorentz boost starting from the rest frame (where W is $(0, mc\mathbf{s}_0)$) to a frame moving at relative velocity $-\mathbf{v}$. Unprimed symbols refer to the situation in the new frame, where the velocity of the particle is $\mathbf{v}$. Recalling eqn (15.18) and using the Lorentz boost eqns (6.35) we find

$$W^0 = \mathbf{s} \cdot \mathbf{p} = \gamma \mathbf{v} \cdot \mathbf{s}_0 m = \mathbf{s}_0 \cdot \mathbf{p},$$

$$W^i = (E/c)\mathbf{s} = mc\mathbf{s}_0 + \frac{\gamma^2 m \mathbf{v} \cdot \mathbf{s}_0}{(1+\gamma)c}\mathbf{v}$$

$$= mc\mathbf{s}_0 + \frac{\mathbf{p} \cdot \mathbf{s}_0}{mc + E/c}\mathbf{p} \tag{15.20}$$

where to obtain the second form we used $\gamma m \mathbf{v} = \mathbf{p}$ and $\gamma mc^2 = E$.

The components of the 3-spin vector parallel and perpendicular to the velocity are

$$s_{\parallel} = s_{0\parallel}, \qquad s_{\perp} = \frac{1}{\gamma}s_{0\perp}. \tag{15.21}$$

These results can be obtained from eqn (15.20), but it is easier to get them directly by applying a Lorentz transform to $(0, \mathbf{s}_0)$ and interpreting the outcome using eqn (15.19). Both results are interesting. The first says that the component of the spin along the particle's velocity direction is given by the amount of spin in the rest frame along that direction. Therefore, when one considers a given particle from the point of view of any one of a set of frames all moving in the same direction relative to the rest frame, this spin projection is invariant. The two together imply that at low velocities the spin direction is almost unaffected by Lorentz boosts (a change in angle only appears at order v^2/c^2, whereas orbital angular momentum is strongly affected by a change of reference frame). Also, in the limit $v \rightarrow c$, $s_{\perp} \rightarrow 0$ so the spin is directed along the velocity: either aligned or anti-aligned.

In the limit $v \rightarrow c$, eqns (15.20) become

$$W \rightarrow \mathbf{s}_0 \cdot \mathbf{p} \, (1, \ \mathbf{p}c/E) = \frac{\mathbf{s} \cdot \mathbf{p}}{p} P, \tag{15.22}$$

(where we have used $E \to pc$). For a massless particle we can no longer interpret $\mathbf{s}_0$ as the spin in the rest frame, since there is no rest frame. Therefore we express the result in terms of $\mathbf{s}$, the spin of the particle in whatever frame is under consideration. We find that for a massless particle the spin 4-vector is proportional to the 4-momentum. This means that the spin must be aligned or anti-aligned with the velocity, which is consistent with the behaviour we already noted in the limit $v \to c$. Therefore $\mathbf{s} \cdot \mathbf{p}/p$ evaluates to either $+s$ or $-s$. We still have $\mathsf{W} \cdot \mathsf{P} = 0$ (see eqn 15.14) since for a massless particle the 4-momentum is null ($\mathsf{P} \cdot \mathsf{P} = 0$).

The projection of the spin along the velocity: i.e.,

$$s_\parallel = \frac{\mathbf{s} \cdot \mathbf{p}}{p}$$

is called the *helicity*. For a massive particle this can depend on reference frame. For example, by adopting a frame that overtakes a particle one could reverse the direction of $\mathbf{p}$ without reversing the direction of $\mathbf{s}$. Therefore the helicity is not a Lorentz-invariant property in general. However, for massless particles there is no reference frame that can overtake the particle. Eqn (15.22) shows that the helicity for a massless particle is a constant of proportionality between two 4-vectors. It follows that it is a Lorentz scalar: i.e., a Lorentz-invariant quantity. Its value is either $+s$ or $-s$.

In the quantum theory the helicity is quantised. It has eigenvalues $m\hbar$ where m is an integer or half-integer depending on the spin of the particle. For electrons m can be $\pm 1/2$, and which value is obtained, $+1/2$ or $-1/2$, is frame-dependent. For photons m can be ± 1, and the value is frame-independent.

15.2.3 Thomas precession revisited

Eqn (15.14) has interesting consequences for the kinematics of spin angular momentum. In the absence of torque, one might be tempted to expect that $d\mathsf{S}/d\tau = 0$. However, if this were so then we could not uphold (15.14). Differentiating that equation with respect to proper time, one finds

$$\mathsf{S} \cdot \dot{\mathsf{U}} + \dot{\mathsf{S}} \cdot \mathsf{U} = 0 \tag{15.23}$$

therefore $\dot{\mathsf{S}}$ cannot be zero in general if the particle is accelerating. This leads one to insist that the evolution of 4-spin in the *absence* of torque must be governed by

$$\frac{d\mathsf{S}}{d\tau} = \frac{\mathsf{S} \cdot \dot{\mathsf{U}}}{c^2} \mathsf{U} \tag{15.24}$$

since then we can guarantee (15.23) and we used the only available 4-vectors. By 'absence of torque' here we mean that the situation in the instantaneous rest frame is $d\mathbf{s}_0/d\tau = 0$.

For convenience let us treat the case of constant rest mass, then $\dot{\mathsf{U}} = \dot{\mathsf{P}}/m$ and the spatial part of eqn (15.24) reads

$$\frac{d}{d\tau}(\gamma \mathbf{s}) = \left(-\frac{\gamma \mathbf{s} \cdot \mathbf{v}}{c}\frac{\dot{E}}{c} + \gamma \mathbf{s} \cdot \dot{\mathbf{p}}\right)\frac{\gamma \mathbf{v}}{mc^2}$$

$$= (-\gamma \mathbf{s} \cdot \mathbf{v}\dot{\gamma} + \gamma \mathbf{s} \cdot (\dot{\gamma}\mathbf{v} + \gamma \dot{\mathbf{v}}))\frac{\gamma \mathbf{v}}{c^2}$$

$$\implies \dot{\gamma}\mathbf{s} + \gamma \dot{\mathbf{s}} = \gamma^3(\mathbf{s} \cdot \dot{\mathbf{v}})\frac{\mathbf{v}}{c^2}. \tag{15.25}$$

Now use $\dot{\gamma} = \gamma^3(\mathbf{v} \cdot \dot{\mathbf{v}})/c^2$ and one finds

$$\frac{d\mathbf{s}}{d\tau} = \frac{\gamma^2}{c^2}\dot{\mathbf{v}} \wedge (\mathbf{v} \wedge \mathbf{s}). \tag{15.26}$$

Since the quantities are all as observed in some given inertial frame, it makes sense to express the result in terms of reference frame time using $d/d\tau = \gamma d/dt$ and therefore $\dot{\mathbf{v}} = \gamma \mathbf{a}$ where $\mathbf{a} = d\mathbf{v}/dt$, giving

$$\frac{d\mathbf{s}}{dt} = \frac{\gamma^2}{c^2}\mathbf{a} \wedge (\mathbf{v} \wedge \mathbf{s}). \tag{15.27}$$

This equation shows that the 3-spin of a particle which accelerates without torque has a constant component along the acceleration, but an evolving perpendicular component. For example, for rectilinear motion the sign of $d\mathbf{s}/dt$ is such as to align the spin more and more onto the direction of motion as the velocity increases.

This is of some interest, but an even more interesting observation emerges if we consider the proper spin, that is, the 3-spin in the rest frame. We should like to discuss the evolution of $\mathbf{s}_0$ for a particle which is accelerating. It is useful to do the analysis in an inertial frame. We shall then have a 'mixed-frame' type of quantity: $d\mathbf{s}_0/dt$ gives the rate of change, with respect to laboratory frame time t, of the proper spin $\mathbf{s}_0$. In case it seems odd to discuss this type of 'mixed' quantity, let us consider some other examples in order to show that it is in fact a sensible thing to do.

Consider the Doppler effect in the case of an accelerating source such as a flying singing bird. We observe waves of changing frequency in some given direction in our inertial frame fixed to the ground. We might well want to know, is the changing frequency wholly owing to the Doppler effect (the bird whistling at fixed frequency, but darting too and fro), or is the bird chirping? In this case an interesting quantity is $f_0(t)$: i.e., the frequency f_0 in the instantaneous rest frame of the bird, evaluated at the bird event whose time is t in our inertial frame.

For a more apt example, suppose there is a magnetic compass fixed in a rally car which is racing down a bumpy twisting track. We might take an interest in the question: is the compass needle maintaining a true indication of north, or is its violent motion throwing it off? The answer to this question depends on the nature of the force between the needle and its pivot (which may not be accurately at the needle's centre) as well as the flow of the surrounding fluid, and the interaction with Earth's magnetic field. We have no way to carry out the calculation in the accelerating frame of the car (unless we 'borrow' techniques from General Relativity), so we much prefer to do the calculation in an inertial frame such as that of the Earth. Nonetheless, the direction we want to know is the one observed by the driver.

I hope I have persuaded you that calculating $d\mathbf{s}_0/dt$ is a worthwhile thing to do, where t is time in some inertial frame, and $\mathbf{s}_0$ is the proper spin of a particle which may be accelerating.

By 'proper spin' we mean, of course, the spin as observed in the instantaneous rest frame. But wait—'*the*' rest frame? Which rest frame? For any given particle at any given event there are an infinite number of rest frames, all related to one another by rotations. For scalar properties such as mass this issue is irrelevant, but for a vector property such as spin we must specify which rest frame we mean. In the following argument we first pick one inertial frame, called the lab frame, which remains fixed throughout, and we study the particle as it moves relative to the lab frame. At any event on the particle's worldline, by 'the instantaneous rest frame' we mean *that instantaneous rest frame which is related to the lab frame purely by a boost.*

Our starting point is eqn (15.24). By dotting eqn (15.24) with S we obtain

$$\mathsf{S} \cdot \frac{d\mathsf{S}}{d\tau} = \frac{1}{2}\frac{d}{d\tau}(\mathsf{S} \cdot \mathsf{S}) = 0 \qquad (15.28)$$

(using eqn (15.14)), therefore S is of fixed size, and therefore, by eqn (15.13), the proper 3-spin is of fixed size s_0 during the motion. Only its direction changes.

Let θ_0 be the angle *in the instantaneous rest frame* between $\mathbf{s}_0$ and the particle's velocity vector $\mathbf{v}$. It will be important to be clear about the definition of this angle. Keep in mind that $\mathbf{v}$ is the relative velocity of the particle and the lab frame, so it is well-defined in both frames and they agree on its angle relative to their respective coordinate axes (since the boost relating them is along $\mathbf{v}$). With this definition the parallel and perpendicular components appearing in eqn (15.21) are

$$s_{0\parallel} = s_0 \cos\theta_0, \qquad s_{0\perp} = s_0 \sin\theta_0.$$

Using eqns (15.19) and (15.21) we therefore have

$$\mathsf{S} = (\gamma s_0(v/c)\cos\theta_0, \ \gamma s_0 \cos\theta_0 \hat{\mathbf{v}} + s_0 \sin\theta_0 \mathbf{n}) \qquad (15.29)$$

where $\hat{\mathbf{v}}$ is a unit vector along $\mathbf{v}$, and $\mathbf{n}$ is a unit vector perpendicular to $\mathbf{v}$ in the plane formed by $\mathbf{v}$ and $\mathbf{s}$.

To find the evolution of θ_0, the method of calculation involves a trick: we express S in terms of two convenient 4-vectors M and N:

$$\mathsf{S} = s_0(\mathsf{M}\cos\theta_0 + \mathsf{N}\sin\theta_0) \qquad (15.30)$$

where in our chosen frame (the lab frame),

$$\mathsf{N} = (0, \ \mathbf{n}) \quad \text{and} \quad \mathsf{M} = (\gamma v/c, \ \gamma\hat{\mathbf{v}}). \qquad (15.31)$$

By noticing that M is the Lorentz-boosted version of $(0, \hat{\mathbf{v}})$ in the rest frame, it is easy to prove that

$$\mathsf{M} \cdot \mathsf{M} = \mathsf{N} \cdot \mathsf{N} = 1$$

and hence

$$\mathsf{M} \cdot \dot{\mathsf{M}} = \mathsf{N} \cdot \dot{\mathsf{N}} = 0.$$

You can also confirm (by evaluating in a convenient frame) that

$$\mathsf{M} \cdot \mathsf{U} = \mathsf{N} \cdot \mathsf{U} = \mathsf{M} \cdot \mathsf{N} = 0.$$

With these preliminaries over, let us calculate $\dot{\mathsf{S}}$:

$$\dot{\mathsf{S}} = s_0 \left(\dot{\mathsf{M}} \cos \theta_0 + \dot{\mathsf{N}} \sin \theta_0 + \dot{\theta}_0 (-\mathsf{M} \sin \theta_0 + \mathsf{N} \cos \theta_0) \right)$$

where we used that s_0 is constant. Substituting into eqn (15.24) and dotting both sides with N gives

$$\mathsf{N} \cdot \dot{\mathsf{M}} \cos \theta_0 + \dot{\theta}_0 \cos \theta_0 = 0 \tag{15.32}$$

Hence

$$\frac{d\theta_0}{d\tau} = -\mathsf{N} \cdot \frac{d\mathsf{M}}{d\tau} = -\mathbf{n} \cdot \frac{d}{d\tau} (\gamma \mathbf{v}/v)$$

$$= -\mathbf{n} \cdot \left(\dot{\gamma} \hat{\mathbf{v}} + \gamma \frac{\dot{\mathbf{v}}}{v} - \gamma \frac{\dot{v}}{v^2} \mathbf{v} \right) = -\gamma \frac{\mathbf{n} \cdot \dot{\mathbf{v}}}{v}. \tag{15.33}$$

Therefore

$$\frac{d\theta_0}{dt} = -\gamma \frac{\mathbf{a} \cdot \mathbf{n}}{v}. \tag{15.34}$$

This equation says that the proper spin 3-vector rotates relative to the velocity whenever the acceleration has a component along $\mathbf{n}$: i.e., perpendicular to the velocity. This is not in itself surprising: classically we would expect the net result to be that the velocity changes direction while the spin does not. However the Lorentz factor γ says that the angle between $\mathbf{s}_0$ and $\mathbf{v}$ is opening up 'too quickly'. Consider, for example, the case of circular motion: then the velocity changes direction at the rate $\omega_v = a/v$ in the lab frame. This is the rotation of the axis relative to which θ_0 was defined, so the proper spin must be rotating relative to a fixed direction, such as one of the lab frame coordinate axes, at the rate

$$\frac{d\theta}{dt} = \frac{d\theta_0}{dt} + \omega_v = -(\gamma - 1)\frac{a}{v} \tag{15.35}$$

(in the simplest case, where $\mathbf{a}$, $\mathbf{v}$ and $\mathbf{s}_0$ are coplanar). This is precisely the Thomas precession that we derived previously in eqn (6.50).

15.2.4   Precession of the spin of a charged particle

We now discuss the motion of a magnetic dipole in a magnetic field. Classically, a dipole $\boldsymbol{\mu}$ in a field $\mathbf{B}$ experiences a torque

$$T = \boldsymbol{\mu} \wedge \mathbf{B}.$$

Therefore, if **s** is the angular momentum of a particle possessing both angular momentum and dipole moment, the equation of motion of the angular momentum is

$$\frac{d\mathbf{s}}{dt} = \boldsymbol{\mu} \wedge \mathbf{B}. \tag{15.36}$$

When the dipole moment is proportional to the angular momentum, as is the case for the intrinsic spin and dipole of a charged particle, for example, the equation of motion becomes

$$\frac{d\mathbf{s}}{dt} = \frac{gq}{2m}\mathbf{s} \wedge \mathbf{B}, \tag{15.37}$$

where q/m is the charge/mass ratio of the particle, and g is the gyromagnetic ratio (equal to 1 in the case of orbital angular momentum). We now generalize to the relativistic case by arguing that eqn (15.37) is the low-velocity limit of a covariant equation. Introduce the 4-spin-vector S whose components are $(0, \mathbf{s}_0)$ in the rest frame (eqn (15.12)), and investigate its product with the field tensor. In the rest frame one obtains

$$\mathbb{F} \cdot \mathsf{S} = \begin{pmatrix} 0 & E_x/c & E_y/c & E_z/c \\ -E_x/c & 0 & B_z & -B_y \\ -E_y/c & -B_z & 0 & B_x \\ -E_z/c & B_y & -B_x & 0 \end{pmatrix} \begin{pmatrix} 0 \\ s_{0x} \\ s_{0y} \\ s_{0z} \end{pmatrix} = \begin{pmatrix} \mathbf{s}_0 \cdot \mathbf{E}/c \\ \mathbf{s}_0 \wedge \mathbf{B} \end{pmatrix}. \tag{15.38}$$

This leads us to suggest that the generalization of eqn (15.37) is

$$\frac{d\mathsf{S}}{d\tau} \overset{?}{=} \frac{gq}{2m}\mathbb{F} \cdot \mathsf{S}. \tag{15.39}$$

There is a problem, however: this equation does not guarantee to preserve the orthogonality between the spin 4-vector and the 4-velocity, eqn (15.14). Dotting with 4-velocity U gives

$$\mathsf{U} \cdot \frac{d\mathsf{S}}{d\tau} = \frac{gq}{2m}\mathsf{U} \cdot \mathbb{F} \cdot \mathsf{S}.$$

For a constant U the left-hand side can be written $(d/d\tau)(\mathsf{U} \cdot \mathsf{S}) = 0$, while the right-hand side is not necessarily zero. However, we can see how to fix the problem: add a term to the right-hand side, so that

$$\frac{d\mathsf{S}}{d\tau} \overset{?}{=} \frac{gq}{2m}\left(\mathbb{F} \cdot \mathsf{S} + \frac{1}{c^2}[\mathsf{U} \cdot \mathbb{F} \cdot \mathsf{S}]\mathsf{U}\right).$$

Now, dotting the right-hand side with U always gives zero, which you can easily see by noticing that the term in square brackets is a scalar. We still have not finished, however, because if the particle is accelerating then we ought not to get zero: we have already discussed this in section 15.2.3, see eqn (15.24), where we saw that an extra term associated with Thomas precession is present. The equation we need is

$$\frac{dS}{d\tau} = \frac{gq}{2m}\left(\mathbb{F}\cdot S + \frac{1}{c^2}[U\cdot\mathbb{F}\cdot S]U\right) + \frac{S\cdot\dot{U}}{c^2}U. \qquad (15.40)$$

We now have a satisfactory covariant equation for the evolution of the spin of a charged particle in an electromagnetic field. It gives the classical result in the rest frame and ensures that S remains perpendicular to U.

If the particle's acceleration is due to the electromagnetic force, then after substituting from eqn (13.9) we have[2]

$$\frac{dS}{d\tau} = \frac{q}{2m}\left(g\mathbb{F}\cdot S - \frac{(g-2)}{c^2}[S\cdot\mathbb{F}\cdot U]U\right) \qquad (15.41)$$

where we have used that $\mathbb{F}$ is antisymmetric, so $S\cdot\mathbb{F}\cdot U = -U\cdot\mathbb{F}\cdot S$. This equation can serve as the starting-point of the treatment of spin-orbit interaction in an atom, and of the precession of the spin of high-velocity particles in high-energy physics experiments. In the case $g = 2$ (a good approximation for electrons) the extra torque term exactly cancels with the Thomas precession term, and we are left with our original conjecture (15.39) after all.

By dotting eqn (15.41) with S you can confirm that $(d/d\tau)(S\cdot S) = 0$ (on the right use $S\cdot U = 0$ and $S\cdot\mathbb{F}\cdot S = 0$, see eqn (12.13)). Therefore the electromagnetic interaction does not change the size of the spin!

The evolution of the spin direction can be obtained by using the method introduced in section 15.2.3 (see exercise 15.5). The result is

$$\frac{d\theta_0}{dt} = \frac{q}{2m}\left[(g-2)(\mathbf{B}\wedge\mathbf{n})\cdot\hat{\mathbf{v}} - \frac{2}{v}\left(1 - \frac{g}{2}\frac{v^2}{c^2}\right)\mathbf{E}\cdot\mathbf{n}\right]. \qquad (15.42)$$

This is the Bargmann, Michel, Telegdi equation. For example, if $\mathbf{E} = 0$ and $\mathbf{B}$ is perpendicular to the velocity then we have circular motion in which the proper spin rotates relative to the velocity at the rate

$$\frac{d\theta_0}{dt} = \frac{q}{2m}(g-2)B.$$

Each time the particle has completed a circle its velocity comes back along the initial direction, whereas its spin has precessed by an amount proportional to $g - 2$. Observation of such a precession offers a precise measure of the difference of g from 2, which constitutes a precise test of quantum field theory.

The case $\mathbf{B} = 0$ applies to an electron in an atom with no externally applied magnetic field. This can be used to calculate the spin-orbit coupling effect, with the Thomas precession fully taken into account.

[2] We here neglect the force on the dipole moment; this is exact in a uniform static B field, but more generally involves approximation.

Exercises

(15.1) (i) Prove that the time rate of change of the angular momentum $\mathbf{L} = \mathbf{r}\wedge\mathbf{p}$ of a particle about an origin O is equal to the couple $\mathbf{r}\wedge\mathbf{f}$ of the applied force about 0.

(ii) If L^{ab} is the particle's 4-angular momentum, and we define the 4-couple $G^{ab} \equiv X^a F^b - X^b F^a$, prove that $(d/d\tau)L^{ab} = G^{ab}$, and that the space-space part of this equation corresponds to the previous 3-vector result.

(15.2) A gyroscope consists of a flat disc rotating about an axis which is fixed in one frame (the CM frame or 'rest frame' of the gyroscope). In that frame the rotation is rigid and the disc is uniform. Show that the intrinsic angular momentum **s** of such a gyroscope is aligned with the axis in the rest frame, but not in most frames.

(15.3) A system is formed of two particles of rest mass m, one lying at $\{x, y, z\} = \{0, -a, 0\}$ in frame S, the other moving on the trajectory $\{x, y, z\} = \{vt, a, 0\}$ with constant speed v. Find the centroid at $t = 0$. Find all components of the tensors $L^{ab}(0)$ (total angular momentum about the origin) and S^{ab} (total angular momentum about the centroid). Show that the centroid in S never coincides with the centroid in the rest frame of the second particle.

(15.4) Let S′ be the rest frame of a gyroscope as described in exercise 15.2. Show that in frame S

(in standard configuration with S′), if the gyroscope's axis is parallel to the y direction then its centroid is not in the xy plane. (A qualitative argument suffices.) Show that a force applied to the axis in the y direction will create a torque about the centroid, with a direction in the correct sense to cause Thomas precession.

(15.5) **Precession of spin in EM field.** Let the 4-spin be written $\mathbf{S} = s_0(\mathbf{M}\cos\theta_0 + \mathbf{N}\sin\theta_0)$, where $\mathbf{M}$ and $\mathbf{N}$ are the 4-vectors introduced in eqn (15.31). From eqn (15.41) deduce

$$\dot{\theta}_0 = \frac{qg}{2m}\mathbf{N} \cdot \mathbb{F} \cdot \mathbf{M} - \mathbf{N} \cdot \dot{\mathbf{M}}.$$

Examine the second term in order to relate it to $\mathbf{N} \cdot \dot{\mathbf{U}}$; hence obtain

$$\dot{\theta}_0 = \frac{qg}{2m}\mathbf{N} \cdot \mathbb{F} \cdot \mathbf{M} - \frac{q}{mv}\mathbf{N} \cdot \mathbb{F} \cdot \mathbf{U}.$$

Thus obtain eqn (15.42).

(15.6) Confirm that if the metric $(+ - --)$ is adopted instead of $(- + ++)$ then $c^2 \to -c^2$ in eqn (15.41) and equations leading up to it.

Energy density

In chapter 6 we introduced the idea of flow, and the 4-current J. Its components, in any reference frame, are density and flux of some Lorentz-invariant quantity such as electric charge or rest mass. Another quantity that can flow, and that we might naturally take an interest in, is energy. In this chapter we take as our starting point the idea of energy per unit volume—energy density. Starting from a frame where everything is static, upon changing reference frame, we find we have to consider the flow of energy, and also momentum, since energy and momentum are partners in Special Relativity. This leads to the idea of both energy transport and momentum transport. The former idea is reasonably intuitive, the second is more challenging. We shall see that it can be connected to, or interpreted as, another way of describing internal forces in a fluid or solid body, such as pressure and stress.

With these concepts in hand, we can then apply the requirement of energy-momentum conservation. We thus obtain equations of motion for internal movements in any continuous body (whether a fluid or a solid—even a solid can vibrate). They describe the relationships between movements in the body and the internal pressures and stresses and applied forces. These are the fundamental equations for relativistic fluid mechanics (called *hydrodynamics*), and they also give some important insight into conservation of energy-momentum in general. This in turn reveals one of the basic ingredients in General Relativity.

16.1 Introducing the stress-energy tensor

Consider a set of inert particles all at rest in some reference frame S_0. Let n_0 be the number of particles per unit volume in that frame, and let m be the rest mass of each particle. The energy density (energy per unit volume) in frame S_0 is then

$$\rho_0 c^2 = n_0 m c^2.$$

In chapter 6 we associated a 4-vector with a charge density, so let us try associating a 4-vector with this energy density, by writing

$$\mathsf{N} \stackrel{?}{=} \rho_0 c \mathsf{U}$$

where U is the 4-velocity of the set of particles (they all have the same 4-velocity in the simple scenario we are considering) and the question-mark here signifies that we are tentatively exploring an idea. Now consider the very same particles, but as observed in a reference frame S moving in the negative x-direction relative to S_0. In the new frame the components of N are

$$N = \rho_0 c(\gamma c, \gamma \mathbf{v}) = (\gamma \rho_0 c^2, \gamma \rho_0 c \mathbf{v}).$$

There is nothing mathematically wrong with these statements, but we find that this 4-vector is of limited usefulness. In the new frame the energy of every particle is increased by a factor γ, and the size of any region containing a fixed number of particles has contracted by a factor γ, so the energy density is now $\gamma^2 \rho_0 c^2$, but this is not equal to the zeroth component of N. This issue did not arise with the 4-current of electric charge because that described a flow of a Lorentz invariant quantity (charge), whereas now we have a quantity that is not invariant (energy). N is a 4-vector, but not a useful one.

To obtain a more useful quantity, we try defining a tensor instead:

$$T^{ab} = \rho_0 \mathsf{U}^a \mathsf{U}^b. \tag{16.1}$$

In the rest frame, this tensor is very simple: it has only one non-zero element, $T^{00} = \rho_0 c^2$, which is the energy density. By evaluating $\Lambda T \Lambda^T$, or simply by writing the general form $\mathsf{U} = (\gamma c, \gamma \mathbf{v})$, we can find out how this tensor appears in a general frame. For simplicity, let us first consider the frame S, in which the particles all move in the positive x direction with speed v. In this frame we find

$$T^{ab} = \begin{pmatrix} \gamma^2 \rho_0 c^2 & \gamma^2 \rho_0 c v & 0 & 0 \\ \gamma^2 \rho_0 c v & \gamma^2 \rho_0 v^2 & 0 & 0 \\ 0 & 0 & 0 & 0 \\ 0 & 0 & 0 & 0 \end{pmatrix} = \gamma^2 \rho_0 c^2 \begin{pmatrix} 1 & v/c & 0 & 0 \\ v/c & v^2/c^2 & 0 & 0 \\ 0 & 0 & 0 & 0 \\ 0 & 0 & 0 & 0 \end{pmatrix} \tag{16.2}$$

Now the T^{00} element is readily interpreted: it has two factors of γ, so it is the energy density in the new frame. We can also recognise $\gamma^2 \rho_0 c v$: this is c times the momentum per unit volume in the new frame, since the momentum of any given particle is $\gamma m v$ and the number of particles per unit volume is $n = \gamma n_0$, which upon multiplying by c gives

$$c(\gamma m v) n = c(\gamma m v)(\gamma n_0) = \gamma^2 \rho_0 c v.$$

This same quantity can also be interpreted as the energy flux divided by c, recall eqn (5.58). Thus both T^{01} and T^{10} are readily interpreted (there is just an ambiguity over whether we should think of them as energy flux or momentum density, or one of each). It remains to interpret $T^{11} = \gamma^2 \rho_0 v^2$. With the hint that flux might be relevant, we interpret this as a momentum flux. This is a more subtle idea: it concerns the idea that momentum itself can be 'transported' from one place to another, like a sack of potatoes. Before going further, we must first stop to appreciate this more thoroughly.

16.1.1   Transport of energy and momentum

Suppose that, in frame S, a particle having energy E and momentum $\mathbf{p}$ moves from $\{x, y, z\} = \{-1, -1, 0\}$ to $\{2, 1, 0\}$. Then it is clear that the energy E, which used to be located at $\{-1, -1, 0\}$, has moved to $\{2, 1, 0\}$. This is energy transport. Equally, the momentum $\mathbf{p}$, which used to be located at $\{-1, -1, 0\}$, is now at $\{2, 1, 0\}$. What this means is that if an observer were to sit at $\{x, y, z\} = \{2, 1, 0\}$ and collect everything that arrives, then he will find himself, after the arrival of the particle, to be in possession of extra energy E and extra momentum $\mathbf{p}$. Unless he somehow delivers an impulse to counter that momentum, he must now be moving. Equally, an observer sitting at $\{-1, -1, 0\}$ and claiming ownership of everything there, will find that, when the particle departs from him, he loses energy E and momentum $\mathbf{p}$.

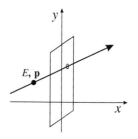

Fig. 16.1 Transporting energy and momentum across a plane.

We can also talk about energy flux and momentum flux. Consider now many particles, uniformly spread with n per unit volume, all with the same energy E and momentum $\mathbf{p}$. We can ask: at what rate does energy cross the plane at $x = 0$ (i.e. the yz plane)? In any given time t, a volume $Av_x t$ of the particle beam moves across an area A of the yz plane, so the amount of energy crossing, per unit area, per unit time—which is what we call the energy *flux*—is

$$S_x = nv_x E. \tag{16.3}$$

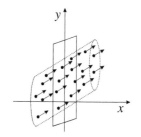

Fig. 16.2

Similarly, the amount of energy crossing the plane $y = 0$, per unit area per unit time, is $S_y = nv_y E$, and crossing the plane $z = 0$ is $S_z = nv_z E$. In total we find the energy flux is a vector, which for a particle beam is given by $\mathbf{S} = nE\mathbf{v}$ (the product of energy density and velocity of the beam).

Next let us ask the same questions about momentum. There are now nine quantities to consider. A particle crossing any given plane, say the plane $x = 0$, carries all three components of its momentum across. The amount of x-momentum crossing the plane, per unit area per unit time, is calculated just as before, except that where previously we had energy E, now we have x-momentum p_x. The flux of x-momentum in the x-direction is therefore

$$nv_x p_x.$$

The flux of y-momentum across this same plane is

$$nv_x p_y$$

and the flux of z-momentum is

$$nv_x p_z.$$

Next consider the rate at which particles cross the plane $y = 0$. This tells us about the flow of p_x and p_y and p_z across the plane $y = 0$. We again obtain three quantities: $nv_y p_x$, $nv_y p_y$, $nv_y p_z$. Upon considering also the plane $z = 0$ we find that in total nine quantities are required to describe momentum flux in a simple particle beam:

$$\sigma^{ij} = nv^i p^j = n \begin{pmatrix} v_x p_x & v_x p_y & v_x p_z \\ v_y p_x & v_y p_y & v_y p_z \\ v_z p_x & v_z p_y & v_z p_z \end{pmatrix} \tag{16.4}$$

The (i, j)th component of σ^{ij} tells us how much j-momentum is being carried along in the i-direction by movement of material (here, particles).

Next, observe that the notion of energy and momentum transport must also occur in a situation where particles are not flying freely but pushing on one another, such as in a solid. If every atom on a solid pushes on its neighbours, then energy and momentum will be moved around just as surely as if the atoms each moved freely. Therefore we can still define an energy flux vector **S** and a momentum flux tensor σ^{ij}, although their expressions in terms of motions of the atoms may be more complicated. In a solid there are, indeed, more possibilities, because the diagonal elements of σ^{ij} do not need to be positive. How shall we interpret that? The idea is to connect momentum flux to force per unit area. Imagine erecting a wall at $x = 0$ in the solid or fluid under consideration, and suppose the wall absorbs or supplies all energy and momentum incident on it or carried away from it. Absorbing energy is easy: just suppose the wall heats up. Absorbing momentum is less easy, but we can imagine that the wall is very massive so it acquires or supplies all the momentum without acquiring any significant velocity. Such a wall finds momentum σ^{0j} brought up to it per unit area per unit time on one side, and carried away on the other. Therefore, on the $x < 0$ side it experiences a force σ^{0j} per unit area, and on the $x > 0$ side it experiences an equal and opposite force per unit area (Fig. 16.3). In short, *momentum flux is force per unit area*—and this is a statement not merely about shared physical dimensions, but about identical physical effects. We have already looked at this idea briefly in section 5.6, and now we are considering it again. The implication is that in the case of a solid or fluid at rest, we can consider the diagonal elements of σ^{ij} to represent *pressure*. A negative value for one of these elements says there is negative pressure, which is *tension*. The off-diagonal elements represent *sheer stress*.

Having used a wall to think about this, now remove the wall and then the forces under consideration are exerted by the material on one side of any given plane on the material on the other side.

For a fluid possessing both internal forces and motion, the total momentum flux tensor is

$$\sigma^{ij} = t^{ij} + g^i u^j \tag{16.5}$$

where t^{ij} is the 3-tensor describing the pressure and stress, **u** is the local flow velocity, and **g** is the momentum density associated with the flow of material and a subtle contribution connected with the rate of doing work, which we shall examine in section 16.2.1.

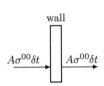

wall

$A\sigma^{00}\delta t$ $A\sigma^{00}\delta t$

Fig. 16.3 A special wall erected in the $x = 0$ plane in a fluid such as to leave the momentum flow in the fluid undisturbed. For $\sigma^{00} > 0$ in a time δt the left face of the wall *absorbs* momentum $A\sigma^{00}\delta t$ and recoils to the right; the right face of the wall *emits* momentum $A\sigma^{00}\delta t$ and recoils to the left. Hence the wall is squeezed. In the absence of the wall, the forces are experienced by each layer of the fluid itself. Other components of σ^{ij} contribute further forces.

In classical mechanics, the tensor σ^{ij} is called the stress tensor. In relativistic mechincs, the tensor T^{ab} that we have begun to investigate is called the **stress-energy tensor**, or the 4-stress.

16.1.2   Ideal fluid

Now let us return to the tensor that we constructed from energy density for particles at rest; eqn (16.1). If we change frame again, but now such that the particles move in the negative x direction, then the tensor will become

$$T^{ab} = \gamma^2 \rho_0 c^2 \begin{pmatrix} 1 & -v/c & 0 & 0 \\ -v/c & v^2/c^2 & 0 & 0 \\ 0 & 0 & 0 & 0 \\ 0 & 0 & 0 & 0 \end{pmatrix}. \tag{16.6}$$

This is to be contrasted with eqn (16.2). Notice that the momentum and energy are now flowing in the other direction, as we should expect, but the T^{11} term has the same sign. This is because negative momentum now passes from right to left (from positive x to negative x), which is the same as positive momentum passing from left to right.

Now suppose that in some reference frame there are two particle beams moving in the x-direction with equal and opposite velocities. By summing eqns (16.2) and (16.6) we find the total stress-energy tensor must be

$$T^{ab} = \rho c^2 \begin{pmatrix} 1 & 0 & 0 & 0 \\ 0 & v^2/c^2 & 0 & 0 \\ 0 & 0 & 0 & 0 \\ 0 & 0 & 0 & 0 \end{pmatrix} \tag{16.7}$$

where $\rho = \gamma^2 \rho_0$ is the total energy density (each beam contributes half of ρ). Finally, consider a set of particles moving with a range of velocities, distributed isotropically (e.g., an ideal gas). The stress-energy tensor will have the form

$$T^{ab} = \begin{pmatrix} \rho c^2 & 0 & 0 & 0 \\ 0 & p & 0 & 0 \\ 0 & 0 & p & 0 \\ 0 & 0 & 0 & p \end{pmatrix} \tag{16.8}$$

where p is the pressure. More generally, any system whose stress-energy tensor has this diagonal form in some frame is called an *ideal fluid*. The frame in which the tensor is diagonal is the rest frame of the fluid (since there is no momentum density or energy flux in that frame). In a general frame such a tensor can be written as shown in table 16.1. This table summarizes the two example systems we have considered. The first system (particles not interacting and all sharing a common 4-velocity near any given event) is called 'dust'.

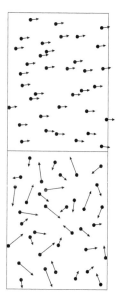

Fig. 16.4 An example of 'dust (top) and an 'ideal fluid' (bottom). Dust can flow but does not have internal pressure or stress; an ideal fluid has pressure but not sheer stress.

Table 16.1 Two example stress-energy tensors. Entries left blank are zero. The examples are necessarily given in some suitably chosen reference frame (the rest frame of the local fluid). The equation gives the form in an arbitrary frame.

Dust	$\begin{pmatrix} \rho_0 c^2 & & & \\ & 0 & & \\ & & 0 & \\ & & & 0 \end{pmatrix}$,	$T^{ab} = \rho_0 U^a U^b$
Ideal fluid	$\begin{pmatrix} \rho_0 c^2 & & & \\ & p & & \\ & & p & \\ & & & p \end{pmatrix}$,	$T^{ab} = (\rho_0 + p/c^2)U^a U^b + p g^{ab}$

16.2 Stress-energy tensor for an arbitrary system

Figure 16.5 summarizes the physical interpretation of the elements of the stress-energy tensor. So far we have explained how this interpretation arises for the case of an ideal fluid and for the case of a system at rest but possessing internal forces. To prove that the interpretation is valid in general one can adopt one of two strategies. The first strategy is to guess that this must be the interpretation in general, in which case one can now skip straight to section 16.3, and then upon examining the physical predictions one is driven to the conclusion that the guess was correct. The second strategy is to start from the rest frame and transform to an arbitrary frame, looking into the physical interpretation in detail of the various quantities that arise in the expression for $\mathbb{T}^{ab}$. This strategy is more laborious, but one can learn something from it, so we shall display it next.

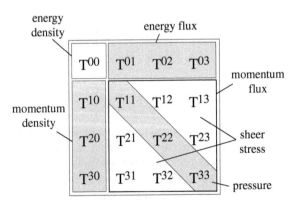

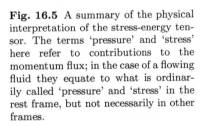

Fig. 16.5 A summary of the physical interpretation of the stress-energy tensor. The terms 'pressure' and 'stress' here refer to contributions to the momentum flux; in the case of a flowing fluid they equate to what is ordinarily called 'pressure' and 'stress' in the rest frame, but not necessarily in other frames.

First consider the pressure on the walls of a cubic chamber of ideal gas. In the rest frame S_0 the pressure p_0 is isotropic. The force on each wall is therefore $p_0 A$ where A is the area of a side of the cube. Now adopt a reference frame in standard configuration with S_0. Applying the equations for the transformation of force, eqn (4.6) (for the case of a pure force), we find the forces on the faces perpendicular to the motion is unchanged, while that on the other faces is reduced by γ. Those faces are also contracted by γ, so the pressure is $p = p_0$ on all faces: uniform motion does not change the pressure (we shall generalize this to arbitrary shapes in the following).

Now consider the stress-energy tensor for some given small portion of a continuous system such as a solid or a fluid. Since we allow that the system may flow, we shall call it a fluid, but the ideas will apply also to a solid which may be regarded as a fluid with restricted flow. The fluid is not necessarily ideal (i.e., it can have stress). We consider a small portion so that all the fluid within it has the same velocity, and therefore we may speak of a rest frame. In the rest frame there is no momentum flux, so the stress-energy tensor of the portion of fluid must take the form

$$\mathbb{T}_0 = \begin{pmatrix} \rho_0 c^2 & 0 & 0 & 0 \\ 0 & p_0^x & \sigma_0^{12} & \sigma_0^{13} \\ 0 & \sigma_0^{21} & p_0^y & \sigma_0^{23} \\ 0 & \sigma_0^{31} & \sigma_0^{32} & p_0^z \end{pmatrix} \tag{16.9}$$

where we wrote $p_0^{x,y,z}$ for the diagonal elements of σ_0, signifying pressure. By applying an inverse Lorentz transformation $\mathbb{T} = \Lambda^{-1} \mathbb{T}_0 \Lambda^{-1}$ (so that $\mathbf{u} = c\boldsymbol{\beta}$ is the velocity of the fluid in the new frame), we find

$$\mathbb{T} = \begin{pmatrix} \gamma^2(\beta^2 p_0^x + c^2 \rho_0) & \beta\gamma^2(c^2\rho_0 + p_0^x) & \beta\gamma\sigma_0^{12} & \beta\gamma\sigma_0^{13} \\ \beta\gamma^2(c^2\rho_0 + p_0^x) & \gamma^2(\beta^2 c^2 \rho_0 + p_0^x) & \gamma\sigma_0^{12} & \gamma\sigma_0^{13} \\ \beta\gamma\sigma_0^{21} & \gamma\sigma_0^{21} & p_0^y & \sigma_0^{23} \\ \beta\gamma\sigma_0^{31} & \gamma\sigma_0^{31} & \sigma_0^{32} & p_0^z \end{pmatrix}$$

$$= \begin{pmatrix} \rho c^2 & \mathbf{g}c \\ \mathbf{g}c & \sigma^{ij} \end{pmatrix}. \tag{16.10}$$

The diagonal element $\mathbb{T}^{11}$ is not the same as in the rest frame, but did we not just establish that the pressure is invariant? The answer is simply that the spatial elements of $\mathbb{T}$ in a frame other than the rest frame do not represent pressure and stress alone. Rather, they represent *total momentum density*. When the fluid is flowing the momentum density along the direction of flow has two contributions: one from the pressure or stress, one from the transport of material, as shown in eqn (16.5), in which $g^i = \mathbb{T}^{0i}/c$. For example, the 11 component of $\mathbb{T}$ is

$$\sigma^{11} = t^{11} + g^1 u^1 \tag{16.11}$$

$$\Rightarrow \gamma^2(u^2\rho_0 + p_0^x) = t^{11} + \frac{u^2}{c^2}\gamma^2(\rho_0 c^2 + p_0^x) \tag{16.12}$$

where we have used eqn (16.10) on the left and also to obtain g^1 on the right. Solving for t^{11} gives $t^{11} = p_0^x$, confirming the interpretation that t^{11} is pressure. In the following we shall prove this physical interpretation more generally.

16.2.1   Interpreting the terms*

We would like to interpret the terms arising in the Lorentz-transformed stress-energy tensor (16.10). As we have just seen, it is necessary to do this in order to avoid misapplying concepts such as pressure. Let S_0 be the rest frame. We are interested in what is observed in some other frame S relative to which the fluid moves at velocity $\mathbf{u}$. In writing down eqn (16.10) we arranged the axes so that $\mathbf{u}$ is along the x direction.

First we *define* the 3-tensor t^{ij} to be that tensor which gives the *forces* per unit area on the sides of any small region within the moving portion of fluid. This is obviously a tensor, since it acts on vectors to produce vectors. That is, the net 3-force per unit area on a boundary normal to the $\mathbf{n}$ direction is

$$f^i = t^{ij} n_j.$$

e.g., for $\mathbf{n}$ along the ith coordinate axis, $\mathbf{f}$ is given by the ith column of t^{ij}. Now, using the transformation of 3-force (4.6) and the Lorentz contraction where appropriate, you can easily confirm that t^{ij} must transform as

$$t^{ij} = \begin{pmatrix} t_0^{11} & \gamma t_0^{12} & \gamma t_0^{13} \\ t_0^{21}/\gamma & t_0^{22} & t_0^{23} \\ t_0^{31}/\gamma & t_0^{32} & t_0^{33} \end{pmatrix}. \tag{16.13}$$

Fig. 16.6 Directions associated with elements of t^{ij}.

For example, the (12) entry gives the force in the x direction, per unit area, on a boundary normal to the y direction. The force is unchanged but the boundary is contracted, yielding γt_0^{12}. The (21) entry gives the force in the y direction, per unit area, on a boundary normal to the x direction. The force is reduced while the boundary is unchanged, yielding t_0^{21}/γ. Note that the result is not symmetric.

Now consider the momentum flowing across a fixed plane. If ρc^2 is the energy density in the moving portion of fluid, we might guess that the momentum density is $\mathbf{g} = \rho \mathbf{u}$ (eqn (5.58)), but by now we are prepared for the fact that the forces may also contribute momentum flow (sections 5.6 and 16.1). For example, suppose I pull on a rope in order to move a heavy object towards me. I am doing work, the heavy object is acquiring kinetic energy. The energy must travel down the rope from me to the object, and therefore (eqn (5.58) again) there is "hidden" momentum in the rope, in the direction from me to the object. If the rate of doing work is fv, then the size of this momentum is fv/c^2 per unit length of rope.

Now consider that any small region of fluid is serving as the 'rope' via which forces are communicated, and energy transported, from one place to another in the fluid. Consider a boundary across the 'rope' (i.e., in the fluid), placed normal to the x^i axis. The matter on the positive side of the boundary experiences a force $\mathbf{t}^{(i)}$ per unit area, where for convenience we wrote $\mathbf{t}^{(i)}$ for the vector (t^{1i}, t^{2i}, t^{3i}). Note that this is a column not a row of t. The matter is moving, so this force does work at the rate $\mathbf{t}^{(i)} \cdot \mathbf{u}$. (Note that it does not matter whether or not you consider that the boundary moves with the fluid; what is important is that the matter on which the force acts moves.) It should be apparent to you that, equally, the matter on the other side of the boundary has work done on it at the rate $-\mathbf{t}^{(i)} \cdot \mathbf{u}$. However, this does not mean there is no energy flow: consider the rope example again, where similar statements apply. In the case of a rope in tension, the direction of energy flow is against the direction of motion of the rope; for a rod in compression the energy flow is along the direction of motion of the rod. In general, we find that the energy flow per unit area in the direction x^i is

$$\mathbf{t}^{(i)} \cdot \mathbf{u} = u_j t^{ji}. \tag{16.14}$$

(not $t^{ij} u_j$—the 3-tensor is not symmetric). Therefore the total momentum density of a small portion of fluid is

$$g^i = \rho u^i + u_j t^{ji}/c^2. \tag{16.15}$$

Using eqns (16.15) and (16.13) we are almost ready to interpret the first row and column of $\mathbb{T}$. First we need an expression for ρ.

In the absence of forces (e.g., for dust) we expect $\rho = \gamma^2 \rho_0$: one factor of γ for the energy, one for the Lorentz contraction of the volume. In the presence of forces we must also consider the work done by those forces. The argument hinges on the relativity of simultaneity. Figure 16.7 shows the relevant region of spacetime. The quantity ρ_0 refers to the energy content (per unit volume) in frame S_0 of a set of particles at the events along the line OA. ρ refers to the energy content (per unit volume) of the same set of particles, but now as observed in frame S, and at the events along the line OB. We have already mentioned that in the absence of forces we would have $\rho = \gamma^2 \rho_0$. It remains to consider the work done, in frame S, between OA and OB. We only need to consider the particle at the end AB of the region of fluid: all other forces are either internal or, at end O, do no work. The other fluid does work on the particles at AB (these are the particles at the boundary) at the rate pAu, where p is pressure and A is the area of the boundary face. If event O is at the origin, then event B is at $t = 0$ in S, and event A is at $t = -\gamma u L_0/c^2$ (by Lorentz transformation from $(0, -L_0, 0, 0)$, where L_0 is the rest length of the portion of fluid). Therefore the work done is

$$pAu^2 \gamma L_0/c^2.$$

The volume of the fluid portion in S is $A L_0/\gamma$, so the total energy per unit volume is

$$\rho c^2 = \gamma^2 \rho_0 c^2 + \gamma^2 \frac{u^2}{c^2} t_0^{11}, \tag{16.16}$$

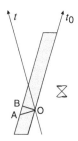

Fig. 16.7 Spacetime diagram to aid the calculation of work done in a fluid. The arrows mark the time axes of frames S and S_0. The shaded region is a small portion of fluid at rest in S_0. OA is a line of simultaneity in S_0; OB is a line of simultaneity in S.

where we have substituted t_0^{11} for p. Now, by combining eqns (16.16), (16.15), and (16.13) one can describe the stress-energy tensor in S in terms of its components in S_0, and by comparison with eqn (16.10) thus confirm eqn (16.5). The net result is to gain a correct interpretation (the one offered by eqn (16.5)) of the stress-energy tensor of a moving fluid: the tensor expresses *the density and flux of energy and momentum.*

We have considered the case where the energy transfer was purely by work and by flow of material; if there was also heat flow then this would contribute a further term to eqn (16.15). This would also change the situation in the rest frame, and a term would correspondingly have to be added to eqn (16.9).

Note that in the case of a fluid moving at the speed of light—for example, a set of electromagnetic waves all moving in the same direction—the concept of 4-stress still applies (as elucidated in section 16.4.3 below), but it is no longer possible to speak of a rest frame, so the decomposition (16.5) is meaningless and t^{ij} becomes irrelevant.

16.3   Conservation of energy and momentum for a fluid

So far we have established what physical quantities are expressed by the elements of $\mathbb{T}$. We have not yet established any constraint on how $\mathbb{T}$ may vary as a function of position and time—from a purely mathematical point of view, it might have any functional form whatsoever. However if $\mathbb{T}$ is to describe a real fluid, then it must be constrained by the laws of physics, and in particular the laws of energy and momentum conservation.

We learned how to apply the idea of conservation to a scalar invariant quantity in chapter 6, where we discussed the **continuity equation** (6.13), which we reproduce here for convenience:

$$\frac{\partial \rho}{\partial t} = -\boldsymbol{\nabla} \cdot \mathbf{j}, \qquad \text{or} \qquad \Box \cdot \mathsf{J} = 0. \tag{16.17}$$

If we now examine the 4-divergence of the stress-energy tensor, then by setting it equal to zero we get four continuity equations: one for energy, and one for each of the components of momentum:

$$\Box \cdot \mathbb{T} \;=\; 0$$

$$\Leftrightarrow \qquad \left(\frac{1}{c}\frac{\partial}{\partial t}, \, \boldsymbol{\nabla}\cdot\right) \left(\begin{array}{c|c} s & c\mathbf{g} \\ \hline c\mathbf{g} & \sigma^{ij} \end{array}\right) \;=\; 0$$

$$\Leftrightarrow \qquad \frac{\partial s}{\partial t} = -\boldsymbol{\nabla} \cdot (c^2 \mathbf{g}) \quad , \quad \frac{\partial g^j}{\partial t} = -\boldsymbol{\nabla} \cdot \boldsymbol{\sigma}^{(j)} \tag{16.18}$$

where s is energy density, $\mathbf{g}$ is momentum density, (so $(c^2 \mathbf{g})$ is energy flux), and $\boldsymbol{\sigma}^{(j)}$ is the jth column of σ^{ij}. In order to physically interpret these equations, integrate each of them over the volume of some region of space. This makes the interpretation easier because most of us find flux through a surface easier to think about than divergence. The integral of $\nabla \cdot (c\mathbf{g})$ over the volume of some region is equal to the net flux of energy out of that region, so the first equation says that the rate of increase of energy in any region is equal to minus the amount of energy flowing out of that region. The second equation says that the rate of increase of x-momentum in any region is equal to minus the amount of x-momentum flowing out of that region, and similarly for y-momentum and z-momentum.

The above is a perfectly legitimate way to interpret the stress-energy tensor. It implies that each column of $\mathbb{T}$ has a density followed by a flux. However, in the literature the following version of the continuity equation for the stress-energy tensor is very often used:

$$\partial_\lambda \mathbb{T}^{a\lambda} = 0 \tag{16.19}$$

This is slightly different, because it takes a sum over a given row rather than a given column of $\mathbb{T}^{ab}$. Since $\mathbb{T}^{ab}$ is symmetric this is equally valid, but it implies a slightly different point of view, in which each *row* of $\mathbb{T}$ is to be interpreted as a density followed by a flux. This is why the components of the stress-energy tensor are usually interpreted as shown in Fig. 16.5.

Now suppose a fluid is flowing in a region where it is subjected to an *external* force (not just the internal forces it generates between parts of itself). For example, think of an electrically charged fluid flowing in a region of externally applied electric field. The external field can 'reach in' and 'grab' any part of the fluid, adding to the forces experienced there. In this case we must write

$$\Box \cdot \mathbb{T} = \mathsf{K} \tag{16.20}$$

where K is the density of external 4-force. If the idea of 'force density' is unfamiliar, get some insight once again by integrating the equation over a small region: the integral of K with respect to volume, over any small region, is the net 4-force on the material inside the region.

For a pure force, K takes the form $(\mathbf{k} \cdot \mathbf{u}/c, \mathbf{k})$ where $\mathbf{u}$ is the local flow velocity and $\mathbf{k}$ is the external 3-force per unit volume. In component form, eqn (16.20) then reads

Conservation of 4-momentum for a fluid

$$\left(\; \mathbf{k} \cdot \mathbf{u}/c, \quad \mathbf{k} \; \right) = \left(\frac{1}{c}\frac{\partial}{\partial t}, \; \nabla \cdot \right) \left(\begin{array}{c|c} \rho c^2 & c\mathbf{g} \\ \hline c\mathbf{g} & \sigma \end{array} \right) \tag{16.21}$$

which gives

$$c^2 \frac{\partial \rho}{\partial t} = -c^2 \nabla \cdot \mathbf{g} + \mathbf{k} \cdot \mathbf{u}, \tag{16.22}$$

$$\frac{\partial g^i}{\partial t} = -\partial_j \sigma^{ij} + k^i. \tag{16.23}$$

We have used $\rho = s/c^2$ here, since this is usually done, but one should keep in mind that the physics here is about energy, not mass. The first of these equations may be called the *continuity equation for energy* or, equivalently, the expression of *conservation of energy*. The second equation may be called any or all of the *continuity equation for momentum*, the expression of *conservation of momentum*, the *equation of motion* (since it relates momentum changes to forces), or the relativistic **Euler equation** for a fluid.

Let us apply eqn (16.20) to the case of an ideal fluid, whose stress-energy tensor takes the form

$$T^{ab} = (\rho_0 + p/c^2)U^a U^b + p\eta^{ab}$$

(see table 16.1) where we used η^{ab} for the Minkowski metric in order to avoid confusion with momentum density. We then find

$$U^\lambda U^b \frac{\partial}{\partial x^\lambda}(\rho_0 + p/c^2) + (\rho_0 + p/c^2)\frac{\partial}{\partial x^\lambda}\left(U^\lambda U^b\right) + \frac{\partial p}{\partial x^\lambda}\eta^{\lambda b} = K^b. \quad (16.24)$$

It is interesting to note that the proper density of the fluid always enters this expression in company with the pressure (divided by c^2). To draw out the implications we shall first manipulate the expression into an exact statement about energy, and then obtain the equation of motion in the case of a slowly-moving fluid.

Absolute derivative. In discussions involving flow or movement along a curve, the *Eulerian derivative* or *absolute derivative* is often useful. For any quantity q associated with a flow or with a curve (e.g., a worldline), the absolute derivative of q is the rate of change of q as it would appear to an observer following along the flow or moving along the curve. In Newtonian mechanics this is

$$\frac{Dq}{Dt} = \frac{\partial q}{\partial t} + \mathbf{v}\cdot\nabla q.$$

In Special Relativity it is

$$\frac{Dq}{D\tau} \equiv U^\lambda \partial_\lambda q \qquad [= U\cdot\Box q. \qquad (16.25)$$

The proof of this is given before eqn (14.25), which is an example of precisely this same idea; we merely postponed naming it until now. In particle mechanics, if q is a property of a particle then $Dq/D\tau \equiv dq/d\tau$: in other words, the absolute derivative is merely a new name for something we have already met—$DX/D\tau$ is 4-velocity, $DU/D\tau$ is 4-acceleration, etc. In the case of fluid flow the notation can be useful in order to clarify which change and which proper time is being referred to.

First, apply the product rule (C.6) of differentiation:

$$\partial_\lambda \left(U^\lambda U^b \right) = (\partial_\lambda U^\lambda) U^b + U^\lambda \partial_\lambda U^b$$

$$= (\partial_\lambda U^\lambda) U^b + \frac{DU^b}{D\tau}$$

where in the second step we used eqn (16.25) (see box above). The second term on the right is the 4-acceleration of a fluid element. Noting this, we can obtain a simpler equation by substituting this expression into eqn (16.24) and then dotting the whole equation with U (i.e., contract with U_b). Then, since $U_b A^b = U \cdot A = 0$, the acceleration term disappears and we have (after also using $\eta^{\lambda b} U_b = U^\lambda$)

$$-c^2 \left(U^\lambda \partial_\lambda (\rho_0 + p/c^2) + (\rho_0 + p/c^2)\partial_\lambda U^\lambda \right) + (\partial_\lambda p) U^\lambda = K^b U_b$$

$$\Rightarrow \qquad c^2 U^\lambda \partial_\lambda \rho_0 + (\rho_0 c^2 + p)\partial_\lambda U^\lambda = -K^b U_b$$

$$\Rightarrow \qquad c^2 \frac{D\rho_0}{D\tau} = - \left(\rho_0 c^2 + p \right) \Box \cdot U + K \cdot U \qquad (16.26)$$

which may be read as a partner to eqn (16.22), giving the rate of change of proper energy density with respect to proper time.

Now let us examine the case of a slowly-moving fluid, such that $U^0 \simeq c$ and $U^i \ll c$. Then, returning to eqn (16.24) and writing out the $b = 0$ and $b = i$ equations separately, we have

$$cU^\lambda \partial_\lambda (\rho_0 + p/c^2) + (\rho_0 + p/c^2)\partial_\lambda \left(U^\lambda c \right) - \frac{\partial p}{\partial ct} = \mathbf{k} \cdot \mathbf{u}/c$$

$$U^i U^\lambda \partial_\lambda (\rho_0 + p/c^2) + (\rho_0 + p/c^2)\partial_\lambda \left(U^\lambda U^i \right) + \frac{\partial p}{\partial x^i} = k^i$$

Now multiply the first equation by U^i/c and subtract it from the second equation:

$$(\rho_0 + p/c^2)U^\lambda \partial_\lambda U^i + \frac{\partial p}{\partial x^i} + \frac{U^i}{c^2}\frac{\partial p}{\partial t} = k^i - (\mathbf{k} \cdot \mathbf{u})\frac{U^i}{c^2} \qquad (16.27)$$

where we have used the product rule (C.6) for the differential $\partial_\lambda \left(U^\lambda U^i \right)$, which then allowed two terms to be cancelled. The first term on the left is $(\rho_0 + p/c^2)Du/D\tau$ by making use of eqn (16.25) again. Upon neglecting terms of order u^2/c^2 (including time dilation: i.e., we write $t = \tau$) we have the overall result

$$\left(\rho_0 + \frac{p}{c^2} \right) \frac{D\mathbf{u}}{Dt} = \mathbf{k} - \nabla p. \qquad (16.28)$$

This is the classical Euler equation for an ideal fluid, also called the **Navier–Stokes equation**, except that the term representing the inertia of the fluid is augmented by the pressure. So, according to Special Relativity, internal pressure as well as mass contributes to inertia! A high-pressure fluid will respond more sluggishly to a given force, compared to a low-pressure fluid of the same density. For ordinary fluids this is a small effect, but for thermal radiation, or a gas at very high temperature, it is significant.

16.4   Electromagnetic energy and momentum

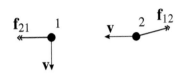

Fig. 16.8 Two charged particles move on orthogonal trajectories. At the moment when one passes directly in front of the other, the forces are not in opposite directions (the electric contributions are opposed, but the magnetic contributions are not). Does this mean that momentum is not conserved?

Now we turn our attention from material fluids to electromagnetic fields.

Consider two identical point charges that are released from rest at the same moment, at some modest distance from one another. They will repel one another, so fly apart with equal and opposite momenta. Thus momentum is conserved. However, consider these events from the perspective of another frame of reference moving along the line between the particles. In the new frame the release events are not simultaneous: one particle starts to accelerate before the other one. It has changed its momentum, but the other has not, so what has happened to conservation of momentum?

Consider another scenario, depicted in Fig. 16.8. Two charges are moving at right angles to one another in a common plane, so that one passes in front of the other. The electric field produced by each particle at the other is directed along the line between them, so the electric forces are in opposite directions. Since a moving charge produces no magnetic field along its line of motion, at the moment when q_2 is moving directly towards q_1, the latter experiences no magnetic field, so experiences no further force: the net force on it is directly away from q_2. However, it produces a non-zero magnetic field at q_2, and the latter is in motion through this field. Therefore q_2 experiences a transverse force: the total force on it is *not* directly away from q_1, but somewhat off to one side. So the forces are not equal and opposite! No momentum conservation again?

In both these examples an attempt was being made to talk about momentum conservation between events at separate locations. However, Relativity teaches us that this is doomed to failure. A conservation law has to be *local*, not just global. That is, the conservation of a substance such as water does not mean merely that the total amount of water in the room is fixed (assuming the door is shut and no chemistry is going on): it means much more than that. If water were to disappear from a vase, it is not enough merely that an equal quantity of water should appear somewhere else such as on the window. We insist that the water *has to get there by flowing across from the one place to the other* (for example, by evaporation and convection). In classical physics we might imagine that a less tangible quantity such as momentum might disappear from one place and appear in another without flowing across the intervening space, but Relativity teaches us that a quantity is conserved locally or not at all. The Principle of Relativity requires that the law, if it is valid, should apply in all reference frames, and the relativity of simultaneity shows that a conservation law that relies on simultaneous behaviour at separate places cannot hold in all reference frames.

Faced with the observed behaviour of charged particles, we must either abandon the principles of conservation of energy and momentum, or else assert that something in addition to the particles, and near to them, can carry energy and momentum. The obvious candidate is the electromagnetic field (or possibly the potentials, but in view of gauge freedom, it would seem less likely that it should work out that way). We have in fact assumed this already when we allowed ourselves to talk about 'the energy carried by a pulse of light', and when we applied to light-pulses concepts such as an energy-momentum 4-vector. Now we shall investigate whether this idea can be made precise and extended to all fields, including static ones. It turns out that it can, and it will lead to a new and more satisfactory way of understanding 'potential energy'.

We shall start with energy, and turn to momentum afterwards, but aim to finish with a covariant treatment in terms of a stress-energy tensor. We could restrict ourselves to covariant 4-tensor notation from the outset, but I think it is easier to understand what is going on in the more familiar language of flow through space and rate of change with time.

In what follows we shall need to discuss the energies both of particles and of fields. It will help if you agree at the outset to abandon all talk of 'potential energy'. You may have been taught that a charged particle 'possesses potential energy $q\phi$' when it is in a static electric field whose scalar potential is ϕ, but we are going to show that this sort of talk is quite muddled and misleading. The only energy a particle has is its rest energy m_0c^2 and its kinetic energy $(\gamma - 1)m_0c^2$, which together make its total energy γm_0c^2. The kinetic energy is the energy a particle has because it is *moving*, not because of where it is, and the rest mass is constant, unchanged by interactions with the electromagnetic field. Note the great contrast with gravitational physics.

Now let us suppose that an electromagnetic field possesses an energy per unit volume (scalar, called u) and an energy current density (vector, $\mathbf{N}$) which is the energy flowing across a small area, per unit area per unit time. These quantities should satisfy the continuity equation

$$\frac{\partial u}{\partial t} = -\boldsymbol{\nabla} \cdot \mathbf{N}.$$

At least, that is what we should expect for fields in free space, when there are no charged particles around. But of course, electromagnetic fields can interact with particles, and presumably exchange energy with them. How does that come about? Only and wholly through the Lorentz force equation, because according to our theory that is the only 'point of impact' of the fields onto matter. Since we have a pure force, we can use $\mathbf{f} \cdot \mathbf{v}$ to get the rate of doing work by the force. To be precise, this is the rate of change of *kinetic energy* of the particle being pushed (that is, the rate of change of its full, relativistic kinetic energy). This is $dW/dt = q\mathbf{E} \cdot \mathbf{v}$ for a particle of charge q. We model a general distribution of charge as many small volumes dV, each containing charge $q = \rho dV$. The combination $q\mathbf{v} = \rho\mathbf{v}dV$ can be recognized as $\mathbf{j}dV$, where $\mathbf{j}$ is the current density, so the rate of doing work at some given point, per unit volume, is $\mathbf{E} \cdot \mathbf{j}$. This work is the energy being given *to* the charged particles (increasing their kinetic energy) and therefore being taken *from* the field.

If the particles are being slowed then this is taken care of by the sign of $\mathbf{E} \cdot \mathbf{j}$. Therefore the conservation of energy is represented by the equation

$$-\frac{\partial u}{\partial t} = \boldsymbol{\nabla} \cdot \mathbf{N} + \mathbf{E} \cdot \mathbf{j}. \tag{16.29}$$

The left-hand side says how much energy is going out of the field in some small volume (per unit volume), and the right-hand side says how much is field energy but is flowing out of the region, and how much is being given to the particles. We have accounted for the total energy of field and particles, and asserted that it is conserved.

You may be concerned that in a typical electric circuit with a constant current, there is inside any resistor a field $\mathbf{E}$ and a constant current density $\mathbf{j}$. The $\mathbf{E} \cdot \mathbf{j}$ says that work is being done, but the constant $\mathbf{j}$ says that the particles are not in fact speeding up—so where is the energy going? Is it 'potential energy' after all? The answer is that the current carriers inside the resistor *are* continually being accelerated by the field, but they immediately suffer collisions with the material of the resistor (nuclei and bound electrons), transferring their new-found kinetic energy to kinetic energy and field energy of the rest of the resistor, in a random form called 'heat'. A detailed model of all these effects must end up confirming eqn (16.29), because it is derived from the only fundamental point of interaction of field and matter.

The following beautiful argument is due to John Henry Poynting (1852–1914).

We should like to find out how u and $\mathbf{N}$ in eqn (16.29) depend on the fields $\mathbf{E}$ and $\mathbf{B}$. To this end, we can use the Maxwell equation M4 to express $\mathbf{j}$ in terms of the fields, giving

$$\mathbf{E} \cdot \mathbf{j} = \epsilon_0 c^2 \mathbf{E} \cdot (\boldsymbol{\nabla} \wedge \mathbf{B}) - \epsilon_0 \mathbf{E} \cdot \frac{\partial \mathbf{E}}{\partial t}.$$

The last term is $(\partial/\partial t)(\frac{1}{2}\epsilon_0 \mathbf{E} \cdot \mathbf{E})$, so it looks as though that this is at least a part of $\partial u/\partial t$. Therefore, we want to turn the first term into the divergence of something.

A divergence involves $\boldsymbol{\nabla}\cdot$ and a vector. The vectors we have available are $\mathbf{E}$, $\mathbf{B}$ and $\mathbf{E} \wedge \mathbf{B}$. The term we are investigating involves both $\mathbf{E}$ and $\mathbf{B}$ so we try taking a look at $\boldsymbol{\nabla} \cdot (\mathbf{E} \wedge \mathbf{B})$. The general result for the divergence of a vector product is

$$\boldsymbol{\nabla} \cdot (\mathbf{E} \wedge \mathbf{B}) = \mathbf{B} \cdot (\boldsymbol{\nabla} \wedge \mathbf{E}) - \mathbf{E} \cdot (\boldsymbol{\nabla} \wedge \mathbf{B}). \tag{16.30}$$

The last term is just what we have. We deduce that

$$\mathbf{E} \cdot \mathbf{j} = -\epsilon_0 c^2 \boldsymbol{\nabla} \cdot (\mathbf{E} \wedge \mathbf{B}) + \epsilon_0 c^2 \mathbf{B} \cdot (\boldsymbol{\nabla} \wedge \mathbf{E}) - \frac{\partial}{\partial t} \left(\tfrac{1}{2}\epsilon_0 \mathbf{E} \cdot \mathbf{E} \right).$$

This is a nice divergence and a time-derivative, plus a part in the middle that is not in the form we want. However, Maxwell's equations will sort it out for us again, this time by using M3 to replace $\boldsymbol{\nabla} \wedge E$, giving

$$\mathbf{E} \cdot \mathbf{j} = -\epsilon_0 c^2 \boldsymbol{\nabla} \cdot (\mathbf{E} \wedge \mathbf{B}) - \frac{\partial}{\partial t} \left(\tfrac{1}{2}\epsilon_0 c^2 \mathbf{B} \cdot \mathbf{B} + \tfrac{1}{2}\epsilon_0 \mathbf{E} \cdot \mathbf{E} \right). \tag{16.31}$$

This beautiful result shows that we can make our energy conservation equation (16.29) apply very nicely to fields and particles together. We just need to *define*

$$u = \tfrac{1}{2}\epsilon_0 c^2 B^2 + \tfrac{1}{2}\epsilon_0 E^2 ,$$

$$\mathbf{N} = \epsilon_0 c^2 \mathbf{E} \wedge \mathbf{B}. \qquad (16.32)$$

We have not proved that eqn (16.32) represents a unique solution: it is possible to define more complicated versions of u and $\mathbf{N}$, such that after differentiating one and taking the divergence of the other, one still obtains eqn (16.31); but this form is the most obvious, and is consistent with all observations in electromagnetism. It is believed to be correct.[1]

$\mathbf{N} = \epsilon_0 c^2 \mathbf{E} \wedge \mathbf{B}$ is called the Poynting vector, after its discoverer. It gives the energy flow per unit area per unit time (also called flux). For an oscillating field such as a light-wave, its time average is the power per unit area, called the intensity. The Poynting vector is often written

$$\mathbf{N} = \mathbf{E} \wedge \mathbf{H}$$

where $\mathbf{H}$ is a field closely related to $\mathbf{B}$, being given in free space by $\mathbf{H} = \mathbf{B}/\mu_0 = \epsilon_0 c^2 \mathbf{B}$.

[1] It is not possible to use electromagnetic theory alone to distinguish this choice of u and $\mathbf{N}$ from other choices that still satisfy eqn (16.31). In General Relativity, however, these quantities enter into the formula for the gravitational field. A precise gravitational experiment could therefore allow a test to distinguish Poynting's choice of u, $\mathbf{N}$ from others that could be made. However, observations to date are not sufficiently accurate to carry out such a test.

16.4.1   Examples of energy density and energy flow

Now we shall explore the physical meaning of u and $\mathbf{N}$ by considering some examples.

Consider a stationary spherical ball of charge. We suppose the ball has a uniform charge density ρ. There is no movement, so no magnetic field. Suppose we had to construct such a ball: we would have to arrange to bring up some charge from a long way away, and push it onto the ball. At any given moment, part way through this construction process, the ball has radius r and therefore total charge

$$q(r) = (4/3)\pi r^3 \rho.$$

The work required to bring up the next little piece dq of charge from infinity to the edge of the ball is

$$dW = -\int_\infty^r f dr' = \frac{q(r)dq}{4\pi\epsilon_0 r}$$

where f is the Coulomb force. Let us write $Q = (4/3)\pi a^3 \rho$ for the total charge on the ball when it reaches its final size a, then

$$q(r) = (r/a)^3 Q, \qquad \Rightarrow \qquad r = (q/Q)^{1/3} a.$$

This allows us to perform the integral for W, obtaining

$$W = \int_0^Q \frac{q}{4\pi\epsilon_0 a} \left(\frac{Q}{q}\right)^{1/3} dq = \frac{3}{5}\frac{Q^2}{4\pi\epsilon_0 a}.$$

Now let us calculate the energy stored in the fields, according to the energy density eqn (16.32i).

Outside the ball the electric field is the same as the field due to a point charge: $E = Q/(4\pi\epsilon_0 r^2)$, radially outwards. Inside the ball the field at any given r is also the same as the field due to a point charge, but the charge in question is now that contained inside the radius r: i.e., $q(r)$. This leads to a field radially outwards again, but increasing linearly with radius: $E = rQ/(4\pi\epsilon_0 a^3)$. The total field energy given by eqn (16.32) is

$$U = \int u\,\mathrm{d}V = \frac{\epsilon_0}{2}\left\{\int_0^a \left(\frac{rQ}{4\pi\epsilon_0 a^3}\right)^2 \mathrm{d}V + \int_a^\infty \left(\frac{Q}{4\pi\epsilon_0 r^2}\right)^2 \mathrm{d}V\right\}$$

$$= \frac{Q^2}{4\pi\epsilon_0 a}\left(\frac{1}{10} + \frac{1}{2}\right) = \frac{3}{5}\frac{Q^2}{4\pi\epsilon_0 a} \tag{16.33}$$

(where the volume element is $\mathrm{d}V = (4\pi r^2)\mathrm{d}r$). This is a standard exercise in elementary electromagnetism, but we displayed it in full in order to comment on the result and raise some more subtle issues later. The amount of work done against the Coulomb repulsion of the charges is found to equal the amount of energy stored in the *whole* field, both inside and outside the ball. So who 'owns' the energy? When one first learns electrostatics, one is usually invited to say that the work done in bringing one charge near to another charge can be described in terms of 'potential energy' of the charge. The work is done, but the charges are not moving at the end, so where did the energy go? In order to preserve energy conservation, this idea of 'potential energy' was introduced. We now see that this was misleading. The charges do not possess any energy beyond their rest energy and kinetic energy. The energy someone provided by doing the work has gone into the *field*. 'Potential energy' is misleading, especially in relativity theory, because it is non-local and does not contribute to the inertia of a particle.

It seems odd at first that the energy is not contained in the ball. That is where it might appear that we put it, but in fact we did not: the forces pushed on the charges throughout their journey from far far away, and they did their work *locally*, putting energy into the electromagnetic field at each place. Again, Special Relativity insists on local conservation if energy is to be conserved at all. Only a small part of the total energy ends up inside the ball. Indeed, if we had constructed instead a thin spherical shell of charge then one could arrange that only a negligible fraction of the total energy is inside the shell.

Next we investigate the Poynting vector $\mathbf{N}$.

Figure 16.9 shows the Poynting vector in the vicinity of a uniformly moving sphere of charge (with no fields present other than its own). We shall call this sphere a 'particle'. However, we shall not allow our 'particle' to be point-like, because we want to avoid the possibility of infinite field energy. Since the $\mathbf{E}$ field is radial and $\mathbf{B}$ circles around, $\mathbf{N}$ is everywhere directed tangential to the surface of a sphere around the particle. The vector is directed from behind the charge to in front of it, representing the movement of field energy as the fields fade away behind

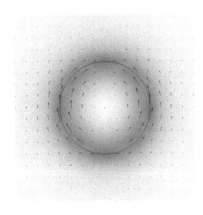

Fig. 16.9 The energy density u and Poynting vector $\mathbf{N}$ in the vicinity of a uniform sphere of charge in uniform motion, with no fields present other than its own. The shading indicates u, and the arrows indicate $\mathbf{N}$ (by their length and direction).

the charge and build up in front of it. If the charge is not accelerating then there is no net influx or outflux of energy towards or away from it. The vanishing field behind the charge provides just enough energy for the increasing field in front to build itself up ... until it fades in turn and passes the energy on.

Fig. 16.10 shows the case of a charged sphere moving at constant velocity in a uniform applied electric field. Again we shall call it a 'particle'. Now there is a net influx of energy into any sphere surrounding the particle. This makes sense because the applied field is doing work on the particle. To have a specific model, imagine that we have a uniform charged sphere, moving in a neutral viscous medium. It has reached its terminal velocity in the medium so moves at constant $\mathbf{v}$, and the net result is that the applied electric field $\mathbf{E}_0$ does work on the charged sphere, which in turn puts energy into the viscous medium, at the rate $\mathbf{f} \cdot \mathbf{v} = qE_0v$, where q is the total charge carried by the sphere.

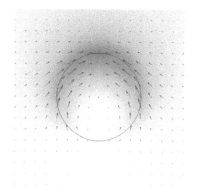

Fig. 16.10 The energy density u and Poynting vector $\mathbf{N}$ in the vicinity of a charged sphere in uniform motion, in a region where there is an applied uniform static electric field in the vertical direction. The Poynting vector now shows a net flux of electromagnetic energy towards the sphere. (Since the sphere is not accelerating there must be other forces on it, and it expends the received electromagnetic energy by doing work against those other forces.)

When is it acceptable to use 'potential energy'?   Having strictly rejected the notion of 'potential energy' as a fundamental property, we can reintroduce it as a calculational device that can be convenient in some circumstances. The idea can be adopted when we have a situation where the field does not 'leak' energy by some process that we are not taking into account. Potential energy is like money in the bank. If I have one million gold sovereigns in the bank, then as I walk around my pockets are not weighed down by one million gold sovereigns, but I am confident that, should I ask the bank for a gold sovereign, I will receive one (and my account will be diminished by 1)—except that the bank may get into financial trouble and disperse 'my' money. In the case of electrostatics, the field acts as a totally trustworthy 'bank' as far as charged particles are concerned, so that a particle can have confidence that by moving away it could pick up the kinetic energy it 'expects' on the basis of a potential energy calculation. We used this idea, for example, in the consideration of motion under a central force in section 4.2.6. The concept begins to fail, however, in dynamic problems when the fields can move energy around by wave motion. Then we have to abandon the idea of potential energy, and trust eqns (16.32).

Now let us calculate the energy flow in the field. We do not care whether we are discussing a point charge or a uniform ball of charge, since we shall only be calculating the fields outside such a ball, but we are assuming the ball is not perturbed by the applied field (i.e., it remains spherical and uniform). We shall calculate the Poynting vector at points on the surface of a sphere $\mathcal{R}$ with radius r centred on the origin, and then integrate over this surface. Note that the surface of integration is fixed in space: it does not move along with the charge. However, it is convenient to perform the calculation of $\mathbf{N}$ at the moment when the charge arrives at the centre of $\mathcal{R}$. At that moment the electric field outside the charged ball is

$$\mathbf{E} = \mathbf{E}_0 + \mathbf{E}_q = E_0\hat{\mathbf{z}} + \frac{q}{4\pi\epsilon_0 r^2}\hat{\mathbf{r}}$$

where, for simplicity, we treat the case of a slowly moving charge, $v \ll c$. The magnetic field is

$$\mathbf{B} = \frac{\mathbf{v} \wedge \mathbf{E}_q}{c^2} = \frac{qv\sin\theta}{4\pi\epsilon_0 c^2 r^2}\hat{\boldsymbol{\phi}}$$

(where $\hat{\boldsymbol{\phi}}$ is a unit vector in the azimuthal direction). The Poynting vector has two contributions:

$$\mathbf{N} = \epsilon_0 c^2 \mathbf{E} \wedge \mathbf{B} = \epsilon_0 c^2 (\mathbf{E}_0 \wedge \mathbf{B} + \mathbf{E}_q \wedge \mathbf{B}).$$

The second of these (the contribution from the charge's own fields) is directed around the sphere of integration, neither in nor out. The first (the applied electric field combining with the charge's magnetic field) is directed towards the z axis. This is the only term that will contribute to the surface integral. It has size

$$\epsilon_0 c^2 E_0 B = \frac{E_0 qv\sin\theta}{4\pi r^2}$$

so the net flux *in* through the surface of $\mathcal{R}$ is

$$\int_{\mathcal{R}} \mathbf{N} \cdot (-\mathbf{n})\, \mathrm{d}\mathcal{S} = \frac{qvE_0}{4\pi r^2} \int_0^{2\pi} \int_0^{\pi} \sin^2\theta\, r^2 \sin\theta \mathrm{d}\theta \mathrm{d}\phi$$

where $\mathbf{n}$ is the unit vector normal to the surface, and we used $\mathbf{N} \cdot (-\mathbf{n}) = N\sin\theta$. Thus there are three factors of $\sin\theta$: one from $\mathbf{E}_0 \wedge \mathbf{B}$, one from $\mathbf{N} \cdot \mathbf{n}$, and one from the surface element $\mathrm{d}\mathcal{S}$. The integral is easily done using $\sin^3\theta = \sin\theta(1 - \cos^2\theta)$, and we obtain

$$-\int_{\mathcal{R}} \mathbf{N} \cdot \mathbf{n}\, \mathrm{d}\mathcal{S} = \frac{2}{3} qE_0 v. \tag{16.34}$$

Thus the net energy flow in through our chosen surface is proportional to qvE_0, the work done on the charge, but there seems to be a mistake: a factor $2/3$ when we expected 1. There is no mistake. It is simply that we have not yet finished. We need to think about the field-energy density u as well. Is it constant or increasing or decreasing, inside region $\mathcal{R}$? At first one might imagine that we have a symmetry, so that $\int u\mathrm{d}V$ would be constant at the moment when the particle reaches the centre of $\mathcal{R}$, but the situation is not symmetrical forward and back. In front of the particle the fields $\mathbf{E}_0$ and $\mathbf{E}_q$ are in the same direction; they reinforce one another to make a big E^2. Behind the particle the fields $\mathbf{E}_0$ and $\mathbf{E}_q$ are opposed; they tend to cancel one another out, leaving a small E^2. So we can picture the energy density $u = \epsilon_0 E^2/2$ as 'heaped up' in front of the particle (see Fig. 16.10). In a given small time interval $\mathrm{d}t$ the particle travels a distance $\mathrm{d}z = v\mathrm{d}t$. If the particle is just passing through the centre of $\mathcal{R}$ then in the next travel distance $\mathrm{d}z$ the distribution of energy density will shift, such that a relatively large 'chunk' of energy is lost from the upper hemisphere of $\mathcal{R}$, while the lower hemisphere gains a smaller amount. Note that here we are not talking about a transport of energy (we have already calculated that) but a rise or fall of energy owing to non-zero values of du/dt. The net effect is calculated (by you) in the exercises: it is found to be just $dU = -(1/3)qE_0dz$, so

$$\frac{d}{dt}\int_{\mathcal{R}} u dV = -\frac{1}{3}qE_0v. \tag{16.35}$$

The net effect, then, is that the energy influx of $(2/3)qE_0v$ plus a power $(1/3)qE_0v$ liberated from the field inside $\mathcal{R}$ combine to provide the qE_0v, which ends up being transferred to the viscous medium (or, more generally, to whatever further object or system the charge is pushing on).

It is noteworthy that whereas the applied force here acts along the direction of travel, the energy flows towards the charge from the *sides*, at right angles to the motion. This is connected to the fact that in this example the momentum of the charged object is not changing.

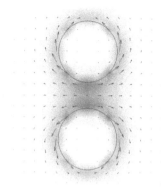

Fig. 16.11 shows the case of two opposite charges being pulled apart at constant speed. Since the charges mutually attract, their outward motion must be being caused by another system (such as a strong man) which does work *on* them. However, their energy is not increasing (their velocity is constant), so they are passing all this energy on to the electromagnetic field. The field energy moves outward from the charges. Although the field $(\mathbf{E}_1 + \mathbf{E}_2)$ between the charges is becoming smaller, so the energy density at any point there is falling, the volume of the region where $(\mathbf{E}_1 + \mathbf{E}_2)^2 > (E_1^2 + E_2^2)$ is becoming larger, with the net effect that the total field energy, integrated over all space, is increasing. Note once again that the direction of the energy flow is at first surprising— at right angles to the forces that do the work. Near either charge the situation is as in Fig. 16.10 but with directions reversed.

Fig. 16.11 Energy density and Poynting vector for a pair of charges being pulled apart, so as to move uniformly away from each other.

Other 'canonical' examples of Poynting's vector are the capacitor, the resistor, and the plane electromagnetic wave. In a plane wave (section 8.1) $\mathbf{N}$ points in the direction of the wave vector $\mathbf{k}$, which makes sense. Its size at any moment is $\epsilon_0 cE^2$ (since the fields are perpendicular and $E = cB$). If E oscillates as $E_0 \cos(\mathbf{k} \cdot \mathbf{r} - \omega t)$, then N oscillates as $N_0 \cos^2(\mathbf{k} \cdot \mathbf{r} - \omega t)$. Thus its direction is fixed but its size oscillates between zero and $\epsilon_0 cE_0^2$. The *intensity* I is defined as the power per unit area, averaged over a cycle: i.e., $I = \langle N \rangle = \epsilon_0 cE_0^2/2$. For such a wave the electric and magnetic contributions to the energy density are equal. Their total is $u = \epsilon_0 E^2$, which also oscillates. Its spatial average is $\epsilon_0 E_0^2/2$, and one can see that the intensity is c times this. The summary is

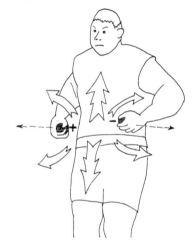

Fig. 16.12 Energy flow. The strong man is pulling the oppositely charged particles *apart*, thus putting energy *into* the electromagnetic field. (The flow of energy in the man's muscles is not shown).

$$I = \langle N \rangle = \langle u \rangle c = \tfrac{1}{2}\epsilon_0 E_0^2 c.$$

Next consider a cylindrical resistor of length d and radius a. If a current I flows and the voltage between the ends is V then the power dissipated in the resistor is VI. The magnetic field at the surface is $B = \mu_0 I/2\pi a = I/(2\pi\epsilon_0 c^2 a)$, directed in loops around the resistor in a right-handed sense with respect to the current. Inside the resistor there is an electric field in the direction of the current flow, of size $E = V/d$. Therefore the size of the Poynting vector is

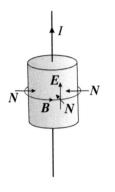

Fig. 16.13 Current-carrying resistor.

$$N = \frac{VI}{2\pi ad}$$

and its direction is radially inwards: i.e., pointing straight in through the curved surface. The area of that surface is $2\pi ad$ (circumference times length), so the total energy flow **in** to the resistor is VI per unit time—exactly matching what we know is dissipated there. This exact balance satisfies us that energy is conserved, but the sign needs a moment's thought. Surely the power VI is leaving the resistor, not coming in? The answer is that there is a conversion of energy going on: electromagnetic field energy enters the resistor and is used up accelerating the charges that carry the current. These charges in turn collide with the material of the resistor, heating it, turning their kinetic energy into heat. This heat subsequently leaves the resistor. Therefore the sign of the flow of field energy is correct: into the resistor.

The location of this flow can seem bizarre at first, however.

The battery is pushing on the charges, which are moving up the wire, so one might think the work is being done right there in the wire. One would expect that that is where the energy is being transported too: down the wire, from the battery to the capacitor. But Poynting says it is not: it is coming in from the sides! The example of a pair of charges (Fig. 16.11) should have prepared you for this. In fact, a moment's reflection should convince you that close to zero work is being done in the wire, because the electric field (and therefore the force) is close to zero there. The work is done in the battery, which draws on energy stored in the fields of its molecules (also called 'chemical energy') to pull apart electrons and positive ions (the very process shown in Fig. 16.11). This 'pumps' energy out of the sides of the battery into the surrounding field. The energy is transported through the field and eventually comes in through the sides of the resistor.

A similar argument can be made for a parallel plate capacitor being charged at a constant rate. The field between the plates grows, and the energy it needs arrives by coming in through the sides, not along the wires.

It turns out that often we do not need to keep track of these energy movements: we can just trust the fields to take care of it without our needing to know the details. However, if we want to hold on to the principle of energy conservation, then Poynting's vector gives a clear and thorough (and correct!) treatment.

16.4.2   Field momentum

The Poynting vector describes the flux of *energy*. We would like to know also about momentum. Does an electromagnetic field carry momentum? The only Lorentz-covariant answer is *yes*. We already presented in section 5.6 the fact that there is a very general relationship between energy flux and momentum density; eqn (5.58). Therefore, we should like to claim that the momentum per unit volume carried by an electromagnetic field is given by

$$\mathbf{g} = \mathbf{N}/c^2. \tag{16.36}$$

This turns out to be correct, but in the course of looking into it we shall begin to uncover the limits of classical electromagnetism.

First let us give some evidence for our claim that we can apply the formula (16.36) to electromagnetic fields. Consider, for example, the phenomenon of radiation pressure. A plane wave incident on a (non-transparent) material body exerts a force on the body. To see how this comes about, consider the motion of a charged particle such as an electron on the surface of such a body. The electric field of the incident wave drives the electron in the *transverse* direction. For example, if the wave is propagating in the z direction and is linearly polarized with its electric field along x, then the electron is pushed in the x direction. This does not give rise to a force in the direction of propagation of the wave. However, the non-zero x component of velocity causes the electron to feel also the magnetic force $q\mathbf{v} \wedge \mathbf{B}$, and this is in the z direction, and causes the radiation pressure. The electron of charge q, velocity $\mathbf{v}$ absorbs energy from the wave at the rate $q\mathbf{E} \cdot \mathbf{v}$. For simplicity let us suppose the motion of the electron is always in the x-direction (the force in the z-direction being opposed by equal and opposite forces from the rest of the body). Then the rate at which energy is being transferred from the field to the body via the electron is qEv, and the Lorentz force component in the z direction is qvB. Therefore the energy and the impulse delivered during some time interval t are[2]

$$\text{energy} = \int qvE\mathrm{d}t, \qquad \text{momentum} = \int qvB\mathrm{d}t.$$

Since for a plane wave the field strengths are related by $E = cB$, we find the ratio of the energy delivered and momentum delivered is c, the same as the ratio of energy and momentum for particles of zero rest mass. It follows that eqn (16.36) can be asserted for electromagnetic plane waves, and therefore (by using Fourier analysis) for electromagnetic waves more generally.

The '4/3 problem'

Next, let us consider another example in which energy is transported by a field. Suppose there exists an electromagnetic field that presents itself as a static electric field in some reference frame. The field possesses energy, the integral of its energy density u over all space. Now consider the situation from the point of view of a reference frame moving with respect to the first. Is the energy content of the field moving in the new frame? The answer is surely 'yes'. Whatever charges gave rise to the field in the first frame are now in motion. The energy of the field must be moving at precisely the same velocity as the charges, because we can imagine deconstructing the charge distribution at some later time and reclaiming the energy stored in the field. For example, think of a capacitor sent on an interstellar journey. It assuredly takes its field energy with it!

[2] If the electron moves freely apart from the forces arising from the wave, then during each cycle of oscillation it undergoes a driven motion but does not on average absorb any energy; a body with only such particles in it would be transparent. If the electron experiences forces from the rest of the body which tend to damp its motion, then the average of $q\mathbf{E} \cdot \mathbf{v}$ over a cycle is non-zero.

Now let us calculate the momentum content of such a 'moving static field'. For the sake of simplicity we will consider a spherically symmetric static electric field, and we will assume the new reference frame moves at low velocity $v \ll c$ relative to the frame containing the static field. Then the electric field E in the new frame is equal to that in the first frame (up to order v^2/c^2) and the magnetic field is $\mathbf{B} = \mathbf{v} \wedge \mathbf{E}/c^2$. The momentum density is

$$\mathbf{g} = \mathbf{N}/c^2 = \epsilon_0 \mathbf{E} \wedge \mathbf{B} = \frac{\epsilon_0}{c^2} \mathbf{E} \wedge \mathbf{v} \wedge \mathbf{E}.$$

Let $\mathbf{v}$ be along the z-axis and let θ be the angle between $\mathbf{r}$ and this axis, then $\mathbf{g}$ has size $g = (\epsilon_0/c^2)E^2 v \sin\theta$ and its z-component is $g_z = g \sin\theta$. When we integrate over all space to determine the total momentum in the field, only the z component survives, and therefore we have

$$
\begin{aligned}
p &= \int_0^\pi \int_0^\infty \frac{\epsilon_0 E^2 v}{c^2} \sin^2\theta \; r^2 \sin\theta \, dr d\theta \, 2\pi \\
&= \frac{4v}{3c^2} \int_0^\infty \frac{1}{2} \epsilon_0 E^2 4\pi r^2 dr
\end{aligned}
\tag{16.37}
$$

Now, for the low velocity under consideration the magnetic field is weak, and contributes negligibly to the energy density compared to the electric field. Therefore, we can recognise the integral on the right-hand side of eqn (16.37) as the total energy content $\mathcal{E}$ of the field (here we write $\mathcal{E}$ for energy to avoid confusion with the electric field strength). We conclude that

$$\frac{\mathbf{p}}{\mathcal{E}} \stackrel{!}{=} \frac{4}{3} \frac{\mathbf{v}}{c^2}. \tag{16.38}$$

This result violates the relation $\mathbf{p} = \mathcal{E}\mathbf{v}/c^2$, which is the universal relationship between energy and momentum for bodies moving at any speed. This '4/3 problem' troubled early workers such as Lorentz. It implies, for example, that the total energy and momentum of this field cannot be considered as a 4-vector.

There is nothing wrong with our calculations of the energy and momentum in the field. Both are correct. The 'problem' is merely that we cannot consider this energy and momentum to be parts of a 4-momentum. The reason is that the field we have considered is not an isolated system. It is in continual interaction with the charges which act as its source. The relation $\mathbf{p} = \mathcal{E}\mathbf{v}/c^2$ applies only to particles or to extended objects that can be considered as isolated entities, free of external influences. We did not encounter this problem for electromagnetic waves because they have a special property: they can be *source-free*. That is to say, although the disturbance which gives rise to electromagnetic waves is usually a charged object in motion, once the source ceases to accelerate the emitted radiation continues to propagate, such that there can exist a source-free volume of space completely containing the electromagnetic radiation field. Such a field *can* be considered to be an isolated system possessing an energy and momentum of its own. Therefore it is legitimate to regard a light-pulse as a single entity with a well-defined energy-momentum 4-vector.

The non-4-vector nature of $\mathcal{E}, \mathbf{p}$ for a static field is also an illustration of the issue we discussed in connection with figure 5.3 and eqn (5.9): one cannot assume that adding up 4-vectors evaluated at different points in space will necessarily give a 4-vector total. It requires something like a conservation law to come into play, to guarantee that the total will give the same 4-vector no matter which time slice is used to calculate it. In the present case we are adding up (i.e., integrating over volume) the energy and momentum of each small region of field; the sum is a 4-vector only if the field is an isolated system, not exchanging energy and momentum with anything else. A static field is not exchanging *energy* with other things, but it is in a state of continuous interaction with its sources, pulling on them. We can think of this, roughly, as a process of continuous elastic collision. If the sources are not accelerating then it must be that some other force is constraining them, and the net result is an interaction between the electromagnetic field and the other force-providing entity, mediated via the charges. A more complete understanding will emerge after we have grappled with this idea in more general terms. That is the subject of the next section.

16.4.3 Stress-energy tensor of the electromagnetic field

In view of our comments in section 16.1 of this chapter, it should not be surprising that the energy density of the electromagnetic field can be understood as one element of a tensor, and that tensor is the stress-energy tensor $\mathbb{T}$ for the electromagnetic field. It describes energy density, energy flux, momentum density, and momentum flux, just as we have already considered for particle beams and fluids, though now we are talking about an electromagnetic field.

Note that we do not construct a 4-vector out of u and the Poynting vector; rather, we find that they form one column of $\mathbb{T}$. Note also that as soon as we allow the idea of energy density, then by Lorentz covariance we must also have not only momentum density but also momentum flux—the σ^{ij} part of the tensor. Electromagnetic waves carry momentum with them, so they can transport momentum across a plane just as surely as can material particles (Fig. 16.1). Furthermore, even static fields exert forces, which means they too transport momentum.

To study the energy and momentum flow in full, a good starting point is the interaction of the fields with the charges, described by the Lorentz force. The conservation of energy and momentum must be expressed by an equation like (16.20), but we shall define the force per unit volume W to be the force exerted *by* the field *on the particles*, so the equation we want is

The 4-force density is defined as

$$W^a \;=\; \mathbb{F}^{a\mu} J_\mu \qquad = (\mathbf{E} \cdot \mathbf{j}/c,\ \rho\mathbf{E} + \mathbf{j} \wedge \mathbf{B}). \qquad (16.40)$$

$$\left[W \;=\; \mathbb{F} \cdot J \right]$$

This is called the *Lorentz force density*. To understand it, multiply by a small volume dV, and then you find that the components are $q\mathbf{E} \cdot \mathbf{v}/c$ and $q(\mathbf{E} + \mathbf{v} \wedge \mathbf{B})$. This is the rate of doing work and the Lorentz force on the charge q contained in the volume dV.[3]

$\mathbb{T}^{ab}$ has the physical dimensions of force per unit area or (equivalently) energy per unit volume.

We can already see that the first column of $\mathbb{T}^{ab}$ should be equal to $(u, \mathbf{N}/c)$, which will yield energy conservation; eqn (16.29). Now we will show how the rest of $\mathbb{T}$ is obtained from the field equations. We shall do this in two ways, both of which provide useful insight. The first method uses 3-vectors and Maxwell's equations to look at just the momentum flow; this results in a suggestion as to how the stress-energy tensor might be formed. The second method uses Lorentz-covariant language (4-vectors and 4-tensors) throughout, and therefore proves that the resultant object is a 4-tensor (i.e., is guaranteed to transform in the right way). This also offers practice in 4-tensor manipulation.

Method 1: 3-vector approach

We examine $\partial\mathbf{g}/\partial t$. This should tell us about the rate of change of momentum, and therefore about the force. Note that although $\mathbf{g}$ is related to the Poynting vector $\mathbf{N}$, it is best to temporarily forget that relation here. In the conservation argument the momentum density $\mathbf{g}$ plays the role, for momentum, which was played by *energy density* u in the Poynting argument. The quantity handling the flow of momentum (i.e., the job done for energy by $\mathbf{N}$) is the *tensor* $\mathbb{T}$.

Conservation of momentum will be achieved if the force on the particles in a small volume dV is equal and opposite to $dV\partial\mathbf{g}/\partial t$, plus another term which signifies the rate at which the field is carrying momentum away. Using eqn (16.36) and Maxwell's equations M3, M4, we have

$$\frac{\partial\mathbf{g}}{\partial t} = \epsilon_0 \left(\mathbf{E} \wedge \frac{\partial\mathbf{B}}{\partial t} + \frac{\partial\mathbf{E}}{\partial t} \wedge \mathbf{B} \right)$$

$$= -\mathbf{j} \wedge \mathbf{B} + \epsilon_0 \left((\boldsymbol{\nabla} \wedge \mathbf{E}) \wedge \mathbf{E} + c^2(\boldsymbol{\nabla} \wedge \mathbf{B}) \wedge \mathbf{B} \right).$$

The first term is the magnetic part of the Lorentz force per unit volume (recall $\mathbf{j}dV = q\mathbf{v}$), with a minus sign as expected. The rest must be either to do with the electric part of the force, or with momentum flow. The

[3] WdV is not a 4-force, because it is missing a factor γ, says eqn (2.75). W is a 4-vector however: it is a 4-force divided by an *invariant* volume dV_0, namely a volume element in the rest frame of the local charge, $dV_0 = \gamma dV$. The 4-force on the charge contained in the laboratory frame volume dV is $WdV_0 = \gamma WdV$. Thus W can be interpreted either as work rate and 3-force per unit volume, or as 4-force per unit proper volume. One does not need to know this, however; the algebraic development will be physically consistent and will lead to an easily interpreted final result in terms of $\mathbf{E}, \mathbf{B}, \rho, \mathbf{j}$.

electric part of the force per unit volume is $\rho\mathbf{E}$, which in terms of the fields alone is $\epsilon_0(\nabla \cdot \mathbf{E})\mathbf{E}$ (using M1). Adding this on, in order to obtain the total force, we have

$$\rho\mathbf{E} + \mathbf{j} \wedge \mathbf{B} = -\frac{\partial \mathbf{g}}{\partial t} + \epsilon_0\left[(\nabla \cdot \mathbf{E})\mathbf{E}\right.$$

$$\left. + (\nabla \wedge \mathbf{E}) \wedge \mathbf{E} + c^2(\nabla \wedge \mathbf{B}) \wedge \mathbf{B}\right]. \qquad (16.41)$$

The term in the square bracket can be written, we hope, as minus the divergence of something. It can, but this argument does not offer an automatic way to see it. Take a look at eqn (16.47) and you will find the answer is

$$\epsilon_0\left[\cdots\right]_i = -\left(\frac{\partial\sigma_{ix}}{\partial x} + \frac{\partial\sigma_{iy}}{\partial y} + \frac{\partial\sigma_{iz}}{\partial z}\right)$$

where

$$\sigma_{ij} = \tfrac{1}{2}\epsilon_0(E^2 + c^2B^2)\delta_{ij} - \epsilon_0(E_iE_j + c^2B_iB_j).$$

You are invited to check this by performing the differentiation. You will find that the B^2 term is needed to give part of $(\nabla \wedge \mathbf{B}) \wedge \mathbf{B}$; it also contributes a $(\nabla \cdot \mathbf{B})\mathbf{B}$ term, but this vanishes by M2.

Method 2: 4-vector calculation

Now for the manifestly covariant approach. We start from the force density, eqn (16.40), and just as we used a Maxwell equation to express $\mathbf{j}$ in terms of the fields in Poynting's argument, now we use the first field equation (13.7) to express J in terms of $\mathbb{F}$:

$$\mathsf{W}^a = -\mathbb{F}^{a\mu}(\epsilon_0c^2)\partial_\lambda\mathbb{F}^\lambda_{\ \mu} \qquad [\,\mathsf{W} = -\,\epsilon_0c^2\mathbb{F}\cdot(\Box\cdot\mathbb{F})]\ \ (16.42)$$

The matrix notation helps to see clearly what we have: it has the structure '$a\partial a$', so it should be possible to relate it to '$\partial(aa)$'. This is the equivalent of step (16.30) in Poynting's argument, and eqn (C.6) (the product rule) contains the result we need:

$$\partial^\lambda\left(\mathbb{F}^{a\mu}\mathbb{F}_{\lambda\mu}\right) = \mathbb{F}^{a\mu}\partial^\lambda\mathbb{F}_{\lambda\mu} + \mathbb{F}_{\lambda\mu}\left(\partial^\lambda\mathbb{F}^{a\mu}\right). \qquad (16.43)$$

The left-hand side of eqn (16.43) is a divergence of a tensor—the very thing we are looking for—so next we concentrate on the last term:

$$\mathbb{F}_{\lambda\mu}\left(\partial^\lambda\mathbb{F}^{a\mu}\right).$$

Just as in Poynting's argument, we need to bring in the other field equation—this time the homogeneous one (the one without a source term). That equation, (13.8) has things such as '$\partial^c\mathbb{F}^{ab}$' in it. Clearly, what we need to do is contract it with $\mathbb{F}_{cb}$:

$$\mathbb{F}_{cb}\left(\partial^c\mathbb{F}^{ab} + \partial^a\mathbb{F}^{bc} + \partial^b\mathbb{F}^{ca}\right) = 0$$

$$\Rightarrow \qquad \mathbb{F}_{cb}\partial^c\mathbb{F}^{ab} = -\mathbb{F}_{cb}\partial^a\mathbb{F}^{bc} - \mathbb{F}_{cb}\partial^b\mathbb{F}^{ca},$$

i.e. $\qquad \mathbb{F}_{\lambda\mu}(\partial^\lambda\mathbb{F}^{a\mu}) = -\mathbb{F}_{\lambda\mu}\partial^a\mathbb{F}^{\mu\lambda} - \mathbb{F}_{\lambda\mu}\partial^\mu\mathbb{F}^{\lambda a}\ \ (16.44)$

where all we did was to carry two terms to the right-hand side, and then relabel dummy indices to make them look more like what we want.

Now, practising the advice given in section 12.2.3 to 'look for scalars', we spot that the first term on the right-hand side is almost a scalar. It is an example of eqn (C.9), except that one pair of indices is the wrong way round. However, since $\mathbb{F}$ is antisymmetric, we can swap them and introduce a minus sign, so we have $+\partial^a D$ where $D = \mathbb{F}_{\lambda\mu}\mathbb{F}^{\lambda\mu}/2$ (see eqn (13.12)).

With the hint that a transpose might be useful, now take a look at the last term in eqn (16.44) and transpose *both* occurrences of $\mathbb{F}$. This makes it look just like the left-hand side, except that the dummy indices are labelled differently. That does not matter, so we have found that

$$\mathbb{F}_{\lambda\mu}(\partial^\lambda \mathbb{F}^{a\mu}) = \partial^a D - \mathbb{F}_{\lambda\mu}(\partial^\lambda \mathbb{F}^{a\mu})$$

$$\Rightarrow \qquad \mathbb{F}_{\lambda\mu}\left(\partial^\lambda \mathbb{F}^{a\mu}\right) = \frac{1}{2}\partial^a D.$$

Substituting this into eqn (16.43), and returning to eqn (16.42), we have

$$\mathsf{W}^a = -\epsilon_0 c^2 \left(\partial^\lambda \left(\mathbb{F}^{a\mu}\mathbb{F}_{\lambda\mu} \right) - \frac{1}{2}\partial^a D \right),$$

$$\Rightarrow \qquad \mathsf{W}^b = -\epsilon_0 c^2 \left(\partial_\lambda \left(\mathbb{F}^{\lambda\mu}\mathbb{F}^b{}_{\mu} \right) - \frac{1}{2}\partial^b D \right). \qquad (16.45)$$

(by reversing the order of the product then using the see-saw rule twice and changing from a to b). We would like to set this equal to $-\partial_\lambda \mathbb{T}^{\lambda b}$, so we want to convert the ∂^b in the second term to ∂_λ. This is easily done by

$$\partial^b = g^{\lambda b}\partial_\lambda$$

Finally, using the antisymmetry of $\mathbb{F}$, we have

Stress-energy tensor [a]

$$\mathbb{T}^{ab} = \epsilon_0 c^2 \left(-\mathbb{F}^{a\mu}\mathbb{F}_\mu{}^b - \frac{1}{2}g^{ab}D \right), \qquad (16.46)$$

where $D = \frac{1}{2}\mathbb{F}_{\mu\nu}\mathbb{F}^{\mu\nu}$.

$$[\text{ i.e. } \mathbb{T} = \epsilon_0 c^2 \left(-\mathbb{F}\cdot\mathbb{F} - \frac{1}{2}gD \right).]$$

Substituting for $\mathbb{F}$ from eqn (7.46), we find

$$\mathbb{T}^{ab} = \left(\begin{array}{c|c} u & \mathbf{N}/c \\ \hline \mathbf{N}/c & \sigma_{ij} \end{array} \right)$$

where

$$\text{energy density} \quad u = \tfrac{1}{2}\epsilon_0 (E^2 + c^2 B^2)$$

$$\text{Poynting vector} \quad \mathbf{N} = \epsilon_0 c^2 \mathbf{E} \wedge \mathbf{B}$$

$$\text{3-stress tensor} \quad \sigma_{ij} = u\delta_{ij} - \epsilon_0 (E_i E_j + c^2 B_i B_j) \tag{16.47}$$

$$= -\sigma_{ij}^M$$

The 'Maxwell stress tensor' σ_{ij}^M is often used in the literature, and its standard definition is such that it is the negative of $\mathbb{T}^{ij}$. By using $\sigma_{ij} = -\sigma_{ij}^M$ we preserve a greater uniformity in the equations describing conservation of energy and momentum below.

$\mathbb{T}$ is fully symmetric. The symmetry of the space–space part is not surprising; the symmetry of the time–space part merits a comment. Suppose N is a 4-vector direction, then $\mathbb{T} \cdot \mathsf{N}$ quantifies the flow of energy and momentum in that direction. The first *row* of $\mathbb{T}$ is used to calculate the flow of *energy*, and the elements of the first *column* are used, together with σ_{ij}, to calculate the flow of *momentum*. That the first row is equal to the first column is an example of the equality of energy flux and momentum density that we noted in section 5.6.

You can now confirm that the time component of the relation $\mathsf{W} = -\Box \cdot \mathbb{T}$ is indeed eqn (16.29) as expected, which represents energy conservation.

Using eqn (16.40) and examining the x-component of $\Box \cdot \mathbb{T}$, we find

$$(\rho\mathbf{E} + \mathbf{j} \wedge \mathbf{B})_x = -\frac{1}{c^2}\frac{\partial N_x}{\partial t} - \nabla_j \sigma_{jx}.$$

To interpret the equation it may be helpful to integrate over a small volume to make the terms more familiar. If the volume is taken small enough that all the charge q in it moves at the same velocity $\mathbf{v}$, then we have

$$q\,(\mathbf{E} + \mathbf{v} \wedge \mathbf{B})_x = -\int \frac{\partial g_x}{\partial t} + \boldsymbol{\nabla} \cdot \boldsymbol{\sigma}_x \mathrm{d}V \tag{16.48}$$

where we wrote $\boldsymbol{\sigma}_x$ for σ_{jx}. This is the flux of x-momentum, and we used eqn (16.36) to convert the Poynting vector into a momentum density. Eqn (16.48) can be 'read' as a statement of Newton's Third Law for the interaction of charge and field. On the left is the force on the charge, and on the right is the force on (i.e., rate of injection of momentum into) the field. The equation states that these forces ('action' and 'reaction' if you like) are equal and opposite. The rate of injection of momentum into the field appears in two parts: $\partial g_x/\partial t$ is the rate at which the momentum of the local field is growing; $\boldsymbol{\nabla} \cdot \boldsymbol{\sigma}_x$ is the rate at which momentum is being supplied to the rest of the field by flowing out of the region under consideration. This confirms the notion of momentum transport that we expected before embarking on the calculation of $\mathbb{T}$.

Gathering the energy and all three momentum components together, the overall conclusion is:

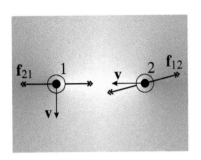

Fig. 16.14 A repeat of Fig. 16.8, but with the effects on the field shown. Each black dot represents a particle, and the attached arrows show the velocity of and force on the particle. Each circle represents a small volume of field, the attached arrows show the rate of injection of momentum from the sources into that volume of field. Conservation of momentum is achieved locally. The stress and momentum density throughout the rest of the field is not shown; it satisfies a continuity equation for each component.

Conservation of 4-momentum of both matter and field together

$$(\mathbf{E} \cdot \mathbf{j}/c, \quad \rho\mathbf{E} + \mathbf{j} \wedge \mathbf{B}) = -\left(\frac{1}{c}\frac{\partial}{\partial t}, \ \boldsymbol{\nabla}\cdot\right)\left(\begin{array}{c|c} u & \mathbf{N}/c \\ \hline \mathbf{N}/c & \sigma \end{array}\right) \quad (16.49)$$

This result is at the heart of all energy-momentum conservation in electromagnetism.

It is sometimes stated that Newton's third law (on action and reaction) breaks down in Special Relativity. It certainly does not, and eqn (16.49) is the proof for the case of electromagnetic interactions. However, it is true to say that Newton's Third Law should not be taken to be a statement about forces at separate locations (e.g., on particles with finite separation); it must be applied locally. What was missing in Fig. (16.8) was a pair of arrows showing the rate of change of momentum in the field. We can now provide those arrows; see Fig. 16.14.

The intuition that 4-momentum should be conserved has been fully born out by the theory. Indeed, the requirement to conserve energy in all reference frames *implies* the conservation of momentum, by the zero-component lemma. We have discovered that in order to make sense of these great conservation principles it is necessary to credit an electromagnetic field not only with energy and momentum, but also with pressure and stress.

Simple examples of stress and pressure

We introduced the stress-energy tensor of the electromagnetic field by using the language of energy and momentum transport. However, in view of the discussion in the first part of this chapter, we know that σ^{ij} can also be interpreted as force per unit area. This helps to get a physical intuition of what it means. Some examples are given in table 16.2.

The part of the stress acting normal to the boundary of an object, in its own rest frame, is the *pressure*, and this is the easiest part to understand. Consider a capacitor, for example. For a parallel plate capacitor aligned along x there is a uniform electric field E in the x direction, and no magnetic field. The force on either plate is $f = QE/2$, where the charge on the plate $Q = \epsilon_0 AE$, so $f = \epsilon_0 E^2 A/2$. Now look at the 3-stress tensor. It is

$$\sigma_{ij} = \frac{1}{2}\epsilon_0 E^2 \begin{pmatrix} -1 & 0 & 0 \\ 0 & 1 & 0 \\ 0 & 0 & 1 \end{pmatrix}. \quad (16.50)$$

A negative pressure is a *tension*; it means that in the x direction the field is pulling its boundary (i.e., the charges on the plate) in towards it (the field). This is the attraction between the capacitor plates that we normally describe as the mutual attraction of opposite charges. The positive xx and yy terms tell us something further: there is an outwards pressure at right angles to the field direction. In general, in the absence of magnetic fields there is always a tension along the electric field lines, providing the 'mechanism' by which opposite charges attract, and there is a pressure at right angles to the field lines, tending to push them apart, and providing the 'mechanism' by which like charges repel.

In a solenoid with a magnetic field B along the x direction, the 3-stress tensor is the same as eqn (16.50), but with E replaced by cB. There is an outward pressure $B^2/2\mu_0$ on the walls of the solenoid, and a tension along the axis.

An electromagnetic wave in free space exerts a force in the direction of travel, and no transverse force. We can always align the x axis with the direction of propagation and write down both $\mathbb{T}$ and the 4-wave-vector $\mathsf{K} = (k, k, 0, 0)$. By spotting that for this case

$$\mathbb{T}^{ab} = \epsilon_0 c^2 \frac{E_0^2}{\omega^2} \cos^2(\mathsf{X}_\mu \mathsf{K}^\mu) \mathsf{K}^a \mathsf{K}^b$$

we deduce that this is the general relationship. By the quotient rule, this implies that *in this context* E_0^2/ω^2 is a scalar, and therefore for a given wave examined in two arbitrarily related reference frames, the energy density, momentum flux, pressure, electric field, and frequency are related by

$$\frac{u'}{u} = \frac{g'}{g} = \frac{p'}{p} = \frac{E_0'^2}{E_0^2} = \frac{\omega'^2}{\omega^2}. \tag{16.51}$$

Table 16.2 Example stress-energy tensors. Entries left blank are zero. The examples are necessarily given in some suitably chosen reference frame; for a covariant expression in each case use eqn 16.46. The capacitor, solenoid and plane wave are aligned along the x axis. 'Point charge' refers to the Coulomb field of a point charge at the origin. All but the last example are stress-free in the chosen frame. Note that for electromagnetic fields, $\mathbb{T}^\lambda_\lambda = 0$ (easily proved from eqn (16.47)).

Capacitor	$\frac{1}{2}\epsilon_0 E^2$	$\begin{pmatrix} 1 & & & \\ & -1 & & \\ & & 1 & \\ & & & 1 \end{pmatrix}$			
Solenoid	$\frac{1}{2}\epsilon_0 c^2 B^2$	$\begin{pmatrix} 1 & & & \\ & -1 & & \\ & & 1 & \\ & & & 1 \end{pmatrix}$			
Plane wave	$\epsilon_0 E_0^2 \cos^2(kx - \omega t)$	$\begin{pmatrix} 1 & 1 & & \\ 1 & 1 & & \\ & & 0 & \\ & & & 0 \end{pmatrix}$	$= \epsilon_0 c^2 (E^2/\omega^2)\mathsf{K}^a\mathsf{K}^b$		
Point charge	$\frac{q^2}{16\pi^2\epsilon_0 r^6}$	$\begin{pmatrix} \frac{r^2}{2} & 0 & 0 & 0 \\ 0 & \frac{r^2}{2} - x^2 & -xy & -xz \\ 0 & -yx & \frac{r^2}{2} - y^2 & -yz \\ 0 & -zx & -zy & \frac{r^2}{2} - z^2 \end{pmatrix}$			

This is the result that we established by less sophisticated methods in chapter 2 (eqn (3.9)).

The point charge (Coulomb field) exhibits a negative pressure and stresses directed towards the origin. What this means is that if you arrange a 'boundary wall' in some region—such as to leave the field on one side of the boundary unaffected but reduced to zero on the other side—then in this case the boundary will be pulled in the direction of the side where the field remains non-zero. For example, place a uniform spherical shell of total charge $-q$ around the point charge q, at some finite radius. Then the field inside the shell is unchanged and the field outside is reduced to zero (from Gauss's theorem). The stress tensor tells us that the spherical shell will experience forces pulling it in towards the origin—which of course we know to be true from the attraction of opposite charges.

What is striking about all this is that the electromagnetic field is behaving like a substantial thing, like a lump of jelly that we could push or pull and be pushed and pulled by in return. It is no wonder that so much time and energy was devoted to the aether model of electromagnetism in the nineteenth century. This time was not wasted: it forced physicists such as Maxwell, Lorentz, and Minkowski to discern and expound these properties. They make the field seem very much like a mechanical entity. Now we have come full circle, and one could say that we do have an 'aether' after all, but *the field itself* is the 'aether'. It is an aether with properties that could not be grasped before the advent of Special Relativity, such as the ability to propagate signals that you cannot catch up with.

Having started in chapter 7 with the idea that electromagnetic fields can seem intangible, it is time to reconsider. Far from being 'hard to see and touch' the electromagnetic field is just about the only thing we ever see or touch! The retinas of our eyes respond to incoming light-waves; the nerve receptors in our fingers respond to the pressure that results when we push the electron clouds of our skin molecules up against the fields supplied by the electron clouds of other objects. The chemical reactions that stimulate our taste buds are a dance of electrons in response to fields in further molecules. Even sound—a pressure wave— relies on electromagnetic fields to allow the air molecules to pass on the pressure as they collide.

Trouton–Noble revisited

We briefly considered the Trouton–Noble experiment in chapter 4; see Fig. 4.1. We there gave an argument sufficient to show that there is no net torque on the system, in any frame, but we did not examine the way by which torque and angular momentum is distributed between the three entities that are involved: rod, charges, and electromagnetic field.

Fig. 4.1 makes clear that the net force on either of the charged particles is zero. Nevertheless, there is a force communicated by each charged particle on to the end of the rod to which it is fixed. Therefore the *rod* does experience non-zero force on each end, and the two forces are not aligned (in frames other than the rest frame), so there is an external torque on the rod, equal to (see Fig. 16.15)

$$f_\| l_0 \cos\theta_0 - f_\perp l_0 \sin\theta_0/\gamma = f l_0 \sin\theta_0 \cos\theta_0 (1 - 1/\gamma^2)$$
$$= f l_0 (v^2/c^2) \sin\theta_0 \cos\theta_0 \quad (16.52)$$

where l_0 is the rest length of the rod, and θ_0 is the angle between the rod and a vertical axis in the rest frame.

It follows that the rod must be acquiring angular momentum (and the electromagnetic field is losing angular momentum). So why does the rod not start to rotate? Or to put it another way, how does it manage to 'push back' in a direction that generates a counter-torque?

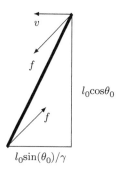

Fig. 16.15 Lengths and directions in Trouton–Noble experiment.

To answer this question, notice that the upper force does positive work $\mathbf{f} \cdot \mathbf{v}$ that provides energy to the top of the rod, and this energy is removed at the bottom where the other applied force does negative work. Therefore there is an energy current $dE/dt = \mathbf{f} \cdot \mathbf{v} = fv \sin\theta_0$ and an associated momentum density. If A is the cross-section through which the energy flows, then the energy flux is $(fv/A) \sin\theta_0$, and therefore the momentum density is $(fv/c^2 A) \sin\theta_0$. The *direction* of this "hidden" momentum is surprising: it is *vertically downwards*. It is located throughout the rod (in equilibrium the internal forces simply pass it on without either augmenting or diminishing it), so the total momentum in the flow is

$$p = \int \frac{fv}{c^2 A} \sin\theta_0 dV = f(v/c^2) l_0 \sin\theta_0 \cos\theta_0$$

(since the vertical height of the rod is $l_0 \cos\theta_0$). Let $d\mathbf{p}$ be the amount of momentum in any given small region; if this region is at position $\mathbf{r}$ then it contributes an angular momentum about the origin of $\mathbf{r} \wedge d\mathbf{p}$. The rate of change of this angular momentum is $\mathbf{v} \wedge d\mathbf{p}$, and therefore the total rate of change of angular momentum is

$$\frac{d\mathbf{L}}{dt} = \mathbf{v} \wedge \mathbf{p} = f(v^2/c^2) l_0 \sin\theta_0 \cos\theta_0 \mathbf{e}$$

where $\mathbf{e}$ is a unit vector in the direction of $\mathbf{v} \wedge \mathbf{p}$. This matches both the size and direction of the applied torque.

16.5   Field and matter pushing on one another

Now for a beautiful result.

In this chapter we first discussed the energy and momentum of a continuous material system, and then discussed the energy and momentum of the electromagnetic field. Now suppose that we have a fluid, and that the external forces on it are wholly electromagnetic in origin. Then the force density K in eqn (16.20) is none other than the Lorentz force density W. Therefore we shall find

$$\square \cdot \mathbb{T}_{\text{matter}} = \mathsf{W},$$

$$\square \cdot \mathbb{T}_{\text{field}} = -\mathsf{W},$$

therefore

Matter interacting with electromagnetic field

$$\square \cdot (\mathbb{T}_{\text{matter}} + \mathbb{T}_{\text{field}}) = 0 \qquad (16.53)$$

The total stress-energy tensor of matter and field together has zero divergence. This result encapsulates energy-momentum conservation for electromagnetic interactions. When other interactions are present, further terms must be added, and the overall result remains of the same form: the total stress-energy tensor of an isolated system is divergence-free. This is a completely general way to express energy-momentum conservation, and it is an important basic idea in General Relativity.

Footnote on the Lorentz force equation

In chapter 13 we obtained Maxwell's equations by starting from the notion of a pure force, with the result that the field tensor was initially defined through the Lorentz force equation (13.2). We then hypothesized field equations, and we had to confirm laboriously that energy and momentum were conserved. Fortunately it turned out that they were. One wonders whether there is another way to get at this connection. There is. By using Lagrangian methods suitable to fields, one can guess or hypothesize Lagrangian densities for the simplest possible fields in vacuum, and by proceeding to a Hamiltonian one obtains eqn (16.47) for the stress-energy tensor of one possible tensor field without even mentioning the idea of charge or particles. One then finds that $\nabla \cdot \mathbf{E}$ is the density of a conserved quantity, so it is named 'charge density'. Upon evaluating $\square \cdot \mathbb{T}$ one finds that it can be non-zero. This leads one to propose the existence of something that the field can push on, and one introduces eqn (16.39) as a definition of 4-force density W (rather than a way to find $\mathbb{T}$). Then one is led inexorably to the fact that the thing pushed by the field (the 'passive charge') is also the source of the field (the 'active charge'). This interesting discovery enables one to write W as in eqn (16.40). Hence the Lorentz force equation may be regarded as a derived result, rather than an axiom, if one places the axiomatic emphasis on 4-momentum conservation.

16.5.1   Resolution of the '4/3 problem' and the origin of mass*

The '4/3 problem' indicated in eqn (16.38) was the problem that the energy and momentum of a field in interaction with sources do not constitute the energy and momentum of an isolated system. Equally, a charged particle (or a charged composite body) cannot be considered to be an isolated entity because it is in continuous interaction with its field. However, we would like to be able to discuss the dynamics of charged particles, and we normally take it for granted that we can do so by furnishing them with a 4-vector momentum—which is what we did in the studies of collisions in chapter 4, for example. We can do this if we can construct a 4-momentum that contains the energy of both the field and the particle together. This extended 'composite' system can be isolated, and therefore the law of conservation of energy-momentum can be applied to it.

For the field part one can always mathematically construct a suitable 4-vector and refer to it as 'the energy-momentum of the field' as long as one is consistent. For example, for a case where there exists a reference frame in which the problem is static, one could *define* the 4-momentum of the field to be $\Lambda^a_\mu \mathsf{P}^\mu_{(0)}$, where $\mathsf{P}^\mu_{(0)} = (\mathcal{E}_{es}/c, 0)$ is its value in the static frame, chosen to agree with the electrostatic energy in that frame. Another reasonable choice would be $(4/3)$ of this. However, we shall describe another, more physically motivated solution, due to Poincaré, that treats the source and field together.

Before constructing the 4-vector we need to beware that we will not be able to apply this idea to point-like particles: the point-like charged particle is a concept fraught with difficulty because it is associated with infinite field strength close to the particle, and infinite associated energy and momentum. Classical physics does not apply in this limit, but similar issues arise in quantum physics. The infinite energies are avoided because in fact no charged particle (such as an electron) ever has its wavefunction concentrated at a point of infinitesimal dimensions—the wavefunctions are always spread out in practice (although for calculational purposes it may be useful to include the highly concentrated wavefunctions in an integral over all 'possible' states of affairs).

In the classical regime we must insist that any distribution of finite charge must have a finite extent, and if a body carries a net charge then there must be binding forces that keep the charges attached to the body. These binding forces cannot all be electromagnetic and classical in origin, because static charge distributions are not stable under electromagnetic forces (a sphere of charge, for example, will fly apart if only electromagnetic forces are present). In the case of the proton, the binding forces are provided by the strong interaction (gluon field) between the quarks inside the proton; in the case of larger objects they may be provided by covalent bonds which rely on quantum theory to allow stable 'orbits' without emission of radiation.

One can keep the binding forces in the energy and momentum accounting without knowing the details of their origin, simply by asserting that they provide whatever forces are needed to make the charge distribution stable. This is done by adding another stress tensor onto the electromagnetic one, such that the total stress tensor describing the 'fields of all kinds' is

$$\mathbb{S}^{ab} = \mathbb{T}^{ab} + \mathbb{B}^{ab}.$$

and this total stress tensor has zero divergence, as in eqn (16.53).

The extra stress $\mathbb{B}^{ab}$ is called the *Poincaré stress tensor*, after Henri Poincaré, who described this resolution. In order to make sure the modification describes a particle that is holding itself together: i.e., that corresponds to what is observed, we pick $\mathbb{B}^{ab}$ in such a way that not only is $\mathbb{S}^{ab}$ divergence-free, but also it describes a system in stable mechanical equilibrium, not imploding nor flying apart. The condition for this is that the self-stress vanishes, meaning that the volume integral of the spatial part of $\mathbb{S}^{ab}$ vanishes in the rest frame:

$$\int \mathbb{S}^{ij}_{(0)} dV_{(0)} = 0$$

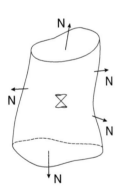

Fig. 16.16 A closed 3-surface and its outward normal N^a. The divergence theorem relates an integral over the interior to an integral over the surface.

Divergence theorem in four dimensions

Gauss's theorem connecting a surface integral to a volume integral in three dimensions reads

$$\oint_{(\mathcal{R})} \mathbf{A} \cdot d\mathbf{S} = \int_{\mathcal{R}} \boldsymbol{\nabla} \cdot \mathbf{A} d^3 x$$

where $(\mathcal{R})$ designates the surface of region $\mathcal{R}$ and the surface element $d\mathbf{S}$ is furnished by convention with a vectorial character, pointing in the outward normal direction. The 4-dimensional version of this connects an integral of a 4-flux over a '3-surface' (i.e., a three-dimensional entity) to an integral of a 4-divergence over a '4-volume':

$$\oint_{(\mathcal{R})} \mathsf{A}_\lambda \mathsf{N}^\lambda dV = \int_{\mathcal{R}} \partial_\lambda \mathsf{A}^\lambda d^4 x. \tag{16.54}$$

where N is a unit 4-vector in the outward normal direction and dV is an invariant volume element. The notation $d\sigma^a \equiv \mathsf{N}^a dV$ for the '3-surface' (i.e., volume) element may also be used. The outward normal is space-like for some parts of the surface, and time-like for others (see Fig. 16.16). Where it is space-like there is a reference frame in which that part of the integral is a spatial surface integral integrated over time. Where it is time-like there is a reference frame in which the 4-surface lies at an instant in time; in that frame $dV = dxdydz$ in rectangular coordinates, and $\mathsf{N} =$ either $(1,0,0,0)$ or $(-1,0,0,0)$. In other frames N is Lorentz-transformed but dV is invariant. If desired, it can be expressed in terms of the new coordinates by $dV = \gamma dx'dy'dz'$ (since the integral in the first frame was over a region such that $dt = 0$ between events in the region).

Gauss's theorem also applies to each row of a tensor (exercise 16.16):

$$\oint_{(\mathcal{R})} \mathsf{M}^{a\lambda} d\sigma_\lambda = \int_{\mathcal{R}} \partial_\lambda \mathsf{M}^{a\lambda} d^4 x. \tag{16.55}$$

(this is an integral over all space at a fixed time; the subscript (0)
denotes the rest frame). Consider, for example, two like charges connected by a rubber band: the electrostatic field has pressure on the
line between the charges, which is compensated by the tension in the
rubber band. The total 4-momentum of the particle is then defined
to be

$$\mathsf{P}^a \equiv \int \mathbb{S}^{a0} dV. \tag{16.56}$$

This is guaranteed to be a 4-vector. Proof: the condition that this P is a
4-vector is that the integral (16.56) should not depend on the orientation
in spacetime of the time-slice used to evaluate it (recall Fig. 5.3). We
prove this by writing the integral

$$\int \mathbb{S}^{a\lambda} \mathsf{N}_\lambda dV, \tag{16.57}$$

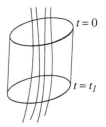

Fig. 16.17

where $\mathsf{N}^a = (1,0,0,0)$ and using a four-dimensional version of the divergence theorem (see box above, eqn (16.55)), as follows.
  Construct a 4-cylinder in spacetime. That is, a region with ends
formed by 'time-slices' at $t = 0$ and $t = t_1$ (in some reference frame), with
t_1 large and negative, and sides at some r (which we will allow to tend to
infinity). Then the integral (16.57) can be understood to be '3-surface'
integral (i.e., an integral over what we normally call a volume), adding
up the amount of $\mathbb{S}^{ab}$ crossing the 3-surface at $t = 0$ in the forward time
direction. This 3-surface is one end of our 4-cylinder. Now complete
the 3-surface integral of $\mathbb{S}^{ab}$ over the whole the 4-cylindrical surface, in
the outward normal direction. Using the divergence theorem, the total
must be zero, because it equals the 4-volume integral of $\Box \cdot \mathbb{S}$, which
is zero. Consequently the amount of $\mathbb{S}$ 'passing through' our time slice
at $t = 0$ is equal to minus the amount passing out of the rest of the
cylinder. By making r large enough, we can argue that the contribution
from the curved sides is negligible, and therefore the amount passing
into the lower end of the cylinder equals the amount passing out of the
top end (this is just another way of talking about energy-momentum
conservation). Finally, allow the top end of the cylinder to vary from
one time-slice to another (corresponding to different inertial frames),
while leaving the bottom unchanged. Since the total flux out remains
zero, and the bottom integral does not change, neither does the top
integral. QED.
  The 4-momentum measured in experiments and discussed in collision
theory is the one given by eqn (16.56). There is no unique answer (offered
by classical electromagnetism) to the question 'how much of the rest
energy comes from the matter, how much from the field?', because one
does not know how much binding energy to associate with the Poincaré
stress part. However, the subject does at least raise the rather wonderful
possibility that we can explain the origin of mass this way: perhaps the

whole rest energy (and therefore rest mass) of fundamental particles such as electrons and protons comes from their electromagnetic field energy?

A distance scale associated with any particle (or composite object) of given charge and mass arises in connection with this question:

$$r_c = \frac{q^2}{4\pi\epsilon_0 mc^2}. \tag{16.58}$$

This is the distance such that the work required to bring two particles of charge q, mass m from far apart to this separation is equal to the rest energy of one of them. Eqn (16.33) shows that if a static uniformly charged sphere of charge q, mass m has radius r_c then its electromagnetic field energy is sufficient to account for 3/5 of its rest energy. More generally, an object of mass m and charge q will generate enough field energy to account for the whole of its rest mass if the physical size is of order r_c (the exact value depending on the distribution of the charge). If q and m are the charge and mass of the electron, then r_c is called the 'classical radius of the electron', equal to 2.818×10^{-15}m. In early work this was envisaged quite literally as a finite radius possessed by the electron (to within a factor of order 1). However, particle collision experiments have shown that electrons are point-like down to much smaller sizes than this. Therefore the 'classical radius of the electron' should not be thought of as a true indication of the 'size' of an electron, but it does indicate a distance scale below which it is difficult to get sensible predictions from classical electromagnetism. The difficulties arise because if a charged sphere had a radius much less than r_c then its rest energy mc^2 would be smaller than its associated field energy, which implies that the matter of the sphere must have negative mass, leading to unphysical behaviour such as self-acceleration.

The Bohr radius and the Compton length of the electron are both much larger than r_c (see appendix B), so in any case classical physics has to be replaced by quantum physics well before one approaches r_c. However, the difficulty of infinite energy persists in the theory of quantum electrodynamics or 'QED', and leads to the famous renormalization problem. In this sense the electrodynamics are indeed crucial to understanding the observed mass. Ultimately QED does not account for the rest mass of the 'bare' electron, but assumes it as a parameter. However, the masses of composite entities such as the pions (139.570 MeV/c^2 for $\pi^{\pm}$, 134.977 MeV/c^2 for π^0) include a small contribution from electromagnetism, and the gluon field can account for most of the rest of mass. (The neutron is heavier than the proton because its different internal strong interactions dominate the electromagnetic ones.) Gluons have zero rest mass. If the quarks also had zero rest mass, the mass of the proton and the neutron would not change by much. In this way Special Relativity, via its prediction of mass-energy equivalence, came close to solving a profound puzzle of the universe: the nature and origin of mass.

Exercises

(16.1) A cable of proper density ρ_0 is subjected to a tension t per unit cross-sectional area. Find the maximum value of t for which $\rho > 0$ in all frames.

(16.2) In this chapter we established that pressure in a fluid is independent of reference frame. However, the fields of an electromagnetic plane wave are frame-dependent, and therefore so is the radiation pressure. Is there a contradiction here? Explain.

(16.3) By applying energy-momentum conservation (eqn (16.19)) to the stress-energy tensor for dust, $T^{ab} = \rho_0 U^a U^b$, prove that, in the absence of external forces, every part of the dust moves uniformly, i.e., that $dU/d\tau = 0$. (Hint: use eqn (16.25); do not assume that ρ_0 is uniform.)

(16.4) §'Incompressible motion' of a fluid is motion in which the proper volume of each fluid element is conserved. Prove that the sufficient and necessary condition for this is that the flow velocity field satisfies $\Box \cdot U = 0$. Prove that, for the case of a pure fluid subject to pure (i.e., rest-mass-preserving) external forces, this condition is equivalent to $d\rho_0/dt = 0$ for each fluid element.

(16.5) **Field energy near a charge moving in an applied electric field**. Obtain eqn (16.35), as follows. Show that for $v \ll c$ the energy density u is dominated by the contribution from the electric field. Obtain $\mathbf{E} \cdot \mathbf{E}$. Do not try to integrate u over the whole sphere, but instead concentrate on that part of the integral which will change over a small time interval dt. The region of integration $\mathcal{R}$ is fixed, but the charge moves. Let $\mathcal{R}$ be a sphere centred at the origin, and consider another sphere centred on the charge, of the same radius a as that of $\mathcal{R}$. Show that when the charge is at $z = vdt$ for small dt, the distance in the radial direction between the two spherical surfaces is $dr = vdt \cos\theta$. Hence in order to find the change dU in field energy over the next time interval dt, it suffices to integrate u over angles θ and ϕ with a volume element $(vdt \cos\theta)a^2 \sin\theta d\theta d\phi$. Perform such an integration and thus obtain eqn (16.35).

(16.6) Show that for a particle at rest possessing both electric charge and magnetic dipole moment (e.g., the electron), the Poynting vector runs in loops around the dipole axis. Could this account for the intrinsic angular momentum of the electron?

(16.7) Investigate the Trouton–Noble experiment using the stress-energy tensor, by writing down the stress-energy tensor for a rod at rest in compression, and hence that of a moving such rod. Calculate $d\mathbf{L}/dt = \int \mathbf{v} \wedge \mathbf{g}dV$ and thus confirm the discussion surrounding Fig. 16.15.

(16.8) Consider the *lever paradox*, as follows. A right-angled lever ABC has its pivot B at the origin, with AB on the x axis and BC on the y axis. The lengths of the arms are equal, $AB = BC = a$. A force f is applied at A in the vertical (y) direction and at C in the horizontal (x) direction. Since the torques are balanced in the rest frame, the lever does not rotate. Show that in another frame in standard configuration with the rest, there is a clockwise torque fav^2/c^2 due to these applied forces. Explain where the consequent steadily increasing angular momentum can be found.

(16.9) Sketch the Poynting vector in the vicinity of the moving pair of charges shown in Fig. 4.1. Does the electromagnetic field contain angular momentum about the origin, and if it does, then in what direction?

(16.10) Find the energy stored in the electromagnetic field inside the capacitor shown in Fig. 7.1, for each of the three different states of motion. Hence show that the field energy in case (c) is larger than in case (b) by $\epsilon_0 E^2 A d\gamma v^2/c^2$. This shows that, for a given capacitor in internal equilibrium and moving at speed v through the laboratory, the energy of the field is smaller when the normal to the capacitor plates is along the direction of motion (case (b)). Does this mean there is a torque tending to rotate the capacitor to this orientation? Explain.

(16.11) **Change of metric**. Confirm that if we change Minkowski metric from $(-+++)$ to $(+---)$ then $\epsilon_0 c^2 \rightarrow -\epsilon_0 c^2$ in eqns (16.42)—(16.46). Thus for given $\mathbf{E}$ and $\mathbf{B}$ fields, $\mathbb{F}$ and g_{ab} change sign but $\mathbb{T}$ does not.

(16.12) Prove that the electromagnetic stress-energy tensor satisfies

$$\mathbb{T}^\lambda_\lambda = 0, \qquad \mathbb{T}^a_\lambda \mathbb{T}^\lambda_b = (\tfrac{1}{2}\epsilon_0 c^2)^2 (D^2 + 4\alpha^2)\delta^a_b,$$

where D and α are the invariants given in eqns (13.12) and (13.13). (Hint: first suppose $\alpha \neq 0$, so there is a frame in which the fields are parallel, and pick axes judiciously, and then infer the other cases.)

(16.13) §Explain why a disordered distribution of radiation ('photon gas') must have the following properties after averaging over time:

(i) $E_x^2 = E_y^2 = E_z^2$, $B_x^2 = B_y^2 = B_z^2$

(ii) $E_x E_y = E_y E_z = E_z E_x = 0$,
$B_x B_y = B_y B_z = B_z B_x = 0$

(iii) $\mathbf{E} \wedge \mathbf{B} = 0$.

Deduce that the stress-energy tensor is diagonal, and matches that of an ideal fluid with $p = \rho_0 c^2 / 3$.

(16.14) §An absorbing plate moves at constant velocity v through a random radiation field as in the previous exercise. Prove that the force per unit area on the front and back surfaces of the plate are

$$p_f = \frac{(1+v/c)^2}{1-v/c} p/2, \qquad p_b = \frac{(1-v/c)^2}{1+v/c} p/2.$$

Verify that $p_f + p_b = \mathbb{T}^{x'x'}$ (assuming the plate moves in the x' direction). (Hint: you must integrate over solid angle, making use of eqn (3.12). The radiation field is isotropic in its rest frame but not in the rest frame of the plate.)

(16.15) Show (e.g., by carrying out the matrix multiplication) that for antisymmetric F^{ab}, with dual $\tilde{F}^{ab}$,

$$g_{\mu\nu} \left(F^{a\mu} F^{b\nu} - \tilde{F}^{a\mu} \tilde{F}^{b\nu} \right) = \tfrac{1}{2} g^{ab} F^{\mu\nu} F_{\mu\nu}$$

and hence that the stress-energy tensor of the electromagnetic field (eqn (16.46)) can alternatively be written

$$\mathbb{T}^{ab} = \tfrac{1}{2} \epsilon_0 c^2 g_{\mu\nu} \left(F^{a\mu} F^{b\nu} + \tilde{F}^{a\mu} \tilde{F}^{b\nu} \right).$$

(16.16) Assuming the four-dimensional Gauss' theorem (16.54), prove eqn (16.55). (Hint: introduce an arbitrary constant 4-vector B and consider $A^a = B_\mu M^{\mu a}$.)

(16.17) Define the total angular momentum of a fluid relative to some origin by

$$L^{ab} = \frac{1}{c} \int \left(x^a \mathbb{T}^{b0} - x^b \mathbb{T}^{a0} \right) dV$$

where $\mathbb{T}^{ab}$ is the stress-energy tensor. Prove that if the fluid is isolated then this quantity is constant in time, and also establish its tensor character. (Proceed as in the argument following eqn (16.56).)

What is spacetime?

Yesterday upon the stair
I met a man who wasn't there.
He wasn't there again today,
Oh, how I wish he'd go away.

after W. H. Mearns

Treatments of Relativity often examine Newton's 'absolute space' and 'Mach's principle' early on, in order to point out some of the issues and subtleties of the idea of absolute motion. We examine them last, because these issues are indeed subtle and not fully understood.

Consider the following thought experiment, proposed by Newton and discussed again by Berkeley and Mach, who in turn influenced Einstein.

A bucket of water is suspended by a rope and set into a spinning motion. The following sequence is observed:

(1) At first, the bucket rotates while the water remains predominantly flat and still.

(2) Owing to viscous forces, the water acquires the rotational motion and rotates at the same angular velocity as the walls of the bucket. Its surface is then curved: it rises up at the the edge and is depressed in the middle.

(3) The bucket is then stopped. For a while the water continues to rotate and continues to have a curved surface.

(4) Finally the water comes to a standstill, and then its surface is flat again.

When the water rotates, its surface is not flat. The question arises: how does the water 'know' to form a curved surface? That is, what determines whether or not the water is rotating, as opposed to merely being observed from a rotating frame? The answer has nothing to do with the bucket (nor anything else nearby), because in stages 1 and 2, and again in stages 3 and 4, the shape of the water differs while the motion of the bucket is the same. Also, in stages 1 and 4, and equally in stages 2 and 3, the state of the water is the same while the motion of the bucket differs.

A related experiment is that of Foucault's pendulum. I strongly urge you to take a chance to visit (or construct) a Foucault pendulum and ponder at length the question: how does the pendulum 'know' not to rotate its plane of swinging along with the Earth?

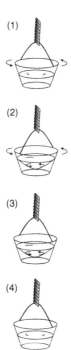

(1)

(2)

(3)

(4)

Fig. 17.1

A pendulum constructed at the North Pole will rotate once, relative to Earth, each 24 hours. Observers on Earth also find that the distant stars move around circular orbits once, relative to Earth, each 24 hours. This seems to be a remarkable coincidence. What has the local pendulum got to do with the distant stars?

These experiments all explore profound questions about *inertial motion*. Ernst Mach, without being completely specific, proposed that the properties of local inertial motion must be in some way connected to the average mass distribution of the distant stars. This idea is called *Mach's principle*. It can be formulated more specifically in various ways, but there is no clear consensus on its precise role or whether it is fully obeyed. It suggests that physics should be primarily relational, which is largely born out by General Relativity and quantum mechanics, and it hints at a connection between inertial motion and matter distribution, which is born out by the connection between matter content and spacetime geometry which underlies General Relativity.

Newton wrote of the concept he called 'absolute space', in the following terms

Absolute space, in its own nature, without regard to anything external, remains always similar and immovable ... Absolute motion is the translation of a body from one absolute place into another; and relative motion, the translation from one relative place into another.

According to Newton's own theory, however, and according to modern physics too, there is no way in practice to detect the distinction he is here making. We can in principle detect every kind of relative motion, but we have no way to say which of all those motions did or did not constitute a displacement relative to 'absolute space'. We cannot even tell which inertial frame is at rest relative to 'absolute space'. Because of these facts, most textbooks advise students to abandon the concept of 'absolute space'. It is too hard to define it in a sensible way, such that it has observable consequences. The same can be said of attempts to define 'absolute time' (although in Newtonian/Galilean physics that idea is sensible).

However, both Special and General Relativity are very elegantly and successfully treatable in terms of geometric concepts such as spacetime and the metric. This seems to suggest that we can assert the existence of something absolute after all: the spacetime which the metric refers to. After all, we can detect absolute properties such as spacetime curvature. If there is a curvature, does that not imply that there is 'something there' that is curved? Or if there might in principle be curvature but none is found, then does that not imply that there is 'something there' that is flat? When gravitational waves propagate from one place to another, surely *something* oscillates and carries the energy and momentum.

One of the important lessons of physics is that on some (not all) questions there can be room for physical pictures that differ without being mutually contradictory. For example, in classical physics one person may assert that each particle moves in response to the net force acting on it at each event, satisfying a second-order differential equation (Newton's Second Law), whereas another person may assert that each particle follows a path of least action between given starting and finishing events. Since these two pictures are mathematically equivalent for reversible motion, one may accept either or both. Since both are also elegant and do not unnecessarily multiply hypotheses, one may give to both a high value and a prominent place in one's thinking.

In the case of spacetime, many physicists and philosophers prefer to 'keep at arm's length' the idea that spacetime is a physical entity having physical properties. One need not abandon it altogether, but one should treat it with caution. We never observe spacetime: we only ever observe particles and fields. 'Spacetime is just a convenient fiction to help organize our mathematical treatment of worldlines,' someone may say, 'the worldlines of particles exist, like straws in a hay bale, but the empty regions between the worldlines are quite literally of no consequence whatsoever.'

It is certain that spacetime, and the gravitational field, are elusive. An 'effect' such as acceleration that can be made to disappear by a mere change of coordinate labelling does not deserve to be called unambiguously 'there' ('on the stair') but a tidal effect that persists in all coordinate systems cannot be ignored. After making such *caveats*, one may say that spacetime is a 'thing', a field on which all other fields are supported, whose chief physical characteristic is to be an 'arena' in which worldlines exist, with meetings governed by covariant equations, and which receives curvature in proportion to the energy and momentum content of those worldlines. All the marvellous physical effects we have studied in this book are, one way or another, consequences of the fact that this arena has Minkowskian geometry in any small region and has the curvature predicted by the Einstein equation, and that physical interactions (such as electromagnetic and nuclear interactions) do not reveal the arena directly: they obey the Strong Equivalence Principle, which is an extension of the Principle of Relativity.

Some basic arguments

A.1 Early experiments

Special Relativity was discovered in part by careful interpretation of the implications of theoretical results coming from the theory of electromagnetism (Maxwell's equations), and in part from counter-intuitive experimental data. Among the important experiments were those on the speed of light in moving water, carried out by Hippolyte Fizeau in 1851, the stellar aberration first observed by James Bradley in 1727 (see chapter 3), the Michelson–Morley experiment of 1887, and the experiments on the velocity-dependence of the charge-to-mass ratio of cathode rays (electrons) performed by Walter Kaufmann between 1901 and 1905.

Bradley's measurements showed that the direction of starlight arriving at Earth is consistent with the idea that the light moves independently of the Earth. From a modern point of view this is not surprising, but it helped to rule out the idea that the light was propagated by a substance ('aether') which could be disturbed by the motion of the Earth.

Fizeau used an interferometer to study the way the speed of light relative to his laboratory was affected by the motion of water through which the light was travelling. If the refractive index of the water is n, then the speed of light relative to the water is c/n, and he found that, relative to his laboratory, the speed w of the light-waves that travelled through moving water was

$$w = \frac{c}{n} + \left(1 - \frac{1}{n^2}\right) v \qquad (A.1)$$

(to within experimental precision), where v is the speed of the water. To interpret this result, Fresnel proposed that the factor $(1 - 1/n^2)$ came about by the water partially dragging a medium ('aether') that conveyed light-waves. In Einstein's theory no such assumption is needed. Rather, as we shall show below, Fizeau's observation was an early example showing (approximately) the way velocities must be added.

Michelson and Morley used an interferometer first developed by Michelson, in which light-waves are split into two paths at right angles, and recombined after reflection from mirrors placed in each 'arm' (Fig. A.1).

If the interferometer is at rest then the round trip time in both arms is $2L_0/c$ where L_0 is the length of either arm. The concept of the Michelson–Morley experiment was that if the interferometer is in motion relative to a medium conveying light-waves, then, according to Newtonian or Galilean notions of distance, time, and velocity, one expects the round trip time in each arm to depend on the orientation of that arm relative to the direction of motion of the interferometer. On this basis the times would be expected to be (exercise for the reader)

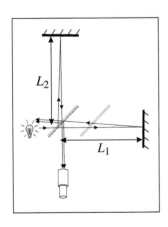

Fig. A.1 Michelson–Morley interferometer.

$$T_1 = \frac{L_1}{c+v} + \frac{L_1}{c-v} = \frac{2L_1/c}{1-v^2/c^2} \tag{A.2}$$

$$T_2 = \frac{2L_2}{\sqrt{c^2-v^2}} = \frac{2L_2/c}{\sqrt{1-v^2/c^2}} \tag{A.3}$$

for an arm oriented along and at right angles to the motion, respectively. By rotating the interferometer, in order to reverse the roles of the arms, any difference between the times can be detected accurately (by looking for a movement of the interference fringes) without requiring the lengths L_1 and L_2 to be known precisely. No such difference was found. This observation is consistent with a Special Relativistic treatment, in which the length L_1 above is predicted to be modified to $L_1(1-v^2/c^2)^{1/2}$.

In Special Relativity we reconsider the very definition of distance and time, in such a way that the Main Postulates of the theory are upheld. The rest of this Appendix presents some simple arguments that adopt and expound such a special relativistic treatment.

A.2   Simultaneity and radar coordinates

If two events happen at the same place in some reference frame, then it is easy to define the time and distance between them. The distance in this case is zero, and the time is determined by the number of ticks of a standard clock situated at the location of both events. (In practice we would also need to agree some sort of standard of time, which is currently done by observing the oscillations of the nucleus of a caesium atom in vacuum, but the details are not necessary in order to study Relativity; we just need to agree that some such standard can be defined. The Principle of Relativity ensures that the definition applies in all reference frames equally.)

For events happening at different places, the time and distance between them has to be defined carefully. A convenient method is first to use 'radar coordinates', and then derive times and distances from those. For any given reference frame F we can pick a position to serve as the spatial origin O of a coordinate system. The particle located at such an origin will have a straight worldline. Now consider an arbitrary event R, not at the origin. We imagine a 'radar echo location' scenario. That is, at time t_1, the particle at the origin of F sends out an electromagnetic pulse, propagating at the speed of light c (think of it as a radio-wave pulse or a flash of light, for example). We suppose that the pulse is reflected off some object present at event R (so R is the event of reflection), and then the pulse propagates back to the particle at the origin, arriving there at time t_2. The times (t_1, t_2) constitute the *radar coordinates* of event R in reference frame F. Together with the direction of travel of the pulses, they suffice to determine the position and time of R in frame F. Let us see how.

First, since the speed of light is independent of processes such as reflection, the outgoing and incoming pulse must have the same speed c in F. It follows that the outgoing pulse takes the same time to get to R as the incoming one takes to come back, so R must occur at a time half way between t_1 and t_2: i.e.,

$$t_R = \frac{t_2 + t_1}{2}$$

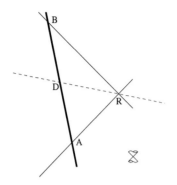

Fig. A.2 Identifying simultaneous events on a spacetime diagram. AB is a straight time-like worldline, so it can be the worldline of a particle in uniform motion, at the spatial origin of an inertial reference frame F. AR and RB are photon worldlines. D is half way between A and B. In the reference frame whose time axis is AB (i.e., frame F), R is simultaneous with D (see text). Applying the argument to further events, one concludes that all events along the dashed line are simultaneous in frame F.

in frame F. Also, since the pulse traveled out and back at the speed c, the distance from O to R must be

$$x_R = \frac{c(t_2 - t_1)}{2}$$

in frame F. This pair of equations allows us to convert easily between radar coordinates and ordinary coordinates. It means that we only need one clock, at the origin, together with the Postulates of Relativity, to define a complete coordinate system for any given reference frame. This radar method is associated with Milne.

The worldline of a particle at the origin of F serves as the 'time axis' of F. The set of events simultaneous with R (according to reference frame F) consists of all those having the same value of the radar echo time $(t_2 + t_1)/2$. On a spacetime diagram these form a line that can be constructed as shown in Fig. A.2. If we choose the scales such that photon worldlines have slope 45° on the diagram, then a line of simultaneity or 'distance axis' for frame F makes the same angle with a photon worldline as the time axis of F does. In other words, the distance axis and time axis of any reference frame are oriented such that the angle between them is bisected by a photon worldline; see Fig. A.3. This is the graphical representation of the fact that the speed of light is the same in all reference frames.

The spacetime diagram construction allows us to see the relativity of simultaneity very easily: sets of events that are simultaneous in one reference frame are not simultaneous in another, because the lines of simultaneity associated with different frames cut through spacetime in different directions.

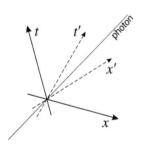

Fig. A.3 Time and position axes for two different reference frames on a spacetime diagram. The photon worldline bisects the angle formed by each axis pair.

A.3   Proper time and time dilation

For any pair of events having a time-like interval between them, the *proper time* is defined to be the time between such a pair of events, as observed in the reference frame in which the events happen at the same place. By definition, the proper time as related to the invariant interval by

$$\tau = \frac{\sqrt{-s^2}}{c} = \frac{|s|}{c} \tag{A.4}$$

(by using eqn (1.7) with $\tau = t_2 - t_1$, and $x_2 = x_1$, $y_2 = y_1$, $z_2 = z_1$).

If a particle or system is in uniform motion at speed v relative to some reference frame F, then the time interval between a pair of events at the particle is related to the proper time by

$$T = \frac{\tau}{\sqrt{1 - v^2/c^2}} = \gamma\tau \tag{A.5}$$

where

$$\gamma(v) \equiv \frac{1}{\sqrt{1 - v^2/c^2}}. \tag{A.6}$$

This is called *time dilation*, because $T \geq \tau$. It means that all systems evolve more slowly if they are moving than if they are at rest. For example, a rabbit carried along in a rocket moving at $v = 0.999c$ would not experience the slightest difference in the laws of physics which determine its metabolism, and yet it would live for 150 years (i.e., 22 times longer than slow rabbits), when the time is measured by clocks relative to which it has this high speed. This effect has not been observed for rabbits, owing to experimental difficulties, but it has been observed for muons and other particles, and for atomic clocks. Its influence has also been inferred in a host of other observations from astronomy and particle physics.

To deduce the time dilation factor, one can argue from the invariance of the interval but the conceptually most simple derivation is arguably the 'photon clock' argument; see Fig. A.4. Note that once the time dilation for one physical phenomenon is known (such as the light-pulse clock), that same factor must apply to all other physical phenomena, by the Principle of Relativity. Otherwise physics would differ from frame to frame. The reader should think through this point carefully. The argument hinges on the fact that 'keeping in synchrony' is frame-independent. For example, if a man steps onto a train at a certain station once each day in the morning, and steps off once each day in the afternoon, then observers in all frames will agree that the man boarded the train, that it was daytime, etc. No observer will find that two rotations of the Earth happened between boarding events, for example. Similarly, two clocks that are situated next to one another and agree in one frame (though their internal workings may differ) will also be found to agree by observers moving quickly past them, because the number of ticks each has made between any given pair of events is merely a matter of counting, which cannot depend on reference frame. Consequently, if one is found to be running slow (relative to the observer's proper clock) then both must be found to be running slow by the same factor. Such clocks could be furnished by physical phenomena such as the hyperfine splitting of hydrogen, or the half-life of the muon, or the heart-beat of a rabbit, etc.

Twin paradox

An important idea related to time dilation is called the 'twin paradox'. A pair of twins are initially the same age; one then travels away from Earth in a high-speed rocket for some time, and returns at similarly high speed. Owing to time dilation, the traveller does not age as much as the twin who remained at home. The 'paradox' arises if one asserts that the situation is symmetric, since Earth travelled away and back relative to the rocket, which suggests the two twins should both age equally. However, the situation is not symmetric, because one twin accelerated (in order to turn back for the return journey), the other did not. This may be illustrated by plotting the worldlines on a spacetime diagram: one worldline is straight, the other is not. The prediction made by Special Relativity is unambiguous, and in truth there is no paradox. However the thought experiment is useful and will be invoked several times in the text.

A.4   Lorentz contraction

If a pair of events has a space-like separation, then there exists a reference frame in which they are simultaneous. The distance between the events, as observed in such a reference frame, is called the *proper distance* L_0. Owing to the fact that we normally study the evolution of particles and systems, the

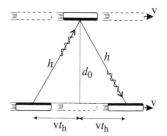

Fig. A.4 The photon clock. A pair of mirrors is attached to a rod of rest length d_0, such that light can bounce between them. In the rest frame, the time for a round trip of a light-pulse between the mirrors is $\tau = 2d_0/c$. The figure shows the situation observed in a reference frame F travelling to the left relative to this 'clock.' In F, the mirrors have a speed v to the right. Let the round trip time in F be t. To complete a round trip the light-pulse must travel a distance $2h$. Clearly, since $h > d_0$ we must find $t > \tau$: time dilation. By Pythagoras' theorem, $h^2 = d_0^2 + (vt_h)^2$ where $t_h = t/2$. Therefore $t = 2h/c = (2/c)(d_0^2 + v^2t^2/4)^{1/2}$. Solving for t one finds $t = \gamma\tau$. The argument hinges on the fact that the events 'pulse leaves' and 'pulse returns' are just that: *events*, so t is the time between the *same two events* as those whose time-separation is τ in the rest frame.

concept of proper time is much more useful in practice than the concept of proper distance. However, distance is also needed in order to obtain a complete description when changing from one reference frame to another.

A physical object can be regarded as a set of worldlines (those of the particles of the object). If these worldlines are straight and parallel then the object has constant velocity and fixed size. The spatial size of an object is defined as the size of the region of space it occupies at any instant of time. Owing to the relativity of simultaneity, this concept is a relative one: i.e., it is well-defined only once a reference frame is specified, and the value obtained for the size can depend on which reference frame is adopted.

Suppose we choose some direction in space, and take an interest in sizes of physical objects along this direction. The length of an object in its own rest frame (i.e., the frame in which its velocity is zero), along the chosen direction, is called its *rest length*. In any reference frame moving with respect to the rest frame, the object has some non-zero velocity $\mathbf{v}$. In such a reference frame the length of the object, along the chosen direction, is given by

$$L = \frac{L_0}{\gamma(v_\parallel)}, \tag{A.7}$$

where $v_\parallel$ is the component of velocity along the chosen direction. Since $L \leq L_0$, this is called *Lorentz contraction* or *space contraction*. It means that an object in motion is contracted along the direction of motion compared to its size when at rest. For example, a rabbit carried along in a rocket moving at $v = 0.999c$ would have physical dimensions approximately 15 cm × 20 cm × 1 cm, when measured by rods relative to which it moves at this high speed.

To deduce this result, consider a clock flying at speed v across a room of rest length L_0. The time taken to cross the room is L_0/v according to clocks at rest in the room. During this period the flying clock registers a smaller time, owing to time dilation: namely, $\tau = (1/\gamma)L_0/v$. An observer on the flying clock finds that the room moves with speed v, and since it takes time τ for the room to pass him, he deduces that the width of the room is $L = v\tau = L_0/\gamma$. Thus he finds the moving room to be contracted by a factor γ compared to its rest length.

It is instructive to consider also another argument based on radar or light-pulse reflection. We use a photon clock once again, but now the clock is oriented along the direction of motion. Thus, suppose a pulse of light is emitted from one end of a rod and travels to the other end, where it is reflected and comes back to the first end. In the rest frame of the rod, the time between emission and final reception is $2L_0/c$ where L_0 is the rod's rest length. The events of emission and final reception happen at the same place in the rest frame of the rod, so their time separation in that frame is the proper time between them, $\tau = t_2 - t_1$. The time between these events in any other frame must therefore be

$$T = \gamma\tau = \gamma\frac{2L_0}{c}.$$

The rod singles out a direction in space by its own axis, and we now consider a reference frame in which it moves in that direction. In such a reference frame the emitted light-pulse moves at speed c and the far end of the rod moves at speed v. It follows that the time taken for the light-pulse to reach the point of reflection is $L/(c-v)$ and the time taken to come back is $L/(c+v)$, where L is the length of the rod in the new reference frame. Hence

$$T = \frac{L}{c-v} + \frac{L}{c+v} = \frac{2cL}{c^2 - v^2} \quad \Rightarrow L = \frac{c^2 - v^2}{c^2}\gamma L_0 = \frac{L_0}{\gamma}.$$

This is eqn (A.7) in the case where $v_{\parallel} = v$. It explains the modification needed to treat the Michelson–Morley interferometer by correcting eqn (A.2). The general case is treated in chapter 2.

A.5 Doppler effect, addition of velocities

Suppose two particles, moving along a line, pass one another at event O, and then at event A the first particle sends a light-signal to the second, where it arrives at event B; see Fig. A.6.

We take O as the origin of position and time. If the relative velocity of the particles is $\mathbf{v}$ then, in the reference frame F of the emitter, the reception event takes place a distance $d = vt_B$ away, so the signal travel time is $d/c = vt_B/c$. It follows that $t_B = t_A + (v/c)t_B$, therefore

$$t_B = \frac{t_A}{1 - v/c}. \tag{A.8}$$

Events O and B take place at the same place in the reference frame F$'$ of the *receiver*, so their time separation t_B' is a proper time in that frame, so $t_B = \gamma t_B'$, hence

$$\frac{t_B'}{t_A} = \frac{1}{\gamma}\frac{1}{1 - v/c} = \sqrt{\frac{1 - v^2/c^2}{(1 - v/c)^2}} = \sqrt{\frac{1 + v/c}{1 - v/c}}. \tag{A.9}$$

We can use this result to deduce the Doppler effect for light-waves. All we need to do is suppose that the emitter emits regularly space signals, once every time t_A in his reference frame, then the above argument applies to all the signals, and the receiver will receive them spaced in time by t_B' as given by (A.9). This set of signals could in fact be one continuous stream of light-waves, with period t_A. Then the event A could be, for example, 'the electric field of the light-wave at the emitter is at a maximum', and the event B could be 'the electric field of the light-wave at the receiver is at a maximum'. Since we define the period of a wave to be the time interval between successive maxima, it follows that eqn (A.9) relates the periods observed at the emitter and receiver. By taking the inverse, we obtain the relationship between the frequencies. Hence if light-waves of frequency ν_0 are emitted by a particle, then the frequency observed by any particle moving directly away from the emitter at speed v is given by

$$\nu = \sqrt{\frac{1 - v/c}{1 + v/c}}\,\nu_0. \tag{A.10}$$

This is called the *longitudinal Doppler effect*, or just Doppler effect. It permits one to deduce, for example, the speed of a star relative to Earth, from the frequency of the received light, if one has independent evidence of what the emitted frequency was.

Now we shall use the Doppler effect to deduce a formula concerning relative velocities. Suppose F$'$ moves relative to F with speed u, and F$''$ moves relative to F$'$ with speed v (i.e., v is the speed of F$''$ *as observed in frame* F$'$), all

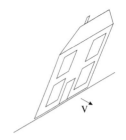

Fig. A.5 An ordinary house subject to Lorentz contraction along its direction of motion relative to some reference body.

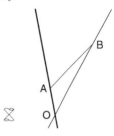

Fig. A.6 A simple set of events suited to reasoning about the Doppler effect. Two particles (with relative velocity less than c) pass one another at the origin O. At event A the first particle sends a light-signal to the second particle, where it arrives at event B.

motions being along the same direction. Then we can calculate the speed w of F″ relative to F by using the Doppler effect formula three times, as follows:

$$\nu' = \sqrt{\frac{1 - u/c}{1 + u/c}}\, \nu, \quad \text{and} \quad \nu'' = \sqrt{\frac{1 - v/c}{1 + v/c}}\, \nu' = \sqrt{\frac{1 - w/c}{1 + w/c}}\, \nu.$$

It follows that

$$\left(\frac{1 - u/c}{1 + u/c}\right)\left(\frac{1 - v/c}{1 + v/c}\right) = \frac{1 - w/c}{1 + w/c}$$

and after a little algebra we obtain

$$w = \frac{u + v}{1 + uv/c^2}. \tag{A.11}$$

This is the formula for 'relativistic addition of velocities'. The generalization to velocities in any direction is presented in chapter 2. The formula predicts that w is never greater than c as long as u and v are both less than or equal to c; this is in agreement with the Light Speed Postulate. The result predicted by classical physics is $w = u + v$; the relativistic formula reproduces this in the limit $uv \ll c^2$.

Eqn (A.11) explains Fizeau's observation (A.1) of the speed of light in moving water, which turns out to be an approximation to the exact result predicted by eqn (A.11):

$$w = \frac{c/n + v}{1 + v/nc}.$$

Note that the velocities in the formula (A.11) are all what we call 'relative velocities', and they concern three different reference frames. There is another type of velocity that can be useful in calculations, which we shall refer to as 'closing velocity'. The concept of 'closing velocity' applies in a *single* reference frame, and it refers to *the rate of change of distance between* two objects, all distances and times being measured in a single reference frame. When both objects are moving relative to the reference frame, a closing velocity is *not* necessarily the velocity of any physical object or signal, and it can exceed the speed of light. For example, an observer standing at the collision point of a modern particle accelerator will observe two bunches of particles coming towards him from opposite directions, both travelling at close to c. The positions of the bunches are, to good approximation, given by $z_1 = L - ct$ and $z_2 = -L + ct$ in the rest frame of such an observer, where L is a constant (equal to half the distance between the bunches at $t = 0$). Hence the distance between the bunches is $z_1 - z_2 = 2L - 2ct$. This distance changes at the rate $-2c$, therefore the observer finds that the two particle bunches have a 'closing velocity' of $2c$. Nevertheless, the *relative velocity* of the bunches is $w = 2v/(1 + v^2/c^2)$, which is less than c. The relative velocity is the velocity which an observer moving along with one bunch will find the other to have.

Constants and length scales

Length scales related to the electron

Name	Symbol	Formula	Value (fm)	Relevance
Bohr radius	a_0	$\hbar/m\alpha c$	53000	require quantum theory
Compton length	λ_C	αa_0	386	require quantum field theory
Classical radius	r_{class}	$\alpha^2 \lambda_C$	0.02	classical electromagnetism consistent

The Bohr radius is, roughly, the distance scale at which low-energy interactions between the electron and other charged particles such as the proton cannot be treated accurately by classical mechanics; quantum mechanics is required. For electrons this can be non-relativistic quantum theory. This limit arises when the combination of distance and momentum uncertainty is of the order of the minimum allowed by Heisenberg's Uncertainty Principle. The Compton length is the distance scale at which Heisenberg's Uncertainty Principle implies the motion of the electron is relativistic: i.e., the speed approaches c, therefore non-relativistic quantum theory breaks down. A relativistically correct quantum theory is required: namely, quantum field theory.

Below both these length scales lies the 'classical radius of the electron'. This is the scale at which all three theories (classical, quantum, quantum field) run into problems in dealing with large field energies close to point-like particles. In principle classical electromagnetism is at least self-consistent all the way down to $r_{\text{classical}}$, although it will not always provide accurate predictions below a_0 owing to quantum effects. It is believed that quantum field theory, such as quantum electrodynamics, handles the field energy divergence satisfactorily through a procedure called *renormalization*. However, it is probable that this and all other current fundamental theories are approximations to an as yet undiscovered approach that, one may hope, avoids divergences in a more elegant manner.

C   Derivatives and index notation

Transformation matrix. For any function f depending on coordinates $\{t, x, y, z\}$ we may write

$$\mathrm{d}f = \frac{\partial f}{\partial t}\mathrm{d}t + \frac{\partial f}{\partial x}\mathrm{d}x + \frac{\partial f}{\partial y}\mathrm{d}y + \frac{\partial f}{\partial z}\mathrm{d}z.$$

In index notation this is written

$$\mathrm{d}f = \frac{\partial f}{\partial x^\lambda}\mathrm{d}x^\lambda = (\partial_\lambda f)\mathrm{d}x^\lambda. \tag{C.1}$$

If we introduce a second coordinate system $\{t', x', y', z'\}$, then as one explores spacetime each of the primed coordinates is some function of all the unprimed coordinates, to which the above result can be applied, and therefore

$$\mathrm{d}x^{a'} = \frac{\partial x^{a'}}{\partial x^\lambda}\mathrm{d}x^\lambda. \tag{C.2}$$

This expression shows how one four-vector, namely $\mathrm{d}x^a$, is expressed in the new coordinate system. In other words, it tells us that the transformation matrix $\Lambda^{a'}_b$ is none other than the set of partial derivatives appearing on the right hand side of this eqn (C.2):

$$\Lambda^{a'}_b \equiv \frac{\partial x^{a'}}{\partial x^b}. \tag{C.3}$$

For Special Relativity, for example, this set constitutes the Lorentz transformation and gives a symmetric matrix. More generally, however (in General Relativity, for example), the transformation matrix need not be symmetric (e.g., consider the case $y' = y + z$, $z' = z$).

When we introduced the gradient operator $\square$ in chapter 6, we proved that it required a minus sign in the time-derivative, in order to ensure that, for scalar f, $\square f$ is a 4-vector. We can now argue that this sign was owing to the fact that $\partial f / \partial x^a$ is covariant. This can be proved in general by using eqn (C.3), as follows.

We have, for any scalar function f,

$$\mathrm{d}f = \frac{\partial f}{\partial x^\lambda}\mathrm{d}x^\lambda.$$

We take an interest in the partial derivatives of f with respect to some other set of coordinates $\{t', x', y', z'\}$. To this end, divide the equation by $dx^{a'}$ while holding $x^{\mu' \neq a'}$ constant:

$$\frac{\partial f}{\partial x^{a'}} = \frac{\partial f}{\partial x^\lambda} \frac{\partial x^\lambda}{\partial x^{a'}}.$$

The RHS can be written $K_{a'}^{\lambda}(\partial f / \partial x^\lambda)$, where the transformation matrix is

$$K_{a'}^{a} \equiv \frac{\partial x^a}{\partial x^{a'}}. \tag{C.4}$$

The transpose of this matrix is the inverse of $\Lambda^{a'}_{a}$, since from eqn (C.2) we have that

$$\frac{\partial x^{a'}}{\partial x^{b'}} = \frac{\partial x^{a'}}{\partial x^\lambda} \frac{\partial x^\lambda}{\partial x^{b'}}. \tag{C.5}$$

But, obviously,

$$\frac{\partial x^{a'}}{\partial x^{b'}} = \delta^{a'}_{b'}$$

so, using eqns (C.3) and (C.4) in (C.5),

$$\Lambda^{a'}_{\lambda} K_{b'}^{\lambda} = \delta^{a'}_{b'}.$$

which shows that $K^T = \Lambda^{-1}$. Thus we have proved that $\partial f / \partial x^a$ transforms like a covariant vector. This justifies the placement of the index in the definition of ∂_a presented in eqn (12.5). The gradient operator $\partial / \partial x^a$ is said to be *naturally covariant* (and the x^a on the *bottom* of the partial derivative gives a reminder that the object one obtains should be exhibited with a *lower* index).

As usual with derivative operators, the order of symbols matters: $\partial_a u^b v^c$ is not the same as $u^b \partial_a v^c$. (A practice that can be useful when a lot of operators are in play is to introduce a comma notation after all the indices: further indices after the comma indicate partial derivatives. Thus a result such as

$$\partial_d(u^a_b v^c) = (\partial_d u^a_b)v^c + u^a_b(\partial_d v^c)$$

would be written

$$(u^a_b v^c)_{,d} = u^a_{b,d} \, v^c + u^a_b \, v^c_{,d}.$$

This notation restores full freedom in the order of writing the symbols. We have not used it in this book, however, because we do not want to require you to learn new notation unnecessarily.)

The partial derivative operators commute among themselves because $\partial^2 f / \partial x \partial y = \partial^2 f / \partial y \partial x$, etc. (assuming the functions are single-valued). So, for example:

$$\partial_a \partial_b u^c = \partial_b \partial_a u^c.$$

The product rule for differentiation reads, for some generic tensors A and B,

$$\partial^\bullet \left(A^{\bullet\bullet} B_{\bullet\bullet\bullet} \right) = B_{\bullet\bullet\bullet} \left(\partial^\bullet A^{\bullet\bullet} \right) + A^{\bullet\bullet} \left(\partial^\bullet B_{\bullet\bullet\bullet} \right) \tag{C.6}$$

where the dots signify any combination of indices, not necessarily repeated. It is just like taking the derivative of a product of scalars, because, after all, each element of a tensor *is* just a scalar (in the sense of a single number, not a Lorentz scalar).

For example, consider a scalar product of two 4-vectors:

$$\partial^a (\mathsf{U}^\lambda \mathsf{V}_\lambda) = (\partial^a \mathsf{U}^\lambda) \mathsf{V}_\lambda + (\partial^a \mathsf{V}^\lambda) \mathsf{U}_\lambda. \tag{C.7}$$

When $\mathsf{U} = \mathsf{V}$ we obtain

$$\partial^a (\mathsf{U}^\lambda \mathsf{U}_\lambda) = 2(\partial^a \mathsf{U}^\lambda) \mathsf{U}_\lambda,$$

i.e. $\qquad \Box (\mathsf{U} \cdot \mathsf{U}) = 2(\Box \mathsf{U}) \cdot \mathsf{U}. \tag{C.8}$

When the *size* of a 4-vector U is constant (i.e., independent of time and space), one has $\Box(\mathsf{U} \cdot \mathsf{U}) = 0$, and then eqn (C.8) says that each row of its gradient tensor $\partial^a \mathsf{U}^\lambda$ is orthogonal to U. This is used in the derivation in appendix D.

Another useful application of the product rule is observed in expressions such as

$$A_{\mu\nu} \partial_a A^{\mu\nu}.$$

Noting the repeated indices, we should like to think we have a scalar somewhere, which is right. We just need to spot that

$$A_{\mu\nu} \partial_a A^{\mu\nu} = \frac{1}{2} \partial_a s \quad \text{where} \quad s = A_{\mu\nu} A^{\mu\nu}. \tag{C.9}$$

(check it by applying eqn (C.6) to the right-hand side). This is the generalization of the familiar $(d/dx)(f^2) = 2f(df/dx)$.

The field of an arbitrarily moving charge

D

This appendix contains two calculations relevant to the problem of finding the electromagnetic field when the charge and current distribution is given.

D.1 Light-cone volume element

We claimed in eqn (8.32) that in the case of integration over a light-cone (e.g., a past light-cone) the combination $\mathrm{d}^3\mathbf{r}_\mathrm{s}/r_\mathrm{sf}$ is Lorentz-invariant, where the notation refers to a volume element divided by spatial distance from source event to the field event, where the source event lies on the cone and the field event is at its apex.

To prove this we can, without loss of generality, place the field event at the origin. We then propose that the variables x_s, y_s, z_s form the spatial part of the null 4-vector

$$\mathsf{R}_{\mathrm{fs}} = \begin{pmatrix} -r_\mathrm{sf} \\ x_s \\ y_s \\ z_s \end{pmatrix} \tag{D.1}$$

where $r_\mathrm{sf} = (x_s^2 + y_s^2 + z_s^2)^{1/2}$ (since now the field event is at the origin). This allows us to understand how the volume element transforms from one reference frame to another, for the case of integration over a light-cone. To reduce clutter, we now drop the subscripts s and sf. When expressed in a new coordinate system (x', y', z') the volume element becomes

$$\mathrm{d}x\,\mathrm{d}y\,\mathrm{d}z = \begin{vmatrix} \frac{\partial x}{\partial x'} & \frac{\partial x}{\partial y'} & \frac{\partial x}{\partial z'} \\ \frac{\partial y}{\partial x'} & \frac{\partial y}{\partial y'} & \frac{\partial y}{\partial z'} \\ \frac{\partial z}{\partial x'} & \frac{\partial z}{\partial y'} & \frac{\partial z}{\partial z'} \end{vmatrix} \mathrm{d}x'\mathrm{d}y'\mathrm{d}z'. \tag{D.2}$$

It is clear by symmetry that this element is invariant under rotations, so it suffices to consider how it varies under the standard Lorentz transformation. We have

$$x = \gamma(x' + vt') = \gamma(x' - vr'/c) \qquad \Rightarrow \qquad \frac{\partial x}{\partial x'} = \gamma\left(1 - \frac{vx'}{r'c}\right)$$

using $\partial r'/\partial x' = x'/r'$. Hence

$$\frac{\partial x}{\partial x'} = \frac{\gamma\,(r' - vx'/c)}{r'} = \frac{r}{r'}$$

where in the final step we used that $-r$ is the time part of the 4-vector R_{fs}, which transforms as $\mathsf{R}'_{fs} = \Lambda\mathsf{R}_{fs}$. Using also $y = y'$ and $z = z'$ we find that the complete Jacobian is

$$\begin{vmatrix} r/r' & 0 & 0 \\ 0 & 1 & 0 \\ 0 & 0 & 1 \end{vmatrix} = \frac{r}{r'}.$$

Hence

$$\frac{dx\,dy\,dz}{r} = \frac{(r/r')dx'dy'dz'}{r} = \frac{dx'dy'dz'}{r'}$$

for the region of integration (i.e., the past light-cone) under consideration. Reinstating finally the subscripts s and sf, we have the result shown in eqn (8.32).

D.2   The field tensor

To obtain the electromagnetic field of an arbitrarily moving point charge, we take as our starting point the Liénard–Wiechert potential, eqn (8.33):

$$\mathsf{A} = \frac{-q}{4\pi\epsilon_0 c}\frac{\mathsf{U}}{\mathsf{R}\cdot\mathsf{U}} \tag{D.3}$$

where $\mathsf{R} = \mathsf{R}_f - \mathsf{R}_s$ is the 4-displacement from the source event to the field event, and U is the 4-velocity at the source event, the source event being retarded, such that R is null. Let

$$s \equiv \mathsf{R}\cdot\mathsf{U},$$

then we are interested in

$$\partial_\mu A^\nu = \frac{s\partial_\mu U^\nu - U^\nu\partial_\mu s}{s^2} \tag{D.4}$$

where we have dropped the factor $-q/4\pi\epsilon_0 c$; we shall reinstate it at the end. Let $a^\nu = d U^\nu/d\tau$ be the 4-acceleration at the source event, and observe (see figure) that since $\mathsf{U} = \mathsf{U}(\tau)$,

$$\frac{\partial U^\nu}{\partial x} = \frac{dU^\nu}{d\tau}\frac{\partial \tau}{\partial x} = a^\nu\frac{\partial \tau}{\partial x}$$

and similarly for other partial derivatives of U^ν. Hence

$$\partial_\mu U^\nu = a^\nu\partial_\mu\tau. \tag{D.5}$$

More generally, for any quantity at the source event,

$$\partial_\mu = \frac{\partial}{\partial x^\mu} = \frac{\partial \tau}{\partial x^\mu}\frac{d}{d\tau} = (\partial_\mu\tau)\frac{d}{d\tau}. \tag{D.6}$$

To find $\partial_\mu\tau$ we use a trick which takes advantage of the fact that R is of fixed size (always null) as we move the field point around. First, since $\mathsf{R} = \mathsf{R}_f - \mathsf{R}_s$ we have

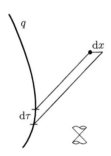

dx

$d\tau$

q

Fig. D.1

$$\partial_\mu R^\nu = \delta_\mu^\nu - \frac{dR_s^\nu}{d\tau}\partial_\mu\tau = \delta_\mu^\nu - U^\nu\partial_\mu\tau. \tag{D.7}$$

But since R is of fixed size, it is orthogonal to its gradient (see eqn (C.8)), so

$$R_\nu\partial_\mu R^\nu \;=\; 0 \;=\; R_\mu - R_\nu U^\nu\partial_\mu\tau$$

$$\Rightarrow \qquad \partial_\mu\tau \;=\; \frac{R_\mu}{s}.$$

After multiplying by a^ν and using eqn (D.5) we obtain a simple expression for the first term in the numerator of eqn (D.4):

$$s\partial_\mu U^\nu = a^\nu R_\mu. \tag{D.8}$$

Now we calculate the second term. We need $\partial_\mu s$, which is

$$\partial_\mu(R_\nu U^\nu) = (\partial_\mu R^\nu)U_\nu + R_\nu(\partial_\mu U^\nu) \tag{D.9}$$

where we have used a see-saw to bring out the relationship to eqn (D.7). Using eqns (D.7) and (D.8) this is

$$\partial_\mu(R_\nu U^\nu) = \left(U_\mu - U^\nu U_\nu\frac{R_\mu}{s}\right) + R_\nu a^\nu\frac{R_\mu}{s}$$

$$= U_\mu + \frac{R_\mu}{s}\left(c^2 + R\cdot a\right)$$

$$\Rightarrow \qquad U^\nu\partial_\mu s = U^\nu U_\mu + \frac{R_\mu}{s}\left(c^2 + R\cdot a\right)U^\nu. \tag{D.10}$$

Substituting eqns (D.10) and (D.8) into eqn (D.4) and gathering terms in R_μ, we find

$$\partial_\mu A^\nu = \frac{-U^\nu U_\mu}{s^2} - \frac{c^2 R_\mu}{s^3}\tilde{U}^\nu$$

where[1]

$$\tilde{U}^\nu \equiv U^\nu - \frac{1}{c^2}\left(a^\nu s - U^\nu a\cdot R\right). \tag{D.11}$$

Upon substituting this into the antisymmetric form $\mathbb{F}^{\mu\nu} = \partial^\mu A^\nu - \partial^\nu A^\mu$ the first term disappears and, after reinstating the factor $-q/4\pi\epsilon_0 c$, we have

$$\mathbb{F}^{\mu\nu} = \frac{qc}{4\pi\epsilon_0}\frac{\left(R^\mu\tilde{U}^\nu - R^\nu\tilde{U}^\mu\right)}{(R_\lambda U^\lambda)^3}. \tag{D.12}$$

This is the field tensor for the electromagnetic field of an arbitrarily moving point charge.

We have presented the calculation for the metric $(-+++)$. The corresponding expressions in the case of the metric $(+---)$ are obtained by substituting $c^2 \to -c^2$ in eqn (D.11).

---

[1] This can also be written $\tilde{U}^\nu \equiv U^\nu - \frac{1}{c^2}\left(a^\nu U^\lambda - U^\nu a^\lambda\right)R_\lambda$.

To find the electric field, use $\mathbf{E} = c\mathbb{F}^{0i}$, keeping in mind

$$\mathsf{R}^\mu = (r, \mathbf{r})$$

$$\mathsf{U}^\mu = \gamma(c, \mathbf{u}) \qquad \Rightarrow \qquad s = \mathsf{R} \cdot \mathsf{U} = \gamma(-cr + \mathbf{u} \cdot \mathbf{r})$$

$$a^\mu = \gamma(\dot\gamma c, \dot\gamma \mathbf{u} + \gamma \mathbf{a}).$$

Upon substituting these into the expression for $\tilde{\mathsf{U}}^\mu$ one finds that the $\dot\gamma$ terms cancel, and one is left with

$$\tilde{\mathsf{U}}^0 = \gamma c + \gamma^3 \mathbf{a} \cdot \mathbf{r}/c,$$

$$\tilde{\mathsf{U}}^i = \gamma \mathbf{u} + \frac{\gamma^2}{c^2}\left(\gamma(\mathbf{a} \cdot \mathbf{r})\mathbf{u} + s\mathbf{a}\right).$$

Putting this into eqn (D.12) leads directly to the expression given in eqn (8.48).

Bibliography

Introductory

Einstein, A. *Relativity: The Special and the General Theory* (Pober Publishing Company, 2010) (first published 1916).

Fayngold, Moses. *Special Relativity and How it Works* (Weinheim: John Wiley, 2008).

Feynman, Richard P. *Six Not-so-Easy Pieces* (London: Penguin Books, 1999).

French, A. P. *Special Relativity* (Wokingham: Van Nostrand Reinhold, 1968).

Mermin, N. D. *It's About Time: Understanding Einstein's Relativity* (Princeton: Princeton University Press, 2009).

Muirhead, H. *The Special Theory of Relativity* (London: Macmillan, 1973).

Rindler, W. *Introduction to Special Relativity* (Oxford: Clarendon Press, 1982).

Rosser, W. G. V. *An Introduction to the Theory of Relativity* (London: Butterworths, 1964).

Steane, A. M. *The Wonderful World of Relativity* (Oxford: Oxford University Press, 2011).

Taylor, E. F. and Wheeler, J. A. *Spacetime Physics* (New York: W. H. Freeman, 1992).

Taylor, J. G. *Special Relativity* (Oxford: Clarendon Press, 1975).

Williams, W. S. C. *Introducing Special Relativity* (London: Taylor and Francis, 2002).

Intermediate or advanced

Cheng, Ta-Pei. *Relativity, Gravitation and Cosmology* (Oxford: Oxford University Press, 2005).

d'Inverno, R. *Introducing Einstein's Relativity* (Oxford: Clarendon Press, 1992).

Jackson, J. D. *Classical Electrodynamics* (Hoboken: John Wiley, 1999).

Landau, L. D. and Lifshitz, E. M. *The Classical Theory of Fields* (London: Butterworth–Heinemann 1987).

Misner, C. W., Thorne, K. S., and Wheeler, J. A. *Gravitation* (New York: W. H. Freeman, 1973).

Rindler, W. *Relativity: Special, General and Cosmological* (2nd edn.) (Oxford: Oxford University Press, 2006).

Schwarz, P. M. and Schwarz, J. J. *Special Relativity, from Einstein to Strings* (Cambridge: Cambridge University Press, 2004).

Index